# Apoptosis and Human Health: Understanding Mechanistic and Therapeutic Potential

Kuladip Jana
Editor

# Apoptosis and Human Health: Understanding Mechanistic and Therapeutic Potential

Springer

*Editor*
Kuladip Jana
Division of Molecular Medicine
Bose Institute
Kolkata, West Bengal, India

ISBN 978-981-97-7904-8 ISBN 978-981-97-7905-5 (eBook)
https://doi.org/10.1007/978-981-97-7905-5



This Springer imprint is published by the registered company Springer Nature Singapore Pte Ltd.
The registered company address is: 152 Beach Road, #21-01/04 Gateway East, Singapore 189721, Singapore

*This Book is dedicated to my respected*
*Parents:*
*Sukanta Kumar Jana & Rina Rani Jana;*
*Dr. Samar Khatua & Krishna Khatua*
*My Teachers;*
*Dr. M. K. Bhattacharya*
*My beloved wife:*
*Moumita Khatua (Jana)*
*&*
*My lovely Son and Daughter:*
*Sharanyo and Shreenika*

# Preface

Cell death represents an irreversible termination of life. It is a fundamental physiological process in all organisms, essential for embryonic development, organ maintenance, and immune function. Recent advancements have deepened our understanding of cell death mechanisms, distinguishing various forms of "programmed cell death" like *apoptosis*, necroptosis, autophagy, pyroptosis, etc. Moreover, critical genes involved in these processes have been identified. Research has demonstrated that cell death is a highly intricate process involving many key participants capable of tipping the delicate balance of cellular environments—from survival to demise and from anti-inflammatory to pro-inflammatory states. As we investigate the complex regulatory mechanisms governing cell death, exciting discoveries are anticipated in this field in the coming years.

When organisms encounter physiological or pathological signals, cells respond by triggering various forms of cell death to maintain normal body functions. Typically, autophagy precedes apoptosis and can initiate it. If apoptotic cells are not promptly cleared, it can lead to programmed necrosis or cell necrosis, where autophagy offers the highest survival advantage, followed by apoptosis and then programmed necrosis, which has the lowest survival advantage. Traditionally, cell death pathways were viewed as independent and parallel processes. However, recent advances reveal that these modes of cell death can interact through interconnected or overlapping signalling pathways. The ultimate fate of cells depends on the interplay between these different cell death modes, closely influencing disease development. Their interactions jointly regulate disease occurrence and outcomes.

Further research is essential to elucidate how different signal pathways interact in cell death, alongside identifying the specific effector molecules in each cell type. In summary, there is an urgent need to investigate the molecular regulatory mechanisms of cell death in various diseases. Future research holds significant scientific importance for advancing fundamental studies on organismal immunity, uncovering disease pathogenesis, and preventing disease onset and progression in connection with the different cell death pathways.

This book aims to outline the transitions between several classical cell death modes in mammals. This seeks to illuminate the role of cell death, particularly apoptosis, in the pathogenesis of diverse diseases, offering fresh insights and fostering new treatment targets to facilitate the exploration and development of therapies.

This book also offers valuable insights into our understanding and provides answers to the following questions:

Is there any evidence of varying sequences of programmed cell death occurrence and progression in different diseases? Can manipulating the expression of critical genes truly facilitate the switch between various forms of programmed cell death, thereby mitigating disease occurrence and progression? How can we achieve a delicate balance between different modes of cell death by regulating the expression of distinct genes during disease onset and progression to achieve optimal therapeutic outcomes?

Over the years, research has elucidated the molecular mechanisms underlying apoptotic signalling and demonstrated that dysregulation of these signalling pathways contributes to human diseases. Indeed, both decreased and increased apoptosis can lead to pathological conditions. Recently, these insights have spurred the development of therapeutic strategies aimed at modulating apoptosis, some of which are undergoing clinical trials or have also been incorporated into medical practice.

# Acknowledgment

I would like to express my deepest gratitude to my teachers for their invaluable guidance and support in editing this book, *Apoptosis and Human Health: Understanding Mechanistic and Therapeutic Potential*. Their expertise and insights have been instrumental in shaping this work, which now stands as an important contribution to the rapidly emerging and highly relevant field of drug design and discovery. Their encouragement and dedication to advancing knowledge in this area have inspired me throughout this journey.

I was able to complete my work with the invaluable support of my Ph.D. student, Satyajit Halder. He provided significant assistance in writing, formatting the chapters, and designing the book's front cover.

I would also like to extend my gratitude to all my friends and colleagues who are the authors of this book for their acceptance of my invitation and timely submission of their chapters. They all kept me going; this book would have been impossible without their support.

Finally, I would like to thank the immense support and love of my family members, parents, and particularly my mother, Rina Rani Jana, for her indefatigable support and inspiration.

I also want to acknowledge my two sisters, Madhumita Jana and Monalisa Jana, for their immense love and support.

I am immensely grateful to my beloved wife, Moumita Khatua (Jana), who is my persistent source of inspiration, motivation, and support.

I am also indebted to my in-laws, Dr. Samar Khatua and Krishna Khatua, and my sister-in-law, Jayeta Khatua, for their enormous love and support.

Last but not the least, I would like to express my heartfelt love and affection to our lovely son and daughter, Sharanyo and Shreenika, for their support and sacrifice.

Kuladip Jana

# Contents

# About the Editor and Contributors

## About the Editor

**Kuladip Jana M.Sc., M.S., Ph.D., Post-doctorate (U.S.A.), FRE, FNESA, FSASS, FARAI, FIARA.** *Dr. Kuladip Jana* **is a** *Principal Scientist* **in the** *Division of Molecular Medicine* **and** *Scientist-In-Charge* **at the** *Centre for Translational Animal Research (CTAR)*, *Bose Institute,* **Kolkata, India. Dr. Jana was trained as a Post-doctoral Research Associate in** *molecular endocrinology and reproduction concerning ageing/Alzheimer's disease at Garrison Institute on Aging, Texas Tech University Health Sciences Center, Texas, USA,* **and in** *vaccine development against cholera and typhoid in National Institute of Cholera and Enteric Diseases, Kolkata,* **and also in the field of** *molecular signalling regulating cell growth/survival and death/apoptosis in breast/cervical cancer cells* **in** *Saha Institute of Nuclear Physics*, **Kolkata, after obtaining his M.S. and Ph.D. degree in** *physiology* **(***microbiology and immunology***;** *molecular and cellular reproductive biology and endocrinology***).**

*Dr. Kuladip Jana* is one of the most outstanding scientist working in the fields of *physiology*, specifically in cancer biology, immunology, reproductive biology and endocrinology, infectious diseases (including *Visceral Leishmania, M. Tuberculosis, Filaria,* etc.), and drug design and development. His research competency is depicted by his international publications in high-impact journals, including Nature Cell Death & Disease, Cell Host & Microbe, Free Radical Biology and Medicine, Science Signaling, PLOS Pathogens, Journal of Biological Chemistry, Journal of Immunology, ACS Applied Materials & Interfaces, ACS Pharmacology & Translational Science, ACS Biomaterials Science and Engineering, among others). Dr. Jana also received several research grants from various government agencies.

Currently, *Dr. Jana's* laboratory is focused on *translational drug developmental research* in diverse directions, i.e. ageing-associated testicular gametogenic and steroidogenic dysfunctions: ameliorating potential of natural product derived small molecules; molecular signalling of germ cell apoptosis by benzo(a)pyrene like environmental toxicants and its protection by natural aryl hydrocarbon receptor (AhR) antagonist; diabetes-associated endocrine and reproductive disorders as well as germ line stem cell apoptosis and its protection by natural flavonoids (sulforaphane analogues); molecular cues related to polycystic ovary syndrome and both type 1 and type 2 diabetes mellitus in female infertility; natural flavonoid-induced amelioration of apoptosis in cervical and breast cancer cells: role of β-catenin and Wnt signalling; and anti-cancer and anti-metastasis potential of trans-stilbene resveratrol derivative and sulforaphane analogue through G2M arrest, microtubule depolarization, and apoptosis: involvement of TGFβ signalling, application of some novel synthetic biomaterials/nano-materials as drug candidates against different diseases.

During the developmental work, *Dr. Jana* established a state-of-the-art preclinical animal research facility "*Centre for Translational Animal Research",* at Bose Institute, Kolkata, funded by DST, Government. of India. *Dr. Jana* also developed a *non-surgical chemical sterilization method for male animals* during his Ph.D. work. *Dr. Jana* serves as an advisor on the scientific advisory boards of various institutes, having been appointed as a Government of India-approved nominee to the *IAEC/IBSC/IHEC.*

Based on his research and fellowship training, *Dr. Jana* received several awards, fellowships, and honours, including *Young Investigator Award, 2011 (USA), Young Scientist Award, 2012 and 2014 (USA), Outstanding Scientist Award, 2015, Eminent Scientist Award, 2022, S Ramanujan Award 2024, Prof. K. C. Bose Gold Medel award, 2024, Fellow of the Scholar Academic & Scientific Society (FSASS), Fellow of the National Environmental Science Academy (FNESA), Fellow of the Reproduction & Endocrinology (FRE), etc.* He serves as an editor and editorial board member of over *35* prestigious international journals. He has also authored many international research articles, books, and book chapters. *Dr. Jana* guided many Ph.D. students, and he is a member of several learned scientific societies like, *the Society for Redox Biology & Medicine, USA, the Science Advisory Board, USA, The Physiological Society of India, The Society of Reproductive Biology & Endocrinology, and National Academy of biological Sciences,* among others. *Dr. Jana* is an Honorary Rosalind Member of the London Journal Press and he visited abroad to deliver invited lectures and chair sessions at many conferences and workshops. *Dr. Jana was also involved in supporting various healthcare programmes for the people living in the urban slums and rural and underserved tribal areas of West Bengal and received significant recognition from the society.*

## Contributors

**Sourav Bar** Coastal Environmental Studies Research Centre of Egra SSB College, Vidyasagar University, Midnapore, West Bengal, India

**Ankita Bhattacharya** Department of Biochemistry, K.P.C Medical College & Hospital, Kolkata, India

**Mihir K. Bhattacharya** ICMR-National Institute of Cholera & Enteric Diseases (NICED), Kolkata, West Bengal, India

**Suresh Challa** Cell Biology Division, National Institute of Nutrition, Hyderabad, India

**Rajarshi Chattopadhyay** Department of Zoology, Ramakrishna Mission Vidyamandira, Belur Math, India

**Deep Sankar Chini** Department of Zoology, Vidyasagar University, Midnapore, West Bengal, India

**Pijush K. Das** Infectious Diseases and Immunology Division, CSIR-Indian Institute of Chemical Biology, Kolkata, India

**Priyanka Das** Polymer-based Medical Devices and Complex Drug Delivery Systems Laboratory, Department of Pharmaceutics, National Institute of Pharmaceutical Education and Research (NIPER), Kolkata, India

**Souhadri Das** Molecular Oncology Laboratory, Department of Biological Sciences, Bose Institute, Kolkata, India

**Tapasi Das** Department of Biochemistry, University of Calcutta, Kolkata, India

**Pallab Datta** Polymer-based Medical Devices and Complex Drug Delivery Systems Laboratory, Department of Pharmaceutics, National Institute of Pharmaceutical Education and Research (NIPER), Kolkata, India

**Aiindrila Dhara** Molecular Oncology Laboratory, Department of Biological Sciences, Bose Institute, Kolkata, India

University of Cambridge, Cambridge, UK

Masaryk University, Brno, Czech Republic

**Soumik Dhara** Coastal Environmental Studies Research Centre of Egra SSB College, Vidyasagar University, Midnapore, West Bengal, India

**Omkar Kailas Gaikwad** Polymer-based Medical Devices and Complex Drug Delivery Systems Laboratory, Department of Pharmaceutics, National Institute of Pharmaceutical Education and Research (NIPER), Kolkata, India

**Srijon Gayen** School of Natural Product Studies, Jadavpur University, Kolkata, West Bengal, India

**Sudipta Kumar Ghorai** Coastal Environmental Studies Research Centre of Egra SSB College, Vidyasagar University, Midnapore, West Bengal, India

**Debidas Ghosh** Molecular Medicine, Nutrigenomics and Public Health Research Laboratory, Department of Bio-Medical Laboratory Science and Management, Vidyasagar University, Midnapore, West Bengal, India

**Noyel Ghosh** Division of Molecular Medicine, Bose Institute, Calcutta, West Bengal, India

**Pratik Ghosh** Department of Zoology, Vidyasagar University, Midnapore, West Bengal, India

**Suparna Ghosh** School of Natural Product Studies, Jadavpur University, Kolkata, West Bengal, India

**Barun Das Gupta** School of Natural Product Studies, Jadavpur University, Kolkata, West Bengal, India

**Pallab Kanti Haldar** School of Natural Product Studies, Jadavpur University, Kolkata, West Bengal, India

**Satyajit Halder** Division of Molecular Medicine, Bose Institute, Unified Academic Campus, Kolkata, West Bengal, India

**Kuladip Jana** Division of Molecular Medicine, Bose Institute, Unified Academic Campus, Kolkata, West Bengal, India

**Shraman Jana** IISER Kolkata, Mohanpur, Haringhata Farm, West Bengal, India

**Vishakha Yogesh Joshi** Polymer-based Medical Devices and Complex Drug Delivery Systems Laboratory, Department of Pharmaceutics, National Institute of Pharmaceutical Education and Research (NIPER), Kolkata, India

**Abhik Kar** Department of Zoology, University of Calcutta, Kolkata, India

**Ramandeep Kaur** Molecular Oncology Laboratory, Department of Biological Sciences, Bose Institute, Kolkata, India

**Sushweta Mahalanobish** Division of Molecular Medicine, Bose Institute, Calcutta, West Bengal, India

**Chirantan Majumder** Division of Molecular Medicine, Bose Institute, Unified Academic Campus, Kolkata, West Bengal, India

**Anirban Manna** Division of Molecular Medicine, Bose Institute, Unified Academic Campus, Kolkata, West Bengal, India

**Arkadeep Mitra** Department of Zoology, City College (Affiliated to University of Calcutta), Kolkata, India

**Niladri Mondal** Department of Biology, Katholic University of Leuven, Leuven, Belgium

**Lokesh Murumulla** Cell Biology Division, National Institute of Nutrition, Hyderabad, India

**Dibya Pal** Molecular Medicine, Nutrigenomics and Public Health Research Laboratory, Department of Bio-Medical Laboratory Science and Management, Vidyasagar University, Midnapore, West Bengal, India

**Shuvam Pal** Molecular Oncology Laboratory, Department of Biological Sciences, Bose Institute, Kolkata, India

**Prasanta Patra** Department of Zoology, Vidyasagar University, Midnapore, West Bengal, India

**Souravi Roy** Department of Biochemistry, University of Calcutta, Kolkata, India

**Sagartirtha Sarkar** Department of Zoology, University of Calcutta, Kolkata, India

**Nirmalya Sen** Molecular Oncology Laboratory, Department of Biological Sciences, Bose Institute, Kolkata, India

**Parames C. Sil** Division of Molecular Medicine, Bose Institute, Calcutta, West Bengal, India

**Nidhi Singh** Polymer-based Medical Devices and Complex Drug Delivery Systems Laboratory, Department of Pharmaceutics, National Institute of Pharmaceutical Education and Research (NIPER), Kolkata, India

**Anindita Ukil** Department of Biochemistry, University of Calcutta, Kolkata, India

# Abbreviations

| | |
|---|---|
| AIF | Apoptosis-inducing factor |
| APAF-1 | Apoptotic protease activating factor 1 |
| ATR | Ataxia telangiectasia and Rad3 related |
| BCL-2 | B-cell lymphoma 2 |
| BAD | Bcl-2-associated death promoter |
| BAK | Bcl-2 antagonist killer 1 |
| BAX | Bcl-2 associated X protein |
| BIK | Bcl-2 interacting killer |
| BIM | Bcl-2-like protein 11 |
| Caspases | Cysteine-aspartic proteases |
| *CARD* | Caspase-recruitment domains |
| *cFLIP* | Cellular FLICE (FADD-like IL-1β-converting enzyme)-inhibitory protein |
| *cIAP* | *C*ellular inhibitor of apoptosis protein |
| CHK1/2 | Checkpoint kinase 1/2 |
| CIAP | Cellular inhibitor of apoptosis |
| *DED* | *D*eath effector domains |
| DISC | Death-inducing signalling complex |
| ERK1/2 | Extracellular signal-regulated kinases |
| FAS | First apoptosis signal |
| FASL | First apoptosis signal ligand |
| FADD | FAS-associated protein with death domain |
| FITC | Fluorescein isothiocyanate |
| ICAD | Inhibitor of caspase-activated DNase |
| Iκβ | Inhibitor of kappa beta |
| MCL-1 | Induced myeloid leukaemia cell differentiation protein |
| *MLKL* | Mixed lineage kinase domain-like protein |
| *PAMPs* | *P*athogen-associated molecular patterns |
| MOMP | Mitochondrial outer membrane permeability |
| NF-κβ | Nuclear factor kappa-light-chain-enhancer of activated B cells |
| P38MAPK | Mitogen-activated protein kinase |
| PARP-1 | Poly(ADP-ribose) polymerase-1 |
| PI3K | Phosphoinositide 3-kinase |
| *PIDD* | p53-inducible protein with a death domain |

| | |
|---|---|
| PKA | Protein kinase A |
| PKB | Protein kinase B |
| PKC | Protein kinase C |
| *RAIDD* | RIP-associated Ich-1/Ced-3 homologous protein with a death domain |
| *RIP1* | Receptor-interacting protein 1 |
| *RIP3* | Receptor-interacting serine-threonine kinase 3 |
| *ROS* | Reactive oxygen species |
| STAT | Signal transducer and activator of transcription |
| TBID | Truncated BH3-interacting domain death agonist |
| *TLR3* | Toll-like receptor 3 |
| *TNFR* | *T*umour necrosis factor receptor |
| TNF-α | Tumour necrosis factor-alpha |
| TRAF-2 | TNF-receptor-associated factor-2 |
| *TRAF* | TNF receptor-associated factors |
| TRAAD | TNF-receptor type-1-associated death domain |
| TRAIL | TNF-related apoptosis-inducing ligand |
| XIAP | X-linked inhibitor of apoptosis |

# 1 Cell Death Pathways: Intricate Connections and Disease Implications

Satyajit Halder and Kuladip Jana

**Abstract**

Cell death is a crucial biological process for an organism's development, growth, and survival, facilitating tissue remodeling and eliminating surplus or harmful cells. This chapter explores the three primary genetically programmed cell death mechanisms: apoptosis, necroptosis, and pyroptosis, each activated by specific conditions and stimuli, and their distinct roles in modulating immune responses. Apoptosis, or programmed cell death, maintains cellular homeostasis with minimal inflammation, which is essential for development and normal physiological functions. Necroptosis is an inflammatory backup when apoptosis is inhibited, which plays a critical role in pathogen defense by releasing damage-associated molecular patterns (DAMPs). Pyroptosis, highly inflammatory and primarily immune-related, responds to intracellular infections by releasing pro-inflammatory cytokines and DAMPs. Dysregulation of these pathways is significant in disease pathogenesis, where inadequate cell death can lead to cancer and autoimmunity, while excessive cell death is linked to degenerative and inflammatory diseases. Understanding the regulation of these cell death mechanisms is imperative for therapeutic intervention, with current research focusing on molecular pathways and potential drug targets, such as apoptosis protein inhibitors for cancer and necroptosis pathway inhibitors for inflammatory diseases.

**Keywords**

Apoptosis · Necroptosis · Pyroptosis · Ferroptosis · Autophagy · Diseases · Therapeutic targets

S. Halder · K. Jana (✉)
Division of Molecular Medicine, Bose Institute, Unified Academic Campus, Kolkata, West Bengal, India
e-mail: kuladip@jcbose.ac.in; kuladip.jana@gmail.com



K. Jana (ed.), *Apoptosis and Human Health: Understanding Mechanistic and Therapeutic Potential*, https://doi.org/10.1007/978-981-97-7905-5_1

## 1.1 Introduction

Cell death is a fundamental aspect of life, intricately involved in developing multicellular organisms and maintaining tissue homeostasis. This process ensures the removal of dispensable cells, thus sculpting tissues during development and maintaining balance within adult tissues (Kerr et al. 1992; Green and Llambi 2015). Beyond its role in development and homeostasis, cell death is crucial in the immune response, aiding in the elimination of infected or damaged cells to protect the organism from pathogens (Shalini et al. 2015). However, the disruption of cell death mechanisms can lead to a variety of diseases, including cancer, where cells evade apoptosis, and neurodegenerative diseases, where excessive cell death leads to loss of neuronal function (Olsson and Zhivotovsky 2011; Orrenius et al. 2011).

Given its pivotal role in health and disease, increasing interest is in modulating cell death pathways for therapeutic purposes. Various forms of cell death have been characterized, with apoptosis, necroptosis, and pyroptosis being the most extensively studied (Fulda et al. 2010). Apoptosis, often termed programmed cell death, involves a tightly regulated sequence of events leading to the orderly dismantling of cellular components without provoking an inflammatory response (Kerr et al. 1992). In contrast, necroptosis is a programmed form of necrosis that acts as a fail-safe mechanism when apoptosis is inhibited, often associated with inflammation due to the release of damage-associated molecular patterns (DAMPs) (Galluzzi et al. 2012). Pyroptosis, predominantly triggered by microbial infections, is characterized by the release of pro-inflammatory cytokines, contributing to the immune response (Bergsbaken et al. 2009).

Recent advances have revealed a complex network of crosstalk and backup mechanisms between these cell death pathways, highlighting the intricacies of cellular demise and survival strategies (Green and Llambi 2015). For instance, pathways that inhibit apoptosis can activate necroptosis, ensuring cell death occurs even when primary mechanisms are blocked (Oberst et al. 2011). This interconnectivity underscores the potential for targeting specific molecules within these pathways to treat diseases characterized by either excessive or insufficient cell death (Fulda 2010).

This chapter aims to provide an in-depth review of the various forms of cell death, their molecular interconnectivity, and the validated therapeutic targets that hold promise for treating diseases associated with dysfunctional cell death signaling.

## 1.2 Apoptosis

Apoptosis, first characterized as programmed cell death in 1972 (Kerr et al. 1992), is integral to maintaining tissue homeostasis and regulating various physiological processes, including cell turnover, immune system function, and embryonic development (Voss and Strasser 2020). It involves distinct morphological changes such as cellular and organelle fragmentation, DNA cleavage, and the formation of apoptotic bodies (Kerr et al. 1992; Zakeri et al. 1993). Despite typically being immunologically silent, apoptosis can contribute to inflammatory conditions under certain pathological circumstances (Rickard et al. 2014; Yang et al. 2015; Singh et al. 2019).

Apoptosis is primarily mediated through two major pathways: intrinsic and extrinsic. The balance of pro-apoptotic and anti-apoptotic members of the Bcl-2 family regulates the intrinsic pathway, whose equilibrium can be disrupted by various cellular stresses, leading to apoptotic cell death. Conversely, the extrinsic pathway involves members of the TNF superfamily binding to death receptors on the cell surface, triggering a signaling cascade that initiates apoptosis (Taylor et al. 2008). Additionally, apoptosis can be induced by cytotoxic T cells recognizing non-self-antigens presented on target cells, where granzymes released by the T cells activate caspases to execute cell death (Voskoboinik et al. 2015a). Regardless of the initiating stimulus, apoptosis pathways converge on caspases, critical proteases that either initiate apoptotic signals (initiator caspases) or execute the final dismantling of the cell (effector caspases), Fig. 1.1.

### 1.2.1 Intrinsic Apoptosis

Intrinsic apoptosis is a fundamental cellular process that occurs in response to a variety of stressors, including growth factor deprivation, DNA damage (e.g., UV radiation), and developmental signals (Fig. 1.1) (Kale et al. 2018). These stressors alter the delicate balance of proteins in the Bcl-2 superfamily, which is critical for controlling mitochondrial integrity and cell survival. This family can be classified into three main subgroups: the anti-apoptotic proteins (e.g., B-cell lymphoma 2, Bcl-2; B-cell lymphoma-extra-large, Bcl-XL; myeloid cell leukemia 1, Mcl-1), the pro-apoptotic BH3-only proteins (e.g., Bcl-2 interacting mediator of cell death, Bim), and the effector proteins Bcl-2-associated X protein (Bax), Bcl-2 homologous antagonist/killer (Bak), and Bcl-2-related ovarian killer (Bok) (K Under normal settings, anti-apoptotic proteins inhibit the pro-apoptotic activity of Bax and Bak, preventing mitochondrial outer membrane permeabilization (MOMP), a vital step in apoptosis.

BH3-only proteins act as cellular stress sensors, blocking anti-apoptotic Bcl-2 proteins or directly activating Bax and Bak. Their regulation is multifaceted: some are transcriptionally controlled (e.g., p53 upregulated modulator of apoptosis, PUMA, in response to DNA damage) (Nakano and Vousden 2001), while others are post-translationally modified (e.g., Bim, BH3-interacting domain death agonist; Bid, BH3-interacting-domain death agonist) (Li et al. 1998; Lei and Davis 2003). Disruption of this delicate equilibrium can tilt the scales favor pro-apoptotic signaling, resulting in MOMP.

Following MOMP, key mediators of intrinsic apoptosis are released from the mitochondria into the cytosol, including cytochrome c (Stein and Hansen 1999) and second mitochondria-derived activator of caspases (SMAC)/direct IAP binding protein with low pI (Diablo), which antagonizes inhibitors of apoptosis proteins (IAPs) such as X-linked inhibitor of apoptosis protein (XIAP), facilitating caspase

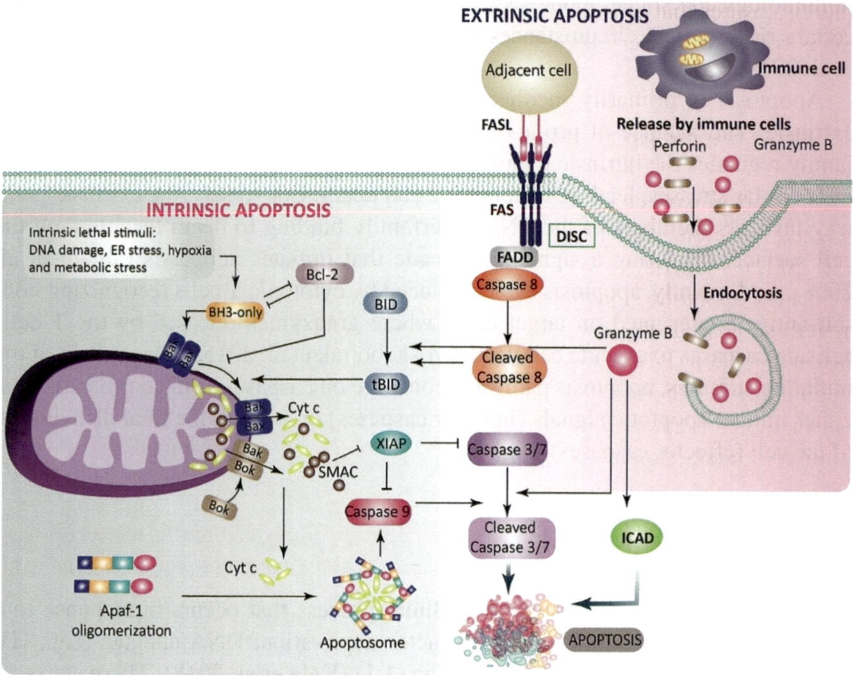


**Fig. 1.1** Intrinsic apoptosis can be triggered by stimuli like growth factor withdrawal, shifting the balance between pro-survival Bcl-2 proteins and BH3-only proteins, leading to Bax and Bak oligomerization, mitochondrial outer membrane permeabilization (MOMP), and the release of cytochrome c and Smac. Cytochrome c binds Apaf-1, forming the apoptosome and activating caspase-9, activating caspase-3/7, leading to apoptosis. Extrinsic apoptosis is initiated by TNF family ligands binding to their receptors, forming the death-inducing signaling complex (DISC) and activating caspase-8, which can directly activate caspase-3 or cleave Bid to trigger intrinsic apoptosis. Caspase activities can be inhibited by XIAP, which is antagonized by Smac. Apoptosis can also be induced by immune cell-released granzyme B and perforin, with granzyme B activating caspase-3 or other apoptosis effectors like CAD

activation within the apoptosome (Du et al. 2000; Verhagen et al. 2000). Apoptotic protease activating factor 1 (Apaf-1) attracts procaspase-9 to form the apoptosome, where caspase-9 is activated, triggering a cascade of caspase activation that includes caspases 3 and 7, resulting in hallmark apoptotic morphological alterations. Caspase-mediated cleavage of specific substrates, such as inhibitor of caspase-activated DNase (ICAD) to active caspase-activated DNase (CAD), causes DNA fragmentation, whereas cleavage of Rho-associated protein kinase I (ROCK-I) promotes cellular blebbing and other apoptotic features.

Anoikis is a type of intrinsic apoptosis caused by the loss of cell-matrix connections via integrins, which communicate through pathways such as phosphoinositide 3-kinase (PI3K) or focal adhesion kinase (FAK) to alter Bcl-2 family proteins.

Anoikis ensures that cells only survive in specific tissue compartments, and its evasion is critical in cancer metastasis and invasiveness.

### 1.2.2 Extrinsic Apoptosis

Extrinsic apoptosis is caused by interactions between members of the TNF family and their receptors, which include TNF, FasL, TRAIL, and TL1A (Fig. 1.2) (Wilson et al. 2009). This route, which focuses on FasL-mediated signaling, is an example of extrinsic apoptotic signaling. FasL interacts with its transmembrane receptor Fas, triggering a cascade that includes Fas-associated protein with death domain (FADD) via death domain (DD) contacts (Chinnaiyan et al. 1995; Boldin et al. 1996). FADD, which comprises both a death domain (DD) and a death effector domain (DED), aids in the formation of the death-inducing signaling complex (DISC) through homotypic interactions with caspase-8 (Boldin et al. 1996; Medema et al. 1997). Caspase-8's p18 and p10 subunits are generated within the DISC by proximity-induced cleavage, which initiates downstream activation of caspase-3 and caspase-7 (type I apoptosis) (Stennicke et al. 1998).

Insufficient caspase-3 activation can result in type II apoptosis, in which caspase-8 cleaves the BH3-only protein known as BH3 interacting domain death agonist (BID) to create its shortened form, truncated BID (tBID) (Li et al. 1998). tBID then activates the intrinsic apoptotic pathway by interacting directly with Bcl-2-associated X protein (Bax) and Bcl-2 homologous antagonist/killer (Bak), causing mitochondrial outer membrane permeabilization (MOMP) and cytochrome c release (Desagher et al. 1999; Wei et al. 2000). The control and activation of these pathways vary between cell types and are influenced by factors such as XIAP expression levels (Jost et al. 2009; Varfolomeev et al. 2009).

### 1.2.3 Granzyme-Mediated Apoptosis

Cytotoxic lymphoid cells, particularly natural killer (NK) cells, and cytotoxic T cells use a variety of ways to cause cell death in their targets, playing an important role in immunological surveillance and protection against infections and cancer. One major mechanism is the release of secretory granules containing perforin and granzyme B in response to identifying infected or altered cells (Voskoboinik et al. 2015b). Perforin enhances granzyme B entry into the target cell cytoplasm, where it can start apoptosis by cleaving different substrates such as caspases and Bid, a pro-apoptotic Bcl-2 family member (Boivin et al. 2009). Importantly, granzyme B can directly cleave an inhibitor of caspase-activated DNase (ICAD), causing DNA fragmentation and cell death in the absence of caspase activation (Thomas et al. 2000).

### 1.2.4 Necroptosis

Necroptosis is a controlled kind of cell death distinguished by its pro-inflammatory characteristics and independence from caspases. It activates receptor-interacting protein kinase 3 (RIPK3) and its downstream effector, mixed lineage kinase domain-like protein (MLKL). RIPK3 activation is triggered by a variety of stimuli, including death receptors (DRs), toll-like receptors (TLRs), and Z-DNA binding protein 1 (ZBP1), according to Newton and Manning (2016). Upon activation, RIPK3 autophosphorylates and then phosphorylates MLKL (Sun et al. 2012; Murphy et al. 2013). Phosphorylated MLKL then forms oligomers that translocate to the cell membrane, producing membrane permeability and beginning cell death (Cai et al. 2014; Chen et al. 2014; Dondelinger et al. 2014; Samson et al. 2020) (Fig. 1.2).

MLKL activation does not irreversibly commit cells to necroptosis; Murphy (2020) carefully reviews the processes that can regulate necroptosis following MLKL activation. Key necroptosis mediators contain RIP homotypic interaction motif (RHIM) domains. RIPK1 is required for death receptor signaling (Sun et al. 2002), TRIF (TIR-domain-containing adapter causing interferon-β) promotes necroptosis through toll-like receptors (Kaiser and Offermann 2005), and ZBP1

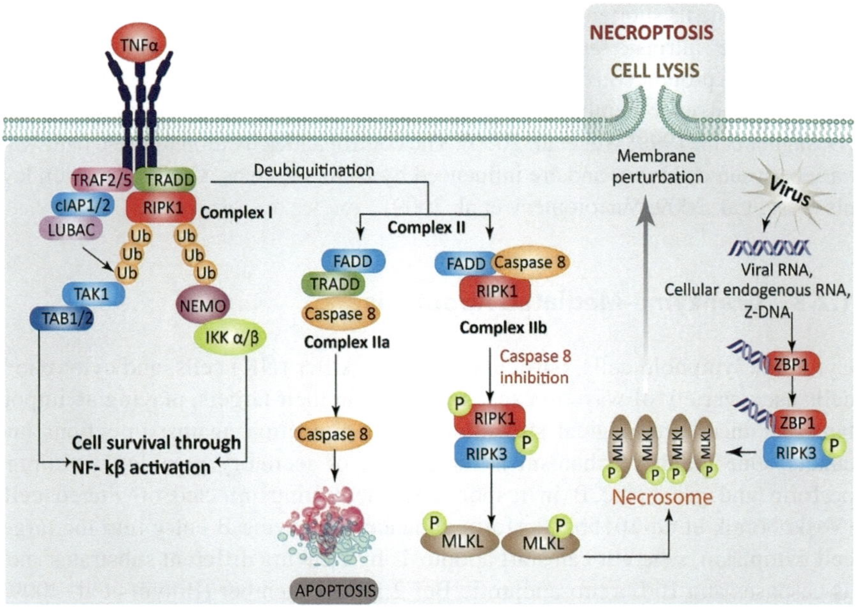


**Fig. 1.2** Necroptosis can be triggered by various stimuli. TNF binding to its receptor forms a complex that activates NF-κB and MAPK-prolonged signaling and caspase inhibition cause RIP1 to move to the cytosol and form complex II, where RIP1 and RIP3 phosphorylate each other, leading to MLKL activation. MLKL oligomerizes, disrupting the membrane and causing cell lysis. Z-DNA sensing by ZBP1 also binds to RIP3 to induce cell death, a process that can be inhibited by RIP1

mediates necroptosis caused by virally encoded nucleic acids (Kaiser et al. 2008). In addition, RIPK3 activates MLKL (Sun et al. 2002).

The TNF receptor-mediated signaling pathway is critical for necroptosis initiation (Fig. 1.2). When TNF binds to TNFR1, complex I (RIPK1, TRADD, TRAF2, and c-IAP1/2) is ubiquitinated, including K11- and K63-linked chains, which activates the NF-κB and MAPK pathways (Bertrand et al. 2008; Varfolomeev et al. 2008). Deubiquitinases reduce these effects by eliminating ubiquitin chains (Lork et al. 2017). RIPK1 can then self-activate via autophosphorylation, boosting necroptosis when caspase-8 activity is insufficient (Newton et al. 2016). FADD, caspase-8, and c-FLIP regulate the balance of apoptosis and necroptosis (Lin et al. 1999; Newton et al. 2019).

TRIF, a key mediator of necroptosis, activates NF-κB signaling via the TRAF2/6-RIPK1 axis when TLR3/4 complexes detect viral RNA or LPS (Kawasaki and Kawai 2014). TRIF can also induce apoptosis or necroptosis depending on the cellular environment and interactions with RIPK1 and caspase-8 (McAllister et al. 2013; Kaiser et al. 2013). ZBP1, when triggered by viral Z-DNA or Z-RNA, causes necroptosis by binding to RIPK3, whereas RIPK1 can block necroptosis signaling via RHIM–RHIM domain interactions (Lin et al. 2016; Newton et al. 2016). Recent research has linked endogenous Z-double-strand RNA to necroptosis under certain conditions (Jiao et al. 2020).

### 1.2.5 Pyroptosis

Pyroptosis is a diligently organized, pro-inflammatory necrotic cell demise that relies on Gasdermin proteins. Activation occurs when caspase-1, caspase-4, caspase-5 (the human counterparts of murine caspase-11), or caspase-11 are stimulated by various inflammasome pathways, ending in their autoprocessing and activation. When activated, these caspases separate Gasdermin-D (GSDMD) into its N-terminal (GSDMD-N) and C-terminal (GSDMD-C) domains (Kayagaki et al. 2015; Shi et al. 2015; Kuang et al. 2017). GSDMD-C acts as an auto-inhibitory domain, suppressing GSDMD-N's cell lytic characteristics until cleavage releases the inhibition (Kayagaki et al. 2015; Shi et al. 2015; Kuang et al. 2017).

Once released, GSDMD-N moves to the cell membrane, where it binds to phosphatidylinositol phosphates and phosphatidylserine and oligomerizes to form holes, causing lytic cell death (Aglietti et al. 2016; Ding et al. 2016; Liu et al. 2016; Sborgi et al. 2016). GSDMD pores allow the release of mature versions of pro-inflammatory cytokines IL-1β and IL-18, which lack secretion sequences and permeate through the extracellular environment (Evavold et al. 2018; Heilig et al. 2018).

The signal for GSDMD cleavage by caspases travels through various inflammasomes via canonical or non-canonical pathways. Canonical inflammasomes typically comprise sensors such as AIM2, NLRP1, NLRP3, NLRC4, and Pyrin, which detect damage-associated molecular patterns (DAMPs) or pathogen-associated molecular patterns (PAMPs). These sensors use Pyrin domains to connect with the adaptor molecule ASC (apoptosis-associated speck-like protein containing a

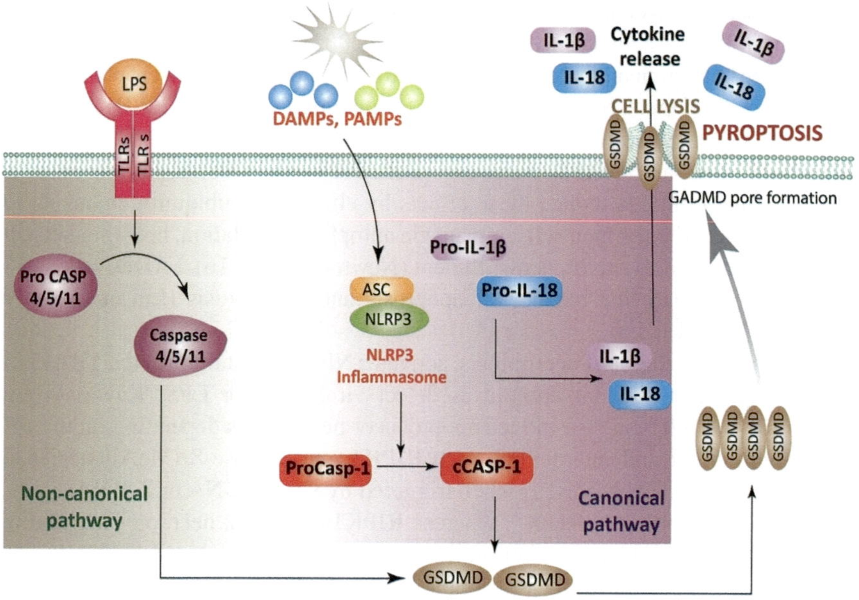


**Fig. 1.3** Pyroptotic cell death is initiated by various stimuli that activate the inflammasome. Activation of NLRP3 leads to its binding with ASC and caspase-1, forming the inflammasome complex. Caspase-1 then processes pro-IL-1β and pro-IL-18 into their active forms and cleaves GSDMD. The N-terminal fragment of GSDMD (GSDMD-N) translocates to the membrane, causing cell lysis and cytokine release. In non-canonical inflammasome activation, intracellular LPS activates caspase-11, which processes GSDMD and caspase-1, leading to cell lysis and further activation of the canonical inflammasome pathway

CARD). ASC then attracts and oligomerizes caspase-1 via its caspase activation and recruitment domains (CARD), activating the protein (Martinon et al. 2002; Srinivasula et al. 2002; Chauhan et al. 2020).

Inflammasome activation is a two-step process. During priming, Toll-like receptors (TLRs) detect PAMPs and activate gene expression through NF-κB and IFN type I, leading to upregulating NLRP3 and caspase-11 (Rathinam et al. 2012). The non-canonical inflammasome is then activated when intracellular lipopolysaccharide (LPS) binds to murine caspase-11 or human caspase-4/5, causing their activation and subsequent cleavage of GSDMD to induce pyroptosis (Kayagaki et al. 2011; Shi et al. 2014; Shi J, Zhao Y, et. al. 2015).

Aside from GSDMD, additional Gasdermin proteins, such as Gasdermin E (GSDME), play important roles in cell death. Caspase-3 and granzyme B can cleave GSDME, which links pyroptosis to apoptotic and cytotoxic pathways (Rogers et al. 2019; Wang et al. 2017; Lee et al. 2018). The interconnectivity of various cell death mechanisms highlights their intricate control and significance in immune responses and disease development (Fig. 1.3).

## 1.3 Other Forms of Regulated Cell Death

Other forms of cell death have been identified in recent years. While some signaling components and pathways are already well understood, others still need further research.

### 1.3.1 Autophagy-Dependent Cell Death (ADCD)

Important for preserving cellular homeostasis, autophagy is the process of cellular breakdown mediated by lysosomal pathways. Usually, the knockout of autophagy mediators results in neonatal mortality (Kuma et al. 2017). This process consists of several phases controlled by protein complexes. A common trigger for autophagy is starvation; Ulk1-ATG13/101 and FIP200 kinase complex start this process (Zachari and Ganley 2017). Under the direction of many protein complexes, nucleation, elongation, and phagophore production follow initiation. LC3 tags the phagophore, which develops into autophagosomes, fusing lysosomes to create autolysosomes for content breakdown. The book by Dikic and Elazar (2018) thoroughly explains this technique.

Certain components of the autophagy pathway also have non-autophagic uses (Galluzzi and Green 2019). Concurrent cell death methods such as apoptosis or necroptosis (Galluzzi et al. 2018) contrast with autophagy-dependent cell death (ADCD), which depends on autophagic machinery—operating through two pathways—one dependent on autophagic flow for organelle degradation (Dasari et al. 2018) and the other, known as Autosis, independent of autophagic flux (Liu et al. 2013). ADCD is both routes that show the development of cytoplasmic vacuole (Bialik et al. 2018). Without triggering death or necroptosis, resveratrol treatment of cancer cells causes autophagy-dependent ADCD (Dasari et al. 2018). Induced by TAT-Beclin-1 peptides, malnutrition, or hypoxia, autosis—mediated by Na+/K+ -ATPase—modulated by cardiac glycosides (Liu et al. 2013) results in cell enlargement and finally plasma membrane rupture. Severe anorexia nervosa patients have been found to contain autonomic cells (Kheloufi et al. 2015). ADCD is involved in several physiological mechanisms and diseases, including malignancies, reperfusion injuries, and other pathologies (Bialik et al. 2018; Denton and Kumar 2019).

### 1.3.2 Mitochondrial Permeability Transition Pore (MPTP)-Mediated Necrosis

Another type of controlled cell death under the influence of intracellular circumstances is mitochondrial permeability transition pore (MPTP)-mediated necrosis. Tracked by oxidative stress or cytosolic/mitochondrial Ca2+ buildup, MPTP opening throws off H+ gradients and stops ATP synthesis (Lemasters et al. 2009; Izzo et al. 2016). Crucially for MPTP activity, cyclophilin D (CypD) controls pore opening, hence influencing ischemia/reperfusion damage in the heart and brain (Baines et al. 2005; Nakagawa et al. 2005; Schinzel et al. 2005). Although other methods propose inadequate suppression by Cyclosporin A (Basso et al. 2005), mitochondrial swelling in CypD mutant animals verifies its function. F1F0 ATP synthase's participation in MPTP highlights its potential as a therapeutic target for several disorders (Bonora et al. 2013; Giorgio et al. 2013; Sileikyte and Forte 2019).

### 1.3.3 Parthanatos

Parthanatos, another regulated cell death mechanism, depends on poly(ADP-ribose) polymerase 1 (PARP1) activity. PARP1 activation in response to severe DNA damage by reactive oxygen or nitrogen species depletes cellular NAD+ and ATP via PAR polymer formation, potentially leading to cell death (Andrabi et al. 2014; Ray et al. 2017). PARP1-induced AIF–MIF interaction facilitates DNA fragmentation in some contexts (Yu et al. 2002; Wang et al. 2016), while AIF-independent pathways also drive retinal degeneration (Jang et al. 2017). RNS associated with parthanatos contributes to neural pathologies (Virag and Szabo 2002; Fatokun et al. 2014).

### 1.3.4 NETosis

NETosis, involving neutrophil extracellular trap (NET) formation, represents a host defense mechanism where neutrophils release DNA containing antimicrobial proteins into the extracellular space (Brinkmann et al. 2004). NADPH oxidase and ROS play pivotal roles in actin depolarization and NET release (Douda et al. 2015). Elevated ROS activate neutrophil elastase (NE), leading to actin depolarization and chromatin decondensation via histone cleavage and citrullination by PAD4, facilitating NET formation. Further research is required to elucidate the molecular mechanisms of non-lytic and lytic NET formation and their interactions with other cell death modalities and autophagy pathways.

### 1.3.5 Ferroptosis

Ferroptosis, another regulated cell death modality, depends on ROS-mediated iron-induced lipid peroxidation (Dixon et al. 2012). Cellular iron and ROS levels contribute to ferroptosis susceptibility, regulated by antioxidant enzymes like glutathione peroxidase 4 (GPX4) and pathways involving lipid hydroperoxide generation. Inhibitors targeting ferroptosis have potential therapeutic implications across various pathologies (Bebber et al. 2020; Belavgeni et al. 2020).

## 1.4 Cell Death Pathways Components as Therapeutic Targets

Cell death pathways are critical for maintaining biological homeostasis, and their dysregulation can lead to many illnesses, including autoimmune disorders, dementia, and cancer. Consequently, these pathways suggest interesting candidates for therapeutic intervention.

The relevance of cell death in human diseases was initially verified by finding Bcl-2 from the genomic region frequently involved in chromosomal translocations in follicular lymphomas (Tsujimoto et al. 1984). The expression of Bcl-2 was discovered to shield cells from apoptosis, resulting in lymphocyte buildup and probable cancer formation (Vaux et al. 1988; McDonnell et al. 1989). The relevance of other Bcl-2 family members, including Bcl-xL and Mcl-1, was subsequently discovered, creating antagonists targeting these anti-apoptotic proteins (Czabotar et al. 2014). These chemicals, known as BH3 mimetics, replicate the cell death-promoting activity of BH3 domains.

The first successful BH3 mimetic, ABT-737, exhibited partial selectivity for Bcl-2, Bcl-xL, and Bcl-w over Mcl-1 and A1, substantially reducing tumor growth in multiple animal cancer models (Oltersdorf et al. 2005). ABT-737 was followed by navitoclax (ABT-263), an orally accessible BH3 mimic (Tse et al. 2008). Navitoclax showed remarkable promise in patients with chronic lymphocytic leukemia (CLL) and indicated excellent potential for combination therapy, particularly those downregulating Mcl-1, a Bcl-2 protein not impacted by ABT-263 (Cragg et al. 2009). However, navitoclax's clinical development was delayed by its adverse effect of producing a rapid decline in platelet count due to Bcl-xL suppression (Roberts et al. 2012). This unfavorable consequence led to a change in drug discovery efforts away from targeting Bcl-xL (Czabotar et al. 2014).

Consequently, the Bcl-2 selective inhibitor venetoclax (ABT-199) emerged as the first BH3 mimetic and cell death-regulating small molecule to earn FDA approval for the treatment of chronic lymphocytic leukemia (CLL) and small lymphocytic lymphoma (Souers et al. 2013). Venetoclax is currently under research in multiple clinical trials, attempting to expand its use across various cancers (Strasser and Vaux 2020). However, Mcl-1 has been recognized as a key resistance factor to Bcl-2 targeting (Gong et al. 2016). Various Mcl-1 selective inhibitors have been discovered to address this and are entering clinical trials (Kotschy et al. 2016; Caenepeel et al. 2018). The availability of selective antagonists for diverse pro-survival Bcl-2 proteins could give oncologists potent and tolerant anti-cancer medicines to treat malignancies successfully.

Several IAP antagonists, such as GDC-0152, TL3271, and SM-164, have exhibited tumor-inhibiting effects in in vivo cancer models without generating considerable toxicity or weight loss in mice (Lu et al. 2008; Flygare and Fairbrother 2010; Flygare et al. 2012; Fulda and Vucic 2012; Morrish et al. 2020a). Encouraged by these preclinical results, many IAP antagonists have progressed to phase I/II clinical studies for various cancers (Fulda and Vucic 2012; Jensen et al. 2020b). Trials with

GDC-0152, LCL161, HGS1029, and TL32711 revealed successful target antagonism, dose-proportional pharmacokinetics, and an absence of dose-limiting toxicity (Morrish et al. 2020a). However, these trials did not indicate significant anti-tumor effectiveness, so further research was not pursued (Morrish et al. 2020a).

Despite these limitations, the capacity of IAP antagonists to promote non-canonical NF-κB signaling has sparked interest in combining them with checkpoint inhibitors (such as anti-PD1 antibodies) and anti-retroviral therapy (Chesi et al. 2016; Nixon et al. 2020; Morrish et al. 2020a). Additionally, IAP antagonists have showed promise in treating liver disorders caused by HBV and Plasmodium infections (Ebert et al. 2020; Morrish et al. 2020b). Future trials will continue to assess the safety and efficacy of IAP antagonists for treating malignancies and infections, seeking to give new treatment alternatives for people in need.

Bcl-2 and IAP antagonists were discovered to trigger cell death in hematological and solid malignancies. However, excessive cell death can be hazardous, leading to tissue damage and neurodegeneration. Therefore, RIP1 has been offered as a safer alternative for treating inflammatory and neurodegenerative illnesses without the danger of immunosuppression (Yuan et al. 2019b; Mifflin et al. 2020). While RIP3 kinase is potentially a target, genetic studies and efforts to block RIP3 have revealed toxicity, making its suppression dangerous (Newton et al. 2014). In contrast, genetic inactivation or chemical inhibition of RIP1 kinase activity is well-tolerated and provides no known hazards (Berger et al. 2014; Newton et al. 2014; Polykratis et al. 2014; Patel et al. 2020; Webster et al. 2020).

Inhibiting RIP1 kinase activity shows promise in treating joint and skin inflammation, ileocolitis, and TNF-induced systemic inflammatory response syndrome (SIRS) (Berger et al. 2014; Vlantis et al. 2016; Newton et al. 2016; Patel et al. 2020; Webster et al. 2020). It also plays a crucial role in neurodegenerative and neuroinflammatory illnesses (Yuan et al. 2019a; Mifflin et al. 2020). However, its therapeutic efficacy in pancreatic tumors, lung metastases, pancreatitis, and certain viral infections remains debatable (Newton et al. 2016; Patel et al. 2020; Webster et al. 2020). Further investigations are necessary to establish the parameters suited for RIP1 inhibition, with clinical trials providing critical confirmation.

RIP1 inhibitors developed by GlaxoSmithKline (GSK) and Denali have shown encouraging safety profiles in clinical settings. Initial studies show that GSK2982772 and DNL104 were generally well tolerated (Harris et al. 2017; Weisel et al. 2017; Grievink et al. 2020; Jensen et al. 2020a; Mifflin et al. 2020). Phase I trials with GSK2982772 found no major adverse events (AEs) or safety issues, prompting GSK to proceed with small phase 2 trials for psoriasis, rheumatoid arthritis, and ulcerative colitis. However, GSK2982772 did not reveal significant therapeutic advantages in psoriasis or rheumatoid arthritis, and findings from the ulcerative colitis trial are forthcoming (ClinicalTrials.gov). GSK also evaluated GSK3145095 in pancreatic and other solid tumors, but the trial was stopped during patient enrollment (Harris et al. 2019). Denali's DNL104, a brain-penetrant RIP1 inhibitor, revealed no detrimental effects on the central nervous system but did observe aberrant liver function in certain healthy subjects (Grievink et al. 2020). Overall, targeting RIP1 is a viable method for treating inflammatory illnesses. Future preclinical and clinical investigations will be critical in determining the best indications and patient populations for RIP1 inhibitors.

An emerging technique for addressing inflammatory illnesses involves preventing inflammasome assembly and the cell death process mediated by proteins, including NLRP3 and GSDMD. GSDMD is vital in pyroptotic cell death, particularly in sepsis or acute respiratory distress syndrome (ARDS) (Chauhan et al. 2020). Despite its potential, efforts to target GSDMD are in the early stages due to its unique nature without known enzymatic activity (Shi et al. 2014). Consequently, no GSDMD inhibitors have progressed to clinical trials yet.

In contrast, established regulators of the NLRP3 inflammasome have gained significant attention in drug discovery. NLRP3 is linked to numerous inflammatory, neurological, and metabolic illnesses. The efficacy of therapeutic targeting of NLRP3 was proven early on with glyburide, which efficiently suppresses IL-1β release and pyroptotic cell death. Another intriguing drug, MCC950/CRID3, targets the NACHT domain of NLRP3 with great potency and selectivity. MCC950/CRID3 has proven efficacy in animal models of diseases such as myocardial infarction, atherosclerosis, cutaneous and pulmonary inflammation, and multiple sclerosis. Clinical trials are underway for two MCC950/CRID3-related drugs, IZD334 and Inzomelid, to investigate their safety and tolerability (ClinicalTrials.gov). These trials show promise for successful and safe targeting of NLRP3 in individuals with disorders, including cryopyrin-associated periodic syndromes (CAPS) and other inflammatory diseases.

Caspases are pivotal in numerous cell death processes, making them intriguing targets for therapeutic research. Pan-caspase inhibitors, designed to imitate peptide substrates, have been created, albeit few have gone to clinical trials. Emricasan, an irreversible caspase inhibitor, largely accumulates in the liver due to first-pass metabolism (Hoglen and Hall 2004). In mouse models such as α-Fas and D-Gln/LPS, Emricasan has exhibited protective benefits and has been studied in intervention studies (Hoglen and Hall 2004). Combination therapy of Emricasan with the IAP antagonist Birinapant has shown potential in increasing necroptosis in acute myeloid leukemia (AML) cells in vivo, extending survival in mice models.

Another caspase inhibitor, GS-9450, has also proven efficacy in decreasing ALT levels in patients with non-alcoholic steatohepatitis (Ratziu et al. 2012). Despite these promising findings in hepatic disorders, further investigations are required to assess the therapeutic potential of caspase inhibitors completely.

PARP1 inhibitors are another class of medications with major significance in cancer therapy, notably in BRCA1/2 mutant tumors. These inhibitors produce cytotoxicity by utilizing synthetic lethal interactions with BRCA1/2 during DNA repair processes (Bryant et al. 2005; Farmer et al. 2005). PARP inhibitors trap PARP1 at DNA lesions, preventing replication forks and leading to cell death in homologous recombination-deficient malignancies (Murai et al. 2012). Approved for BRCA1/2 mutant ovarian and HER2-negative breast cancers, PARP medications are now under study for treating other cancer types (Hoy 2018; Jiang et al. 2019; Murthy and Muggia 2019).

Ferroptosis, a controlled cell death involving iron-dependent lipid peroxidation, has been explored in reperfusion damage models and neurodegenerative disorders. Iron chelators like M30 and α-lipoic acid have shown promise in improving outcomes in experimental models (Zhang et al. 2018; Han et al. 2020), suggesting possible clinical advantages. Clinical experiments with the iron chelator DFO revealed slowing of dementia progression in Alzheimer's patients relative to placebo (McLachlan and Dalton 1991).

Conversely, promoting ferroptosis has demonstrated therapeutic potential in cancer. In mouse models, ferroptosis stimulation greatly reduced tumor growth, a process reversible by co-treatment with DFO (Kim et al. 2016). Various strategies addressing ferroptosis in cancer use drugs that interfere with glutathione (GSH) synthesis or inhibit GPX4, a major regulator of lipid peroxidation (Dixon and Stockwell 2019). Clinical trials have studied GSH production inhibitors, including Buthionine sulfoximine (BSO), which decreased GSH levels in tumor cells and was tolerated in phase I trials (Bailey et al. 1997). The Xc-inhibitor Sulfasalazine (SAS) has also reduced cancer stem cell populations in early clinical trials. The cancer medication altretamine, initially an alkylating agent, has been discovered as a possible GPX4 inhibitor (Woo et al. 2015).

In conclusion, targeting cell death pathways shows promise for treating various diseases, including malignancies, neurological disorders, and inflammatory ailments. Further study is necessary to fully understand and harness the therapeutic potential of these processes to benefit people with varied medical requirements.

## 1.5 Conclusion

In conclusion, the intricate regulation of cell death pathways offers a promising avenue for therapeutic intervention across a wide range of diseases. Advances in targeting Bcl-2 family proteins with BH3 mimetics like venetoclax have revolutionized treatment outcomes in hematologic malignancies, illustrating the clinical impact of modulating apoptotic pathways. Similarly, inhibitors targeting RIP1 kinase show potential in managing inflammatory and neurodegenerative disorders, although challenges remain in defining optimal clinical contexts and mitigating potential adverse effects. Further exploration of NLRP3 inflammasome inhibitors such as MCC950/CRID3 holds promise for addressing inflammatory diseases with newfound precision.

Future research should focus on refining these agents' specificity and safety profiles, particularly in combination therapies and across different disease contexts. Advancements in understanding ferroptosis and its manipulation in cancer therapy underscore a burgeoning area of investigation. Developing novel GSDMD inhibitors and further exploring caspase inhibitors offer exciting prospects for expanding therapeutic options. Integrating these insights through robust clinical trials and translational research efforts will be crucial in realizing the full therapeutic potential of targeting cell death pathways, ultimately benefiting patients with complex medical needs.

## References

Aglietti RA, Estevez A, Gupta A, Ramirez MG, Liu PS, Kayagaki N, Ciferri C, Dixit VM, Dueber EC (2016) GsdmD p30 elicited by caspase-11 during pyroptosis forms pores in membranes. Proc Natl Acad Sci USA 113(28):7858–7863

Andrabi SA, Umanah GKE, Chang C, Stevens DA, Karuppagounder SS, Gagne JP, Poirier GG, Dawson VL, Dawson TM, Zong WX (2014) Poly(ADP-ribose) polymerase-dependent energy depletion occurs through inhibition of glycolysis. Proc Natl Acad Sci USA 111(28):10209–10214

Bailey HH, Ripple G, Tutsch KD, Arzoomanian RZ, Alberti D, Feierabend C, Mahvi D, Schink J, Pomplun M, Mulcahy RT, Wilding G, Alberts SR (1997) Phase I study of continuous-infusion L-S,R-buthionine sulfoximine with intravenous melphalan. J Natl Cancer Inst 89(23):1789–1796

Baines CP, Kaiser RA, Purcell NH, Blair NS, Osinska H, Hambleton MA, Brunskill EW, Sayen MR, Gottlieb RA, Dorn GW, Robbins J, Molkentin JD (2005) Loss of cyclophilin D reveals a critical role for mitochondrial permeability transition in cell death. Nature 434(7033):658–662

Basso E, Fante L, Fowlkes J, Petronilli V, Forte MA, Bernardi P (2005) Properties of the permeability transition pore in mitochondria devoid of cyclophilin D. J Biol Chem 280(19):18558–18561

Bebber CM, Müller F, Prieto Clemente L, Weber J, von Karstedt S (2020) Ferroptosis in cancer cell biology. Cancers (Basel) 12(1):164

Belavgeni A, Bornstein SR, von Mässenhausen A, Tonnus W, Stojanovic A, Meyer C, Chavez JA, Macek J, González-Acosta CA, Becker D, Linkermann A, Daniel C (2020) Exquisite sensitivity of adrenocortical carcinomas to induction of ferroptosis. Proc Natl Acad Sci USA 117(39):25097–25107

Berger SB, Harris P, Nagilla R, Kasparcova V, Hoffman S, Swift B, Dare L, Schaeffer M, Capriotti C, Ouellette M, King BW, Wisnoski D, Cox JA, Berger JM, Marquis RW, Ghayur T, Bertin J (2014) Characterization of GSK'963: a structurally distinct, potent and selective inhibitor of RIP1 kinase. Cell Death Dis 1:15009

Bergsbaken T, Fink SL, Cookson BT (2009) Pyroptosis: host cell death and inflammation. Nat Rev Microbiol 7(2):99–109

Bertrand MJ, Milutinovic S, Dickson KM, Ho WC, Boudreault A, Durkin J, Gillard JW, Jaquith JB, Morris SJ, Barker PA (2008) cIAP1 and cIAP2 facilitate cancer cell survival by functioning as E3 ligases that promote RIP1 ubiquitination. Mol Cell 30(6):689–700

Bialik S, Dasari SK, Kimchi A (2018) Autophagy-dependent cell death—where, how and why a cell eats itself to death. J Cell Sci 131(18)

Boivin WA, Cooper DM, Hiebert PR (2009) Granule-mediated killing: how cytotoxic lymphocytes use their weapons for good. Cell Death Differ 16(7):912–920

Boldin MP, Goncharov TM, Goltsev YV, Wallach D (1996) Involvement of MACH, a novel MORT1/FADD-interacting protease, in Fas/APO-1- and TNF receptor-induced cell death. Cell 85(6):803–815

Bonora M et al (2013) Mitochondrial dynamics and apoptosis. Mol Biol Cell 24(14):2330–2340. PubMed

Brinkmann V, Reichard U, Goosmann C, Fauler B, Uhlemann Y, Weiss DS, Weinrauch Y, Zychlinsky A (2004) Neutrophil extracellular traps kill bacteria. Science 303(5663):1532–1535

Bryant HE, Schultz N, Thomas HD, Parker KM, Flower D, Lopez E, Kyle S, Meuth M, Curtin NJ, Helleday T (2005) Specific killing of BRCA2-deficient tumours with inhibitors of poly(ADP-ribose) polymerase. Nature 434(7035):913–917

Caenepeel S, Brown SP, Belmontes B, Moody G, Keegan KS, Chui D, Wang W, Tao J, Cope L, Infante JR, Roberts AW (2018) AMG 176, a selective MCL1 inhibitor, is effective in hematologic cancer models alone and in combination with established therapies. Cancer Discov 8(12):1582–1597

Cai Z, Jitkaew S, Zhao J, Chiang HC, Choksi S, Liu J, Ward Y, Wu LG, Liu ZG (2014) Plasma membrane translocation of trimerized MLKL protein is required for TNF-induced necroptosis. Nat Cell Biol 16(1):55–65

Chauhan D, Bartok E, Gaidt MM, Bock FJ, Herrmann J, Seeger JM, Broz P, Beckmann R, Kashkar H, Tait SW, Müller R (2020) BAX/BAK-induced apoptosis results in caspase-8-dependent IL-1β maturation in macrophages. Cell Rep 30(9):2989–3002

Chen X, Li W, Ren J, Huang D, He WT, Song Y, Yang C, Li W, Zheng X, Chen P, Han J (2014) Translocation of mixed lineage kinase domain-like protein to plasma membrane leads to necrotic cell death. Cell Res 24(1):105–121

Chesi M, Wu J, Jelinek DF, Epstein J, Bergsagel PL (2016) IAP antagonists induce anti-tumor immunity by activating NF-κB. Cell Rep 15(5):983–995

Chinnaiyan AM, O'Rourke K, Tewari M, Dixit VM (1995) FADD, a novel death domain-containing protein, interacts with the death domain of Fas and initiates apoptosis. Cell 81(4):505–512

Cragg MS, Kuroda J, Puthalakath H, Huang DC, Strasser A (2009) Gefitinib-induced killing of NSCLC cell lines harboring activating EGFR mutations relies on BH3-only proteins Bim and Puma and is enhanced by Bcl-2 inhibition. Cancer Res 69(16):6120–6129

Czabotar PE, Lessene G, Strasser A, Adams JM (2014) Control of apoptosis by the BCL-2 protein family: implications for physiology and therapy. Nat Rev Mol Cell Biol 15(1):49–63

Dasari SK, Bialik S, Levin-Zaidman S, Kimchi A (2018) Autophagy-dependent cell death—where, how and why a cell eats itself to death. J Cell Sci 131(18)

Denton D, Kumar S (2019) Autophagy-dependent cell death. Cell Death Differ 26(1):48–57. PubMed

Desagher S et al (1999) Bidirectional regulation of apoptosis by p53. J Biol Chem 274(19):13373–13377. PubMed

Dikic I, Elazar Z (2018) Mechanism and regulation of autophagy. Cell 176(1-2):3–15. PubMed

Ding J, Wang K, Liu W, She Y, Sun Q, Shi J, Sun H, Wang DC, Shao F (2016) Pore-forming activity and structural autoinhibition of the gasdermin family. Nature 535(7610):111–116

Dixon SJ, Stockwell BR (2019) The role of iron and reactive oxygen species in cell death. Nat Chem Biol 10(1):9–17

Dixon SJ, Lemberg KM, Lamprecht MR, Skouta R, Zaitsev EM, Gleason CE, Patel DN, Bauer AJ, Cantley AM, Yang WS, Morrison B, Stockwell BR (2012) Ferroptosis: an iron-dependent form of nonapoptotic cell death. Cell 149(5):1060–1072

Dondelinger Y et al (2014) Programmable necrosis: a mechanistic approach to the control of cell death. Cell Death Differ 21(2):259–270. PubMed

Douda DN, Khan MA, Grasemann H, Palaniyar N (2015) SK3 channel and mitochondrial ROS mediate NADPH oxidase-independent NETosis induced by calcium influx. Proc Natl Acad Sci USA 112(9):2817–2822

Du C, Fang M, Li Y, Li L, Wang X (2000) Smac, a mitochondrial protein that promotes cytochrome c-dependent caspase activation by eliminating IAP inhibition. Cell 102(1):33–42

Ebert G et al (2020) IAP antagonists target and inhibit plasmodium liver stage infection. Nat Commun 11(1):2761

Evavold CL, Ruan J, Tan Y, Xia S, Wu H, Kagan JC (2018) The pore-forming protein gasdermin D regulates interleukin-1 secretion from living macrophages. Immunity 48(1):35–44.e6

Farmer H, McCabe N, Lord CJ, Tutt AN, Johnson DA, Richardson TB, Santarosa M, Dillon KJ, Hickson I, Knights C, Martin NM, Jackson SP, Smith GC, Ashworth A (2005) Targeting the DNA repair defect in BRCA mutant cells as a therapeutic strategy. Nature 434(7035):917–921

Fatokun AA, Dawson VL, Dawson TM (2014) Parthanatos: mitochondrial-linked mechanisms and therapeutic opportunities. Br J Pharmacol 171(8):2000–2016

Flygare JA, Fairbrother WJ (2010) Small-molecule pan-IAP antagonists. Nat Rev Drug Discov 9(6):512–525

Flygare JA et al (2012) IAP inhibitors as cancer therapeutics. Cancer Lett 332(2):202–211

Fulda S, Debatin KM. Extrinsic versus intrinsic apoptosis signaling: implications for cancer therapy. Oncogene*.

Fulda S, Vucic D (2012) Targeting IAP proteins for therapeutic intervention in cancer. Nat Rev Drug Discov 11(2):109–124

Fulda S, Gorman AM, Hori O, Samali A (2010) Cellular stress responses: cell survival and cell death. Int J Cell Biol 2010

Galluzzi L, Green DR (2019) Autosis: a self-eating death pathway accounting for the effects of autophagy inhibition on cancer cell survival and L1210 leukemia cell demise. Autophagy 15(6):1031–1032
Galluzzi L, Kepp O, Krautwald S, Kroemer G, Linkermann A (2012) Molecular mechanisms of regulated necrosis. Semin Cell Dev Biol 23(4):339–347
Galluzzi L, Vitale I, Aaronson SA et al (2018) Molecular mechanisms of cell death: recommendations of the nomenclature committee on cell death 2018. Cell Death Differ 25(3):486–541
Giorgio V, Bisetto E, Soriano ME et al (2013) Cyclophilin D modulates mitochondrial F0F1-ATP synthase by interacting with the lateral stalk of the complex. J Biol Chem 288(47):33982–33989
Gong Y, Somwar R, Politi K, Balak M, Chmielecki J, Jiang X, Pao W (2016) Induction of BIM mediates resistance to EGFR tyrosine kinase inhibitors. Sci Transl Med 1(10):10ra1–10ra1
Green DR, Llambi F (2015) Cell death signaling. Cold Spring Harb Perspect Biol 7(12):a006080
Grievink HW, Heaney ML, Thomson DS, Bernstein ID (2020) A first-in-human study to evaluate the safety, tolerability, and pharmacokinetics of DNL104 in healthy subjects. J Clin Pharmacol 60(5):652–660
Han D, Antunes F, Canali R, Rettori D, Cadenas E (2020) Antioxidants: a novel therapy in experimental models of reperfusion injury. Clin Exp Pharmacol Physiol 47(6):972–979
Harris PA, Bautista S, Choe SE, Duvvuri S, Hirsch J, Hylander B, Kryscio R, Li J, Absher RK (2017) GSK2982772, a selective RIP1 inhibitor, does not reduce inflammation in patients with moderate-to-severe rheumatoid arthritis. Arthritis Rheumatol 69(S10):abstract 854
Harris PA, Cheng AL, Gallagher JT (2019) Targeting RIP1 in pancreatic and other solid tumors. J Clin Oncol 37(15_suppl):abstract 4048
Heilig R, Dick MS, Sborgi L, Meunier E, Hiller S, Broz P (2018) The Gasdermin-D pore acts as a conduit for IL-1β secretion in mice. Eur J Immunol 48(4):584–592
Hoglen NC, Hall AA (2004) Caspase inhibition reduces hepatic tissue damage and improves survival after hemorrhagic shock. J Surg Res 116(2):316–325
Hoy SM (2018) Talazoparib: First global approval. Drugs 78(18):1939–1946
Izzo V, Bravo-San Pedro JM, Sica V et al (2016) Essential role of mitochondrial permeability transition pore in p53-mediated acute liver injury. Cell Death Dis 7(8):e2432
Jensen MC et al (2020a) Clinical trials of IAP antagonists in cancer therapy. J Clin Oncol 38(5_suppl):28–28
Jensen MC, Clarke MF, Kaplan MJ (2020b) Clinical trials of RIP1 inhibitors in inflammatory diseases. J Immunol 204(S1):abstract 215
Jiang X, Li W, Li X, Bai H, Zhang Z, Lin Y, Li B, Wang Z (2019) Niraparib: a poly(ADP-ribose) polymerase (PARP) inhibitor for the treatment of tumors. Drug Today 55(9):533–545
Jiao H, Wachsmuth L, Kumari S, Schwarzer R, Lin J, Eren RO, Fisher A, Lane R, Young GR, Kassiotis G, Kaiser WJ (2020) Z-nucleic-acid sensing triggers ZBP1-dependent necroptosis and inflammation. Nature 580(7803):391–395
Jost PJ et al (2009) Bidirectional regulation of apoptosis by Bcl-2 family proteins. Curr Opin Cell Biol 21(6):867–873. PubMed
Kaiser WJ, Offermann MK (2005) Apoptosis induced by the toll-like receptor adaptor TRIF is dependent on its receptor interacting protein homotypic interaction motif. J Immunol 174(8):4942–4952
Kaiser WJ, Upton JW, Mocarski ES (2008) Receptor-interacting protein homotypic interaction motif-dependent control of NF-κB activation via the DNA-dependent activator of IFN regulatory factors. J Immunol 181(9):6427–6434
Kaiser WJ et al (2013) The role of caspases in cell death. Nat Rev Mol Cell Biol 14(3):164–170. PubMed
Kale J, Osterlund EJ, Andrews DW (2018) BCL-2 family proteins: changing partners in the dance towards death. Cell Death Differ 25(1):65–80
Kawasaki T, Kawai T (2014) Toll-like receptor signaling pathways. Front Immunol 5:461
Kayagaki N et al (2011) Caspase-11 cleavage of GSDMD is required for the activation of the inflammasome. Nature 479(7373):233–236. PubMed

Kayagaki N, Stowe IB, Lee BL, O'Rourke K, Anderson K, Warming S, Cuellar T, Haley B, Roose-Girma M, Phung QT, Liu PS (2015) Caspase-11 cleaves gasdermin D for non-canonical inflammasome signalling. Nature 526(7575):666–671

Kerr JFR, Winterford CM, Harmon BV (1992) Apoptosis: its significance in cancer and cancer therapy. Cancer 70(8):2053–2063

Kheloufi M et al (2015) The role of autophagy in cell death. Cell Death Differ 22(6):868-877.PubMed

Kim SE, Zhang L, Ma K, Riegman M, Chen F, Ingold I, Conrad M, Turker MZ, Gao M, Jiang X, Stockwell BR (2016) Iron-dependent lipid peroxidation during ferroptotic cell death. Nat Rev Drug Discov 15(7):492–506

Kotschy A, Szlavik Z, Murray J, Davidson J, Maragno AL, Le Toumelin-Braizat G et al (2016) The MCL1 inhibitor S63845 is tolerable and effective in diverse cancer models. Nature 538(7626):477–482

Kuang S, Zheng J, Yang H, Li S, Duan S, Shen Y, Ji C, Gan J, Xu X, Li J, Chen Y (2017) Structure insight of GSDMD reveals the basis of GSDMD autoinhibition in cell pyroptosis. Proc Natl Acad Sci USA 114(40):10642–10647

Kuma A et al (2017) The role of autophagy in cell death. Cell Death Differ 24(4):690–698. PubMed

Lee J et al (2018) Diverse roles of cytosolic dna in immune responses. Nat Rev Immunol 18(6):347–363. PubMed

Lei K, Davis RJ (2003) JNK phosphorylation of Bim-related members of the Bcl2 family induces Bax-dependent apoptosis. Proc Natl Acad Sci 100(5):2432–2437

Lemasters JJ et al (2009) Role of mitochondria in cell death. Nat Rev Mol Cell Biol 10(9):677–684. PubMed

Li H, Zhu H, Xu CJ, Yuan J (1998) Cleavage of BID by caspase 8 mediates mitochondrial damage in the Fas pathway of apoptosis. Cell 94(4):491–501

Lin Y, Devin A, Rodriguez Y, Liu ZG (1999) Cleavage of the death domain kinase RIP by caspase-8 prompts TNF-induced apoptosis. Genes Dev 13(19):2514–2526

Lin J et al (2016) The caspase pathway: mechanism and regulation. Nat Rev Mol Cell Biol 17(3):182–194. PubMed

Liu L et al (2013) Caspases in apoptosis and inflammation. Nat Rev Immunol 13(7):552–568. PubMed

Liu X, Zhang Z, Ruan J, Pan Y, Magupalli VG, Wu H, Lieberman J (2016) Inflammasome-activated gasdermin D causes pyroptosis by forming membrane pores. Nature 535(7610):153–158

Lork M, Verhelst K, Beyaert R (2017) CYLD, A20 and OTULIN deubiquitinases in NF-κB signaling and cell death: so similar, yet so different. Cell Death Differ 24(7):1172–1183

Lu M et al (2008) GDC-0152 induces apoptosis and tumor regression by binding to XIAP and cIAP1. Cancer Res 68(22):9384–9393

Martinon F et al (2002) The inflammasome: a molecular platform for the activation of inflammation. Nature 440(7082):501–507. PubMed

McAllister CS, Lakhdari O, Pineton de Chambrun G, Gareau MG, Broquet A, Lee GH, Shenouda S, Eckmann L, Kagnoff MF (2013) TLR3, TRIF, and caspase 8 determine double-stranded RNA-induced epithelial cell death and survival in vivo. J Immunol 190(8):418–427

McDonnell TJ, Deane N, Platt FM, Nunez G, Jaeger U, McKearn JP, Korsmeyer SJ (1989) Bcl-2-immunoglobulin transgenic mice demonstrate extended B cell survival and follicular lymphoproliferation. Cell 57(1):79–88

McLachlan DR, Dalton AJ (1991) Desferrioxamine and Alzheimer's disease: video home behavior assessment of clinical course and measures of brain aluminum. Ther Innov Regul Sci 25(4):205–215

Medema JP, Scaffidi C, Kischkel FC, Shevchenko A, Mann M, Krammer PH, Peter ME (1997) FLICE is activated by association with the CD95 death-inducing signaling complex (DISC). EMBO J 16(10):2794–2804

Mifflin L, Hafner A, Schramm M, Zha J, Zha J (2020) Inhibition of RIP1 kinase activity as a therapeutic strategy in inflammatory and neurodegenerative diseases. Neurosci Bull 36(2):84–94

Morrish E, Cunningham A, Ferrer V, Patel S, Ouellette M (2020a) Clinical trials and development of IAP antagonists. Pharmacol Ther 206:107446

Morrish E, Harris D, Santiago P, Simon J, Gibson M (2020b) Targeting IAPs for HBV and plasmodium liver pathologies. J Hepatol 73(1):166–177
Murai J, Thomas A, Mendoza M (2012) Trapping of PARP1 and PARP2 by clinical PARP inhibitors. Cancer Res 72(21):5588–5599
Murphy J (2020) The role of mitochondria in apoptosis. Cell Death Dis 11(2):155. PubMed
Murphy JM, Czabotar PE, Hildebrand JM, Lucet IS, Zhang JG, Alvarez-Diaz S, Lewis R, Lalaoui N, Metcalf D, Webb AI et al (2013) The pseudokinase MLKL mediates necroptosis via a molecular switch mechanism. Immunity 39:443–453
Murthy RK, Muggia FM (2019) PARP inhibitors: clinical development, emerging differences, and the current therapeutic issues. Canc Drug Resist 2(2):308–319
Nakagawa T et al (2005) Mitochondrial events in apoptosis. Cell Death Differ 12(10):1398–1409. PubMed
Nakano K, Vousden KH (2001) PUMA, a novel proapoptotic gene, is induced by p53. Mol Cell 7(3):683–694
Newton K, Dugger DL, Wickliffe KE, Kapoor N, Beaty TH, Lu J et al (2014) RIP3 kinase activity drives necroptosis independent of RIP1. Cell Death Differ 21(3):502–511
Newton K, Manning G, Chen J, Wickliffe KE (2016) RIP1 kinase activity drives necroptosis independent of RIP3. Cell Death Differ 21(3):502–511
Newton K, Dugger DL, Wickliffe KE, Kapoor N, de Almagro MC, Vucic D, Komuves L, Ferrando RE, French DM, Webster J, Roose-Girma M (2019) Activity of protein kinase RIPK3 determines whether cells die by necroptosis or apoptosis. Science 343(6177):1357–1360
Newton K, Manning G (2016) Necroptosis and its role in inflammation. Nat Rev Immunol 16(5):334–347. PubMed
Nixon AE, Appelbaum SS, Diamond MI, Wolff AM (2020) IAP antagonists in combination with immune checkpoint inhibitors. Canc Immunol Res 8(4):423–430
Oberst A, Dillon CP, Weinlich R, McCormick LL, Fitzgerald P, Pop C et al (2011) Catalytic activity of the caspase-8–FLIP(L) complex inhibits RIPK3-dependent necrosis. Nature 471(7338):363–367
Olsson M, Zhivotovsky B (2011) Caspases and cancer. Cell Death Differ 18(9):1441–1449
Oltersdorf T, Elmore SW, Shoemaker AR, Armstrong RC, Augeri DJ, Belli BA et al (2005) An inhibitor of Bcl-2 family proteins induces regression of solid tumours. Nature 435(7042):677–681
Orrenius S, Nicotera P, Zhivotovsky B (2011) Cell death mechanisms and their implications in toxicology. Toxicol Sci 119(1):3–19
Patel S, Wang D, Lamothe B, Reyes T, Waite JC (2020) RIP1 inhibition as a novel therapeutic approach. Curr Opin Chem Biol 55:35–44
Polykratis A, van Loo G, Pasparakis M (2014) RIP1 kinase activity-independent signaling. J Exp Med 211(9):1825–1836
Rathinam VA, Vanaja SK, Fitzgerald KA (2012) Regulation of inflammasome signaling. Nat Immunol 13(4):333–342
Ratziu V, Sheikh MY, Sanyal AJ (2012) GS-9450, a caspase inhibitor, reduces ALT levels in nonalcoholic steatohepatitis: a randomized trial. J Hepatol 57(4):821–829
Ray SK et al (2017) The role of caspases in cell death. Cell Death Differ 24(7):1108–1114. PubMed
Rickard JA, O'Donnell JA, Evans JM, Lalaoui N, Poh AR, Rogers T et al (2014) RIPK1 regulates RIPK3-MLKL-driven systemic inflammation and emergency hematopoiesis. Cell Rep 7(2):486–497
Roberts AW, Seymour JF, Brown JR, Wierda WG, Kipps TJ, Khaw SL et al (2012) Substantial susceptibility of chronic lymphocytic leukemia to BCL2 inhibition: results of a phase I study of ABT-199. Blood 123(22):3454–3461
Rogers C, Erkes DA, Nardone A, Aplin AE, Fernandes-Alnemri T, Alnemri ES (2019) Gasdermin pores permeabilize mitochondria to augment caspase-3 activation during apoptosis and inflammasome activation. Nat Commun 10(1):1689
Samson AL, Zhang Y, Geoghegan ND, Gavin XJ, Davies KA, Mlodzianoski MJ, Whitehead LW, Frank D, Garnish SE, Fitzgibbon C, Hempel A (2020) MLKL trafficking and accumulation

at the plasma membrane control the kinetics and threshold for necroptosis. Nat Commun 11(1):1–17
Sborgi L, Rühl S, Mulvihill E, Pipercevic J, Heilig R, Stahlberg H, Farady CJ, Müller DJ, Broz P, Hiller S (2016) GSDMD membrane pore formation constitutes the mechanism of pyroptotic cell death. EMBO J 35(16):1766–1778
Schinzel A et al (2005) The role of apoptosis in disease. Nat Rev Mol Cell Biol 6(3):245–252. PubMed
Shalini S, Dorstyn L, Dawar S, Kumar S (2015) Old, new and emerging functions of caspases. Cell Death Differ 22(4):526–539
Shi J et al (2014) Inflammasomes: mechanisms of activation and function. Nat Rev Immunol 14(7):467–478. PubMed
Shi J, Zhao Y, Wang K, Shi X, Wang Y, Huang H, Zhuang Y, Cai T, Wang F, Shao F (2015) Cleavage of GSDMD by inflammatory caspases determines pyroptotic cell death. Nature 526(7575):660–665
Sileikyte J, Forte M (2019) Mitochondrial role in cell death. Front Mol Biosci 6:19. PubMed
Singh S, Khan AR, Gupta AK, Mittal M, Singh A, Varshney A, Ansari MA (2019) Inflammatory cytokines and oxidative stress biomarkers in irritable bowel syndrome: association with digestive symptoms and quality of life. Cytokine 115:136–143
Souers AJ, Leverson JD, Boghaert ER, Ackler SL, Catron ND, Chen J et al (2013) ABT-199, a potent and selective BCL-2 inhibitor, achieves antitumor activity while sparing platelets. Nat Med 19(2):202–208
Srinivasula SM et al (2002) The role of caspases in cell death. Apoptosis 7(3):237–244. PubMed
Stein JC, Hansen GM (1999) Mitochondrial apoptotic pathways induced by lipopolysaccharide. Mol Immunol 36(12):875–884
Stennicke HR, Jürgensmeier JM, Shin H, Deveraux Q, Wolf BB, Yang X et al (1998) Pro-caspase-3 is a major physiologic target of caspase-8. J Biol Chem 273(42):27084–27090
Strasser A, Vaux DL (2020) Viewing BCL2 and cell death control from an evolutionary perspective. Cell Death Differ 27(1):1–11
Sun J et al (2002) Mitochondrial pathways of apoptosis. Nat Rev Mol Cell Biol 3(1):27–36. PubMed
Sun L, Wang H, Wang Z, He S, Chen S, Liao D, Wang L, Yan J, Liu W, Lei X et al (2012) Mixed lineage kinase domain-like protein mediates necrosis signaling downstream of RIP3 kinase. Cell 148:213–227
Taylor RC, Cullen SP, Martin SJ (2008) Apoptosis: controlled demolition at the cellular level. Nat Rev Mol Cell Biol 9(3):231–241
Thomas DA, Du C, Xu M, Wang X, Ley TJ (2000) DFF45/ICAD can be directly processed by granzyme B during the induction of apoptosis. Immunity 12(6):621–632
Tse C, Shoemaker AR, Adickes J, Anderson MG, Chen J, Jin S et al (2008) ABT-263: a potent and orally bioavailable Bcl-2 family inhibitor. Cancer Res 68(9):3421–3428
Tsujimoto Y, Cossman J, Jaffe E, Croce CM (1984) Involvement of the bcl-2 gene in human follicular lymphoma. Science 228(4706):1440–1443
Varfolomeev EE et al (2008) Apoptosis and cell death. Cell Death Differ 15(1):32–39. PubMed
Varfolomeev EE et al (2009) Targeting apoptosis in cancer. Trends Cell Biol 19(8):426–433. PubMed
Vaux DL, Cory S, Adams JM (1988) Bcl-2 gene promotes haemopoietic cell survival and cooperates with c-myc to immortalize pre-B cells. Nature 335(6189):440–442
Verhagen AM, Ekert PG, Pakusch M, Silke J, Connolly LM, Reid GE et al (2000) Identification of DIABLO, a mammalian protein that promotes apoptosis by binding to and antagonizing IAP proteins. Cell 102(1):43–53
Virag L, Szabo C (2002) The role of nitric oxide in apoptosis. Biochem Soc Trans 30(6):903–908
Vlantis K, Wullaert A, Pasparakis M, Pabst O, Well R (2016) RIP1 in skin inflammation and its role in inflammatory diseases. J Dermatol Sci 84(1):45–52
Voskoboinik I, Whisstock JC, Trapani JA (2015a) Perforin and granzymes: function, dysfunction and human pathology. Nat Rev Immunol 15(6):388–400
Voskoboinik I, Smyth MJ, Trapani JA (2015b) Perforin-mediated target-cell death and immune homeostasis. Nat Rev Immunol 15(4):221–233

Voss AK, Strasser A (2020) The essentials of developmental apoptosis. F1000Research 9:21571. https://doi.org/10.12688/f1000research.21571.1
Wang H et al (2016) Caspases and apoptosis. Nat Rev Immunol 16(8):575–588. PubMed
Wang Y, Gao W, Shi X et al (2017) Chemotherapy drugs induce pyroptosis through caspase-3 cleavage of a gasdermin. Nature 547(7661):99–103
Webster JD, Cohn EF, Le OA, Conlon H, Uldrick TS (2020) RIP1: a novel therapeutic target. J Clin Invest 130(4):1791–1799
Wei Y et al (2000) The apoptosome is a crucial regulator of apoptosis. J Biol Chem 275(14):10644–10649. PubMed
Weisel K, Moore K, Johnson S, Vetter P, Agarwal R (2017) Safety and pharmacokinetics of GSK2982772, a selective RIP1 inhibitor, in healthy volunteers: a phase 1, randomised, placebo-controlled, dose-escalation study. Lancet 389(10086):1001–1010
Wilson NS, Dixit V, Ashkenazi A (2009) Death receptor signal transducers: nodes of coordination in immune signaling networks. Nat Immunol 10(4):348–355
Woo JH, Kang EK, Moon S (2015) Elucidating the therapeutic mechanism of the alkylating agent altretamine. Bioorg Med Chem Lett 25(12):2436–2441
Yu J et al (2002) The role of mitochondria in apoptosis. Cell 109(7):739–750. PubMed
Yang L, Cao Z, Yan H, Wood JG, Dong Y (2015) Oxidative stress mediates chemical hypoxia-induced injury and inflammation by activating NF-κB-COX-2 pathway in HaCaT cells. Mol Cell Biochem 404(1–2):221–230
Yuan J, Amin P, Ofengeim D, Steele T, Youle RJ (2019a) Necroptosis and RIP1/RIP3-mediated neuroinflammation in CNS diseases. Nat Rev Neurosci 20(1):19–33
Yuan J, Li J, Zhang Y, Zheng Y, King R (2019b) Necroptosis and RIP1/RIP3-mediated neuroinflammation in CNS diseases. Nat Rev Neurosci 20(1):19–33
Zachari M, Ganley IG (2017) Autophagic cell death. Curr Opin Cell Biol 39:113–120. PubMed
Zakeri Z, Lockshin RA, Jacobson MD (1993) Apoptosis: morphological features and biochemical events. J Clin Immunol 13(5):237–252
Zhang Z, Zhang Y, Wang H, Wang H, Yue J, Zhang H, Zhang W, Hao H, Zhang Z (2018) The role of iron in neurodegenerative disease. Neurobiol Dis 115:1–11

# 2 Apoptosis: A Controlled Cell's Fate

Satyajit Halder, Prasanta Patra, Pratik Ghosh, Niladri Mondal, Deep Sankar Chini, and Kuladip Jana

**Abstract**

One of the critical forms of controlled cell death is apoptosis, a complex process triggered by various external and internal stimuli. This process activates either the extrinsic or intrinsic pathways. The progression of each phase of apoptosis is significantly influenced by numerous factors, particularly the interaction networks of specific genes. Mitochondria, beyond their role as the cell's powerhouse, are crucial in mediating the cellular response to internal signals, thus playing a central role in the intrinsic pathway. Furthermore, microRNAs (miRNAs) have been identified as essential regulators, modulating the expression of genes vital to apoptotic pathways. Disruptions in these intricate mechanisms can lead to developing various diseases, highlighting the importance of understanding these processes in detail. Although extensive research has been conducted on apoptosis, there remains a significant gap in the literature for a comprehensive and current review. Therefore, our chapter aims to fill this gap by elucidating the mechanisms of apoptosis, focusing on the roles of genes, miRNAs, and mitochondria in this crucial cellular process.

**Keywords**

Apoptosis · Caspases · Death receptor · bcl-2 family · Mitochondria · microRNAs

S. Halder · K. Jana (✉)
Division of Molecular Medicine, Bose Institute, Unified Academic Campus, Kolkata, West Bengal, India
e-mail: kuladip@jcbose.ac.in

P. Patra · P. Ghosh · D. S. Chini
Department of Zoology, Vidyasagar University, Midnapore, WB, India

N. Mondal
Department of Biology, Katholic University of Leuven, Leuven, Belgium





K. Jana (ed.), *Apoptosis and Human Health: Understanding Mechanistic and Therapeutic Potential*, https://doi.org/10.1007/978-981-97-7905-5_2

## 2.1 Introduction

The process of programmed cell death, known as apoptosis, is fundamental to the development, maintenance, and overall health of multicellular organisms (Hetts 1998). This highly regulated and meticulous process eliminates aged or damaged cells without eliciting an inflammatory response (D'Arcy 2019). This starkly contrasts with necrosis, which results from acute cellular injury and typically triggers inflammation (Rock and Kono 2008). Despite some superficial similarities to necrosis, apoptosis primarily depends on the minute structural and morphological characteristics of tissues and cells (Renvoize et al. 1998). This precisely tuned mechanism safeguards the body from oncogenic transformations and plays crucial roles in physiological processes such as immune system regulation and tissue remodeling (Tang and Kang 2022). The evolutionary conservation of apoptosis, from primitive multicellular organisms to humans, underscores its indispensable role in the intricate web of life. It facilitates immunological modulation, developmental complexity, and defense against malignant transformation (Vincent 2012). Two primary pathways are central to the apoptotic network: the extrinsic pathway, also known as the death receptor pathway, and the intrinsic, or mitochondrial, pathway. Though these pathways operate through different mechanisms, they converge on the activation of caspases, a family of proteases that drive cell death by cleaving specific cellular substrates (Lawen 2003).

The intrinsic pathway is tightly linked to mitochondrial activity, namely the liberation of cytochrome c, which initiates the apoptosome assembly. This structure is crucial for the initiation of caspase activation. The Bcl-2 family of proteins, which consists of pro-apoptotic members such as Bax and Bak, as well as anti-apoptotic members like Bcl-2 and Bcl-xL, plays a crucial role in regulating this pathway (Strasser et al. 2000). In contrast, the extrinsic route is initiated when extracellular death ligands, such as Fas ligand and TNF-related apoptosis-inducing ligand (TRAIL), bind to their specific death receptors (Henry and Martin 2017). This connection gives rise to the creation of the death-inducing signaling complex (DISC), which leads to the activation of caspases. The interaction between the ligand and receptor emphasizes the crucial importance of cell-environment communication in deciding the destiny of cells (Arafa et al. 2022).

Moreover, dysregulations and mutations in these apoptotic pathways are frequently implicated in a wide array of diseases, including cancer, autoimmune disorders, and neurodegenerative conditions (Elmore 2007). The caspase cascade, a series of proteolytic events, plays a critical role in the execution phase of apoptosis, leading to the cleavage of vital cellular components and the orderly disassembly of the cell. The regulation of this cascade, and apoptosis in general, highlights the complex genetic control that dictates cellular life and death (Shi 2004).

By understanding the detailed mechanisms governing apoptosis, including the roles of the Bcl-2 family in the intrinsic pathway and the ligand–receptor interactions in the extrinsic pathway, we can better comprehend how dysregulation in these processes contributes to disease. This knowledge is crucial for developing therapeutic strategies to modulate apoptosis in various pathological conditions.

## 2.2 Apoptotic Signaling Pathway

There are two main significant processes of cell death: necrosis and apoptosis (Fig. 2.1). By the effect of external injury, they undergo necrosis. They are induced to commit programmed suicide with the generation of internal or external stimuli and undergo apoptosis or programmed cell death. However, Apoptosis is primarily activated through multi-signal pathways and regulated via extrinsic and intrinsic ligands. Further, apoptosis pathways are classified with the help of the dependences of caspase proteins; whether it involved or not in the pathway regulation. In the

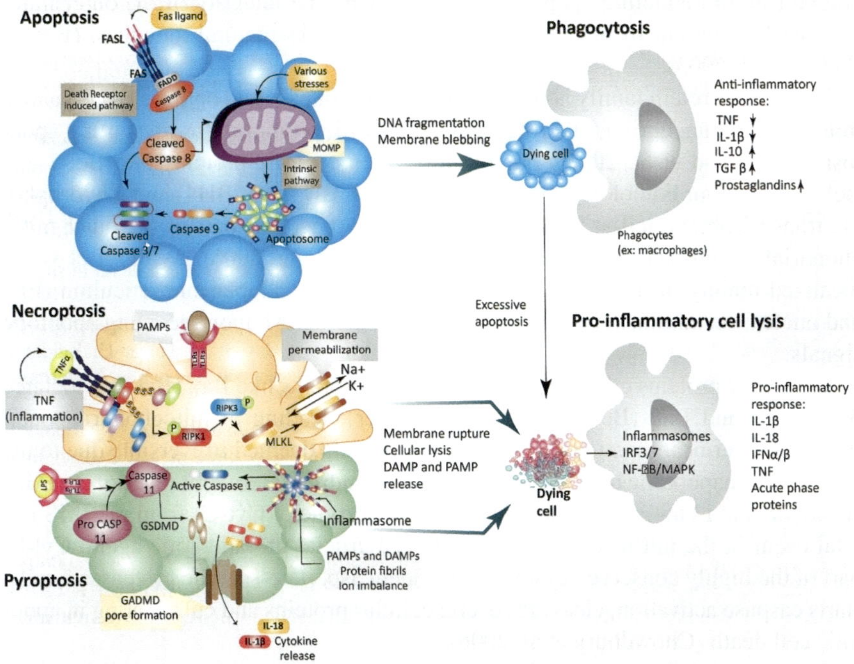


**Fig. 2.1** A schematic diagram of the types of cell death based on morphological and biochemical studies. Apoptosis can be triggered through two primary pathways: the extrinsic (death receptor) pathway and the intrinsic (mitochondrial) pathway. The extrinsic pathway is initiated when ligands, such as FAS ligands, bind to death receptors like FAS, activating caspase 8. On the other hand, the intrinsic pathway is activated by stress stimuli such as DNA damage. This leads to the permeabilization of the mitochondrial outer membrane, the release of cytochrome c, the formation of the apoptosome, and the activation of caspase 9. Both pathways ultimately converge on the effector caspases 3 and 7, responsible for the breakdown of cellular components, fragmentation of DNA, and the exposure of phosphatidylserine on the cell membrane. This exposure acts as a signal for phagocytes to engulf and clear the apoptotic cells. However, excessive apoptosis can overwhelm phagocytes, potentially resulting in secondary necrosis and inflammation. Other forms of programmed cell death, such as necroptosis and pyroptosis, are pro-inflammatory and involve cell lysis and the release of inflammatory molecules

apoptosis pathway, mitochondria connect the pathway with the presence of multi-complicated cascade protein.

### 2.2.1 Intrinsic Pathway (Mitochondrial)

The intrinsic mechanism of apoptosis, also known as the mitochondrial pathway, is triggered when cells experience internal stress (Estaquier et al. 2012) (Fig. 2.3). Apoptosis, in mammals, is mainly initiated by activating caspases in response to certain stimuli. However, it should be noted that not all caspase activities lead to cell death (Earnshaw et al. 1999). The BCL2 gene encodes a protein that can be either antiapoptotic or proapoptotic (Chipuk et al. 2010). The Bcl-2 family proteins play a crucial role in regulating apoptosis and preserving the integrity of the outer mitochondrial membrane by managing its permeability and structural soundness (Bossy-Wetzel et al. 2003).

The Bcl-2 protein family is split into three subgroups based on their BH domains, mitochondrial anchoring, and roles associated with promoting or preventing apoptosis (Mohamad et al. 2005). For instance, antiapoptotic proteins like Bcl-2 and Bcl-xL have four homology domains (BH1, BH2, BH3, and BH4) and a hydrophobic transmembrane tail at the end, which usually becomes part of the outer mitochondrial membrane (OMM) (Van Delft and Huang 2006). While Bcl-xL is localized mainly in the OMM, Bcl-2 is found in the endoplasmic reticulum (ER) and nuclear membranes (NM), translocating to the OMM upon receiving apoptotic signals.

BH3-only proteins can activate proapoptotic proteins like Bax (Bcl-2-associated X protein) and Bak (Bcl-2 antagonist or killer), leading to mitochondrial outer membrane permeabilization (MOMP). This event is deemed irreversible due to the release of proapoptotic proteins from the mitochondrial intermembrane space (IMS) (Brunelle and Letai 2009). The release of cytochrome c (Cyt-C) from the IMS is a vital event in the intrinsic route, as Cyt-C is a crucial pro-apoptotic agent. Cyt-C, part of the highly conserved cysteine-dependent aspartate-specific proteases family, starts caspase activation, cleaving several cellular proteins and culminating in apoptotic cell death (Chowdhury et al. 2006).

Once released into the cytoplasm, Cyt-C promotes the construction of the apoptosome, a multiprotein complex in the cytosol that includes Cyt-C, apoptotic protease activating factor 1 (Apaf-1), and procaspase-9 (Pinton et al. 2001). Apaf-1 has many structural domains, including caspase recruitment domains (CARDs), a nucleotide-binding domain (NBD), a helical domain (HD1), a winged helix domain (WHD), a second helical domain (HD2), and 15 WD40 repeats in the C-terminal region. The core structure of Apaf-1 contains seven copies of the NBD, HD1, and WHD, while the 15 WD40 repeats form two β propellers connected to the central hub by the HD2 domain (Cain et al. 2002; Shakeri et al. 2017). This route is critical for maintaining cellular homeostasis and responding to internal stresses, resulting in programmed cell death by caspase activation and apoptosome formation (Fig. 2.2).

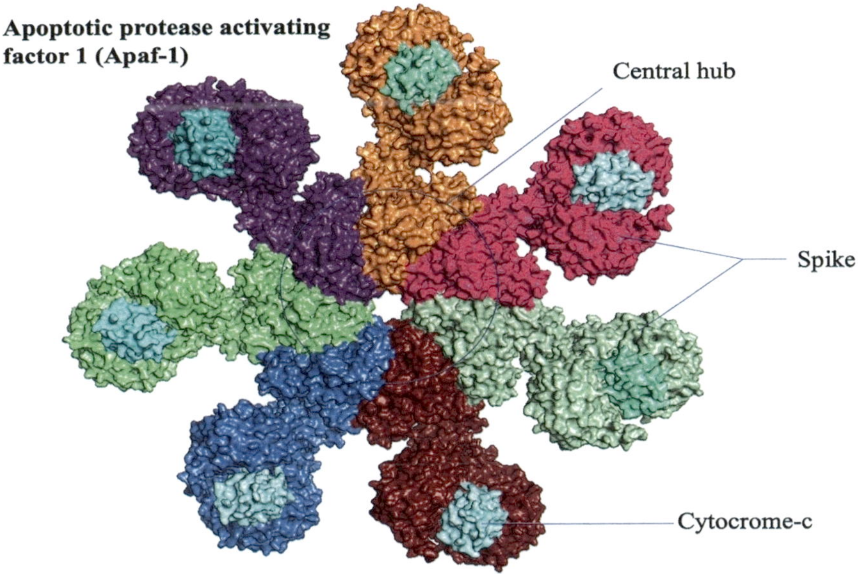


**Fig. 2.2** An overall view 3D structure of the Apoptotic protease activating factor 1 (Apaf-1) apoptosome

The activated apoptosome initiates the CASP9 cleaves and activates effector downstream caspases, such as CASP3, CASP6, and CASP7, through proteolytic cleavage that rearranges a specific internal protein loop in the formation of active sites by the binding of auto-processed ProCASP9 and CARD (Budihardjo et al. 1999). Activating effector caspases initiates proteolytic cleavage in a broad range of cellular targets that eventually force cell death. Recent research demonstrated that PIDDosome, apoptosome, FADDosome, and apoptosome are involved in the specific mechanism in the intrinsic and extrinsic apoptotic pathways to stimulate various initiator caspases (Thompson et al. 2015). Out of the three caspase-activation platforms, PIDDosome is involved in the intrinsic pathway (Berube et al. 2005).

### 2.2.2 Extrinsic Pathway of Apoptosis

The extrinsic signaling cascade of apoptosis is activated by external death signals (Fig. 2.3). This occurs when particular signaling molecules reach a crucial threshold that promotes cell death or when death receptors (DRs) connect to their corresponding death ligands (Obeng 2020). These death receptors have a cytosolic death domain and an external cysteine-rich domain that transfer the death signal from the cell surface to the inside (Papenfuss et al. 2008). Most death receptors belong to the tumor necrosis factor (TNF) superfamily, including Fas, DR4/DR5, TNF-R, and TNF-related apoptosis-inducing ligand receptors (TRAIL-R) (Kashyap et al. 2021). Some receptors are engaged in apoptotic processes depending on p53 and MYC.

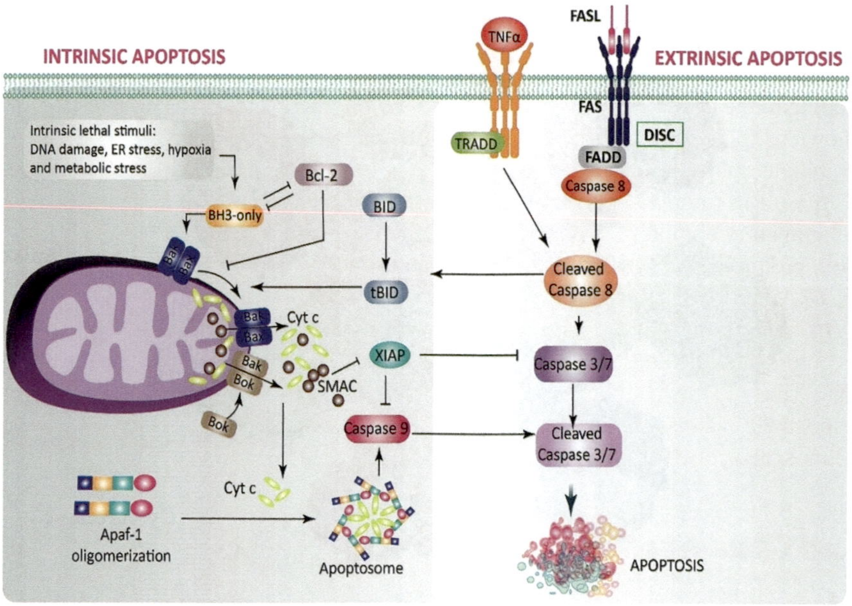


**Fig. 2.3** Schematic overview of the apoptotic pathways. Intrinsic apoptosis can be initiated by stimuli such as growth factor withdrawal. This shift affects the balance between pro-survival Bcl-2 proteins and BH3-only proteins, resulting in the oligomerization of Bax and Bak, mitochondrial outer membrane permeabilization (MOMP), and the release of cytochrome c and Smac. Cytochrome c then binds to Apaf-1 to form the apoptosome, which activates caspase-9, subsequently activating caspase-3 and caspase-7, leading to apoptosis. Extrinsic apoptosis begins with TNF family ligands binding to their receptors, forming the death-inducing signaling complex (DISC) and activating caspase-8. Caspase-8 can directly activate caspase-3 or cleave Bid to initiate intrinsic apoptosis. The activities of caspases can be inhibited by XIAP, which Smac counteracts

Death receptors can be classed depending on their adaptor proteins (Lee et al. 2012). The first group contains receptors linked with the FAS-associated death domain (FADD), which assist in binding caspase-8 to activate the death signal, producing the Death-Inducing Signaling Complex (DISC) (Obeng 2020). DRs such as DR2, DR4, and DR5 initiate apoptotic signals through this route.

The second group contains receptors linked with the TNF-associated death domain (TRADD), which enables the binding of receptor-interacting protein kinase (RIP1), TNF receptor-associated factors 2 and 5 (TRAF2 and TRAF5), and cellular inhibitors of apoptosis proteins (cIAPs). This complex is called Complex I in apoptosis signaling (Xiao et al. 2023). DRs such as DR1, DR3, DR6, and EDAR trigger apoptotic signaling via this route. Once DISC is formed, active caspase-8 progressively activates other caspase-3, caspase-6, and caspase-7 (Ramirez and Salvesen 2018). The creation of DISC can be blocked by regulators such as the FLICE-like inhibitory protein (cFLIP), which mimics caspase-8 but lacks the catalytic domain (Tummers and Green 2017). TNFR1 can also indirectly start apoptosis through TNFR1-linked death domain proteins via the FADD-caspase-8 route while increasing anti-apoptotic signaling through the TRF2/cIAP pathway, resulting in NF-kB and JNK/AP-1 activation (Hui et al. 2004).

During genotoxic stress, the activity of XIAP, cIAP1, and cIAP2 is reduced, leading to the development of a complex containing RIPK1, FADD, and caspase-8, known as the ripoptosome. This complex can convert pro-inflammatory cytokines to initiate apoptosis and necrosis with the assistance of RIPK1. Thus, IAPs that decrease the ripoptosome regulate RIPK1 (Bertheloot et al. 2021). Another pro-inflammatory complex, known as the FADDosome, can activate NF-kB. This complex, produced by FADD, RIPK1, TRAF2, caspase-8, and caspase-10, is triggered by TRAIL-R and DNA damage (Henry and Martin 2017).

In essence, the extrinsic apoptosis pathway contains a complex network of death receptors and adaptor proteins that regulate cell death in response to external signals. These pathways play crucial roles in maintaining cellular homeostasis and responding to stress signals, ultimately assuring the controlled removal of damaged or unneeded cells.

### 2.2.3 Role of Caspases in the Execution of Apoptosis

Caspases are specialized cysteine proteases that play key roles in the intrinsic and extrinsic pathways of apoptosis. They are involved in numerous apoptotic processes, leading to the activation or inactivation of target proteins to carry out cell death. Caspases are divided into initiators, effectors, and inflammatory caspases. Initiator caspases, such as Caspase-2, Caspase-8, Caspase-9, and Caspase-10, activate effector caspases (Tang and Kang 2022). Effector caspases, including Caspase-3, Caspase-6, and Caspase-7, trigger other proteins to enhance apoptosis, while inflammatory caspases, including Caspase-1, Caspase-4, Caspase-5, Caspase-12, Caspase-13, and Caspase-14, initiate inflammatory responses (Sahoo et al. 2023).

Initiator caspases have autoprocessed pro-domains that engage with specific signaling complexes. They activate effector caspases through TRAIL in intrinsic and extrinsic apoptotic pathways (McIlwain et al. 2013). Once activated, effector caspases launch a series of processes that destroy nuclear materials and the cytoskeletal structure via cellular endonucleases and proteases, respectively (Alenzi et al. 2010). Caspase-3, for instance, cleaves the inhibitor of caspase-activated DNase (CAD), hence activating CAD to promote DNA damage via the p53 signaling cascade in apoptotic cells (Orth et al. 2012). Caspase-3 also attaches to and encourages the fragmentation of gelsolin, which is involved in actin assembly and disassembly. This fragmentation affects nuclear and cytoskeletal structures (Kusano et al. 2000).

Active cell death mechanisms are not specific to animals; they have also been reported in plants, yeast, protozoa, and bacteria. In certain circumstances, these events mirror apoptosis, suggesting an evolutionary relationship. Although this approach is contested, investigating cell death in non-animal creatures can provide insights into the life and death of these organisms and show general principles of active cell death.

## 2.3 Exploring Apoptosis Beyond Animals

Caspase proteases are peculiar to animals, meaning apoptosis, when properly defined by the activation of caspases, does not occur in other kingdoms. However, related proteases known as metacaspases have been linked to cell death in plants and yeast (Choi and Berges 2013). Additionally, the paracaspase family, found in mammals and Dictyostelium, does not play a role in cell death, even in Dictyostelium, which lacks an apoptosis-like cell death pathway (Aravind et al. 1999). Several cell death pathways in plants seem to require proteins that mimic the adaptor APAF1 (Lam and Zhang 2012). Unlike caspases, which cleave substrates following an aspartate residue, metacaspases cleave after an arginine residue. Interestingly, a substrate cleavable by both protease types has been found and may be implicated in cell death (Watanabe and Lam 2005). This discovery shows that apoptosis and cell death mechanisms in non-metazoan creatures might share a similar ancestry (Uren et al. 2000).

However, this hypothesis faces substantial obstacles. Despite certain sequence and structural similarities between caspases and metacaspases, their biochemical characteristics are fundamentally different. The notion of induced proximity, needed for the activation of initiator caspases by their adapter platforms, does not apply to metacaspases, which are monomers and not triggered by dimerization (Madeo et al. 2009). Thus, a metacaspase would not be triggered even if recruited to an adapter protein like APAF1. Consequently, the functional processes of apoptosis, requiring interactions between adapters, initiator caspases, and executioner caspases, are unlikely to be recreated by homologs in plants or yeast. Although caspases and metacaspases might share a common molecular ancestor, there is no evidence that their activation mechanisms are the same. Future research may clarify if molecular cousins of APAF1 could activate metacaspases in plant cell death, but currently, no data supports this notion (Van Hautegem et al. 2015).

### 2.3.1 Variations in the Mitochondrial Pathway of Apoptosis Across Different Animal Phyla

Efforts to find homologs of apoptosis-related proteins have been fruitful across numerous animal taxa. Homologs of chordate BCL-2 proteins have been discovered in a range of taxa, including poriferans (sponges) (Wiens et al. 2000), cnidarians (Hydra) (Cikala et al. 1999), platyhelminths (planaria) (Bender et al. 2015), arthropods (flies) (Hay et al. 1994), and nematodes (Horvitz 1999). Cytochrome c stands out as one of the most conserved proteins, occurring in all mammals, eukaryotic, and even bacteria. Homologs of APAF1 have been described in various animal phyla, including placozoans (Trichoplax), potentially the most primitive animal phylum (though this is contested) (Signorovitch et al. 2007). Various caspases, including those with prodomains having protein interaction motifs similar to caspase-9, have been found throughout different animal phyla. It is usually thought that the mitochondrial mechanism of apoptosis, or at least its constituent proteins, started with mammals. However, the molecular actions of these proteins in apoptosis reveal a more complicated picture.

One prominent example is the connection between cytochrome c, APAF1, and caspase-9. Functional investigations have established that the interaction between APAF1-like molecules and caspase-9-like molecules stimulates caspase activation, as seen in arthropods, worms (Horvitz 1999), and chordates (Li et al. 1997). However, the role of cytochrome c is more complicated. In chordates, cytochrome c is needed to activate APAF1 and generate the activation platform for caspase-9. This is not the case for nematodes, where the APAF1 homolog, CED-4, lacks the area necessary for cytochrome c interaction (Zou et al. 1997; Seshagiri and Miller 1997). In arthropods, such as *Drosophila*, the APAF1 homolog has this area; however, the role of cytochrome c in activating this APAF1 homolog is contested. Some investigations have revealed cytochrome c mutants defective in apoptosis (Kanuka et al. 1999), while in vitro experiments in *Drosophila* cells have failed to uncover a role for cytochrome c in caspase activation (Rodriguez et al. 1999; Dorstyn et al. 2002). Conversely, cytosolic extracts from platyhelminths (planaria) and echinoderms (sea urchins and sand dollars) have shown significant caspase activation in response to cytochrome c addition (Bender et al. 2015), demonstrating that APAF1 in these animals is activated by cytochrome c.

In chordates, BCL-2 proteins regulate mitochondrial outer membrane permeabilization (MOMP), which is important for releasing cytochrome c into the cytosol and activating APAF1. Although BCL-2 proteins are ubiquitous throughout the animal kingdom, the relevance of MOMP in apoptosis varies among species. In platyhelminths, cytochrome c release during apoptosis shows the existence of MOMP (Bender et al. 2015). In contrast, no MOMP is identified during apoptosis in worms (Chen et al. 2000). Correspondingly, a BCL-2 family protein from planaria demonstrates MOMP activity (Bender et al. 2015), whereas the sole nematode BCL-2 protein in Caenorhabditis worms lacks this function and instead has an anti-apoptotic impact unrelated to MOMP (Conradt and Horvitz 1998). In *Drosophila*, cytochrome c release occurs post-apoptosis and relies on caspase function, unlike most MOMP processes in chordate cells (Dorstyn et al. 2002).

Arthropod BCL-2 proteins have not been discovered to directly regulate apoptosis (Quinn et al. 2003). Instead, apoptosis in arthropods seems to follow a distinct mechanism. The *Drosophila* APAF1 homolog, ARK, is constitutively active, while the activation of the initiator caspase, DRONC, is blocked by DIAP1 (similar to X-IAP in mammals). Upon the onset of apoptosis, proteins antagonizing DIAP1 are released, freeing active DRONC, which subsequently activates executioner caspases, leading to cell death (Kaiser et al. 1998). These antagonistic proteins work similarly to mammalian X-IAP antagonists but are not sequestered in the mitochondrial intermembrane gap as in mammals.

A conundrum occurs when comparing the functions of BCL-2 proteins in worms and chordates. In *Caenorhabditis elegans*, the anti-apoptotic BCL-2 protein, CED-9, binds directly to the APAF1 homolog, CED-4, inhibiting its activation. The BH3-only protein EGL-1 breaks this connection, allowing CED-4 to bind and activate the caspase CED-3 (Chen et al. 2000). In contrast, anti-apoptotic BCL-2 proteins in chordates do not inhibit APAF1 or caspase-9 activation directly; they hinder mitochondrial outer membrane permeabilization (MOMP), necessary for cytochrome c release and subsequent APAF1 activation (Conradt and Horvitz 1998). Despite the similarities in overall effect—both anti-apoptotic BCL-2 proteins block APAF1 activation—the underlying molecular pathways are unique. This discovery is not intrinsically paradoxical, as proteins can adopt diverse functions in different animals. What is striking is that the inhibition of caspase activation by BCL-2 proteins appears to be mediated through fundamentally different biochemical mechanisms: inhibition of MOMP in chordates versus direct inhibition of APAF1 in worms (Fig. 2.4).

When examining APAF1 homologs across other animal species, it seems possible that nematodes have lost the ability of their APAF1 protein to interact with cytochrome c. This might also apply to arthropods, some of which retain the

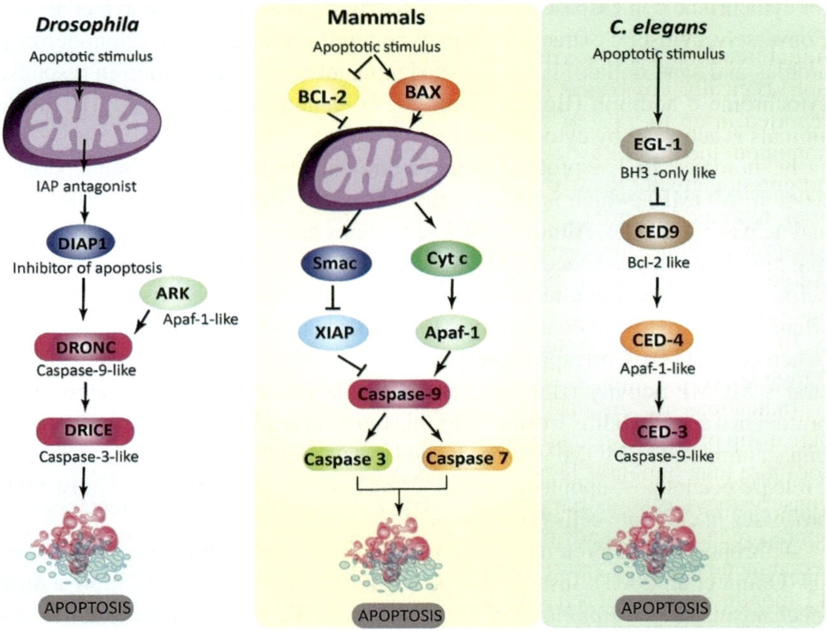


**Fig. 2.4** Variations in apoptosis across different animal phyla. Comparison of apoptosis-inducing pathways in mammals, *Drosophila melanogaster*, and *Caenorhabditis elegans*. In mammals, apoptosis can be triggered via the intrinsic pathway, which involves mitochondrial activation. This process leads to the activation of caspase-9, which in turn activates downstream effector caspases such as caspase-3 and caspase-7. In contrast, while *Drosophila melanogaster* is suspected to use mitochondria similarly in apoptosis, this has yet to be definitively confirmed. In *Caenorhabditis elegans*, mitochondria do not play a role in the initiation of apoptosis. The initiator caspases vary among these organisms: mammals use caspase-9, *Drosophila* utilizes a caspase-9 homolog known as Dronc, and *C. elegans* employs CED-3, which is also similar to caspase-9. The effector caspases differ as well: mammals have caspase-3 and caspase-7, *Drosophila* features caspase-3-like DrICE and Dcp-1, and *C. elegans* relies on CED-3 for effector functions. Activating these effector caspases is responsible for the degradation of cellular proteins, a hallmark of apoptosis

cytochrome c-binding (WD) domain but possess an APAF1-like protein that does not seem to interact with cytochrome c (Quinn et al. 2003). This hypothesis holds if the common ancestor of platyhelminths, nematodes, and arthropods had an APAF1 triggered by cytochrome c. Alternatively, cytochrome c-mediated APAF1 activation might have developed after these groups split.

### 2.3.2 Variations in the Apoptosis Across Different Plant Phyla

Much like animals, plants have developed sophisticated systems to govern programmed cell death (PCD) during numerous developmental stages, such as xylem production, seed maturation, leaf senescence, and reproductive functions. PCD in plants also plays a vital role in responses to pathogen infections from organisms such as *Sclerotinia sclerotiorum, Cochliobolus victoriae*, tobacco mosaic virus, and *Botrytis cinerea*. Environmental stressors, including drought, heat, cold, and hypersalinity, can also promote PCD in plants (van Doorn and Woltering 2005; Lam 2004). According to Fukuda et al. plant PCD can be categorized into three types: apoptotic-like cell death, senescence-associated mortality, and vacuole-mediated cell death (Fukuda 2000). Unlike animals, plants do not possess real caspases but instead have caspase-like proteases such as vacuolar processing enzymes (VPEs), cathepsin B, caspase, phytase, aspartyl proteases, and metacaspases (Bonneau et al. 2008; Watanabe and Lam 2011; Hatsugai et al. 2004). The roles of these proteases are somewhat analogous to animal caspases, with evidence supporting similar activities. For instance, animal caspase-3 and plant metacaspases can cleave the Tudor staphylococcal nuclease (Sundström et al. 2009). In Norway spruce (Picea abies), the metacaspase mcII-Pa is important for suspensor PCD during embryogenesis, and its silencing impairs embryogenesis (Bozhkov et al. 2005). However, the stiff cell walls of plant cells limit the development of apoptotic bodies common in animal PCD, as they hinder cell disintegration and phagocytosis (van Doorn and Woltering 2005). Yet, under specific situations, such as exposure to the mycotoxin Fumonisin B1 or heat stress, apoptotic-like structures and membrane blebbing have been detected in plant cells (Kerr et al. 1972). These findings show that the type and intensity of signals determine the form of cell death, but apoptotic bodies have not been identified in complete plant tissues (Zhang and Lam 2011).

Research on anti-apoptotic genes in animal viruses has shown that viral genomes can generate cytoprotective proteins with minimal primary amino-acid sequence similarity to their cellular counterparts yet fulfill the same function. For example, the E1B 19 K protein of adenovirus functions similarly to cytoprotective Bcl-2 family proteins but lacks sequence homology (White 2001). E1B 19 K interacts with and neutralizes pro-apoptotic Bcl-2 family members, such as Bax (Bcl-2-associated x), and can substitute for Bcl-2 in several in vitro cell death scenarios (Han et al. 1996). It is speculated that plant genomes might potentially contain genes producing functional analogs of anti-apoptotic Bcl-2 family members that have little or no sequence similarity to those in animal cells. Structural similarities, particularly in functional sections like catalytic domains, may explain this functional similarity despite the lack of sequence homology.

Bax generally sits in the cytosol in an inactive state but translocates to the mitochondrial membranes upon receiving an apoptotic signal or through overexpression. This translocation is accompanied by conformational changes that enable Bax to embed in the membrane, resulting in cytochrome c release (Goping et al. 1998). Notably, when murine Bax is produced in plants, it is constantly localized to the mitochondria (Lacomme and Santa Cruz 1999). In plants, cytochrome c release occurs under diverse stress circumstances, but it does not directly activate apoptotic-like properties (Balk et al. 1999; Balk and Leaver 2001; Tiwari et al. 2002). This suggests that an old system of pore-forming proteins, which regulate cell death in mammals, might possibly exist and operate similarly in plants.

The identification of the plant Bcl-2-associated athanogene (BAG) gene family was based on structural conservation rather than sequence similarity (Doukhanina et al. 2006). Initially found in animals, BAG proteins are characterized by a conserved BAG domain (BD) and their role in controlling apoptosis while working as co-chaperones with Hsp70 (Takayama and Reed 2001; Briknarova et al. 2001). Plant BAG proteins similarly give cellular protection, bind to Hsp70, and reduce apoptotic-like cell death under stress circumstances (Doukhanina et al. 2006). There is remarkable conservation between human Bag4 and *Arabidopsis* BAG family members, particularly in important functional domains (Doukhanina et al. 2006; Williams et al. 2010). However, *Arabidopsis* BAG proteins exhibit multiple subcellular localizations unlike their animal counterparts, including the nucleus, cytosol, mitochondria, vacuole, and endoplasmic reticulum. At least three members of the *Arabidopsis* BAG family (AtBAG4, AtBAG6, and AtBAG7) are implicated in autophagy induction and demonstrate cytoprotective effects in response to cold, drought, and heat stress (Doukhanina et al. 2006; Yan et al. 2003; Kabbage et al. 2013).

There is emerging evidence that plant vacuoles serve a critical role in the removal of excess or damaged cells, compensating for the absence of a phagocytic system. The hallmark of vacuolar cell death (VCD) is the engulfment of cytoplasmic components by the vacuole (Hara-Nishimura and Hatsugai 2011). This system has prompted similarities to animal autophagy (van Doorn and Woltering 2005; van Doorn 2011; Kabbage et al. 2013), however the analogy is complex. Notably, the ultimate stages of VCD resemble apoptotic-like cell death. For example, during leaf perforation in the lace plant, VCD engulfs the protoplast and destroys the plant cell wall (Wright et al. 2009). Another connection to apoptotic-like cell death is the participation of caspase-like proteases, notably the proteasome subunit beta type 1 (PBA1), which operates similarly to caspase-3, and the vacuolar processing enzyme (VPE), akin to caspase-1 (Hatsugai et al. 2004). The role of the proteasome in targeted protein degradation adds a new dimension to the regulation of plant PCD. In Nicotiana benthamiana, the defense response to TMV triggers vacuole rupture, destroying cell contents and consequent cell death (Hara-Nishimura and Hatsugai 2011). Thus, VCD may be an efficient way for plants to eradicate growing viruses.

The collapse of vacuoles can release hydrolytic enzymes, such as papain-like cysteine proteases (PLCPs), into the cytoplasm. One major PLCP, RD21, is crucial in oxalic acid (OA)-induced apoptotic-like cell death (Lampl et al. 2013). In mutants missing RD21, OA-induced cell death does not occur, but an overabundance of RD21 accelerates the process. Interestingly, RD21's protease activity can be blocked by creating a covalent compound with AtSerpin1, a serine protease inhibitor conserved across plants and animals that can also suppress metacaspase-1-induced cell death (Lampl et al. 2010, 2013). Cells lacking in AtSerpin1 exhibit increased OA-induced cell death, while those overexpressing AtSerpin1 ameliorate the OA response. Thus, the balance between serpin inhibitors and released PLCPs ultimately determines whether a cell lives or undergoes death. In plants, the expression of ced-9 and Bcl-xl has been shown to reduce cell death, potentially by stabilizing mitochondrial membrane potential through the reduction of cytoplasmic Ca2+ levels (Xu and Reed 1998). Additionally, the transgenic expression ced-9 in plants prevents salinity-induced death by modulating K+ and H+ transport across membranes (Kawasaki et al. 2001). Further evidence of ced-9's role in ion homeostasis is observed in soybeans, where transgenic expression enhances stress tolerance but reduces nodule formation, likely by affecting the ion signatures required for rhizobium perception (Kuwana et al. 2002). Studies also indicate that Bcl-2 homologues of ced-9 regulate autophagy pathways, vesicular trafficking, and organogenesis in animals through interactions with Beclin 1 (Kihara et al. 2001). The expression of ced-9 in soybeans is proposed to impact nodule formation by interfering with conserved autophagy and vesicular trafficking pathways, suggesting potential crosstalk between apoptotic-like and autophagy pathways (Kihara et al. 2001).

The inhibitors of apoptosis proteins (IAPs) represent another class of programmed cell death (PCD) regulators in plants that have minimal sequence homology with their animal counterparts. IAPs, which were first discovered in baculoviruses, are distinguished by the presence of C-terminal RING domains and baculoviral IAP repeat (BIR) domains. For the purpose of reducing caspase activity, the BIR domain is structurally unique and essential (Birnbaum et al. 2000; Tenev et al. 2005). BIR domains have not yet been discovered in plants despite being widely distributed in IAPs across a variety of taxa (Vaux and Silke 2005). Rather, plants encode IAP-like proteins (ILPs) with a C-terminal RING domain with the E3 ligase activity necessary for controlling proteasomes but lacking BIR domains (Yang et al. 2000). One such is the *Arabidopsis* IAP-like protein (AtILP), which when treated with fumonisin B1 decreases PCD and suppresses caspase-like activity (Kabbage and Dickman 2008). The N-terminus of AtILP can reduce apoptosis and lower caspase-3 activity in animal cells, despite its lack of BIR domains and little similarity to animal IAPs (Kabbage and Dickman 2008).

Numerous parallels and divergences between the mechanisms of animal and plant programmed cell death (PCD) have been brought to light by research on PCD. In order to control cell death throughout development, in reaction to infections, and in stressful environmental conditions, plants employ a variety of routes. Certain proteases, like as metacaspases and vacuolar processing enzymes (VPEs), function similarly to animal caspases. The presence of proteins, such as IAP-like proteins (ILPs) and plant Bcl-2-associated athanogene (BAG) proteins, that are functionally similar to animal apoptotic regulators but do not share sequence homology highlights the evolutionary conservation and diversity of cell death processes. Furthermore, the significance of vacuolar cell death (VCD) and the secretion of hydrolytic enzymes such as RD21 underscore the distinct tactics that plants have developed to handle cellular demise and stress reactions. Investigating the interactions between plant autophagy and apoptotic-like pathways may lead to the development of novel agricultural biotechnology approaches for controlling cell death, which would ultimately improve food security and sustainable crop production.

## 2.4 Genes in the Apoptotic Pathways

### 2.4.1 APAF1 (Apoptotic Peptidase Activating Factor 1)

In the presence of pro-apoptotic molecules such as Bax and Bid and anti-apoptotic molecules such as Bcl-X and Bcl-2, the APAF1 gene (Apoptotic peptidase activating factor 1) encodes the Apaf1 protein (Kaman et al. 2019) (Fig. 2.5). The Apoptosome complex plays a critical role in regulating programmed cell death during mammalian development and in orchestrating oncogene and p53-dependent apoptosis. Its activation of CASP9 initiates a cascade that ultimately leads to the activation of CASP3, pivotal in the process (Shi 2001). Proceeding to apoptosome complex construction, Apaf1 protein resides in the cytoplasm as a monomeric isoform with co-ordinates with Cyt C, then binds its critical cofactor, dATP or ATP, forming a multimeric complex called the apoptosome (Kim et al. 2005). Apaf-1 is central in initiating the intrinsic pathway and governing crucial aspects of apoptosis. It also serves as a point of contention, with some perspectives suggesting its activation of caspase-9 through the apoptosome. Recent research demonstrated the role of Apaf1 expressed in a human embryonic stem cell assay, which remarked that low levels of APAF1 expression in the cellular were possibly indicated for preventing cell death in neuronal development (Karimzadeh et al. 2018).

### 2.4.2 BCL2 (B-Cell Lymphoma Protein 2)

The BCL2 (B-cell lymphoma protein 2, apoptosis regulator) gene encodes the BCL-2 protein, which is an anti-apoptotic protein (Deng et al. 2020). The aberrant expression of this protein is implicated in tumor formation. A family of at least nine proteins regulates the intrinsic apoptosis pathway, collectively known as the BCL-2 family. Within this family, certain members, such as BAX, BAK, and BOK, are pro-apoptotic, while others, including BCL-B, BCL-2, BCL-XL, MCL-1, BCL-w, and BFL-1/A1, are anti-apoptotic. Additionally, there are proteins with similar functions, such as BCL2L13, BCL2L14, BCL2L15, and BID (Chao and Korsmeyer 1998).

The balance between pro-apoptotic and anti-apoptotic BCL-2 proteins determines a cell's sensitivity to apoptotic stimuli (Siddiqui et al. 2015). A higher

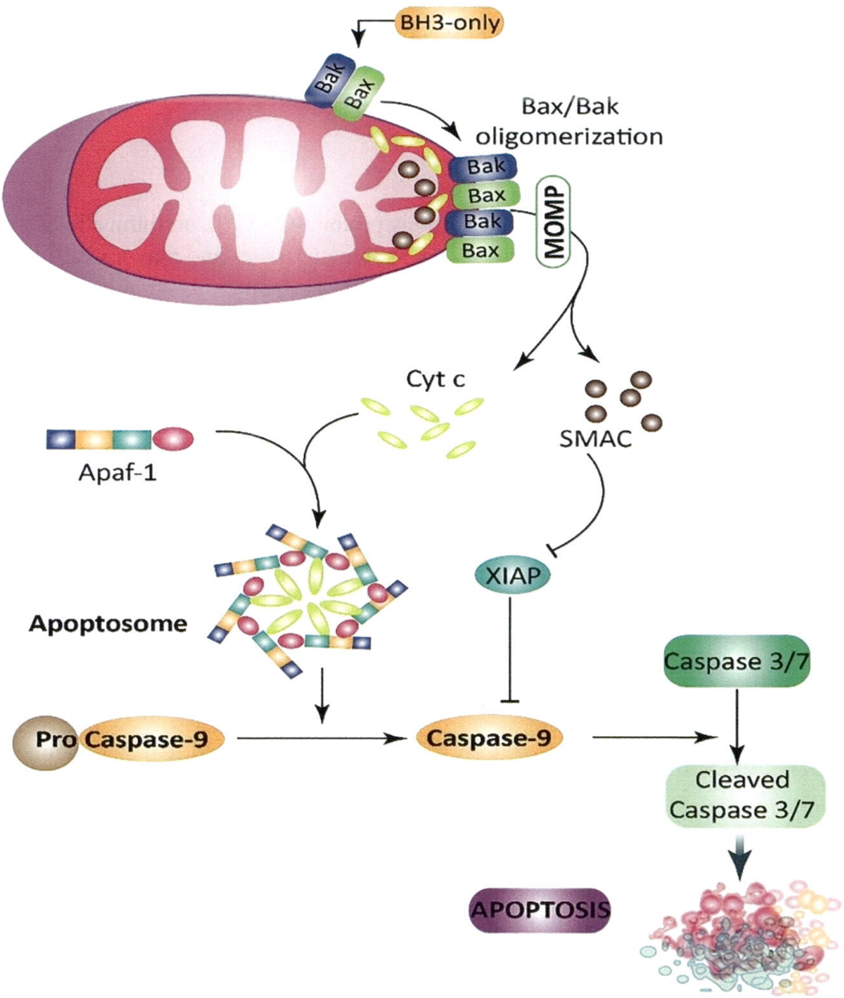


**Fig. 2.5** The mitochondrial pathway of apoptosis. Shifting the balance between pro-survival Bcl-2 and BH3-only proteins leads to Bax and Bak oligomerization, mitochondrial outer membrane permeabilization (MOMP), and the release of cytochrome c and Smac. Cytochrome c binds Apaf-1, forming the apoptosome and activating caspase-9, which then activates caspase-3/7, leading to apoptosis

concentration of pro-apoptotic proteins makes cells more susceptible to apoptosis, whereas an excess of anti-apoptotic proteins renders cells more resistant. BCL-B, an anti-apoptotic member of the BCL-2 family, contains all four homologous domains (BH), from which anti-apoptotic BH3-only proteins such as BAD, BCL-XS, BLK, and BLD are derived (Robert et al. 2012). Activator BH3-only proteins, such as BID or BIM, are activated through transcriptional or post-translational modifications in response to cellular damage. These proteins are crucial for initiating mitochondrial permeabilization and programmed cell death processes (Perini et al. 2018). Their balance is essential for cellular homeostasis and response to apoptotic signals.

### 2.4.3 CASP9 (Caspase 9)

The CASP9 gene encodes the homonymous protein (CASP9), an intracellular protease found to be constitutively expressed in human tissues and play a vital role in the intrinsic apoptotic pathway, which is involved in numerous stimuli, including stress agents, chemotherapies, and radiation (Johnson and Jarvis 2004). It has an inactive form called ProC9 (416 amino acids) and comprises three distinct parts: An N-terminal CARD, a large catalytic subunit, and a small catalytic subunit (Shakeri et al. 2017). CASP9 is activated on the apoptosome complex with the help of kinases like Akt and p21-Ras induced phosphorylation of proC9 and initiates catalytic activeness (Johnson and Jarvis 2004). Inactivation of caspase-9 has a profound physiological effect, leading to the development of cancer. The apoptosome consists of relative concentrations of caspase-9 pro-domain with Apaf-1 and cytochrome C and mediates the formation of CASP9 homodimers and CASP9 with APAF1 heterodimers (Cui et al. 2021).

## 2.5 Genes Involved in Extrinsic Pathway

### 2.5.1 CASP8 Gene

Translation CASP8 gene produces Caspase-8 protein, which is an initiator caspase and is activated through DISC in the extrinsic pathway of apoptosis (Ivanisenko and Lavrik 2019). A higher concentration of Caspase-8 directly cleaves Caspase-3, while a lower concentration of Caspase-8 cleaves BID and releases CytC, thereby triggering caspase-3 through caspase-9 (Kim et al. 2017). Caspase-8 is also a potential biomarker for diabetes mellitus because it is involved in pancreatic β-cell apoptosis (Rojas et al. 2018). So, plasma Caspase-8 level can point out the future progression of diabetes mellitus disease before the emergence of its symptoms (Feltham et al. 2017).

### 2.5.2 MORT1 Gene

The MORT1 gene encodes the FADD protein, which functions as the adapter protein for the extrinsic pathway. FADD is the principal adapter protein of DR-induced apoptosis that activates DISC (Wang et al. 2010). FADD has a death domain that binds Fas to initiate apoptosis through the DISC signaling cascade. Another domain of FADD is called the Death Effectors Domain (DED), which interacts with DEDs upon Caspase-8 or Caspase-10 (Yang 2015). FADD plays a role in T-cell development, and its activation also has a significant contribution to lipid metabolism (Zhuang et al. 2016). FADD also has immunoreactivity characteristics, which shows the involvement of FADD in dementia and Alzheimer 's-like diseases (Ramos-Miguel et al. 2017).

### 2.5.3 FAS Gene

The FAS gene encodes Fas transmembrane protein, one of the DR present on the cell surface. Fas combines with FasL to initiate apoptotic signaling (Papenfuss et al. 2008). Upon FasL/Fas binding, Fas forms a trimeric structure that activates FADD. Fas has an established role in fibroblast proliferation through ERK1/2 signaling (Wajant et al. 2003). Fas have also a significant role in NF-kB activation, which can regulate anti-apoptotic factors. Besides, it is also involved in tumor growth and invasion, lung inflammation, eosinophilia, etc. Non-apoptotic Fas signaling is helpful in B-cell maturation and is also involved in the activation of mTOR (Staniek et al. 2024).

## 2.6 Genes Involved in Intrinsic and Extrinsic Pathway

### 2.6.1 TP53

TP53 or Tumor Protein 53 translates p53 protein, a tumor suppressor, and several functions of cells from transcription to apoptosis. Around 900 genes involved in cell signaling are directly or indirectly regulated by the p53 protein (Nguyen et al. 2018). p53 regulates the extrinsic pathway of apoptosis by influencing the Fas-dependent pathway. p53 induces the promoter region of the FAS gene and enhances the transcription, a promising cancer treatment pathway (Müller et al. 1998). p53 also has a great influence on the intrinsic pathway of apoptosis. It regulates the transcription of PUMA, BID, and Noxa genes by enhancing their expression to promote apoptosis (Chipuk and Green 2006). Upon cell stress and DNA damage, p53 activates Caspase-8 and Caspase-10 to facilitate cell death. Mitochondrial Bax protein releases cyt-C activated by influx of p53, which in turn initiates caspase-9-dependent apoptosis (Mishra and Kumar 2005). Further, p53 can enhance the secretion of Apaf1 by inducing the promoter of this gene, thereby boosting the formation of apoptosome.

### 2.6.2 MYC

MYC is a gene family that regulates cell maturation, growth, and death. The MYC gene has oncogenic and pro-apoptotic functions depending on the expression level (Meyer et al. 2006). MYC performs its role in apoptosis through an extrinsic pathway by involving the p53 protein. MYC blocks a p53 inhibitor hMDM2 through higher expression of p14ARF (Zindy et al. 1998). This p53 is widely involved in intrinsic and extrinsic pathways. High expression of MYC-activated Bax mediated intrinsic pathway through Bim (BH3-only protein) (Jiang et al. 2007). The higher expressions of MYC also correlate with tumorigenesis and genomic instability (Cavalcante et al. 2019).

### 2.6.3 Role of CASP3 Gene in the Execution of Apoptosis

The CASP3 gene expresses the Caspase-3 protein, the crucial caspase involved in apoptosis. Caspase-3 is an effector's caspase that degrades the inhibitor of CAD, which results in the activation of CAD. CAD is an endonuclease that destroys the nuclear DNA of an apoptotic cell. Caspase-3 can also disrupt the cytoskeleton of apoptotic cells with the help of Gelsolin protein. Deregulated caspase-3 promotes metastasis and tumor invasion in colorectal cancer (Liu et al. 2018). Caspase-3 also promotes Alzheimer's disease (AD) by the cleavage of amyloid-beta 4A precursor protein (Stone et al. 2002).

## 2.7 Mitochondrial Epigenetic Function

Mitochondria, vital organelles in cells, primarily produce ATP via oxidative phosphorylation (OXPHOS) perform crucial functions like calcium regulation and lipid balance, and promote cell signaling and death (Singh and Mabalirajan 2021). However, OXPHOS also leads to generating reactive oxygen species (ROS) through ETC, which, when accumulated, can harm mitochondrial DNA (mtDNA). Dysfunctional cells undergo apoptosis, where mitochondria play a central role in apoptosis (intrinsic) and also deliver energy for the execution of the process (Nissanka and Moraes 2018).

In the intrinsic apoptosis pathway, pro-apoptotic proteins (Bcl-2 family) cause mitochondrial outer membrane permeabilization, releasing specific molecules (IAP inhibitors and Cyt-c) and initiating cell death. Mitochondrial dynamics, including organelle's fission (by Dynamin-related protein 1), outer membrane fusion (by Optic atrophy 1), lipid composition (ceramides), and interaction with the endoplasmic reticulum, influence this process (Kroemer et al. 2007; Wang and Youle 2009).

Mitochondria can induce apoptosis even without caspase (by Bax/Bak-induced MOMP) involvement, leading to a pro-inflammatory type of cell death by Bax/Bak-induced MOMP by the release of mtDNA, which leads to IFN expression. They also participate in the extrinsic apoptosis pathway, influenced by death receptor proteins and the activation of death receptor proteins in some cells by CASP8 activation (a type of Bcl-2) (Bock and Tait 2020).

Mutations in mtDNA genes can influence cell death processes, particularly in the presence of heteroplasmic variations, where different genotypes of a mutation coexist within the cell (Pereira et al. 2021). The mtDNA methylation, controlled by DNA methyltransferases (DNMTs), regulates gene expression and is implicated in various diseases (Cui and Xu 2018). Non-coding RNAs (ncRNAs), categorized into long and small types, play regulatory roles at multiple levels within mitochondria, possibly influencing apoptosis through interactions with DNA, RNA, and proteins. Some ncRNAs, like MEG3 and specific miRNAs, have been associated with apoptosis regulation within mitochondria, suggesting their involvement in cell death mechanisms (Dong et al. 2017).

ASncmtRNA-1 and ASncmtRNA-2 are antisense long ncRNAs originating from mtDNA, which are heavily investigated for their roles in triggering apoptosis in diverse tumor cell lines and contributing to aging mechanisms in healthy human cells (Cavalcante et al. 2020).

### 2.7.1 Role of Micro RNAs (miRNAs) in Apoptosis

MiRNAs are small non-coding RNAs known for regulating gene expression and various cellular processes, including apoptosis. They exert their influence on both intrinsic and extrinsic pathways: (i) acting as anti-apoptotic molecules by targeting pro-apoptotic mRNAs or positive regulators of pro-apoptotic mRNAs, and (ii) serving as pro-apoptotic miRNAs by targeting anti-apoptotic mRNAs or their positive regulators (Subramanian and Steer 2010).

Several miRNAs regulate Bcl-2 expression in the intrinsic apoptosis pathway. miR-15 and miR-16 negatively regulate Bcl-2 expression, leading to activation of intrinsic apoptosis (Xia et al. 2008). Downregulation of miR-204 results in Bcl-2 overexpression and increased cell proliferation in gastric cancer cell lines. Moreover, miR-4779 targets PAK2 and CCND3, inhibiting cancer cell growth by inducing apoptosis and cell cycle arrest (Cavalcante et al. 2019). Similarly, upregulation of miR-25 protects tumor cells against TRAIL-induced apoptosis in cholangiocarcinoma (Razumilava et al. 2012). Several other miRNAs target proteins in the TRAIL pathway, such as miR-182 and miR-96 targeting FADD and CASP3, and miR-145 and miR-216 targeting TRAIL receptors DR4 and DR5 (Park and Peter 2008).

The intricate regulation of cell death metabolism encompasses a variety of complex mechanisms, featuring multiple checkpoints that dictate the destiny of cells in reaction to diverse stimuli and microenvironments. Depending on the genes they target, miRNAs can function as either anti-apoptotic agents, fostering conditions conducive to epithelial–mesenchymal transition, cell invasion, and circumvention of cell cycle arrest, or as pro-apoptotic agents, prompting cell death.

## 2.8 Apoptosis Role in the Developmental Process

Cell death is an indispensable and strictly regulated process fundamental to both the development and prevention of disease. Normal conditions see low amounts of pro-apoptotic proteins in the cytoplasm, but their expression or activity must quickly rise upon planned cell death. Both transcriptional and post-translational processes control several genes and proteins associated with death. APAF-1, for example, is under control by interactions with protein modifiers, subcellular localization, and proteolytic cleavage. Important in the evolution of the mammalian central nervous system is the transcriptional control of Apaf-1. Analogous modulation of members of the Bcl-2/CED-9/EGL-1 family is accomplished via both transcriptional and post-translational pathways. In mammals, multiple Bcl-2 family members are transcriptionally controlled, in *C. elegans*, the BH3-only protein EGL-1 is suppressed by the transcription factor Tra-1A in hermaphrodites but not in males. The p53 transcription factor, for instance, increases Noxa and Puma in response to DNA damage; the Jun kinase pathway induces Bim and Hrk/DP5 expression during development. Activation of Bcl-2 family proteins depends on post-translational control including changes in protein structure brought about by phosphorylation. The EGFR-Ras-Raf-MAPK pathway controls hid in *Drosophila* using both transcriptional and post-transcriptional processes. *Drosophila* models' genetic research revealed that the MAPK phosphorylation sites in Hid are essential for cell survival, underscoring the complex regulating systems controlling death in development and disease.

Studies of mutant animals show the significance of Bcl-2 family homologs. Targeting the bcl-2 gene specifically in mice produces phenotypic features, including lymphocytopenia, gray hair, and polycystic kidney disease. On the other hand, homozygous disruption of the anti-apoptotic bcl-x gene causes substantial death in post-mitotic immature neurons of the brain and spinal cord as well as in the hematological system and is embryonic fatal. This disturbance reduces lymphocyte maturation, lowering immature lymphocytes' lifetime while not appreciably harming adult lymphocytes. This suggests the part Bcl-x plays in maintaining the survival of immature cells during development.

Less severe phenotypes of Bax null animals include lymphocyte proliferation and ovarian and testicular degeneration. Studies further show that Bax is essential for cell death in the nervous system deprived of neurotrophic substances. Like bcl-2 overexpression, Bax deficiency promotes cancer, underscoring the significance of Bcl-2 family proteins in preserving tissue homeostasis and preventing neoplastic transformation.

Often dying in utero or soon after birth, Apaf-1 null animals show a range of developmental abnormalities. Indicating impaired death of apoptotic cells, these anomalies include severe craniofacial deformities, neural and brain enlargement, and retention of interdigital webs. Though its function in other types of developmental death is poorly known, Apaf-1 is essential for appropriate brain development and central nervous system death. Unlike mice, *Drosophila* mutants devoid of Apaf-1 expression are live, recommending caution when contrasting developmental death mechanisms across species.

Emphasizing the critical function of the FADD/caspase-8 system in the formation of particular cell types, caspase-8 null mice die in utero due to poor heart muscle development and show diminished hematopoietic progenitor cells. Analogous early developmental death and deficiencies in apoptosis in the nervous system brought on by mutations in caspase-3 and caspase-9 produce brain enlargement. These results underline the critical participation of cell death in the evolution of the nervous, brain, and cardiovascular systems, stressing the need for more research to clarify the exact function of dying cells in developing these structures and systems.

## 2.9 Apoptosis Role in Disease Pathogenesis

Apoptosis, or programmed cell death, is crucial in maintaining tissue homeostasis by eliminating unwanted or damaged cells. Dysregulation of apoptosis contributes to various diseases across different organ systems:

### 2.9.1 Apoptosis Dysregulation and Cancer

Maintaining tissue homeostasis by removing damaged or extraneous cells depends on apoptosis, sometimes called programmed cell death. Along with resistance to traditional treatments, dysregulation of death is intimately linked to cancer genesis and spread. Changes in both intrinsic and extrinsic death pathways help cancer cells survive and avoid immune system destruction and therapeutic agent damage.

An often occurring change in cancer is the overexpression of anti-apoptotic Bcl-2 family proteins, including Bcl-2 and Bcl-xL. By preventing mitochondrial outer membrane permeabilization (MOMP), these proteins improve cell viability and hence stop the release of cytochrome c and consequent caspase activation driving death. Linked to several cancers, including lymphomas, breast cancer, and lung carcinoma, where it corresponds with tumor aggressiveness and resistance to chemotherapy and radiotherapy, is overexpression of Bcl-2. On the other hand, mutations or lack of pro-apoptotic Bcl-2 family members like Bak and Bax also help to generate tumors. Mutations in Bax and Bak have been documented in colon and stomach carcinomas, and these proteins are essential for activating MOMP and starting the apoptotic cascade, therefore acting as tumor suppressors.

Made upon cytochrome c release, the protein complex known as the apoptosome activates caspase-9 and starts the caspase cascade. A key element of the apoptosome, apaf-1, is often turned off or suppressed in malignancies like melanoma, leukemia, and glioblastoma via epigenetic processes. This loss reduces the death response to biological stresses, encouraging cancer cell survival and proliferation.

Furthermore, flaws in death receptor pathways are important for cancer development. When death receptors—like TRAIL receptors and CD95 (Fas)—bind to their respective ligands, they send death signals. As demonstrated in hepatocellular carcinoma and esophageal cancer, downregulation or mutation of these receptors lets cancer cells avoid immune surveillance and withstand treatments meant to induce death.

Moreover, dysregulated caspase activity under the impact of mutations or aberrant expression of caspase inhibitors such as cFLIP and inhibitors of apoptosis proteins (IAPs) can compromise the execution of death in cancer cells. Initially found to be a caspase inhibitor, IAPs also control inflammatory pathways and cell movement, promoting tumorigenesis and metastases.

One intriguing approach for cancer treatment is to target these death routes. Clinical trials are underway on small molecule inhibitors of Bcl-2 family proteins and IAPs as well as agonists of death receptors such as TRAIL. These treatments seek to induce death in cancer cells, sensitizing them to conventional therapies and maybe overcoming resistance mechanisms.

### 2.9.2 Apoptosis Dysregulation and Neurological Disorders

Key roles in both neurodevelopment and disease pathogenesis within the central nervous system (CNS) are played by apoptosis, a method of programmed cell death. Neuronal death is linked in neurodegenerative diseases such as Alzheimer's disease (AD) to the pathogenicity evolution. Amyloid-β peptides accumulating together activate caspase-3, which causes neuronal death and cognitive decline (Rohn 2010; Shang et al. 2012). Furthermore, worsening neuronal degeneration is tau protein hyper-phosphorylation generated by caspase-3 cleavage, which stimulates neurofibrillary tangle development (Noble et al. 2009). Dopaminergic neuron loss in the substantia nigra is linked to Parkinson's disease (PD) with caspase-1 and caspase-3 activation-driven death (Samantaray et al. 2007; Valente et al. 2004). Mutations in PINK1 improve mitochondrial dysfunction and apoptotic signaling, contributing to neuronal death (Wood-Kaczmar et al. 2008). Moreover, caspase-8 activation by the HIP-1-Hippi complex induces death in PD models (Gervais et al. 2002). Defined by mutant huntingtin protein aggregation, Huntington's disease (HD) consists of caspase-mediated proteolytic cleavage of mutant huntingtin and HIP-1 dysregulation, exacerbating neurodegeneration (Goldberg et al. 1996; Bhattacharyya et al. 2008). Crucially in HD development, caspase-6 activation offers an early diagnostic and therapeutic target. Through mechanisms including oxidative stress, inflammation, and caspase activation, mutations in SOD1 cause motor neuron degeneration in amyotrophic lateral sclerosis (ALS), (Pasinelli et al. 2000; Meissner et al. 2010). Indicating their role in ALS pathogenesis, caspase-1, and caspase-3 activation adds to neuronal death (Pasinelli et al. 2000). Acute central nervous system injuries like stroke set up necrotic and dead pathway reactions enhanced by inflammation and oxidative stress (Yuan 2009). Through EP2 receptors, prostaglandin E2 activates caspase-3 in hippocampal neurons, therefore causing ischemic brain injury (Takadera et al. 2004). Furthermore, important roles in ischemic neuronal death are played by JNK activation and Fas/FasL signaling (Martin-Villalba et al. 2001). Reperfusion damage following stroke results from cytochrome C release and reactive oxygen species (ROS)-mediated mitochondrial malfunction, activating apoptotic processes (Pan et al. 2007; Jung et al. 2010). Deleting caspase-3 and caspase-1 in animal models reduces neuronal death, implying possible therapeutic targets (Le et al. 2002).

### 2.9.3 Apoptosis Dysregulation and Cardiovascular Disorders

Adult cardiomyocytes are largely post-mitotic, reducing their ability to repair following damage. Acute assaults like ischemia result in various forms of cell death, but chronic stress often leads to hypertrophy and remodeling. Although the normal myocardium shows low apoptosis (0.001–0.002%) (Whelan et al. 2010), its incidence increases considerably in heart diseases, playing critical roles in circumstances like ischemia/reperfusion damage (I/R) and chronic heart failure (CHF) (Okuno et al. 2004; Le et al. 2002). I/R, characterized by oxidative stress and inflammation during reperfusion, induces necrotic, apoptotic, and autophagic cell deaths. Both intrinsic and extrinsic apoptotic pathways are implicated in I/R, with therapies targeting pathways like Fas, TNF, and p53 showing promise (Pan et al. 2007; Jung et al. 2010). In CHF, low apoptosis rates (0.08–0.25%) contribute to left ventricular remodeling, increased by stressors such as mechanical strain and neurohormones (Whelan et al. 2010). Strategies targeting apoptotic pathways, such as caspase inhibitors or Akt activation, show therapeutic potential against myocardial apoptosis and heart failure development (Whelan et al. 2010; Okuno et al. 2004).

## 2.10 Conclusion

Apoptosis is an energy-dependent, meticulously regulated process characterized by distinct morphological and biochemical features, with caspase activation playing a central role. While numerous essential apoptotic proteins and their roles in apoptotic pathways have been identified, a comprehensive understanding of their molecular mechanisms remains incomplete, necessitating ongoing research. Significant advancements have been made in understanding apoptosis, and it is crucial to continue exploring its mechanisms and interactions with other biological systems. This knowledge can lead to identifying biomarkers that may influence the clinical progression of various diseases. To address this, we delve into the mechanisms of apoptosis and emphasize the critical cellular and genetic components involved. Looking to the future, it is essential to investigate additional epigenetic factors beyond DNA methylation and miRNA activity, which have yet to be fully explored in the context of apoptosis and mitochondrial functions. Studying genetic variants within apoptotic pathways from a population genetics perspective could provide valuable insights into disease development. Understanding the molecular mechanisms underlying apoptosis and other forms of programmed cell death is crucial for gaining profound insights into diverse disease processes. This knowledge can also inform the development of therapeutic strategies. Future research should identify novel epigenetic regulators and genetic variants involved in apoptosis and explore their potential as therapeutic targets. Future research should further elucidate the molecular mechanisms and regulatory networks governing plant PCD. Understanding the structural and functional dynamics of plant PCD regulators could provide insights into their evolutionary origins and roles in plant immunity and stress tolerance. Moreover, integrating multi-omics approaches and advanced computational models could enhance our understanding of apoptosis and its role in health and disease, ultimately leading to improved diagnostic and therapeutic interventions.

## References


Alenzi FQ, Lotfy M, Wyse R (2010) Swords of cell death: caspase activation and regulation. Asian Pac J Cancer Prev 11(2):271–280

Arafa KK, Ibrahim A, Mergawy R, El-Sherbiny IM, Febbraio F, Hassan RY (2022) Advances in cancer diagnosis: bio-electrochemical and biophysical characterizations of cancer cells. Micromachines 13(9):1401

Aravind L, Dixit VM, Koonin EV (1999) The domains of death: evolution of the apoptosis machinery. Trends Biochem Sci 24(2):47–53

Balk J, Leaver CJ (2001) The PET1-CMS mitochondrial mutation in sunflower is associated with premature programmed cell death and cytochrome c release. Plant Cell 13(8):1803–1818

Balk J, Leaver CJ, McCabe PF (1999) Translocation of cytochrome c precedes the visible signs of cell death in UV-C-irradiated leaves of wild-type and lesion mimic mutants of maize. Proc Natl Acad Sci 96(20):11621–11626

Bender CE, Fitzgerald P, Tait SW, Llambi F, McStay GP, Tupper DO, Pellettieri J, Saghafi N, Saenz HM, Wasserman SA, Salvesen GS (2015) Mitochondrial pathway of apoptosis is ancestral in metazoans. Proc Natl Acad Sci 112(19):6088–6093

Bertheloot D, Latz E, Franklin BS (2021) Necroptosis, pyroptosis and apoptosis: an intricate game of cell death. Cell Mol Immunol 18(5):1106–1121

Berube C, Boucher L-M, Ma W, Wakeham A, Salmena L, Hakem R, Yeh W-C, Mak TW, Benchimol S (2005) Apoptosis caused by p53-induced protein with death domain (PIDD) depends on the death adapter protein RAIDD. Proc Natl Acad Sci 102(40):14314–14320

Bhattacharyya NP, Banerjee M, Majumder P (2008) Huntington's disease: roles of huntingtin-interacting protein 1 (HIP-1) and its molecular partner HIPPI in the regulation of apoptosis and transcription. FEBS J 275:4271–4279

Birnbaum MJ, Clem RJ, Miller LK (2000) An apoptosis-inhibiting gene from a nuclear polyhedrosis virus encoding a polypeptide with Cys/His sequence motifs. J Virol 68(4):2521–2528

Bock FJ, Tait SW (2020) Mitochondria as multifaceted regulators of cell death. Nat Rev Mol Cell Biol 21(2):85–100

Bonneau L, Ge Y, Drury GE, Gallois P (2008) What happened to plant caspases? J Exp Bot 59(3):491–499

Bossy-Wetzel E, Barsoum MJ, Godzik A, Schwarzenbacher R, Lipton SA (2003) Mitochondrial fission in apoptosis, neurodegeneration and aging. Curr Opin Cell Biol 15(6):706–716

Bozhkov PV, Suarez MF, Filonova LH (2005) Programmed cell death in plant embryogenesis. Curr Top Dev Biol 67:135–179

Briknarova K, Takayama S, Brive L, Havert ML, Knee DA, Velasco J, Homma S, Cabezas E, Stuart J, Hoyt DW, Satterthwait AC, Llinás M, Reed JC (2001) Structural analysis of BAG1 cochaperone and its interactions with Hsc70 heat shock protein. Nat Struct Biol 8(4):349–352

Brunelle JK, Letai A (2009) Control of mitochondrial apoptosis by the Bcl-2 family. J Cell Sci 122(4):437–441

Budihardjo I, Oliver H, Lutter M, Luo X, Wang X (1999) Biochemical pathways of caspase activation during apoptosis. Annu Rev Cell Dev Biol 15(1):269–290

Cain K, Bratton SB, Cohen GM (2002) The Apaf-1 apoptosome: a large caspase-activating complex. Biochimie 84(2–3):203–214

Cavalcante GC, Schaan AP, Cabral GF, Santana-da-Silva MN, Pinto P, Vidal AF, Ribeiro-dos-Santos Â (2019) A cell's fate: an overview of the molecular biology and genetics of apoptosis. Int J Mol Sci 20(17):4133

Cavalcante GC, Magalhães L, Ribeiro-dos-Santos Â, Vidal AF (2020) Mitochondrial epigenetics: non-coding RNAs as a novel layer of complexity. Int J Mol Sci 21(5):1838

Chao DT, Korsmeyer SJ (1998) BCL-2 family: regulators of cell death. Annu Rev Immunol 16(1):395–419

Chen F, Hersh BM, Conradt B (2000) Transcriptional regulation of cell death genes in *Caenorhabditis elegans*: distinct roles of ces-1 and ces-2 bHLH proteins. Cell 99(3):293–305

Chipuk J, Green D (2006) Dissecting p53-dependent apoptosis. Cell Death Differ 13(6):994–1002

Chipuk JE, Moldoveanu T, Llambi F, Parsons MJ, Green DR (2010) The BCL-2 family reunion. Mol Cell 37(3):299–310

Choi CJ, Berges JA (2013) New types of metacaspases in phytoplankton reveal diverse origins of cell death proteases. Cell Death Dis 4(2):e490

Chowdhury I, Tharakan B, Bhat GK (2006) Current concepts in apoptosis: the physiological suicide program revisited. Cell Mol Biol Lett 11:506–525

Cikala M, Wilm B, Hobmayer E, Bottger A, David CN (1999) Identification of caspases and apoptosis in the simple metazoan Hydra. Curr Biol 9(17):959–962

Conradt B, Horvitz HR (1998) The *C. elegans* protein EGL-1 is required for programmed cell death and interacts with the Bcl-2-like protein CED-9. Cell 93(4):519–529

Cui D, Xu X (2018) DNA methyltransferases, DNA methylation, and age-associated cognitive function. Int J Mol Sci 19(5):1315

Cui J, Zhao S, Li Y, Zhang D, Wang B, Xie J, Wang J (2021) Regulated cell death: discovery, features and implications for neurodegenerative diseases. Cell Commun Signal 19:1–29

D'arcy MS (2019) Cell death: a review of the major forms of apoptosis, necrosis and autophagy. Cell Biol Int 43(6):582–592

Deng H, Yue JK, Zusman BE, Nwachuku EL, Abou-Al-Shaar H, Upadhyayula PS, Okonkwo DO, Puccio AM (2020) B-cell lymphoma 2 (Bcl-2) and regulation of apoptosis after traumatic brain injury: a clinical perspective. Medicina 56(6):300

Dong Y, Yoshitomi T, Hu J-F, Cui J (2017) Long noncoding RNAs coordinate functions between mitochondria and the nucleus. Epigenet Chromatin 10:1–11

Dorstyn L, Colussi PA, Quinn LM, Richardson H, Kumar S (2002) DRONC, an ecdysone-inducible *Drosophila* caspase. Proc Natl Acad Sci 96(8):4307–4312

Doukhanina EV, Chen S, van der Zalm E, Godzik A, Reed J, Dickman MB (2006) Identification and functional characterization of the BAG protein family in *Arabidopsis thaliana*. J Biol Chem 281(28):18793–18801

Earnshaw WC, Martins LM, Kaufmann SH (1999) Mammalian caspases: structure, activation, substrates, and functions during apoptosis. Annu Rev Biochem 68(1):383–424

Elmore S (2007) Apoptosis: a review of programmed cell death. Toxicol Pathol 35(4):495–516

Estaquier J, Vallette F, Vayssiere J-L, Mignotte B (2012) The mitochondrial pathways of apoptosis. Adv Mitochond Med:157–183

Feltham R, Vince JE, Lawlor KE (2017) Caspase-8: not so silently deadly. Clin Transl Immunol 6(1):e124

Fukuda H (2000) Programmed cell death of tracheary elements as a paradigm in plants. Plant Mol Biol 44(3):245–253

Gervais FG, Singaraja R, Xanthoudakis S et al (2002) Recruitment and activation of caspase-8 by the Huntingtin-interacting protein Hip-1 and a novel partner Hippi. Nat Cell Biol 4:95–105

Goldberg YP, Nicholson DW, Rasper DM et al (1996) Cleavage of huntingtin by apopain, a proapoptotic cysteine protease, is modulated by the polyglutamine tract. Nat Genet 13:442–449

Goping IS, Gross A, Lavoie JN, Nguyen M, Jemmerson R, Roth K, Korsmeyer SJ, Shore GC (1998) Regulated targeting of BAX to mitochondria. J Cell Biol 143(1):207–215

Han J, Sabbatini P, Perez D, Rao L, Modha D, White E (1996) The E1B 19K protein blocks apoptosis and downregulates Bax oligomerization in the absence of E1B 55K. Cell Growth Differ: The Mol Biol J Am Assoc Cancer Res 7(6):801–808

Hara-Nishimura I, Hatsugai N (2011) The role of vacuole in plant cell death. Cell Death Differ 18:1298–1304

Hatsugai N, Kuroyanagi M, Yamada K, Meshi T, Tsuda S, Kondo M, Nishimura M, Hara-Nishimura I (2004) A plant vacuolar protease, VPE, mediates virus-induced hypersensitive cell death. Science 305(5685):855–858

Hay BA, Wolff T, Rubin GM (1994) Expression of baculovirus P35 prevents cell death in *Drosophila*. Development 120(8):2121–2129

Henry CM, Martin SJ (2017) Caspase-8 acts in a non-enzymatic role as a scaffold for assembly of a pro-inflammatory "FADDosome" complex upon TRAIL stimulation. Mol Cell 65(4):715–729. e715

Hetts SW (1998) To die or not to die: an overview of apoptosis and its role in disease. JAMA 279(4):300–307

Horvitz HR (1999) Genetic control of programmed cell death in the nematode *Caenorhabditis elegans*. Cancer Res 59(7):1701s–1706s

Hui H, Dotta F, Mario UD, Perfetti R (2004) Role of caspases in the regulation of apoptotic pancreatic islet beta-cells death. J Cell Physiol 200(2):177–200

Ivanisenko N, Lavrik I (2019) Mechanisms of procaspase-8 activation in the extrinsic programmed cell death pathway. Mol Biol 53:732–738

Jiang X, Tsang YH, Yu Q (2007) c-Myc overexpression sensitizes Bim-mediated Bax activation for apoptosis induced by histone deacetylase inhibitor suberoylanilide hydroxamic acid (SAHA) through regulating Bcl-2/Bcl-xL expression. Int J Biochem Cell Biol 39(5):1016–1025

Johnson C, Jarvis W (2004) Caspase-9 regulation: an update. Apoptosis 9(4):423–427

Jung JE, Kim GS, Chen H et al (2010) Reperfusion and neurovascular dysfunction in stroke: from basic mechanisms to potential strategies for neuroprotection. Mol Neurobiol 41:172–179

Kabbage M, Dickman MB (2008) The BAG proteins: a ubiquitous family of chaperone regulators. Cell Mol Life Sci 65(9):1390–1402

Kabbage M, Williams B, Dickman MB (2013) Cell death control: the interplay of apoptosis and autophagy in the context of plant immunity. J Mol Biol 429(3):595–605

Kaiser WJ, Vucic D, Miller LK (1998) The *Drosophila* cell death gene hid functions by antagonizing IAPs, but does not require caspase activation or binding to the IBM motif. Mol Cell 1(3):289–299

Kaman T, Karasakal ÖF, Oktay EÖ, Ulucan K, Konuk M (2019) In silico approach to the analysis of SNPs in the human APAF1 gene. Turk J Biol 43(6):371–381

Kanuka H, Sawai T, Gibson MD, Tojo K, Miura M (1999) JNK and caspase-9-mediated induction of apoptosis in *Drosophila* S2 cells. J Biol Chem 274(49):35538–35541

Karimzadeh S, Hosseinkhani S, Fathi A, Ataei F, Baharvand H (2018) Insufficient Apaf-1 expression in early stages of neural differentiation of human embryonic stem cells might protect them from apoptosis. Eur J Cell Biol 97(2):126–135

Kashyap D, Garg VK, Goel N (2021) Intrinsic and extrinsic pathways of apoptosis: role in cancer development and prognosis. Adv Protein Chem Struct Biol 125:73–120

Kawasaki T, Henmi K, Ono E, Hatakeyama S, Iwano M, Satoh H, Shimamoto K (2001) The small GTP-binding protein rac is a regulator of cell death in plants. Proc Natl Acad Sci 96(19):10922–10926

Kerr JFR, Wyllie AH, Currie AR (1972) Apoptosis: a basic biological phenomenon with wide-ranging implications in tissue kinetics. Br J Cancer 26(4):239–257

Kihara A, Noda T, Ishihara N, Ohsumi Y (2001) Two distinct Vps34 phosphatidylinositol 3-kinase complexes function in autophagy and carboxypeptidase Y sorting in Saccharomyces cerevisiae. J Cell Biol 152(3):519–530

Kim H-E, Du F, Fang M, Wang X (2005) Formation of apoptosome is initiated by cytochrome c-induced dATP hydrolysis and subsequent nucleotide exchange on Apaf-1. Proc Natl Acad Sci 102(49):17545–17550

Kim W-S, Lee K-S, Kim J-H, Kim C-K, Lee G, Choe J, Won M-H, Kim T-H, Jeoung D, Lee H (2017) The caspase-8/Bid/cytochrome c axis links signals from death receptors to mitochondrial reactive oxygen species production. Free Radic Biol Med 112:567–577

Kroemer G, Galluzzi L, Brenner C (2007) Mitochondrial membrane permeabilization in cell death. Physiol Rev 87(1):99–163

Kusano H, Shimizu S, Koya RC, Fujita H, Kamada S, Kuzumaki N, Tsujimoto Y (2000) Human gelsolin prevents apoptosis by inhibiting apoptotic mitochondrial changes via closing VDAC. Oncogene 19(42):4807–4814

Kuwana T, Smith JJ, Muzio M, Dixit V, Newmeyer DD, Kornbluth S (2002) Apoptosis induction by caspase-8 is amplified through the mitochondrial release of cytochrome c. J Biol Chem 273(26):16589–16594

Lacomme C, Santa Cruz S (1999) Bax-induced cell death in tobacco is similar to the hypersensitive response. Proc Natl Acad Sci 96(14):7956–7961

Lam E (2004) Controlled cell death, plant survival and development. Nat Rev Mol Cell Biol 5(4):305–315

Lam E, Zhang Y (2012) Regulation of cell death by mitochondrial nucleases in plants. Front Plant Sci 3:88

Lampl N, Budai-Hadrian O, Davydov O, Joss TV, Haran-Marozevich I, Komaniecki D, Szeder K, Aviezer-Hagai K, Varatharajoo S, Fluhr R (2010) *Arabidopsis* AtSerpin1, crystal structure and in vivo interaction with its target protease RD21. J Biol Chem 285(3):13550–13560

Lampl N, Alkan N, Davydov O, Fluhr R (2013) Set-point control of RD21 protease activity by AtSerpin1 controls cell death in *Arabidopsis*. Plant J 74(3):498–510

Lawen A (2003) Apoptosis—an introduction. BioEssays 25(9):888–896

Le DA, Wu Y, Huang Z et al (2002) Caspase activation and neuroprotection in caspase-3-deficient mice after in vivo cerebral ischemia and in vitro oxygen glucose deprivation. Proc Natl Acad Sci USA 99:15188–15193

Lee E-W, Seo J-H, Jeong M-H, Lee S-S, Song J-W (2012) The roles of FADD in extrinsic apoptosis and necroptosis. BMB Rep 45(9):496–508

Li P, Nijhawan D, Budihardjo I, Srinivasula SM, Ahmad M, Alnemri ES, Wang X (1997) Cytochrome c and dATP-dependent formation of Apaf-1/caspase-9 complex initiates an apoptotic protease cascade. Cell 91(4):479–489

Liu F, Wang X, Liu H, Wang Y, Liu X, Hao X, Li H (2018) LncRNA BX357664 inhibits cell proliferation and invasion and promotes cell apoptosis in human colorectal cancer cells. Oncol Lett 15(6):8237–8244

Madeo F, Carmona-Gutierrez D, Ring J, Büttner S, Eisenberg T, Kroemer G (2009) Apoptosis in yeast: a new model system with applications in cell biology and medicine. Cell Death Differ 16(8):1118–1124

Martin-Villalba A, Hahne M, Kleber S et al (2001) Therapeutic neutralization of CD95-ligand and TNF attenuates brain damage in stroke. Cell Death Differ 8:679–686

McIlwain DR, Berger T, Mak TW (2013) Caspase functions in cell death and disease. Cold Spring Harb Perspect Biol 5(4):a008656

Meissner F, Molawi K, Zychlinsky A (2010) Mutant superoxide dismutase 1-induced IL-1beta accelerates ALS pathogenesis. Proc Natl Acad Sci USA 107:13046–13050

Meyer N, Kim SS, Penn LZ (2006) The Oscar-worthy role of Myc in apoptosis. Semin Cancer Biol

Mishra NC, Kumar S (2005) Apoptosis: a mitochondrial perspective on cell death. Indian J Exp Biol 43(01):25–35

Mohamad N, Gutiérrez A, Núñez M, Cocca C, Martín G, Cricco G, Medina V, Rivera E, Bergoc R (2005) Mitochondrial apoptotic pathways. Biocell 29(2):149–161

Müller M, Wilder S, Bannasch D, Israeli D, Lehlbach K, Li-Weber M, Friedman SL, Galle PR, Stremmel W, Oren M (1998) p53 activates the CD95 (APO-1/Fas) gene in response to DNA damage by anticancer drugs. J Exp Med 188(11):2033–2045

Nguyen T-AT, Grimm SA, Bushel PR, Li J, Li Y, Bennett BD, Lavender CA, Ward JM, Fargo DC, Anderson CW (2018) Revealing a human p53 universe. Nucleic Acids Res 46(16):8153–8167

Nissanka N, Moraes CT (2018) Mitochondrial DNA damage and reactive oxygen species in neurodegenerative disease. FEBS Lett 592(5):728–742

Noble W, Garwood C, Stephenson J et al (2009) Minocycline reduces the development of abnormal tau species in models of Alzheimer's disease. FASEB J 23:739–750

Obeng E (2020) Apoptosis (programmed cell death) and its signals-A review. Braz J Biol 81:1133–1143

Okuno S, Saito A, Hayashi T, Chan PH (2004) The c-Jun N-terminal protein kinase signaling pathway mediates Bax activation and subsequent neuronal apoptosis through interaction with Bim after transient focal cerebral ischemia. J Neurosci 24:7879–7887

Orth JD, Loewer A, Lahav G, Mitchison TJ (2012) Prolonged mitotic arrest triggers partial activation of apoptosis, resulting in DNA damage and p53 induction. Mol Biol Cell 23(4):567–576

Pan J, Konstas AA, Bateman B et al (2007) Reperfusion injury following cerebral ischemia: pathophysiology, MR imaging, and potential therapies. Neuroradiology 49:93–102

Papenfuss K, Cordier SM, Walczak H (2008) Death receptors as targets for anti-cancer therapy. J Cell Mol Med 12(6b):2566–2585

Park S-M, Peter ME (2008) microRNAs and death receptors. Cytokine Growth Factor Rev 19(3–4):303–311

Pasinelli P, Houseweart MK, Brown RH Jr, Cleveland DW (2000) Caspase-1 and -3 are sequentially activated in motor neuron death in Cu, Zn superoxide dismutase-mediated familial amyotrophic lateral sclerosis. Proc Natl Acad Sci USA 97:13901–13906

Pereira CV, Gitschlag BL, Patel MR (2021) Cellular mechanisms of mtDNA heteroplasmy dynamics. Crit Rev Biochem Mol Biol 56(5):510–525

Perini GF, Ribeiro GN, Pinto Neto JV, Campos LT, Hamerschlak N (2018) BCL-2 as therapeutic target for hematological malignancies. J Hematol Oncol 11:1–15

Pinton P, Ferrari D, Di Virgilio F, Pozzan T, Rizzuto R (2001) Molecular machinery and signaling events in apoptosis. Drug Dev Res 52(4):558–570

Quinn LM, Dorstyn L, Mills K, Colussi P, Chen P, Coombe M, Abrams J, Kumar S, Richardson H (2003) An essential role for the caspase Dronc in developmentally programmed cell death in *Drosophila*. J Biol Chem 278(50):47685–47693

Ramirez MLG, Salvesen GS (2018) A primer on caspase mechanisms. Semin Cell Dev Biol

Ramos-Miguel A, García-Sevilla JA, Barr AM, Bayer TA, Falkai P, Leurgans SE, Schneider JA, Bennett DA, Honer WG, García-Fuster MJ (2017) Decreased cortical FADD protein is associated with clinical dementia and cognitive decline in an elderly community sample. Mol Neurodegener 12:1–14

Razumilava N, Bronk SF, Smoot RL, Fingas CD, Werneburg NW, Roberts LR, Mott JL (2012) miR-25 targets TNF-related apoptosis inducing ligand (TRAIL) death receptor-4 and promotes apoptosis resistance in cholangiocarcinoma. Hepatology 55(2):465–475

Renvoize C, Biola A, Pallardy M, Breard J (1998) Apoptosis: identification of dying cells. Cell Biol Toxicol 14:111–120

Robert G, Gastaldi C, Puissant A, Hamouda A, Jacquel A, Dufies M, Belhacene N, Colosetti P, Reed JC, Auberger P (2012) The anti-apoptotic Bcl-B protein inhibits BECN1-dependent autophagic cell death. Autophagy 8(4):637–649

Rock KL, Kono H (2008) The inflammatory response to cell death. Annu Rev Pathol Mech Dis 3:99–126

Rodriguez A, Oliver H, Zou H, Chen P, Wang X, Abrams JM (1999) Dark is a *Drosophila* homolog of Apaf-1/CED-4 and functions in an evolutionarily conserved death pathway. Nat Cell Biol 1(5):272–279

Rohn TT (2010) The role of caspases in Alzheimer's disease; potential novel therapeutic opportunities. Apoptosis 15:1403–1409

Rojas J, Bermudez V, Palmar J, Martínez MS, Olivar LC, Nava M, Tomey D, Rojas M, Salazar J, Garicano C (2018) Pancreatic beta cell death: novel potential mechanisms in diabetes therapy. J Diabetes Res 2018

Sahoo G, Samal D, Khandayataray P, Murthy MK (2023) A review on caspases: key regulators of biological activities and apoptosis. Mol Neurobiol 60(10):5805–5837

Samantaray S, Knaryan VH, Guyton MK et al (2007) The parkinsonian neurotoxin rotenone activates calpain and caspase-3 leading to motoneuron degeneration in spinal cord of Lewis rats. Neuroscience 146:741–755

Seshagiri S, Miller LK (1997) *Caenorhabditis elegans* CED-4 stimulates CED-3 processing and CED-3-induced apoptosis. Curr Biol 7(7):455–460

Shakeri R, Kheirollahi A, Davoodi J (2017) Apaf-1: regulation and function in cell death. Biochimie 135:111–125
Shang YC, Chong ZZ, Wang S, Maiese K (2012) Prevention of beta-amyloid degeneration of microglia by erythropoietin depends on Wnt1, the PI 3-K/mTOR pathway, Bad, and Bcl-xL. Aging (Albany NY) 4:187–201
Shi Y (2001) A structural view of mitochondria-mediated apoptosis. Nat Struct Biol 8(5):394–401
Shi Y (2004) Caspase activation: revisiting the induced proximity model. Cell 117(7):855–858
Siddiqui WA, Ahad A, Ahsan H (2015) The mystery of BCL2 family: Bcl-2 proteins and apoptosis: an update. Arch Toxicol 89:289–317
Signorovitch AY, Dellaporta SL, Buss LW (2007) Molecular signatures for sex in the Placozoa. Proc Natl Acad Sci 102(44):15518–15522
Singh S, Mabalirajan U (2021) Mitochondrial calcium in command of juggling myriads of cellular functions. Mitochondrion 57:108–118
Staniek J, Kalina T, Andrieux G, Boerries M, Janowska I, Fuentes M, Díez P, Bakardjieva M, Stancikova J, Raabe J (2024) Non-apoptotic FAS signaling controls mTOR activation and extrafollicular maturation in human B cells. Sci Immunol 9(91):eadj5948
Stone JR, Okonkwo DO, Singleton RH, Mutlu LK, Helm GA, Povlishock JT (2002) Caspase-3-mediated cleavage of amyloid precursor protein and formation of amyloid β peptide in traumatic axonal injury. J Neurotrauma 19(5):601–614
Strasser A, O'Connor L, Dixit VM (2000) Apoptosis signaling. Annu Rev Biochem 69(1):217–245
Subramanian S, Steer CJ (2010) MicroRNAs as gatekeepers of apoptosis. J Cell Physiol 223(2):289–298
Sundström JF, Vaculova A, Smertenko AP, Savenkov EI, Golovko A, Minina E, Tiwari BS, Rodriguez-Nieto S, Zamyatnin AA, Välineva T, Saarma M, Ståhl U, Hussey PJ, Silke J, Zhivotovsky B, Bozhkov PV (2009) Tudor staphylococcal nuclease is an evolutionarily conserved component of the programmed cell death degradome. Nat Cell Biol 11(11):1347–1354
Takadera T, Shiraishi Y, Ohyashiki T (2004) Prostaglandin E2 induced caspase-dependent apoptosis possibly through activation of EP2 receptors in cultured hippocampal neurons. Neurochem Int 45:713–719
Takayama S, Reed JC (2001) Molecular chaperone targeting and regulation by BAG family proteins. Nat Cell Biol 3(10):237–241
Tang D, Kang R (2022) Cell death: machinery and regulation. In: Mechanisms of cell death and opportunities for therapeutic development. Elsevier, pp 47–64
Tenev T, Zachariou A, Wilson R, Ditzel M, Meier P (2005) IAPs are functionally non-equivalent and regulate effector caspases through distinct mechanisms. Nat Cell Biol 7(1):70–77
Thompson R, Shah RB, Liu PH, Gupta YK, Ando K, Aggarwal AK, Sidi S (2015) An inhibitor of PIDDosome formation. Mol Cell 58(5):767–779
Tiwari BS, Belenghi B, Levine A (2002) Oxidative stress increased respiration and generation of reactive oxygen species, resulting in ATP depletion, opening of mitochondrial permeability transition, and programmed cell death. Plant Physiol 128(4):1271–1281
Tummers B, Green DR (2017) Caspase-8: regulating life and death. Immunol Rev 277(1):76–89
Uren AG, O'Rourke K, Aravind L, Pisabarro MT, Seshagiri S, Koonin EV, Dixit VM (2000) Identification of paracaspases and metacaspases: two ancient families of caspase-like proteins, one of which plays a key role in MALT lymphoma. Mol Cell 6(4):961–967
Valente EM, Abou-Sleiman PM, Caputo V et al (2004) Hereditary early-onset Parkinson's disease caused by mutations in PINK1. Science 304:1158–1160
Van Delft MF, Huang D (2006) How the Bcl-2 family of proteins interact to regulate apoptosis. Cell Res 16(2):203–213
van Doorn WG (2011) Classes of programmed cell death in plants, compared to those in animals. J Exp Bot 62(14):4749–4761
Van Doorn WG, Woltering EJ (2005) Many ways to exit? Cell death categories in plants. Trends Plant Sci 10(3):117–122
Van Hautegem T, Waters AJ, Goodrich J, Nowack MK (2015) Only in dying, life: programmed cell death during plant development. Trends Plant Sci 20(2):102–113

Vaux DL, Silke J (2005) IAPs, RINGs and ubiquitylation. Nat Rev Mol Cell Biol 6(4):287–297

Vincent M (2012) Cancer: a de-repression of a default survival program common to all cells? A life-history perspective on the nature of cancer. BioEssays 34(1):72–82

Wajant H, Pfizenmaier K, Scheurich P (2003) Non-apoptotic Fas signaling. Cytokine Growth Factor Rev 14(1):53–66

Wang C, Youle RJ (2009) The role of mitochondria in apoptosis. Annu Rev Genet 43:95–118

Wang L, Yang JK, Kabaleeswaran V, Rice AJ, Cruz AC, Park AY, Yin Q, Damko E, Jang SB, Raunser S (2010) The Fas–FADD death domain complex structure reveals the basis of DISC assembly and disease mutations. Nat Struct Mol Biol 17(11):1324–1329

Watanabe N, Lam E (2005) Two *Arabidopsis* metacaspases AtMCP1b and AtMCP2b are arginine/lysine-specific cysteine proteases and activate apoptosis-like cell death in yeast. J Biol Chem 280(16):14691–14699

Watanabe N, Lam E (2011) Calcium-dependent activation and autolysis of *Arabidopsis* metacaspase 2a. FEBS Lett 585(2):225–230

Whelan RS, Kaplinskiy V, Kitsis RN (2010) Cell death in the pathogenesis of heart disease: mechanisms and significance. Annu Rev Physiol 72:19–44

White E (2001) Regulation of the cell cycle and apoptosis by the oncogenes of adenovirus. Oncogene 20(54):7836–7846

Wiens M, Belikov SI, Kaluzhnaya OV, Krasko A, Thakur NL, Perovic-Ottstadt S, Muller WE (2000) Homeobox-containing gene associated with axial differentiation in the sponge Suberites domuncula. J Biol Chem 275(35):35953–35959

Williams B, Kabbage M, Britt R, Dickman MB (2010) AtBAG7, an *Arabidopsis* Bcl-2-associated athanogene, resides in the endoplasmic reticulum and is involved in the unfolded protein response. Proc Natl Acad Sci 107(13):6088–6093

Wood-Kaczmar A, Gandhi S, Yao Z et al (2008) PINK1 is necessary for long term survival and mitochondrial function in human dopaminergic neurons. PLoS One 3:e2455

Wright KM, Duncan GH, Pradel KS, Roberts MR (2009) Age-dependent programmed cell death in the lace plant: a model system for the study of leaf perforation. J Exp Bot 60(8):2129–2142

Xia L, Zhang D, Du R, Pan Y, Zhao L, Sun S, Hong L, Liu J, Fan D (2008) miR-15b and miR-16 modulate multidrug resistance by targeting BCL2 in human gastric cancer cells. Int J Cancer 123(2):372–379

Xiao S-Q, Cheng M, Wang L, Cao J, Fang L, Zhou X-P, He X-J, Hu Y-F (2023) The role of apoptosis in the pathogenesis of osteoarthritis. Int Orthop 47(8):1895–1919

Xu Q, Reed JC (1998) Bax inhibitor-1, a mammalian apoptosis suppressor identified by functional screening in yeast. Mol Cell 1(3):337–346

Yan S, Tang Z, Su W, Sun W (2003) Proteomic analysis of salt stress-responsive proteins in rice root. Proteomics 5(9):235–244

Yang JK (2015) Death effecter domain for the assembly of death-inducing signaling complex. Apoptosis 20:235–239

Yang Y, Fang S, Jensen JP, Weissman AM, Ashwell JD (2000) Ubiquitin protein ligase activity of IAPs and their degradation in proteasomes in response to apoptotic stimuli. Science 288(5467):874–877

Yuan J (2009) Neuroprotective strategies targeting apoptotic and necrotic cell death for stroke. Apoptosis 14:469–477

Zhang Y, Lam E (2011) Shedding light on the role of metacaspases. Plant Signal Behav 6(12):2029–2032

Zhuang H, Wang X, Zha D, Gan Z, Cai F, Du P, Yang Y, Yang B, Zhang X, Yao C (2016) FADD is a key regulator of lipid metabolism. EMBO Mol Med 8(8):895–918

Zindy F, Eischen CM, Randle DH, Kamijo T, Cleveland JL, Sherr CJ, Roussel MF (1998) Myc signaling via the ARF tumor suppressor regulates p53-dependent apoptosis and immortalization. Genes Dev 12(15):2424–2433

Zou H, Li Y, Liu X, Wang X (1997) An APAF-1·cytochrome c multimeric complex is a functional apoptosome that activates procaspase-9. J Biol Chem 272(29):18556–18558

# 3 Apoptosis: Novel Insights into the Role of Death Receptors and Caspases in Cell Fate

Satyajit Halder, Anirban Manna, and Kuladip Jana

**Abstract**

Caspases, a group of cysteine-aspartic proteases, have a critical function in carrying out apoptosis, a highly regulated process of programmed cell death that is essential for maintaining tissue balance and promoting development. Caspases are produced as inactive zymogens and can be divided into two groups: initiator caspases (caspase-2, -8, -9, and -10) and effector caspases (such as caspase-3, caspase-6, and caspase-7). Initiator caspases are triggered by apoptotic signals, either through death receptors or mitochondrial pathways, leading to their oligomerization and subsequent activation. This activation then triggers effector caspases, which break down cellular components by means of proteolytic cleavage. These series of events result in distinct changes in the structure and chemistry of the cell, such as the breaking down of DNA, the production of small protrusions on the cell membrane, and the creation of apoptotic bodies. In addition to their traditional function in programmed cell death, recent research suggests that caspases can play a role in non-apoptotic processes such as inflammation, cell differentiation, and tissue remodeling. This chapter consolidates the latest information on the processes of caspase activation, their specificity in targeting substrates, the factors that regulate their activity, and their wider significance in both health and illness. Gaining knowledge about the function of caspases in programmed cell death not only reveals essential aspects of cellular biology but also presents new opportunities for developing treatments for conditions marked by either excessive or insufficient apoptosis.

S. Halder · A. Manna · K. Jana (✉)
Division of Molecular Medicine, Bose Institute, Unified Academic Campus, Kolkata, West Bengal, India
e-mail: kuladip@jcbose.ac.in



K. Jana (ed.), *Apoptosis and Human Health: Understanding Mechanistic and Therapeutic Potential*, https://doi.org/10.1007/978-981-97-7905-5_3

**Keywords**

Apoptosis · Death receptor · Caspases · Inflammation · Immunity · Cancer

## 3.1 Introduction

Apoptosis, also known as programmed cell death, is a crucial mechanism that helps maintain tissue balance and stability. The process guarantees the removal of impaired cells and the reutilization of cellular constituents essential for specialization and growth (Eskandari and Eaves 2022). Traditionally, cell death has been classed based on the morphological changes found in dying cells, with apoptosis and necrosis being the most well-known kinds. However, a new study has uncovered additional forms of cell death. Cysteine-aspartic proteases, popularly known as caspases, are key cell death and inflammation regulators. The discovery of caspases dates back over two decades, with the identification of ced-3, a caspase implicated in a specific form of cell death during the metamorphosis of Caenorhabditis elegans (Boice and Bouchier-Hayes 2020). The earliest mammalian homolog of ced-3 was initially designated "interleukin-1-beta-converting enzyme (ICE)" and later renamed caspase-1, while Nedd2 was reclassified as caspase-2.

Currently, there are 18 recognized mammalian caspases, albeit discrepancies in name have produced some confusion. Newly found caspases such as -15, -17, and -18 are often lacking in placental animals, with the exception of caspase-16. Additionally, some caspases, such as caspase-5, are not found in mice, while murine caspases include caspase-11 and -13, and bovine orthologues include caspase-4 (Orning and Lien 2021). In humans, truncated and full-length alleles of caspase-12 exist, but rodents solely contain full-length caspase-12. Caspase-14, expressed in the epidermis, is vital for safeguarding the underlying skin through cornification.

The mammalian caspases can be divided into three broad categories termed executioner caspases, initiator caspases, and inflammatory caspases. The executioner caspases include caspase-3 and caspase-7. Caspase-6 is also in this group, but it has a less clearly defined role in apoptosis. Instead, it plays a unique role in neurodegenerative disease. The executioner caspases all have in common a short prodomain of 20–30 residues that lacks any defined motif or protein interaction site. Executioner caspases are so-called because they bring about the execution phase of apoptosis by cleaving many different structural and regulatory proteins that shut down the functions of the cell and contribute to the morphological hallmarks of apoptosis (Kesavardhana et al. 2020). The executioner caspases are activated when cleaved by an upstream initiator caspase or additional protease. Caspase-2, -8, -9, and -10 comprise the initiator caspases. They each have a long prodomain that encodes a protein: a protein interaction motif used to facilitate dimerization of the catalytic domains. These motifs include the caspase recruitment domain (CARD) found in caspase-2 and caspase-9 and the death effector domain (DED) found in the prodomains of caspase-8 and caspase-10. These regions enhance the dimerization and recruitment of caspases into bigger complexes, crucial for their activation (Vigneswara and Ahmed 2020).

The autoactivation of the caspase zymogen into the active protease is driven by proximity-induced dimerization, leading to conformational changes that result in the proteolytic excision of flexible linker regions separating the prodomain and the large and small catalytic subunits (Lamkanfi et al. 2002). In contrast, executioner caspases, which lack a long amino-terminal prodomain, require cleavage by initiator caspases for their activation (Ramirez and Salvesen 2018). Historically, caspases have been predominantly connected with apoptosis, a regulated and non-lytic form of cell death required for the coordinated disassembly and elimination of old and damaged cells (Ramirez and Salvesen 2018). More recently, caspase-mediated cleavage events have been demonstrated to decrease necroptosis, a lytic-controlled cell death phase driven by receptor-interacting protein (RIP) kinases (Pasparakis and Vandenabeele 2015). Furthermore, research has identified ways by which inflammatory caspases promote pyroptosis, another important lytic cell death mode defined by the release of inflammatory cytokines interleukin (IL)-1β and IL-18 (Vande Walle and Lamkanfi 2016).

Over the past three decades, considerable evidence has implicated dysregulated caspase activation as a causative mechanism in several diseases, including cancer, autoimmunity, autoinflammation, and infectious disorders (Halaby 2012; Van Gorp et al. 2019). This chapter concisely explores the activities and signaling processes of apoptotic caspases before focusing on recent findings into the activation mechanisms of inflammatory caspases and their various roles in cell death and inflammation. Additionally, it addresses developing evidence of the molecular interplay between inflammatory and apoptotic caspases and examines innovative techniques for therapeutic control of caspase activation in a spectrum of communicable, inflammatory, neoplastic, metabolic, and neurodegenerative illnesses (Fig. 3.1).

## 3.2 Caspases in Apoptotic Cell Death

It is crucial to understand that during the activation of apoptosis, initiator caspases are regulated by two principal apoptotic pathways: the intrinsic (also known as the mitochondrial or Bcl-2 regulated) pathway and the extrinsic (death receptor-mediated) pathway. The intrinsic pathway is typically activated in response to cellular stress, such as exposure to cytotoxic drugs or DNA damage (Dhani et al. 2021). This pathway involves the Bcl-2 family of proteins and requires the activation of pro-apoptotic effectors BAX and BAK. These effectors facilitate mitochondrial outer membrane permeabilization (MOMP), leading to the release of cytochrome c. Cytochrome c then interacts with Apaf-1 (Apoptotic Protease-Activating Factor 1), forming a large complex known as the apoptosome, which subsequently activates caspase-9.

Apoptosis is mediated by death receptors through ligand binding and activation of the tumor necrosis factor receptor (TNFR) superfamily, which includes TNFR, Fas, and TRAIL receptors. These receptors contain the death domain, leading to the recruitment and activation of caspase-8 or caspase-10 via the death-inducing signaling complex (DISC). DISC comprises FAS-associated death domain protein (FADD)

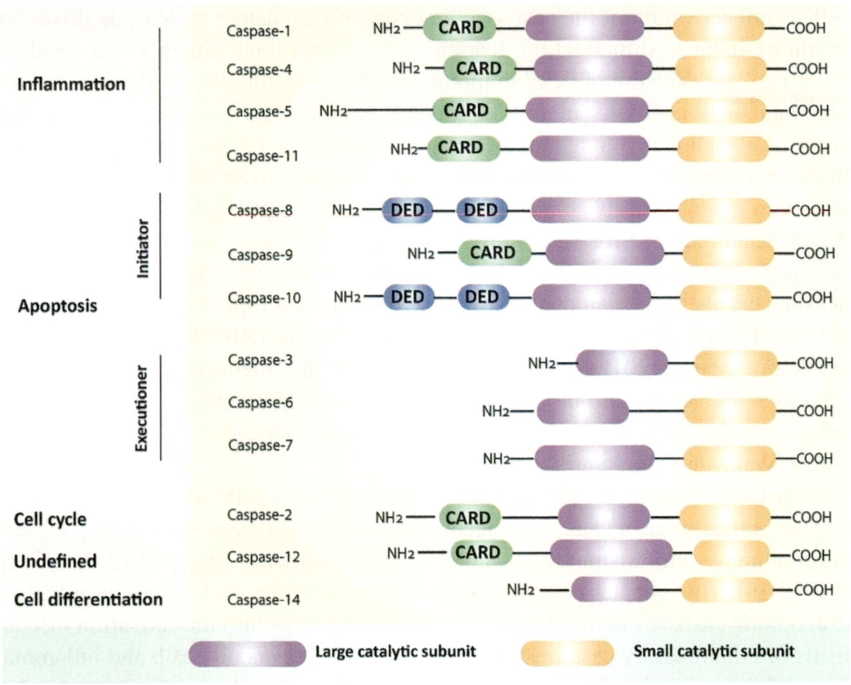


**Fig. 3.1** Domain architecture and functional roles of caspases. Members of the caspase family are classified into inflammatory and apoptotic categories based on their functions and domain architectures. Inflammatory caspases, including caspases-1, -4, -5, and -11, are characterized by the presence of a caspase activation and recruitment domain (CARD) at their N-terminal end. Apoptotic caspases are further divided into initiator and executioner caspases. Initiator caspases, such as caspases-8, -9, and -10, share similar domain architectures with inflammatory caspases but are primarily involved in initiating apoptosis by activating executioner caspases like caspases-3, -6, and -7. Caspase-2, although it shares structural similarities with inflammatory caspases, is primarily associated with cell cycle regulation. Murine caspase-12, while structurally related to inflammatory caspases, has an undefined function to date

or TNFR-associated death domain protein (TRADD) (Lu et al. 2020). Caspase-8 also processes BID into its truncated form, tBID, which subsequently activates the mitochondrial pathway and enhances apoptosis. Effector caspases such as caspase-3, caspase-6, and caspase-7 are then activated by initiator caspases following either the extrinsic or intrinsic pathways.

Caspases are implicated in apoptosis, necroptosis, and autophagy. Apoptosis itself can proceed via two pathways: intrinsic and extrinsic. Intrinsic apoptosis is triggered by internal stimuli, including activating p53 and Bcl-2 family proteins like Bax and Bak (Jiang et al. 2021). This activation leads to mitochondrial outer membrane permeabilization (MOMP) and apoptosome formation, where caspase-9 activates effector caspases such as caspase-3. Caspase-2 is also activated by DNA damage, cleaving Bid, and promoting MOMP. Conversely, extrinsic apoptosis involves the DISC, leading to the activation of caspase-8, which subsequently activates effector caspases (Nirmala and Lopus 2020). Caspase-8 also cleaves Bid, releasing cytochrome c and promoting apoptosis via the mitochondrial pathway.

Various complexes are formed depending on ligands such as TNF and Fas. TNF binds TNFR1, leading to the recruitment of TRADD, IAPs, TRAFs, and RIP1, forming a "TRADD-dependent complex I." This complex undergoes RIP1 de-polyubiquitination, disassembling and transitioning to "complex IIa/DISC," which includes TRADD, procaspase-8, FADD, and cFLIP. Activation of caspase-8 within this complex initiate apoptosis (Chen et al. 2020). Under conditions where caspase activity is inhibited, RIP1, FADD, and the FLIPL–caspase-8 heterodimer complex can form a large structure known as the "apoptosome" in response to genotoxic stress. Loss of caspase activity stabilizes RIP1 and RIP3, forming "complex IIb" or necrosomes in microfilament-like structures, leading to necroptotic cell death. At the plasma membrane, recruitment of the MLKL complex occurs, further driving necroptosis. Active caspase-8 can also inactivate RIPK1 and RIPK3 (Araya et al. 2021).

TNF receptors can also stimulate Toll-like receptors, which contribute to necroptosis induction. Additionally, certain caspases, including caspase-3, regulate autophagy by cleaving specific proteins essential for this process. Caspase-2 provides an example of a protein activated both before and after mitochondrial outer membrane permeabilization (MOMP) in response to various apoptotic signals (Yi 2020). Its activation involves CARD formation and subsequent dimerization through cleavage. Interestingly, RAIDD and PIDD are not mandatory for caspase-2 activation in vivo, as evidenced by its activation in RAIDD and PIDD-deficient mice.

## 3.3 Caspases in Non-apoptotic Cell Death

### 3.3.1 Necroptosis

Necroptosis is a distinct form of programmed cell death that occurs when caspase activity is inhibited. It is characterized by cellular swelling and eventual plasma membrane rupture, leading to cell lysis. This mode of cell death can be triggered by various stimuli such as TNF, Toll-like receptor 3 (TLR3) activation by pathogen-associated molecular patterns (PAMPs), or genotoxic stress. Among these triggers, TNF-induced necroptosis is extensively studied and serves as a prominent model (Yan et al. 2022).

A complex signaling cascade is initiated upon TNF binding to its receptors, particularly TNFR1 on the cell surface. TNFR1 recruits several key factors including TNFR-associated death domain (TRADD), TNF receptor-associated factors (TRAFs), receptor-interacting protein kinase 1 (RIP1), and inhibitor of apoptosis proteins (IAPs). Initially, this complex activates the NFκB pathway, which is crucial for inflammatory responses and cell survival. However, if caspase-8 activity is blocked or insufficient, RIP1 and RIP3 form a complex known as the necrosome. The necrosome subsequently recruits mixed lineage kinase domain-like protein (MLKL), leading to necroptotic cell death characterized by cellular swelling, membrane rupture, and leakage of intracellular contents (Liu and Jiao 2020). In addition to its role in regulating cell death pathways like necroptosis, RIP1 also has signaling functions that promote cell survival under certain conditions.

### 3.3.2 Pyroptosis

Pyroptosis is a regulated form of cell death mediated by caspase-1 and caspase-11, typically observed in macrophages infected with intracellular pathogens such as Shigella flexneri or Salmonella Typhimurium. Unlike apoptosis, pyroptosis involves the formation of membrane pores, cell swelling, and osmotic lysis, releasing pro-inflammatory factors (Yu et al. 2021). This process plays a crucial role in the immune response by facilitating the removal of intracellular pathogens. Upon induction of pyroptosis, cytokines are secreted, and bacteria are expelled from the dying cell, subsequently being targeted and eliminated by neutrophils. Importantly, pyroptosis is characterized by DNA cleavage and nuclear condensation, although it lacks the typical DNA fragmentation observed in apoptosis. Pyroptosis represents a defense mechanism against intracellular infections, contributing to the elimination of pathogens while promoting an inflammatory response essential for immune activation and resolution of infection (Yu et al. 2021) (Fig. 3.2).

### 3.3.3 Autophagy

Caspases can cleave many human ATG (autophagy-related) proteins, suppressing autophagy and potentially promoting death. For example, hATG3 can be cleaved by caspase-3, -6, and -8. The cleavage event hinders the role of hATG3 in the creation of autophagosomes, leading to a decrease in autophagy activity and a preference for apoptosis (Ichimiya et al. 2020).

Likewise, the protein Beclin-1 (also known as ATG-6), which plays a crucial role in starting the process of autophagy, can be divided into smaller parts by an enzyme called caspase-3. The process of cleaving Beclin-1 hinders its ability to perform autophagy, causing a shift in the cellular equilibrium towards increased apoptosis (Ichimiya et al. 2020). In addition, caspase-6 and -8 have a role in the fragmentation of p62, which is an essential adaptor protein for selective autophagy. The action of these caspases on p62 inhibits autophagy processes, hence enhancing apoptotic pathways (Ichimiya et al. 2020). Caspase-2 is acknowledged as a controller of autophagy, frequently functioning as an inhibitor. Suppressing or blocking caspase-2 has been demonstrated to augment autophagy, therefore facilitating cell survival during periods of stress (Ichimiya et al. 2020). In addition to hATG3 and Beclin-1, caspases also target Atg4 and AMBRA1 as substrates in the control of autophagy. This emphasizes the intricate relationship between apoptosis and autophagy pathways in maintaining cellular balance and responding to stress.

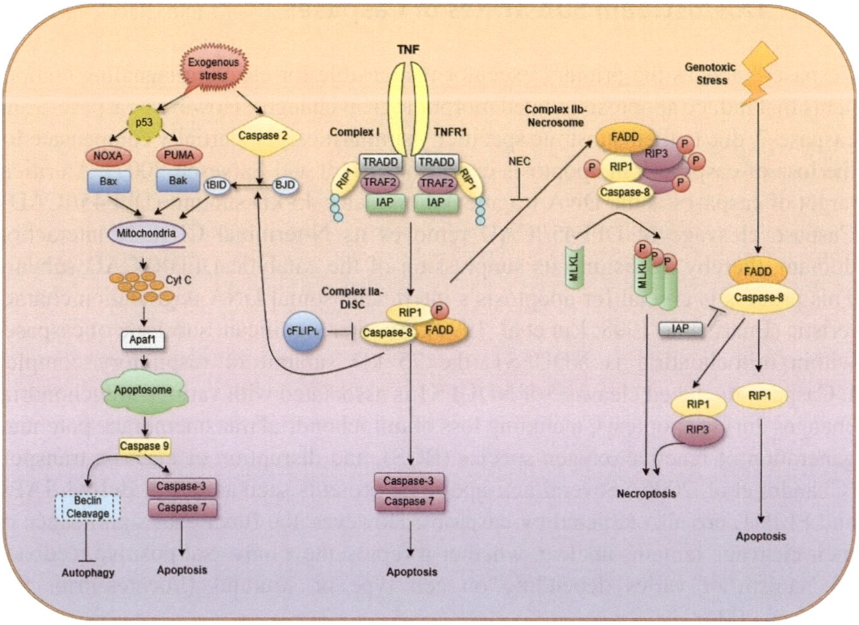


**Fig. 3.2** Cell death pathway of Caspases. Caspases are crucial in cell death pathways, including apoptosis, necroptosis, and autophagy. Apoptosis can occur intrinsically or extrinsically, with intrinsic apoptosis initiated by a death-inducing stimulus. Intrinsic apoptosis involves the activation of p53 and pro-apoptotic Bcl-2 proteins, leading to mitochondrial outer membrane permeabilization. Extrinsic apoptosis involves the formation of complexes by death-domain-containing proteins, such as DISC and caspase-8, which directly activate effector caspases. Different complexes are formed depending on ligands, such as TNF and Fas, and can result in necroptotic cell death. Some caspases, like caspase-3, indirectly control autophagy by cleaving crucial autophagic proteins

## 3.3.4 Mitotic Catastrophe

Mitotic catastrophe is a form of cell death triggered by the malfunction of cell cycle checkpoints during mitosis, often associated with significant DNA damage (Sazonova et al. 2021). This process involves targeting cell-cycle kinases and checkpoint proteins, which fail to properly regulate mitotic progression. As a consequence, cells undergo aberrant mitosis characterized by chromosome segregation errors, multipolar spindles, and ultimately, mitotic arrest.

In response to these mitotic abnormalities, mitotic catastrophe activates various cellular responses, including the activation of caspases such as caspase-2. Caspase-2, in turn, initiates apoptotic pathways by cleaving key substrates involved in cell survival and integrity. Additionally, mitotic catastrophe can lead to the dysregulation of trafficking proteins and organelle dynamics, further contributing to the demise of the affected cells.

## 3.4 Downstream Substrates of Caspases

Caspase-3 acts as the primary executor responsible for cleaving signaling components that induce apoptosis-related morphological changes. However, caspase-6 and caspase-7, due to their substrate specificity similarities, can partially compensate for the loss of caspase-3 in apoptosis execution (Riedl and Salvesen 2007). A critical target of caspases is the DNA fragmentation factor 45 kD subunit (DFF45/ICAD). Caspase cleavage of DFF45/ICAD removes its N-terminal CIDE-N interaction domain, thereby abolishing its suppression of the catalytic DFF40/CAD subunit. This process is crucial for apoptosis's internucleosomal DNA degradation characteristic (Enari et al. 1998; Liu et al. 1997). Another significant substrate of caspases within mitochondria is NDUFS1, the 75 kD subunit of respiratory complex I. Caspase-mediated cleavage of NDUFS1 is associated with various mitochondrial changes during apoptosis, including loss of mitochondrial transmembrane potential, generation of reactive oxygen species (ROS), and disruption of electron transport (Chandra et al. 2006). Several anti-apoptotic proteins such as Bcl-2, Bcl-Xl, IAPs, and FLIP L are also targeted by caspases. However, the functional significance of their cleavage remains unclear, whether it represents a universal positive feedback mechanism or varies depending on cell type or stimulus (Fuentes-Prior and Salvesen 2004).

Caspases also target various protein kinases involved in both pro-apoptotic and anti-apoptotic roles. Among prominent anti-apoptotic kinases, AKT is cleaved during matrix detachment-induced apoptosis in epithelial cells, leading to its inactivation (Thompson 2012). RIP kinase, a death-inducing signaling complex (DISC) component, is cleaved by caspase-8, thereby suppressing its anti-apoptotic function and NF-κB induction (O'Donnell et al. 2011). Focal adhesion kinase (FAK), which transduces survival signals from the extracellular matrix, is cleaved and inactivated during apoptosis induced by various stimuli such as TRAIL, resulting in the loss of survival signals and detachment from the matrix (Cai et al. 1998).

Conversely, caspases can cleave and activate several pro-apoptotic kinases. For instance, ROCK1 kinase is cleaved and activated in response to TNF, facilitating apoptotic membrane blebbing (Sebbagh et al. 2001). Caspase-3-mediated cleavage of PAK2 generates a constitutively active form of the kinase, promoting the formation of apoptotic bodies (Kumar et al. 2006). MEKK1 activation during genotoxic stress and Fas-induced apoptosis contributes to apoptotic signaling through JNK activation (Wang and Liu 1996).

Caspase-mediated cleavage of specific substrates plays a crucial role in several characteristic features of apoptosis. For example, cleavage of fodrin and gelsolin disrupts the actin filament network, contributing to the loss of overall cell shape and detachment from the matrix during apoptosis. Cleavage of nuclear lamins is essential for nuclear shrinkage and budding, a hallmark of apoptotic morphology. Additionally, intermediate filament proteins such as keratins 18 and 19 and vimentin are targeted and cleaved by caspases during apoptosis.

Furthermore, caspase cleavage of adherence junction proteins like β-catenin and plakoglobin disrupts cell–cell interactions, potentially attenuating their pro-survival signaling. This caspases dismantling of cellular architecture is critical for the completion of apoptosis. DNA repair is a highly energy-demanding process that competes for cellular resources such as ATP, essential for the execution of apoptosis. The shutdown of DNA repair pathways can facilitate apoptosis, as these processes typically involve cell cycle arrest, which inhibits apoptosis (Nagai and Pleasance 2007). Caspases target DNA repair proteins, including DNA-dependent protein kinase (DNA-PK), Rad51, and ATM, during apoptosis, impacting their function and potentially hindering repair processes (Sakamoto et al. 2007). Poly(ADP-ribose) polymerase (PARP) is another critical enzyme involved in DNA repair. Activated by DNA breaks, PARP catalyzes the attachment of ADP-ribose polymers to nuclear factors to facilitate repair. This process consumes large amounts of NAD+, indirectly depleting cellular ATP stores (Soldatenkov et al. 2002). Loss of ATP has been shown to prevent apoptosis and promote necrosis, underscoring the importance of cellular energy balance in apoptosis (Andrabi et al. 2006). Caspase-mediated cleavage and inactivation of PARP during apoptosis may serve to modulate this energy balance and promote apoptotic cell death.

During apoptosis, several cell cycle inhibitory factors are cleaved, including Cdc27, Wee1, and two cyclin-dependent kinase (Cdk) inhibitors, p21^CIP1 and p27^KIP1 (Lee et al. 2000; Yu et al. 2003). Furthermore, death receptor signaling has been shown to cause caspase-dependent inactivation of the proliferation inhibitor Rb (Khwaja et al. 2006). Apoptosis induction activates multiple cell cycle-associated factors through cleavage and inactivation of their inhibitors. As illustrated by recent studies, the cleavage of cell cycle regulators during apoptosis serves multiple roles. Loss of Rb promotes the accumulation of procaspases, enhancing apoptotic signaling pathways (Khwaja et al. 2006). In another example, Cdc2 phosphorylates the pro-apoptotic Bcl-2 family member Bad, inhibiting its interaction with AKT and thereby activating its proapoptotic activity (del Peso et al. 1997). These cleavage events may have evolved as a tumor suppressor mechanism, ensuring cell death induction in response to aberrant cell cycle signaling by oncogenes.

## 3.5 Regulation of Caspases

The regulation of caspase activity involves proteolytic activation and diverse regulatory pathways. Apoptosis, while typically independent of protein synthesis, can be influenced by the gene expression of caspases, altering cellular sensitivity to apoptotic signals. For instance, dysregulation of E2F by adenovirus E1A, loss of Rb, or enforced expression of E2F-1 can lead to the accumulation of caspase proenzymes, enhancing p53-mediated apoptosis (Nahle et al. 2002; Maclachlan and El-Deiry 2002). Additionally, caspase-6 and caspase-10 have been identified as transcriptional targets of p53, further linking their regulation to cellular responses (Konishi et al. 2002; Rikhof et al. 2003).

Beyond transcriptional control, post-translational modifications such as nitrosylation, oxidation, ubiquitination, and phosphorylation also play crucial roles in modulating caspase activity. For example, phosphorylation by AKT can inactivate mouse caspase-9, highlighting a mechanism for fine-tuning apoptotic pathways (Cardone et al. 1998). Recently, a family of apoptotic inhibitors known as CARPs was discovered, which binds to and negatively regulates DED caspases through ubiquitin-mediated proteolysis, underscoring another layer of regulatory complexity (Nahle et al. 2002). Maintaining the delicate balance of caspase activity is essential for cell survival. Several key negative regulators of caspases have been identified in viruses and humans alike, exploited by viruses to evade host defenses by inhibiting apoptosis. These regulators include crmA, P35, v-FLIP, and the inhibitor of apoptosis proteins (IAPs) (Irusta et al. 2003).

Among the IAPs, XIAP is particularly notable for its potent inhibition of caspases in humans. Featuring Baculovirus IAP repeat (BIR) domains at their N-termini and RING finger domains at their C-termini, IAPs like XIAP, c-IAP1, c-IAP2, and Survivin inhibit apoptosis induced by diverse stimuli, including death ligands, through direct caspase binding and inhibition (Deveraux et al. 1997; Takahashi et al. 1998; Chai et al. 2001). XIAP's BIR3 domain specifically targets caspase-9, while its linker region near BIR2 inhibits caspases 3 and 7, demonstrating its multifaceted regulatory capabilities (Takahashi et al. 1998; Chai et al. 2001). Moreover, certain IAPs act as E3 ubiquitin ligases, promoting the ubiquitination and subsequent degradation of caspases to further dampen their activity (Suzuki et al. 2001; Huang et al. 2000). The dynamic regulation of IAP activity during apoptosis includes caspase-mediated cleavage of XIAP and cIAP-1, ensuring their specific activation to promote apoptosis (Deveraux et al. 1999; Creagh et al. 2004). Furthermore, mitochondrial proteins like Smac/DIABLO are released upon apoptotic stimulation to inhibit IAPs, thereby facilitating caspase activation and apoptotic cell death (Du et al. 2000; Creagh et al. 2004).

## 3.6 Inflammation and Immunity Connected to Caspases

Initially, the role of caspases in inflammation was elucidated through the function of caspase-1, which processes and matures interleukins IL-1β and IL-18, crucial cytokines in inflammation (Lee et al. 2020). In *Drosophila*, a caspase-8-like protein named Dredd regulates the innate immune response by interacting with the processed form of NF-κB, leading to immune activation against Gram-negative bacteria. Dredd enhances susceptibility to these bacteria, suggesting a key role in immune modulation (Fig. 3.3).

In mammals, several caspases are classified as inflammatory caspases essential for the innate immune response. These include human caspase-1, -4, -5, -12, and their mouse counterparts caspase-1, -11, -12. For instance, caspase-1 is activated by caspase-4, -5, and -11, acting as cytoplasmic receptors for LPS that bind with their CARD domain, inducing inflammasome assembly and activation (Andrade and Zamboni 2020). Caspase-12, however, exhibits anti-inflammatory properties by inhibiting caspase-1 activation, thereby protecting against bacterial endotoxin-induced lethality or sepsis in mice (Hou et al. 2021). Interestingly, full-length human caspase-12 is predominantly found in about 20% of sub-Saharan African

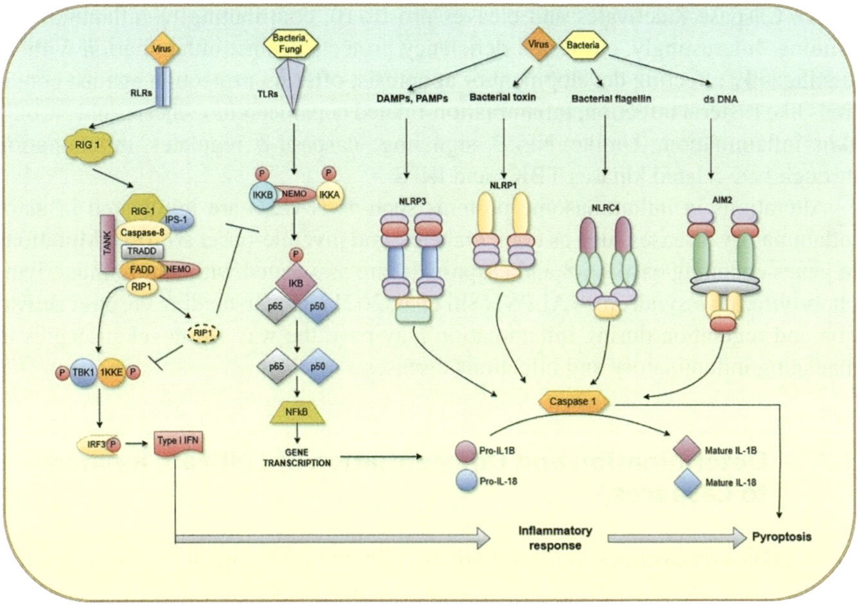


**Fig. 3.3** Inflammatory response mediated with caspases. Caspase-8 restricts RIG-1 signaling, promoting anti-viral immunity. It recruits the caspase-8 DISC complex to the RIG-1 complex, inhibiting IRF3 and NFκB, reducing IFN and IL-1β-/IL-18-mediated inflammatory responses. Inflammasome complex formation is triggered by pathogen-associated molecular patterns and Toll-like receptors. Caspase-1 activates inflammasome complexes, mediating the cleavage and maturation of pro-inflammatory cytokines

descendants, leading to reduced inflammatory responses and increased susceptibility to sepsis.

Inflammasomes, large multi-protein structures, are activated in response to cellular damage from pathogens, toxins, or other stressors. Various pattern-recognition receptors (PRRs) stimulate different inflammasomes, including Toll-like receptors (TLRs) for bacterial, viral, and fungal PAMPs, RIG-like receptors (RLRs) for viral infections, and NOD-like receptors (NLRs) for bacterial components. Specific inflammasomes include NLRP1 (targets anthrax lethal toxin), NLRP3 (responds to a wide range of PAMPs and pathogens), NLRC4 (activated by bacterial flagellin), and AIM2 (detects cytosolic dsDNA). Caspase-1 is activated by NLRP1 and NLRC4, while NLRP3 and AIM2 require ASC for activation.

Caspase-1 within inflammasomes processes pro-IL-1β and pro-IL-18 into their mature forms, leading to pyroptosis, a pro-inflammatory form of cell death (Zheng et al. 2020). Inflammasome activation involves priming, initiated by TLR or RIG activation, which induces transcription of NALP3 and inflammatory cytokines. Subsequently, caspase-1 is recruited and activated again, facilitating cytokine maturation and release (Fig. 3.3).

Further research indicates that caspase-8 is involved in regulating inflammasomes and cytokine processing independently of its role in apoptosis (Tsuchiya

2021). Caspase-8 activates and cleaves pro-IL-1β, contributing to inflammasome priming. Interestingly, caspase-8 deficiency protects against inflammation without significantly affecting development or apoptosis, offering protection against conditions like Listeria infection, inflammation-related hepatocarcinogenesis, and chronic skin inflammation. Unlike NFκB signaling, caspase-8 regulates inflammation through IκK-related kinases TBK1 and IRF3.

Alterations in inflammasome proteins such as NALP3 are implicated in autoinflammatory diseases such as gout, diabetes, and juvenile-onset arthritis. Mutations in genes encoding caspase-8 and caspase-10 are associated with autoimmune lymphoproliferative syndrome (ALPS) (Shi et al. 2021). Understanding caspase activation and regulation during inflammation may pave the way for novel strategies in managing inflammatory and infectious diseases.

## 3.7 Determination and Differentiation of Cell-Fate Related to Caspases

Early studies on caspases in *Drosophila* highlighted their critical role in cell differentiation. For instance, research showed that mutations in the Apaf-1 orthologue weaken Dronc, an initiator caspase, thereby inhibiting developmental cell death (Bianchi and Mezzapelle 2020). Furthermore, dark-mutant flies expressing a dominant-negative form of Dronc displayed developmental defects, suggesting that Dronc has functions beyond apoptosis. These include the regulation of neural precursor cell formation through the cleavage of Shaggy, a process controlled by DmIKK. In higher animals, caspases also contribute significantly to the terminal differentiation of various cell types (Li et al. 2020). Caspase-3, for example, is involved in non-apoptotic functions such as lens fiber formation and enucleation during erythroid differentiation. Additionally, it plays roles in cell proliferation and migration during neuronal differentiation.

In muscle progenitor cells, caspase-3 and caspase-9 are crucial for differentiation processes. Their absence in mesenchymal stem cells (MSCs) affects osteogenic differentiation, leading to reduced bone mineral density. Notably, caspase-7 has been identified as playing a critical role in dental mineralization processes (Farooq et al. 2021). Caspase-3 also plays a significant role in embryonic and hematopoietic stem cells, influencing monocyte differentiation into macrophages through its enzymatic activity. Caspase-8, on the other hand, is responsible for cleaving RIP1, which suppresses NEMO during macrophage differentiation. This regulatory function underscores its importance in immune cell differentiation processes. Caspase-14, expressed in all terrestrial mammals, is essential for epidermal keratinocytes. It contributes to differentiation, cornification, skin hydration, and protection against UVB radiation (Patergnani et al. 2020). Caspase-14 interacts with caspase-3 and is implicated in keratinocyte differentiation, Notch1 signaling, and the regulation of PKC-δ. Its dysregulation is associated with skin inflammatory conditions like psoriasis.

## 3.8 Proliferation of Cells and Regeneration of Tissue Associated with Caspase

Apoptosis is closely connected with the regeneration process; regeneration is conservative, which is seen in various kinds of animals. Hydra, which is an invertebrate, and Xenopus, an amphibian, present increased apoptosis in regeneration. Initiated within the dying cells, MAPK signaling responsible for activating Wnt-ligand synthesis escalates the level of compensating proliferation in the neighboring cells (Loftus et al. 2022). In *Drosophila*, compensatory proliferation supplements for cell death due to injury and keeps the size and shape of the tissue in balance. In flies, Dronc, an initiator caspase, constructs mitogens, for example, Wingless (Wg) and Decapentaplegic (Dpp). Hedgehog signaling in effector caspases also evokes cell proliferation. Caspases were also noted to regulate cell dying and also the site of cell duplication as shown by the research.

In mammals, the regenerating ability of some tissues is fairly well-marked notably in the liver and skin. Lack of caspase-3 and caspase-7 in mice results in poor liver regeneration and wound healing by calcium-independent phospholipase A2 (iPLA2). Caspase cleaves iPLA2, which releases arachidonic acid and *lysophosphocholine*, which in turn creates PGE2 for stem cell proliferation and tissue repair (Sladky and Villunger 2020). Therefore, it is necessary to include that PGE2 binds to the Wnt signaling to regulate the formation of hematopoietic stem cells. Caspases do not have to be involved in cell death to affect the regulation of paracrine factors. That means in caspase-8 knockout mice, the criteria of enhanced liver regeneration have been associated with augmented NFκB signaling. Caspase-7 is important in osteogenesis, and Casp-8exhibits mice reduced bone mineral density and suppressed bone-forming genes, Smad1 and Msx1.

Caspase activation during radiotherapy or tissue injury makes cancer cells proliferate through iPLA2 and PGE2 activation and thus limits an effective treatment. Apoptosis cells release small microparticles to activate the neighboring cells and increase regeneration (reg) genes (He et al. 2022). It has also been observed that PGE2 acts in conjunction with the Wnt signaling pathway regarding stem cell maintenance. Caspases also regulate cell proliferation in an execution point manner, meaning they bring about cell proliferation in a cell type-specific manner. This is indicated by the fact that caspase-8 deficiency affects lymphocyte proliferation and reduces peripheral T cells, as well as an impaired response. Human mutations of caspase-8 cause lymphocyte and natural killer cell dysfunction (Özkan et al. 2021). Necrostatin and RIPK3 inhibited experiments indicate that loss of caspase-8 and FADD result in necroptosis of cells or hematopoietic progenitor cell proliferation requires FADD. The DISC complex also regulates the cell cycle of hepatocytes. Caspase-1 and -6 are implicated in B-cell proliferation and lymphocyte hypo-cellularity.

On the other hand, caspase-3-deficient B cell exhibits hyper-proliferation associated with the cyclin-dependent kinase inhibitor p21 (Medina et al. 2020). Caspase-3 cleaves the cell-cycle regulators, including p21 and p27, and thus inhibits cell division.

## 3.9 Tumorigenesis Related to Caspases

Cell division and cell death are intimately linked processes in biology. The failure of cells to undergo programmed cell death, or apoptosis, can lead to uncontrolled cell division, a hallmark of cancer. Many genes involved in cell death pathways also function as tumor suppressor genes (Bi et al. 2022). For example, in cancers such as non-Hodgkin's lymphomas and small cell lung cancers, overexpression of anti-apoptotic proteins like Bcl-2 can prevent caspase activation through phosphorylation, thereby inhibiting apoptosis. Inhibitors of apoptosis proteins (IAPs) also play a critical role in promoting cell survival by either directly inhibiting caspases or altering cell growth pathways mediated by NFκB. Despite their primary role in apoptosis, caspases influence various cellular processes unrelated to cell death, such as proliferation, invasion, and migration, all of which are crucial aspects of tumorigenesis.

### 3.9.1 Caspase-1

Caspase-1 has been identified in some studies as a potential tumor suppressor. Overexpression of caspase-1 increases the sensitivity of prostate cancer cells to radiation-induced cell death. Reduced expression of caspase-1 is observed in many human cancers, and its restoration through demethylation has been shown to enhance apoptosis and inhibit tumor formation (Ball et al. 2020). In experimental models, caspase-1-deficient mice show increased tumor development in conditions like colitis-associated cancer (CAC), accompanied by heightened epithelial cell proliferation and reduced apoptosis. Interestingly, lethality in caspase-1-deficient animals due to lipopolysaccharide (LPS) exposure is attributed to mutations in the caspase-11 locus rather than caspase-1 itself.

### 3.9.2 Caspase-2

Caspase-2 is also implicated in tumor suppression. It is located on chromosome 7q, a region commonly lost in leukemia. Low levels of caspase-2 are associated with poor prognosis in certain cancers, while its overexpression correlates with remission and improved survival in acute leukemias. Caspase-2 knockout mice demonstrate increased tumor susceptibility in lymphoma and mammary carcinoma models, suggesting its role in inhibiting tumorigenesis (Green 2022). Caspase-2 deficiency has been shown to impair DNA damage response, reduce p53 activation and function, and enhance proliferation rates in KRas-driven lung cancer, affecting chemotherapy efficacy and overall survival negatively. Despite its synergy with p53, caspase-2 does not significantly influence the manifestation of thymomas or fibrosarcomas in mice. Notably, caspase-2 promotes a tumor suppressor profile in high c-myc expressing cells, where loss of its activator Pidd has minimal impact. Elevated levels of caspase-2 in the serum of neuroblastoma patients correlate with lower survival rates, highlighting its complex role as both a potential tumor suppressor and promoter depending on the tissue context.

### 3.9.3 Caspase-3

The risk associated with polymorphisms in CASP3 and CASP7 genes varies across different cancers. CASP3 variant alleles have been linked to increased risk in some cancers but decreased risk in lung cancer. In breast cancer, caspase-3 is often down-regulated overall but up-regulated in malignant tissues where it correlates with apoptosis (Pisani et al. 2020). Interestingly, in tumors undergoing programmed cell death, caspase-3 can enhance tumor repopulation post-radiotherapy through the release of PGE2, a factor supported by studies in head and neck as well as breast carcinoma patients, where higher caspase-3 levels are associated with increased recurrence rates.

### 3.9.4 Caspase-6 and -7

CASP6 and CASP7 mutations are relatively rare but are implicated in decreasing apoptosis in cancers where they occur. CASP6 mutations have been reported in colon and gastric cancers, while CASP7 mutations are found in colon, esophageal, and head/neck carcinomas. Specific CASP7 polymorphisms increase the risk of lung and endometrial cancers (Ahmad et al. 2022).

### 3.9.5 Caspase-8

Caspase-8 plays diverse roles in carcinogenesis. Loss of caspase-8 is rare but observed in invasive colorectal carcinomas, hepatocellular carcinomas (HCCs), and advanced gastric cancers, where it contributes to decreased apoptosis. However, certain CASP8 polymorphisms have been associated with reduced breast cancer risk (Pereira et al. 2020). Caspase-8 deficiency can enhance metastatic potential in neuroendocrine tumors and neuroblastoma models, promoting tumor progression. Conversely, high expression of caspase-8 is observed in advanced tumors and is involved in metastasis through interactions with the focal adhesion complex, promoting cell migration. Caspase-8 also exhibits anti-necrop totic activity, and its deletion specifically in hepatocytes is beneficial in preventing hepatocarcinogenesis.

### 3.9.6 Caspase-10

Inactivating mutations in CASP10 are found in non-Hodgkin lymphoma (NHL), T-cell leukemia, multiple myeloma, and various solid tumors, highlighting its role in preventing apoptosis in these malignancies (Consonni et al. 2024).

### 3.9.7 Caspase-14

Caspase-14 exhibits complex regulation in cancer. It is down-regulated in advanced stages of colon, cervical, and ovarian cancers but up-regulated in ductal carcinomas. Overexpression of caspase-14 in salivary gland adenocarcinoma suppresses tumor growth and metastasis, indicating its potential role as a tumor suppressor in certain contexts (Markiewicz et al. 2021).

## 3.10 Activation of Caspase While Aging

Aging is characterized by the progressive deterioration of organ function due to accumulated cellular damage within tissues (Brahadeeswaran et al. 2022). Caspases, pivotal in the apoptosis process, play a dual role in aging: they efficiently eliminate abnormal cells but also contribute to aging by targeting essential post-mitotic cells like neurons and cardiac myocytes.

Genetic variants that enhance caspase-1 activity and its effect on interleukin-1 beta and interleukin-18 release are implicated in systemic inflammatory diseases that promote aging. Caspase-1 is associated with two biomarkers of aging: increased oxidative stress and chronic low-grade inflammation. Studies manipulating caspase-1 in rats have suggested potential applications to delay certain manifestations of aging (Flores et al. 2022). Caspase-6 is linked to the pathogenesis of Alzheimer's disease, an age-associated condition. Sarcopenia, characterized by muscle degeneration, shows increased apoptosis and elevated expressions of CASP2, CASP6, CASP7, and CASP9 in satellite cells, leading to reduced stem cell numbers and worsening muscle function.

Age-related hearing loss involves heightened generation and apoptosis of cochlear cells in aged animals. Calorie restriction (CR) has been shown to reduce oxidative stress and mitochondrial damage, thereby mitigating age-related changes. The unfolded protein response (UPR), crucial for protein quality control, declines with age, particularly evident in stressed aged mice compared to their younger counterparts, where increased caspase-12 activity and apoptosis are observed (Flores et al. 2020). Caspase-2 knockout mice exhibit decreased antioxidant activity, increased oxidative stress, DNA oxidation, and a shortened lifespan, indicating a non-apoptotic role of caspase-2 in oxidative stress and aging, linked to mitochondrial ROS production associated with complex-III activity. These findings underscore the intricate involvement of caspases in aging processes beyond their classical role in apoptosis, influencing various age-related diseases and physiological changes.

## 3.11 Neural Development Related to Caspases

Apoptosis plays a crucial role in neural development during both embryonic and adult stages, involving key molecules such as caspase-3, caspase-9, and Apaf-1. Their absence leads to severe neural malformations like exencephaly and neural hyperplasia due to increased apoptosis in cultured neural progenitor cells (NPCs) and neurons. Beyond apoptosis, caspases are actively involved in various non-apoptotic processes essential for brain development.

One significant non-apoptotic role is dendrite pruning, crucial for eliminating damaged dendrites and refining axonal contacts within neural networks (Nguyen et al. 2021). Studies in *Drosophila* have highlighted the role of Dronc, a caspase, in dendrite pruning during larval stages, which is vital for developing the mature nervous system. Dronc initiates pruning through ubiquitin-proteasome degradation of DIAP1, leading to morphological changes such as cytoskeletal disruption, cell fragmentation, and phagocytic removal, all crucial for neuronal remodeling without inducing cell death.

Research using caspase-3 exon 3 knockout models has demonstrated its role in regulating spine density and dendrite morphology apart from apoptosis. Knockout mice exhibit supernumerary spines, indicating caspase-3's involvement in spine pruning (Gupta et al. 2021). Similarly, caspase-2 knockout in J20 APP transgenic mice preserves cognitive abilities with age, attributed to maintained dendritic spine density. Caspase-2 facilitates this by modulating the RhoA or ROCK-II signaling pathway, promoting dendritic spine shrinkage. Caspases also contribute to axonogenesis and synaptogenesis. Knockout mice lacking caspase-9 and Apaf1 exhibit aberrant axonal pathfinding, defective synaptogenesis, and sensory neuronopathy, particularly in the olfactory system, despite no increase in apoptosis (Voss and Strasser 2020). Caspase-9 is involved in semaphorin 7A cleavage, critical for proper axonal guidance. Other caspases like caspase-3 and caspase-6 also play roles in axonal guidance processes.

In the context of synaptic plasticity, which underlies learning and memory, caspase-3 plays a crucial role in long-term depression (LTD) in the hippocampus. It mediates the endocytosis of AMPA receptors located in the post-synaptic density during LTD (Pemberton et al. 2021). This process, essential for normal brain function, involves rapid and transient activation of caspase-3 without inducing cell death. Peptide inhibitors of caspase-3 and caspase-9 have been shown to prevent LTD, highlighting the specific role of caspases in synaptic plasticity mechanisms.

## 3.12 Caspases in Therapeutic Disease Management

### 3.12.1 Caspase Activators in Therapeutic Disease Management

Over the past ten years, numerous attempts have been made to create compounds that can directly activate caspase-3 for use in cancer treatment. One target that has been recommended for intervention is the "safety catch" sequence found in procaspase-3 that is not active (Roy et al. 2001). This sequence, which keeps procaspase-3 inactive in resting cells, is made up of a triplet of aspartic acid residues that sustain intramolecular electrostatic contacts (Roy et al. 2001). Numerous molecules, such as gambonic acid, α-(trichloromethyl)-4-pyridineethanol (PETCM), and the gambonic acid derivative MX-2060, have been found through high-throughput screening (HTS) initiatives to effectively activate caspase-3 in vitro (Jiang et al. 2003; Zhang et al. 2004; Fischer and Schulze-Osthoff 2005). Although there has been little information on the clinical development of these drugs, this series has demonstrated potential in triggering apoptosis in cancer cell lines. The first procaspase-activating molecule (PAC-1), which has a zinc-chelating motif essential to its capacity to activate caspase-3, is another intriguing caspase-3 activator discovered by HTS (Putt et al. 2006; Peterson et al. 2009). A "next-generation" chemical (S-PAC-1) has been the subject of successful in vivo canine trials that have resulted in partial tumor remission (Peterson et al. 2009). But the mechanism of PAC-1-mediated caspase activation is debatable because Putt et al.'s results could not be verified by another group (Putt et al. 2006; Denault et al. 2007).

Substances that indirectly activate caspases are the subject of ongoing research in another field. While some of these drugs are analogues of the endogenous IAP inhibitor Smac (Chen and Huerta 2009), others block endogenous caspase inhibitors such as the Bcl-2 and IAP proteins (Vogler et al. 2009). Activators and antibodies that bind to death receptors are among the remaining ones (Lu 2011). Clinical trials are now being conducted to examine a number of these compounds.

Apoptosis-inducing substances have long been used to treat cancer, and most of these drugs have the same transport and specificity issues as conventional chemotherapeutics. But when it comes to cancer cells that use antiapoptotic proteins to avoid dying, some caspase activators—specifically, those that block antiapoptotic molecules like Bcl-2—show an improved therapeutic index (Certo et al. 2006).

In graft versus host disease (GVHD), inducing apoptosis is crucial to eliminate autoreactive lymphocytes. A phase I/II clinical trial is testing a modified caspase-9 "safety switch." This engineered caspase-9 lacks the CARD domain and includes a forkhead protein binding sequence. Activated by a specific small molecule, it triggers apoptosis in allodepleted T cells transplanted post-T-cell-depleted stem cell transplants (Tey et al. 2007; Straathof et al. 2005).

### 3.12.2 Caspase Inhibitors in Therapeutic Disease Management

Caspase activation typically yields pronounced therapeutic benefits, whereas caspase inhibition has proven effective in mitigating conditions marked by excessive apoptotic cell death. In chronic hepatitis C infection, hepatocyte apoptosis is detrimental, prompting exploration of caspase inhibitors in clinical trials (Manns et al. 2010). Recent research underscores caspase inhibitors as promising therapeutic candidates for various neurological conditions. Traumatic injuries to the central nervous system (CNS), such as traumatic brain injury (TBI) and spinal cord injury (SCI), trigger apoptosis, contributing significantly to neuronal loss and long-term neurological deficits (Citron et al. 2008; Clark et al. 2000; Akdemir et al. 2008). In animal models, caspase-3 inhibitors like Z-DEVD-FMK and Ac-DMQD-CHO have demonstrated neuroprotective effects by attenuating apoptosis and improving functional outcomes post-injury (Citron et al. 2008). Similarly, caspase-9 inhibition with Z-LEHD-FMK reduces lesion expansion in SCI models, suggesting a role in limiting secondary injury mechanisms (Clark et al. 2000). Broad-spectrum caspase inhibitors such as Q-VD-OPh and Boc-D-FMK also show promise in mitigating CNS trauma effects (Hernandez et al. 2022).

Elevated IL-1β levels correlate with seizure onset in epilepsy, where caspase-1 inhibitors VX-740 and VX-765 have effectively suppressed IL-1β production in animal models, delaying seizures and improving outcomes (Devinsky et al. 2018; Aronica and Crino 2011). VX-765 has demonstrated clinical potential with reduced seizure frequency observed in human trials, though safety concerns remain (Devinsky et al. 2018). Neurodegenerative diseases like Alzheimer's disease (AD), amyotrophic lateral sclerosis (ALS), Huntington's disease (HD), and Parkinson's disease (PD) involve progressive neuronal loss mediated by caspase-driven apoptosis (Rohn 2010; Toulmond et al. 2004). Caspase inhibitors such as Q-VD-OPh and Z-DEVD-FMK have shown efficacy in preclinical models by preventing neurodegeneration in AD and ALS (D'Amelio et al. 2011; Rohn 2010). VX-765 exhibits promise in preserving cognitive function in AD models, suggesting its potential as a disease-modifying therapy (Garcia and Adams 2017). In models of HD and PD, caspase inhibitors have protected neurons by mitigating apoptosis and preserving neuronal integrity (Turmel et al. 2001; Tatton et al. 2003). While these findings underscore the therapeutic potential of caspase inhibitors across neurological diseases, further clinical trials are essential to validate their efficacy and safety in human patients, addressing critical gaps between preclinical promise and clinical application.

Caspase inhibitors have emerged as crucial tools in studying and potentially treating various inflammatory disorders. In psoriasis, characterized by chronic skin inflammation, elevated levels of caspase-1 and proinflammatory cytokines like IL-1β and TNF-α contribute to disease pathogenesis, highlighting caspase-1 as a promising therapeutic target (Lowes et al. 2014; Parisi et al. 2013). Clinical trials using VX-765, a caspase-1 inhibitor, have shown initial promise in mitigating psoriasis symptoms, though further outcomes are awaited (Lowes et al. 2014; Parisi et al. 2013; Boxer et al. 2010). Similarly, in rheumatoid arthritis (RA) and osteoarthritis (OA), caspase inhibitors like pralnacasan have demonstrated efficacy in preclinical models by reducing joint inflammation and cartilage damage (Le and Abbenante 2005; D'Lima et al. 2006). Concerns over long-term liver toxicity have stalled the further development of some caspase inhibitors despite promising results in early clinical trials for RA (Le and Abbenante 2005). In sepsis, characterized by systemic inflammation and organ dysfunction, caspase inhibitors such as Z-VAD-FMK and VX-166 have shown potential in animal models by preventing lymphocyte apoptosis and improving survival rates (Hotchkiss et al. 2000; Oberholzer et al. 2006; Li et al. 2019). However, challenges remain in translating these findings to clinical settings due to issues with pharmacokinetics and toxicity profiles observed in animal studies (Oberholzer et al. 2006). Further optimization of caspase inhibitors is crucial to enhance their efficacy and safety for broader clinical application across these diverse inflammatory conditions.

Insulin resistance and inflammation result from adipocyte hypertrophy in obesity, triggering pyroptosis, necroptosis, and apoptosis—three types of programmed cell death (PCD). This inflammatory response is facilitated by Caspase-1 activation through inflammasomes, affecting insulin sensitivity. Pralnacasan and other caspase inhibitors have the ability to increase insulin sensitivity in obese mice (Tinahones et al. 2013). According to Vincent and Mohr (2007), caspase-3 is linked to retinal cell apoptosis in diabetes, a defining feature of diabetic retinopathy. This suggests that caspase inhibitors, such as minocycline, may hold therapeutic promise. Caspase inhibitors (Z-VAD-FMK, EP1013, IDN-6556) have been investigated in islet transplantation to improve graft survival and function; however, results have been inconsistent, indicating issues with specificity and long-term efficacy (Emamaullee et al. 2007). Caspases are involved in inflammation and hepatocyte apoptosis in non-alcoholic fatty liver disease (NAFLD), progressing to non-alcoholic steatohepatitis (NASH). While the caspase inhibitor emricasan initially showed promise in preclinical models, clinical studies did not improve liver histology, indicating challenges in translating preclinical findings to human outcomes (Shiffman et al. 2019). Despite the potential shown by other inhibitors, such as VX-166 and GS-9450, in preclinical research, concerns over selectivity and safety in clinical settings persist (Witek et al. 2009; Ratziu et al. 2012). Overall, caspase inhibitors hold therapeutic potential for treating metabolic illnesses, but further study is needed to ascertain their specificity, long-term effects, and alternative pathways of cell death to determine their clinical utility.

## 3.13 Conclusions

Since their discovery over a decade and a half ago, caspases have been shown to govern a wide array of cellular processes. While much is understood about their mechanisms and activation pathways in cell death and apoptosis, their non-apoptotic roles remain largely enigmatic. Caspases primarily function through the proteolytic cleavage of target proteins, dictating various cellular processes with substrate selectivity. Caspases are implicated in complex biological phenomena such as tissue differentiation, disorders of cell proliferation, tumor progression, and metastasis. They also play crucial roles in immune responses, synaptic plasticity, DNA damage response, stress signaling, metabolism, and aging. While some functions of caspases lead to apoptosis and cell death, others operate independently of apoptosis. This dual nature of caspases underscores unresolved questions and highlights the potential significance of further research into these processes. A deeper understanding of caspase functions could pave the way for improved strategies in managing conditions associated with inflammation, autoimmune diseases, and cancer. By elucidating the multifaceted roles of caspases in cellular processes, new hypotheses can be formulated, opening avenues for innovative treatments and enhanced prognostic approaches in these fields.

## References

Ahmad I, Dera AA, Irfan S, Rajagopalan P, Beg MMA, Alshahrani MY, Mir MA, Abohashrh M, Alam MM, Wahab S, Verma AK (2022) BV6 enhances apoptosis in Lung cancer cells by ameliorating caspase expressions through attenuation of XIAP, cIAP-1, and cIAP-2 proteins. J Cancer Res Ther 18(6):1651–1657

Akdemir O, Berksoy I, Karaoğlan A, Barut Ş, Bilguvar K, Çirakoğlu B, Özcan M (2008) Therapeutic efficacy of Ac-DMQD-CHO, a caspase 3 inhibitor, for rat spinal cord injury. J Clin Neurosci 15:672–678

Andrabi SA, Dawson TM, Dawson VL (2006) Mitochondrial and nuclear cross talk in cell death: parthanatos. Ann N Y Acad Sci 1010(1):178–183

Andrade WA, Zamboni DS (2020) NLRC4 biology in immunity and inflammation. J Leucocyte Biol 108(4):1117–1127

Araya LE, Soni IV, Hardy JA, Julien O (2021) Deorphanizing caspase-3 and caspase-9 substrates in and out of apoptosis with deep substrate profiling. ACS Chem Biol 16(11):2280–2296

Aronica E, Crino PB (2011) Inflammation in epilepsy: clinical observations. Epilepsia 52:26–32

Ball DP, Taabazuing CY, Griswold AR, Orth EL, Rao SD, Kotliar IB, Vostal LE, Johnson DC, Bachovchin DA (2020) Caspase-1 interdomain linker cleavage is required for pyroptosis. Life Sci Alliance 3(3)

Bi Q, Wu JY, Qiu XM, Zhang JD, Sun ZJ, Wang W (2022) Tumor-associated inflammation: the tumor-promoting immunity in the early stages of tumorigenesis. J Immunol Res 2022(1):3128933

Bianchi ME, Mezzapelle R (2020) The chemokine receptor CXCR4 in cell proliferation and tissue regeneration. Front Immunol 11:560971

Boice A, Bouchier-Hayes L (2020) Targeting apoptotic caspases in cancer. Biochimica et Biophysica Acta (BBA)-Mol Cell Res 1867(6):118688

Boxer MB, Quinn AM, Shen M, Jadhav A, Leister W, Simeonov A, Thomas CJ (2010) A highly potent and selective caspase 1 inhibitor that utilizes a key 3-cyanopropanoic acid moiety. ChemMedChem 5(5):730–738

Brahadeeswaran S, Sivagurunathan N, Calivarathan L (2022) Inflammasome signaling in the aging brain and age-related neurodegenerative diseases. Mol Neurobiol 59(4):2288–2304

Cai D, Shames BD, Klemke RL (1998) Regulation of matrix detachment-induced apoptosis in human epithelial cells. J Cell Biol 144(3):601–610. https://doi.org/10.1083/jcb.144.3.601

Cardone MH, Roy N, Stennicke HR, Salvesen GS, Franke TF, Stanbridge E, Frisch S, Reed JC (1998) Regulation of cell death protease caspase-9 by phosphorylation. Science 282(5392):1318–1321

Certo M, Del Gaizo Moore V, Nishino M, Wei G, Korsmeyer S, Armstrong SA, Letai A (2006) Mitochondria primed by death signals determine cellular addiction to antiapoptotic BCL-2 family members. Cancer Cell 9(5):351–365

Chai J, Shiozaki E, Srinivasula SM, Wu Q, Datta P, Alnemri ES, Shi Y, Datta P (2001) Structural basis of caspase-7 inhibition by XIAP. Cell 104(5):769–780

Chandra D, Liu JW, Tang DG, Earnshaw WC (2006) Mitochondrial pathway: caspase-independent pathways of apoptosis. Cold Spring Harb Perspect Biol 2(9):a000581. https://doi.org/10.1101/cshperspect.a000581

Chen L, Huerta S (2009) Smac mimetics as new cancer therapeutics. Anti-Cancer Drugs 20(8):646–658

Chen KW, Demarco B, Broz P (2020) Beyond inflammasomes: emerging function of gasdermins during apoptosis and NETosis. The EMBO J 39(2):e103397

Citron BA, Arnold PM, Haynes NG, Ameenuddin S, Farooque M, Santacruz K, D'Arcangelo G (2008) Neuroprotective effects of caspase-3 inhibition on functional recovery and tissue sparing after acute spinal cord injury. Spine 33:2269–2277

Clark RS, Kochanek PM, Watkins SC, Chen M, Edward Dixon C, Seidberg NA, Loeffert JE (2000) Caspase-3 mediated neuronal death after traumatic brain injury in rats. J Neurochem 74:740–753

Consonni F, Moreno S, Vinuales Colell B, Stolzenberg MC, Fernandes A, Parisot M, Masson C, Neveux N, Rosain J, Bamberger S, Vigue MG (2024) Study of the potential role of CASPASE-10 mutations in the development of autoimmune lymphoproliferative syndrome. Cell Death Disease 15(5):315

Creagh EM, Murphy BM, Duriez PJ, Duckett CS, Martin SJ (2004) Smac/Diablo antagonizes ubiquitin ligase activity of inhibitor of apoptosis proteins. J Biol Chem 279(26):26906–26914

D'Amelio M, Cavallucci V, Middei S, Marchetti C, Pacioni S, Ferri A, Marie H (2011) Caspase-3 triggers early synaptic dysfunction in a mouse model of Alzheimer's disease. Nat Neurosci 14:69–79

D'Lima D, Hermida J, Hashimoto S, Colwell C, Lotz M (2006) Caspase inhibitors reduce severity of cartilage lesions in experimental osteoarthritis. Arthritis Rheum 54(5):1814–1821

del Peso L, González-García M, Page C, Herrera R, Nuñez G (1997) Interleukin-3-induced phosphorylation of BAD through the protein kinase Akt. Science 278(5338):687–689

Denault JB, Salvesen GS, Alnemri ES (2007) Apoptotic caspase activation and activity. Methods Enzymol 322:233–244

Deveraux QL, Takahashi R, Salvesen GS, Reed JC (1997) X-linked IAP is a direct inhibitor of cell-death proteases. Nature 388(6639):300–304

Deveraux QL, Leo E, Stennicke HR, Welsh K, Salvesen GS, Reed JC (1999) Cleavage of human inhibitor of apoptosis protein XIAP results in fragments with distinct specificities for caspases. EMBO J 18(19):5242–5251

Devinsky O, Vezzani A, O'Brien TJ, Jette N, Scheffer IE, de Curtis M, Wright D (2018) Epilepsy. Nat Rev Dis Prim 3:1–24

Dhani S, Zhao Y, Zhivotovsky B (2021) A long way to go: caspase inhibitors in clinical use. Cell Death Dis 12(10):949

Du C, Fang M, Li Y, Li L, Wang X (2000) Smac, a mitochondrial protein that promotes cytochrome c-dependent caspase activation by eliminating IAP inhibition. Cell 102(1):33–42

Emamaullee JA, Stanton L, Schur C, Shapiro AMJ (2007) Caspase inhibitor therapy enhances marginal mass islet graft survival and preserves long-term function in islet transplantation. Diabetes 56:1289–1298

Enari M, Sakahira H, Yokoyama H, Okawa K, Iwamatsu A, Nagata S (1998) A caspase-activated DNase that degrades DNA during apoptosis, and its inhibitor ICAD. Nature 391(6662):43–50

Eskandari E, Eaves CJ (2022) Paradoxical roles of caspase-3 in regulating cell survival, proliferation, and tumorigenesis. J Cell Biol 221(6):e202201159

Farooq M, Khan AW, Kim MS, Choi S (2021) The role of fibroblast growth factor (FGF) signaling in tissue repair and regeneration. Cells 10(11):3242

Fischer U, Schulze-Osthoff K (2005) New approaches and therapeutics targeting apoptosis in disease. Pharmacol Rev 57(2):187–215
Flores J, Noël A, Foveau B, Beauchet O, LeBlanc AC (2020) Pre-symptomatic Caspase-1 inhibitor delays cognitive decline in a mouse model of Alzheimer disease and aging. Nature Commun 11(1):4571
Flores J, Fillion ML, LeBlanc AC (2022) Caspase-1 inhibition improves cognition without significantly altering amyloid and inflammation in aged Alzheimer disease mice. Cell Death Dis 13(10):864
Fuentes-Prior P, Salvesen GS (2004) The protein structures that shape caspase activity, specificity, activation and inhibition. Biochem J 384(2):201–232
Garcia JA, Adams J (2017) VX-765 alleviates cognitive impairment and delays cognitive decline in Alzheimer's disease mouse models. J Alzheimers Dis 56:775–785
Green DR (2022) Caspases and their substrates. Cold Spring Harbor Persp Biol 14(3):a041012
Gupta R, Ambasta RK, Kumar P (2021) Autophagy and apoptosis cascade: which is more prominent in neuronal death? Cell Mol Life Sci 78(24):8001–8047
Halaby MJ (2012) Caspase activation and diseases. J Clin Invest 122(5):1761–1765
He S, Chakraborty R, Ranganathan S (2022) Proliferation and apoptosis pathways and factors in oral squamous cell carcinoma. Int J Mol Sci 23(3):1562
Hernandez M, Brown D, Martinez J, Wilson S (2022) Broad-spectrum caspase inhibitors as potential neuroprotective agents in CNS trauma: current perspectives. Neurosci Rev 45:112–125
Hotchkiss RS, Chang KC, Swanson PE, Tinsley KW, Hui JJ, Klender P et al (2000) Caspase inhibitors improve survival in sepsis: a critical role of the lymphocyte. Nat Immunol 1(6):496–501
Hou J, Hsu JM, Hung MC (2021) Molecular mechanisms and functions of pyroptosis in inflammation and antitumor immunity. Mol Cell 81(22):4579–4590
Huang H, Joazeiro CA, Bonfoco E, Kamada S, Leverson JD, Hunter T (2000) The inhibitor of apoptosis, cIAP2, functions as a ubiquitin-protein ligase and promotes in vitro monoubiquitination of caspases 3 and 7. J Biol Chem 275(35):26661–26664
Ichimiya T, Yamakawa T, Hirano T, Yokoyama Y, Hayashi Y, Hirayama D, Wagatsuma K, Itoi T, Nakase H (2020) Autophagy and autophagy-related diseases: a review. Int J Mol Sci 21(23):8974
Irusta PM, Chen YB, Hardwick JM (2003) Viral modulators of cell death provide new links to old pathways. Curr Opin Cell Biol 15(6):700–705
Jiang X, Wang X, Cytochrome C, New Y, York N (2003) Apaf-1, a human protein homologous to *C. elegans* CED-4, participates in Cytochrome C-dependent activation of Caspase-3. Cell 94(1):339–352
Jiang M, Qi L, Li L, Wu Y, Song D, Li Y (2021) Caspase-8: a key protein of cross-talk signal way in "PANoptosis" in cancer. Int J Cancer 149(7):1408–1420
Kesavardhana S, Malireddi RS, Kanneganti TD (2020) Caspases in cell death, inflammation, and pyroptosis. Annu Rev Immunol 38:567–595
Khwaja A, Rodriguez-Viciana P, Wennström S et al (2006) Matrix, and that the expression of HSPs, that was. Oncogene 17(22):2865–2875
Konishi Y, Lehtinen M, Donovan N, Bonni A (2002) Cdc2 phosphorylation of Bad links the cell cycle to the cell death machinery. Mol Cell 9(5):1005–1016. https://doi.org/10.1016/S1097-2765(02)00500-2
Kumar S, Jiang MS, Adams JL (2006) Caspase 3-cleaved PAK2 as a novel PAK2 kinase regulator. J Biol Chem 281(32):21475–21482. https://doi.org/10.1074/jbc.M603197200
Lamkanfi M, Festjens N, Declercq W, Vanden Berghe T, Vandenabeele P (2002) Caspases in cell survival, proliferation and differentiation. Cell Death Differ 9(4):358–367
Le GT, Abbenante G (2005) Inhibitors of TACE and caspase-1 as anti-inflammatory drugs. Curr Med Chem 12(26):2963–2977
Lee JY, Brafford PA, Herlyn M (2000) Functional implications of mutations in the BRAF gene in human cancer. Oncogene 19(5):5449–5454
Lee S, Channappanavar R, Kanneganti TD (2020) Coronaviruses: innate immunity, inflammasome activation, inflammatory cell death, and cytokines. Trends Immunol 41(12):1083–1099

Li X, Yao X, Zhu Y, Zhang H, Wang H, Ma Q, Li W (2019) The caspase inhibitor Z-VAD-FMK alleviates endotoxic shock via inducing macrophages necroptosis and promoting MDSCs-mediated inhibition of macrophages activation. Front Immunol 10:1–14

Li M, Liao L, Tian W (2020) Extracellular vesicles derived from apoptotic cells: an essential link between death and regeneration. Front Cell Develop Biol 8:1063

Liu ZG, Jiao D (2020) Necroptosis, tumor necrosis and tumorigenesis. Cell Stress 4(1):1

Liu X, Zou H, Slaughter C, Wang X (1997) DFF, a heterodimeric protein that functions downstream of caspase-3 to trigger DNA fragmentation during apoptosis. Cell 89(2):175–184

Loftus LV, Amend SR, Pienta KJ (2022) Interplay between cell death and cell proliferation reveals new strategies for cancer therapy. Int J Mol Sci 23(9):4723

Lowes MA, Suárez-Fariñas M, Krueger JG (2014) Immunology of psoriasis. Annu Rev Immunol 32:227–255

Lu Y (2011) Death receptors in apoptosis and cancer. Cancer Sci 102(9):1661–1666

Lu F, Lan Z, Xin Z, He C, Guo Z, Xia X, Hu T (2020) Emerging insights into molecular mechanisms underlying pyroptosis and functions of inflammasomes in diseases. J Cell Physiol 235(4):3207–3221

Maclachlan TK, El-Deiry WS (2002) Apoptotic threshold is lowered by p53 transactivation of caspase-6. Proc Natl Acad Sci USA 99(14):9492–9497. https://doi.org/10.1073/pnas.152327299

Manns MP, Wedemeyer H, Singer A, Khomutjanskaja N, Dienes HP, Roskams T et al (2010) Inhibition of hepatitis C virus by caspase inhibitors. Lancet 376(9742):1447–1454

Markiewicz A, Sigorski D, Markiewicz M, Owczarczyk-Saczonek A, Placek W (2021) Caspase-14—from biomolecular basics to clinical approach. A review of available data. Int J Mol Sci 11:5575

Medina CB, Mehrotra P, Arandjelovic S, Perry JS, Guo Y, Morioka S, Barron B, Walk SF, Ghesquière B, Krupnick AS, Lorenz U (2020) Metabolites released from apoptotic cells act as tissue messengers. Nature 580(7801):130–135

Nagai MA, Pleasance ED (2007) Sequence analysis of the SRGAP2 gene locus reveals that it was truncated in early primates. Genomics 89(6):744–751

Nahle Z, Polakoff J, Davuluri RV, McCurrach ME, Jacobson MD, Narita M, Zhang MQ, Lazebnik Y, Bar-Sagi D, Lowe SW (2002) Direct coupling of the cell cycle and cell death machinery by E2f. Nat Cell Biol 4(11):859–864

Nguyen TTM, Gillet G, Popgeorgiev N (2021) Caspases in the developing central nervous system: apoptosis and beyond. Front Cell Develop Biol 9:702404

Nirmala JG, Lopus M (2020) Cell death mechanisms in eukaryotes. Cell Biol Toxicol 36(2):145–164

O'Donnell MA, Perez-Jimenez E, Oberst A, Ng A, Massoumi R, Xavier R, Green DR (2011) Caspase 8 inhibits programmed necrosis by processing CYLD. Nat Cell Biol 13(12):1437–1442. https://doi.org/10.1038/ncb2362

Oberholzer C, Tschoeke SK, Moldawer LL, Oberholzer A (2006) Local thymic caspase-9 inhibition improves survival during polymicrobial sepsis in mice. J Mol Med 84(5):389–395

Orning P, Lien E (2021) Multiple roles of caspase-8 in cell death, inflammation, and innate immunity. J Leucocyte Biol 109(1):121–141

Özkan İ, Koçak P, Yıldırım M, Ünsal N, Yılmaz H, Telci D, Şahin F (2021) Garlic (Allium sativum)-derived SEVs inhibit cancer cell proliferation and induce caspase mediated apoptosis. Sci Rep 11(1):14773

Parisi R, Symmons DPM, Griffiths CEM, Ashcroft DM (2013) Global epidemiology of psoriasis: a systematic review of incidence and prevalence. J Invest Dermatol 133(2):377–385

Pasparakis M, Vandenabeele P (2015) Necroptosis and its role in inflammation. Nature 517(7534):311–320

Patergnani S, Danese A, Bouhamida E, Aguiari G, Previati M, Pinton P, Giorgi C (2020) Various aspects of calcium signaling in the regulation of apoptosis, autophagy, cell proliferation, and cancer. Int J Mol Sci 21(21):8323

Pemberton JM, Pogmore JP, Andrews DW (2021) Neuronal cell life, death, and axonal degeneration as regulated by the BCL-2 family proteins. Cell Death Different 28(1):108–122

Pereira LM, Assis PA, de Araújo NM, Durso DF, Junqueira C, Ataíde MA, Pereira DB, Lien E, Fitzgerald KA, Zamboni DS, Golenbock DT (2020) Caspase-8 mediates inflammation and disease in rodent malaria. Nature Commun 11(1):4596

Peterson QP, Hsu DC, Novotny CJ, West DC (2009) Procaspase-activating compound (PAC-1) and cancer treatment. The Rockefeller University Press 2009(1):75–84

Pisani C, Ramella M, Boldorini R, Loi G, Billia M, Boccafoschi F, Volpe A, Krengli M (2020) Apoptotic and predictive factors by Bax, Caspases 3/9, Bcl-2, p53 and Ki-67 in prostate cancer after 12 Gy single-dose. Sci Rep 10(1):7050

Putt KS, Chen GW, Pearson JM, Sandhorst JS, Hoagland MS, Kwon JT et al (2006) Small-molecule activation of procaspase-3 to caspase-3 as a personalized anticancer strategy. Nature 61(2):748–752

Ramirez MLG, Salvesen GS (2018) A primer on caspase mechanisms. Semin Cell Dev Biol 82:79–85

Ratziu V, Sheikh MY, Sanyal AJ, Lim JK, Conjeevaram H, Chalasani N et al (2012) A phase 2, randomized, double-blind, placebo-controlled study of GS-9450 in subjects with nonalcoholic steatohepatitis. Hepatology 55:419–428

Riedl SJ, Salvesen GS (2007) The apoptosome: signalling platform of cell death. Nat Rev Mol Cell Biol 8(5):405–413. https://doi.org/10.1038/nrm2163

Rikhof B, Corn PG, El-Deiry WS (2003) Caspase 10 levels are increased following DNA damage in a p53-dependent manner. Cancer Biol Ther 2(6):707–712. https://doi.org/10.4161/cbt.2.6.606

Rohn TT (2010) The role of caspases in Alzheimer's disease; potential novel therapeutic opportunities. Apoptosis 15:1403–1409

Roy N, Deveraux QL, Takahashi R, Salvesen GS (2001) Caspase family proteases and apoptosis. Cell Death Different 6(5):362–369

Sakamoto S, Imai K, Tamamura Y et al (2007) Distinct roles of HDAC2 and HDAC11 in cell proliferation and chemoresistance in breast cancer. Biochem Biophys Res Commun 354(1):21–26. https://doi.org/10.1016/j.bbrc.2006.12.188

Sazonova EV, Petrichuk SV, Kopeina GS, Zhivotovsky B (2021) A link between mitotic defects and mitotic catastrophe: detection and cell fate. Biol Direct 16(1):25

Sebbagh M, Renvoizé C, Hamelin J, Riché N, Bertoglio J, Bréard J (2001) Caspase-3-mediated cleavage of ROCK I induces MLC phosphorylation and apoptotic membrane blebbing. Nat Cell Biol 3(4):346–352. https://doi.org/10.1038/35070029

Shi Z, Yuan S, Shi L, Li J, Ning G, Kong X, Feng S (2021) Programmed cell death in spinal cord injury pathogenesis and therapy. Cell Proliferat 54(3):e12992

Shiffman M, Freilich B, Vuppalanchi R, Watt K, Chan JL, Spada A et al (2019) Randomised clinical trial: emricasan versus placebo significantly decreases ALT and caspase 3/7 activation in subjects with non-alcoholic fatty liver disease. Aliment Pharmacol Ther 49:64–73

Sladky VC, Villunger A (2020) Uncovering the PIDDosome and caspase-2 as regulators of organogenesis and cellular differentiation. Cell Death & Differentiation 27(7):2037–2047

Soldatenkov VA, Chasovskikh S, Potaman VN, Trofimova I, Smulson ME, Dritschilo A (2002) A chemosensitizing anticancer drugs of caged adenovirus-mediated UV irradiation and sodium butyrate treatment. Cancer Res 62(2):307–312

Straathof KC, Pulè MA, Yotnda P, Dotti G, Vanin EF, Brenner MK, Heslop HE (2005) An inducible caspase 9 safety switch for T-cell therapy. Blood 105(11):4247–4254

Suzuki Y, Nakabayashi Y, Takahashi R (2001) Ubiquitin-protein ligase activity of X-linked inhibitor of apoptosis protein promotes proteasomal degradation of caspase-3 and enhances its anti-apoptotic effect in FAS-induced cell death. Proc Nat Acad Sci USA 98(15):8662–8667

Takahashi R, Deveraux Q, Tamm I, Welsh K, Assa-Munt N, Salvesen GS, Reed JC (1998) A single bir domain of XIAP sufficient for inhibiting caspases. J Biol Chem 273(13):7787–7790

Tatton WG, Chalmers-Redman R, Brown D, Tatton N (2003) Apoptosis in Parkinson's disease: signals for neuronal degradation. Ann Neurol 53:S61–S72

Tey SK, Dotti G, Rooney CM, Heslop HE, Brenner MK (2007) Inducible caspase 9 suicide gene to improve the safety of allodepleted T cells after haploidentical stem cell transplantation. Biol Blood Marrow Transplant 13(8):913–924

Thompson JE (2012) AKT regulation of cell proliferation and apoptosis. Biol Reprod 86(3):1–12. https://doi.org/10.1095/biolreprod.111.098442

Tinahones FJ, Araguez LCI, Murri M, Olivera WO, Torres MDM, Barbarroja N et al (2013) Caspase induction and BCL2 inhibition in human adipose tissue. Diabetes Care 36:513–521

Toulmond S, Tang K, Bureau Y, Ashdown H, Degen S, O'Donnell R et al (2004) Neuroprotective effects of M826, a reversible caspase-3 inhibitor, in the rat malonate model of Huntington's disease. Br J Pharmacol 141:689–697

Tsuchiya K (2021) Switching from apoptosis to pyroptosis: gasdermin-elicited inflammation and antitumor immunity. Int J Mol Sci 22(1):426

Turmel H, Hartmann A, Parain K, Douhou A, Srinivasan A, Agid Y et al (2001) Caspase-3 activation in 1-methyl-4-phenyl-1,2,3,6-tetrahydropyridine (MPTP)-treated mice. Mov Disord 16:185–189

Van Gorp H, Delanghe T, Roelandt R, Nerinckx W, Verrijken A, Georgy S et al (2019) Caspase-1 inhibition decreases adipose tissue inflammation and fibrosis in mice and humans. Cell Death Dis 10(6):472. https://doi.org/10.1038/s41419-019-1716-3

Vande Walle L, Lamkanfi M (2016) Pyroptosis. Curr Biol 26(13):R568–R572

Vigneswara V, Ahmed Z (2020) The role of caspase-2 in regulating cell fate. Cells 9(5):1259

Vincent JA, Mohr S (2007) Inhibition of caspase-1/interleukin-1β signaling prevents degeneration of retinal capillaries in diabetes and galactosemia. Diabetes 56:224–230

Vogler M, Walczak H, Stadel D (2009) Small-molecule drugs targeting the intrinsic apoptosis pathway. Semin Cancer Biol 19(1):1–6

Voss AK, Strasser A (2020) The essentials of developmental apoptosis. F1000Research 9

Wang Z, Liu Y (1996) Activation of MEK kinase 1 by the c-Jun N-terminal kinase in response to DNA damage. J Biol Chem 271(16):9358–9363. https://doi.org/10.1074/jbc.271.16.9358

Witek RP, Stone WC, Karaca FG, Syn WK, Pereira TA, Agboola KM et al (2009) Pan-caspase inhibitor VX-166 reduces fibrosis in an animal model of nonalcoholic steatohepatitis. Hepatology 50:1421–1430

Yan J, Wan P, Choksi S, Liu ZG (2022) Necroptosis and tumor progression. Trends Cancer 8(1):21–27

Yi YS (2020) Functional crosstalk between non-canonical caspase-11 and canonical NLRP3 inflammasomes during infection-mediated inflammation. Immunology 159(2):142–155

Yu D, Coomber D, Riggins GJ (2003) Lack of telomerase activity in human myeloma cells and primary. Nat Genet 33(4):415–420

Yu P, Zhang X, Liu N, Tang L, Peng C, Chen X (2021) Pyroptosis: mechanisms and diseases. Signal Transduct Targeted Ther 6(1):128

Zhang C, Cai T, Zhu H, Hu Y (2004) Pseudoginsengenin DQ induces caspase-independent cell death of human colon carcinoma HT-29 cells. Phytother Res 18(7):593–598

Zheng M, Karki R, Vogel P, Kanneganti TD (2020) Caspase-6 is a key regulator of innate immunity, inflammasome activation, and host defense. Cell 181(3):674–687

# Apoptosis: Mechanisms and Clinical Implications

**4**

Sourav Bar, Soumik Dhara, Satyajit Halder, Kuladip Jana,
and Sudipta Kumar Ghorai

**Abstract**

Tissue homeostasis is efficiently maintained by the fine balance between cell death and cell survival. The major players in this complex molecular process are caspases, which depend on aspartate-directed proteases. These molecular players control the whole apoptosis for programmed cell death. Initiator caspases are functionalized through ligand–receptor interaction or mitochondrial signaling. The activated initiator caspases start a downstream cascade to activate effector caspases. The interplay between cellular survival and death plays a crucial role in embryonic tissue development, regulation of immune response, and tissue remodeling. Dysregulation of the apoptotic pathways is responsible for the initiation and progression of numerous diseases like neurodegenerative diseases and cancer—these cellular dynamics at a molecular level help to develop targeted therapeutic strategies. Modern clinical approaches help to maintain cellular equilibrium in pathological conditions by modulating apoptosis.

**Keywords**

Apoptosis · Programmed cell death · Caspases · Diseases · Therapeutics · Clinical diagnosis

S. Bar · S. Dhara · S. K. Ghorai
Coastal Environmental Studies Research Centre of Egra SSB College, Vidyasagar University, Midnapore, West Bengal, India

S. Halder · K. Jana (✉)
Division of Molecular Medicine, Bose Institute, Unified Academic Campus, Kolkata, West Bengal, India
e-mail: kuladip@jcbose.ac.in



K. Jana (ed.), *Apoptosis and Human Health: Understanding Mechanistic and Therapeutic Potential*, https://doi.org/10.1007/978-981-97-7905-5_4

## 4.1 Introduction

Apoptosis is a meticulously regulated and intricately organized form of programmed cell death. This biological process is crucial for maintaining development, immune responses, and tissue homeostasis in humans and other multicellular organisms. Apoptosis is characterized by a series of unique cellular morphological changes, including cell shrinkage, chromatin condensation, and the formation of apoptotic bodies. Neighboring phagocytes efficiently recognize and engulf these apoptotic bodies, removing the dying cells without causing inflammation (Kerr et al. 1972; Elmore 2007).

Apoptosis is executed through a series of molecular events within two principal pathways: the intrinsic (or mitochondrial) pathway and the extrinsic (or death receptor) pathway. The Bcl-2 family of proteins, which includes both anti-apoptotic and pro-apoptotic members, plays a pivotal role in maintaining mitochondrial integrity (Adams and Cory 2007). Apoptotic signaling is driven by a cascade of caspases, activated by the release of cytochrome c from the mitochondria, leading to the systematic dismantling of the cell. A finely tuned, complex interaction among various regulatory factors balances cell survival and death. Proteins such as the Inhibitor of Apoptosis (IAP) critically modulate the molecular mechanisms of apoptosis (Adams and Cory 2007).

Apart from its crucial role in normal physiological processes, the dysregulation of apoptotic pathways carries significant clinical implications. Apoptotic dysfunction is implicated in various diseases, including neurodegenerative disorders, cancer, cardiovascular diseases, and autoimmune diseases. In cancer, for instance, defects in apoptotic pathways lead to uncontrolled cell proliferation and tumor formation (Hanahan and Weinberg 2000). Neurodegenerative conditions such as Parkinson's and Alzheimer's disease exhibit abnormal apoptosis, resulting in neuronal cell death and cognitive decline (Mattson and Kroemer 2003). Alzheimer's disease is characterized by amyloid-β protein aggregation in the brain, contributing to senile plaque formation; these proteins activate apoptosis via caspases and inflammation pathways (Takada et al. 2020). Autoimmune diseases involve impaired apoptosis of self-reactive immune cells, perpetuating autoimmunity (Nagata 2018). Apoptotic dysfunction also contributes to cardiovascular diseases like myocardial infarction and heart failure (Jassem and Heaton 2004).

Mutations in the tumor suppressor gene TP53 are infrequent in acute myeloid leukemia (AML), yet disruptions in the p53 pathway are rapid, leading to dysregulated apoptotic pathways. ASPP1 and ASPP2 proteins are notable in AML, with reduced ASPP2 levels acting as tumor suppressors and influencing chemotherapeutic resistance. Decreased ASPP1 expression in AML holds functional and predictive value in guiding therapeutic approaches (Schittenhelm et al. 2024).

Understanding the molecular intricacies of apoptosis is crucial for developing targeted therapies. Research aims to manipulate apoptotic pathways, from inducing apoptosis in cancer cells to neuroprotective strategies against neurodegenerative diseases. These efforts underscore the therapeutic potential of modulating apoptosis for diverse clinical applications.

## 4.2 Mechanism

### 4.2.1 Molecular Pathways of Apoptosis

Apoptosis is a tightly regulated cellular process that orchestrates programmed cell death. It plays pivotal roles in diverse physiological processes, including immune system modulation, tissue homeostasis, and embryonic development. The intricate molecular pathways of apoptosis involve a network of signaling molecules such as Bcl-2 family proteins, caspases, and regulators of mitochondrial permeability transition pores (MPTP). Recent research has unveiled complex interconnected mechanisms governing apoptosis execution and regulation. For instance, BNIP3L/NIX, a Bcl-2 family protein, activates the mitochondrial apoptotic pathway by inducing MPTP activation and cytochrome c release (Yuan et al. 2017). Another significant finding is the novel pathway where GSK3β phosphorylates the anti-apoptotic protein Mcl-1, influencing apoptosis regulation (Wakatsuki et al. 2017).

Key events in apoptosis include the activation of caspases, a family of cysteine proteases that cleave target proteins at specific aspartic acid residues, leading to controlled cellular breakdown. The crucial role of caspases in initiating and executing apoptosis is underscored in subsequent discussions. Apoptotic caspases encompass initiator caspases (e.g., caspase-8 & 9) and executioner caspases (e.g., caspase-3 & 7), with the latter activated by the former to initiate the cascade culminating in programmed cell death (Yuan and Yankner 2000). Effector caspases then execute the final steps by cleaving various cellular substrates. Moreover, caspases are further categorized into distinct subgroups, such as inflammatory caspases (e.g., caspase-1), implicated in inflammatory responses and cytokine processing (Alberts et al. 2014).

Two key regulatory pathways are present in the cell for the proper execution of apoptosis. Those are the intrinsic or mitochondrial pathway of apoptosis and extrinsic or death receptor-mediated apoptosis pathways. The activation of caspase-9 initiates the intrinsic pathway. In contrast, the activation of caspase-8 is involved in the extrinsic pathways. The complexity of apoptotic signaling can be well understood by studying the interplay between these two primary pathways and their regulation by caspase-8 (Green and Llambi 2015). One of the very crucial aspects of apoptosis is the regulation of mitochondrial membrane disruption. The Bcl-2 family proteins play a central role in this process. So, it is very interesting to understand how caspase-8 can control mitochondrial membrane dynamics through interactions with Bcl-2 family proteins and further activate caspase-9 (Adams and Cory 2007).

Inflammasomes are protein complexes that are crucial in regulating immune response and inflammation. NLRP3 is a special type of inflammasome that consists of the NOD, LRR, and pyrin domain-containing protein 3. Among all other inflammasomes, particularly NLRP3 (NOD, LRR, and Pyrin domain-containing protein 3) plays a vital role in apoptosis. The activation of Toll-like receptors such as TLR-4 and TLR-6, along with the NFκB signaling pathways and NLRP3 inflammasomes, is mediated in response to toxic amyloid β-protein fibrils formed on neuronal cells. These toxic Amyloid β-protein fibrils are found in neurodegenerative disorders such as Alzheimer's disease, triggering a cellular response involving this pathway. This activation of protein complexes, which shows a relationship between inflammatory signaling and neurodegenerative processes, is eventually associated with apoptosis (Takada et al. 2020).

#### 4.2.1.1 Intrinsic (Mitochondrial) Pathway

The intrinsic or mitochondrial pathway of apoptosis plays a central role in maintaining cellular fate, ensuring the intricate balance between programmed cell death and cell survival. The interconnected molecular events linked to the mitochondria are involved in this pathway, which is crucial for maintaining cellular homeostasis. This pathway deals with the regulatory mechanisms, molecular players, and physiological functions of the intrinsic pathway.

The intrinsic pathway of apoptosis plays an important role in many physiological processes. During the gestational period, the intrinsic pathway helps to complete the embryonic development and, in this meantime, controlled cell death by the intrinsic pathway of apoptosis shapes embryonic tissues and organs (Danial and Korsmeyer 2004). The intrinsic pathway also helps to eliminate undesired or defective cells to prevent autoimmunity in the immune system (Green and Llambi 2015). Furthermore, the removal of senescent cells is mediated by the intrinsic apoptosis pathway, ensuring the proper maintenance of tissue homeostasis (D'arcy 2019). This mechanism also takes care of the DNA damage response to prevent the multiplication of defective cells and maintain genomic integrity (Chipuk et al. 2012). The intrinsic or mitochondrial apoptosis pathway is a delicately balanced process with multifaceted outcomes for cellular homeostasis.

##### Pro-apoptotic Activation

The B-cell lymphoma 2 (Bcl-2) family of proteins is a central molecular player in the intrinsic pathway of apoptosis. After receiving apoptotic signals, conformational changes take place in Bax and Bak, and therefore, they translocate to the outer membrane of mitochondria for oligomer formation (Youle and Van Der Bliek 2012). BH3-only proteins such as Bim and Puma tightly regulate these events and act as inducers or direct activators to promote the activation of Bax and Bak (Westphal et al. 2011). Mitochondrial Outer Membrane Permeabilization (MOMP), a major step in the initiation of apoptosis, is promoted by Pro-apoptotic proteins of apoptosis such as Bax and Bak (Czabotar et al. 2014). Anti-apoptotic proteins, including Bcl-2 and Bcl-xL, hinder this process by preventing MOMP (Chen et al. 2005). These proteins' structural and functional complexity are pivotal in apoptotic regulations (Westphal et al. 2011).

##### MOMP and Cytochrome c Release

The intrinsic apoptosis pathway is a complex process with several interconnected molecular mechanisms. Mitochondrial Outer Membrane Permeabilization (MOMP) (Fig. 4.2) is one of the primary events responsible for the initiation of the intrinsic pathway of apoptosis. This crucial step is tightly regulated by the interplay between pro-apoptotic and anti-apoptotic members of the Bcl-2 family proteins. The initiation of this pathway starts with the conformational change and oligomerization of pro-apoptotic members, such as Bax and Bak, in response to cellular stress (Czabotar et al. 2014). The anti-apoptotic proteins tightly maintain a fine balance with their anti-apoptotic counterparts, including Bcl-2 and Bcl-xL (Chen et al. 2005). Molecular interactions between these pro-apoptotic and anti-apoptotic proteins decide the activation and inhibition of MOMP.

In the intrinsic apoptosis pathway, the commitment to programmed cell death is marked by the cytochrome c release from the mitochondria into the cytoplasm. This cellular process is highly controlled and organized by the Bcl-2 family of proteins. Pro-apoptotic proteins such as Bax and Bak are pivotal in this cellular process. After the activation of Bax and Bak, oligomeric complexes form on the outer membrane of mitochondria. The oligomerization activates Bax and Bak, and activating these pro-apoptotic members results in the mitochondrial outer membrane permeabilization. The oligomeric complex formed by Bax and Bak on the outer membrane of mitochondria creates pores, and this critical event is known as Mitochondrial Outer Membrane Permeabilization (MOMP), which causes the release of pro-apoptotic factors from the mitochondrial intermembrane space with cytochrome c (Czabotar et al. 2014). The release of cytochrome c has been marked as the irreversibility of the apoptotic process, where cells are committed to programmed cell death (Youle and Van Der Bliek 2012). MOMP is the point of no return from the intrinsic pathway of apoptosis. After the release of cytochrome c into the cytoplasm, the formation of apoptosome, which is a multiprotein complex containing Apaf-1, is started. This complex further activates caspase-9, which initiates the downstream cascade of caspase and is responsible for the systematic accomplishment of apoptosis (Czabotar et al. 2014).

### Apoptosome Formation and Caspase Activation

A tightly regulated molecular pathway that helps to make cellular decisions is controlled by apoptosome (Fig. 4.1) formation and caspase-9 activation. In the cytoplasm, the released Cytochrome c binds to apoptotic protease-activating factor 1 (Apaf-1) for apoptosome formation (Czabotar et al. 2014). The caspase-9 is activated by this apoptosome complex to initiate the caspase cascade downstream pathways. The cleavage and subsequent activation of effector caspases like caspase-3 and caspase-7 are catalyzed by this complex. This ultimately leads to the execution of apoptosis by deconstructing cellular components (Czabotar et al. 2014).

The apoptosome acts as a crucial network in the intrinsic apoptosis pathway and organizes the activation of whole downstream caspase cascades for the ultimate execution of apoptosis. Apoptosome-mediated activation of caspase-9 is important for determining cellular fate. The apoptosome has a unique structure with the entangled arrangement of cytochrome c. A main component of the apoptosome is Apaf-1, which generally resides in inactive form as a monomer in the absence of apoptotic signals. The conformational change in Apaf-1 after the release of cytochrome c is responsible for oligomerization to form a heptameric structure (Pang et al. 2015). The heptameric complex of apoptosome is responsible for the recruitment and activation of procaspase-9. This induces a conformational change to reveal the Caspase

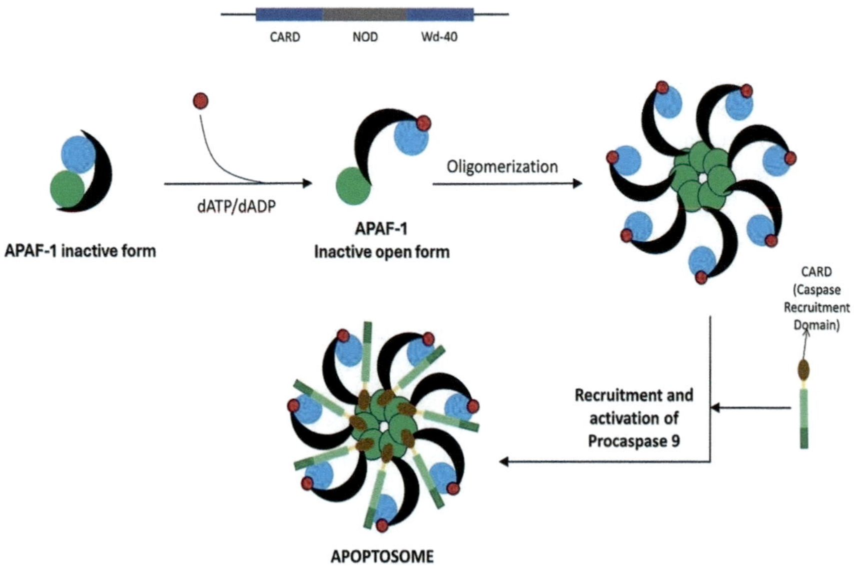


**Fig. 4.1** Formation of apoptosome includes APAF-1, cytochrome c, and caspase-9

Recruitment Domain (CARD) on Apaf-1. The recruitment process of procaspase-9 begins with its association with the WD-40 repeats present in Apaf-1. This interaction is facilitated by the exposure of the CARD region due to a conformational change, thus enabling the progression of the process. The activation of pro-caspase-9 is accomplished by the exposed CARD, which subsequently leads to the autocatalytic cleavage of pro-caspase-9 and converts it into the active caspase-9 (D'Amelio et al. 2008; Czabotar et al. 2014). The apoptotic signal is amplified by the active Caspase-9, which acts as a molecular switch. Therefore, the activation of downstream effector caspases, specifically caspase-3 and caspase-7, is initiated. The final execution of apoptosis is executed by these effector caspases, which cleave a great number of cellular substrates, leading to programmed cell death.

### Crosstalk with Mitochondrial Dynamics

The mitochondrial dynamics and quality control mechanisms are intensely linked to the intrinsic apoptosis pathway. Some studies have shown that mitochondrial fission and fusion events affect the sensitivity of mitochondria to MOMP (Youle and Van Der Bliek 2012). On the other hand, the removal of damaged mitochondria is regulated by mitophagy, a mitochondrial quality control mechanism that influences the initiation of apoptosis (Twig et al. 2008). The interplay between the intrinsic pathway of apoptosis and mitochondrial dynamics helps to intersect cellular health and stress responses.

## Effect of Cellular Metabolism on the Intrinsic Pathway

Metabolic rearrangement has come out as a key determinant of cell fate. Cellular metabolism has a significant effect on the intrinsic pathway of apoptosis. The activity of pro-apoptotic and anti-apoptotic Bcl-2 family members is directly influenced by metabolic outcomes such as Reactive Oxygen Species (ROS) and ATP (Chipuk et al. 2012). Another layer of complexity to the intrinsic pathway is added by this metabolic regulation, and a connection between cellular energy status and apoptotic susceptibility is established. Increased ROS level in cells indicates cellular stress and damage that can directly influence the activation of pro-apoptotic proteins. Conformational changes in Bax can be induced by oxidative stress, and this promotes oligomerization of Bax, which is subsequently inserted into the mitochondrial outer membrane, causing the release of cytochrome c (Korsmeyer 2007; Hockenbery et al. 1993). However, the detrimental effects of ROS are neutralized by anti-apoptotic Bcl-2 proteins, which act as scavengers for free radicals, emphasizing the complex interaction between apoptosis regulation and cellular redox status (Yu et al. 2004). Similarly, ATP levels are critical indicators of cellular energy status, which is critical in modulating apoptosis. ATP levels can significantly affect both pro-apoptotic and anti-apoptotic Bcl-2 family of proteins. However, ATP depletion in cells activates the pro-apoptotic protein Bak which promotes mitochondrial outer membrane permeabilization and apoptotic signaling (Wang and Wan 2023). Conversely, cellular survival can be supported by anti-apoptotic Bcl-2 proteins which preserve mitochondrial integrity against metabolic challenges (Fig. 4.2).

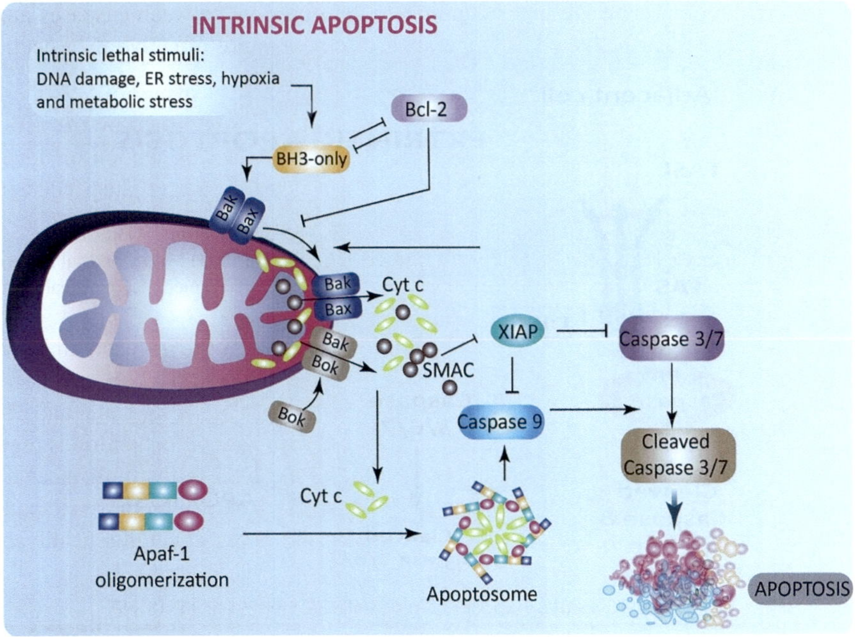


**Fig. 4.2** Intrinsic (mitochondrial) pathway of apoptosis

#### 4.2.1.2 Extrinsic (Death Receptor) Pathway

The extrinsic apoptosis pathway is also known as the death receptor pathway, which is tightly regulated by the attachment of extracellular signal or death ligands to death receptors on the cell surface (Fig. 4.3). The Tumor Necrosis Factor (TNF) superfamily is critical in initiating this pathway. TNF superfamily consists of TNF-α, Fas Ligand (FasL), and TNF-Related Apoptosis-Inducing Ligand (TRAIL) (Ashkenazi and Dixit 1998; Schneider-Brachert et al. 2004). After binding of the death ligand, conformational changes occur within the death receptors such as Fas (CD95) and Tumor Necrosis Factor Receptor 1 (TNFR1), recruiting the adaptor proteins such as Fas-Associated Death Domain (FADD) and TNFR1-Associated Death Domain (TRADD). Therefore, the Death-Inducing Signaling Complex (DISC) is formed, and it acts as an activation platform for procaspase-8 (Kischkel et al. 1995; Nagata 1994). The downstream effector caspases, like caspase-3, which begins the execution phase of apoptosis, can be cleaved and activated by active caspase-8 (Fig. 4.3) (Boatright et al. 2003). The extrinsic pathway responds to external signals to remove damaged or unwanted cells, acting as a crucial part of the surveillance mechanism. Impairment of this pathway can result in various diseases (Ashkenazi 2002).

##### Death Receptor Activation

The extrinsic pathway of apoptosis, which plays a critical role in tissue homeostasis and immune system surveillance, is initiated by the attachment of the TNF superfamily death ligands to their corresponding receptors such as Fas (CD95), TNFR1,

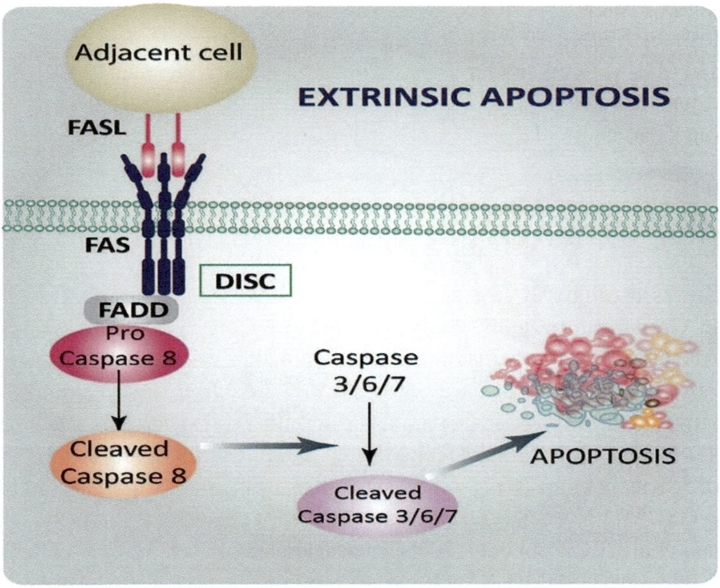


**Fig. 4.3** Extrinsic or death receptor pathway of apoptosis

or DR4/DR5 for TRAIL. This death receptor pathway responds to external signals that trigger cell death (Ashkenazi 2002; Schneider-Brachert et al. 2004).

After the formation of DISC, proteolytic cleavage activates procaspase-8 and generates active caspase-8. This initiator caspase has a dual role, such as it can directly act upon downstream effector caspases like caspase-3 for cleavage and subsequent activation to initiate the execution of apoptosis (Boatright et al. 2003). The activation of effector caspases, specifically caspase-3, can cause characteristic morphological and biochemical changes associated with apoptosis by cleaving a multitude of specific substrates, including structural and repair proteins, and ultimately results in the fragmentation of DNA and membrane blebbing (Ashkenazi and Dixit 1998). Moreover, caspase-8 acts upon the BH3-only protein called Bid and cleaves it into the truncated form known as tBid. This tBid triggers the mitochondrial outer membrane permeabilization (MOMP) to activate the intrinsic (mitochondrial) apoptotic pathway (Li et al. 1997; Luo et al. 1998). The truncated Bid enters into the mitochondria and stimulates the intrinsic pathway of apoptosis by activating Bak and Bax. Thus, the connection between the extrinsic and intrinsic pathways is established (Li et al. 1998; Luo et al. 1998).

One of the critical targets of effector caspase-3 is the DNA repair enzyme Poly (ADP-ribose) polymerase (PARP), which belongs to a family of enzymes that can catalyze the transfer of ADP-ribose to target proteins. PARP is cleaved by the Caspase-3 into distinct fragments that cannot repair DNA damage, resulting in genomic instability (Lazebnik et al. 1994). Nuclear DNA fragmentation is one of the hallmark events in apoptosis. The internucleosomal DNA fragments are generated from intact DNA by the action of endonucleases, which are activated by caspases. This produces the typical ladder pattern, which can be observed during gel electrophoresis (Liu et al. 1996).

The activation of caspase-3 leads to characteristic changes in the morphology of cells, which are distinct features of apoptotic cells. The cleavage of structural and cytoskeletal proteins is mediated by caspase, resulting in the transformation of cell shape (Coleman et al. 2001). These alterations in cell shape induce the packaging of cellular components into apoptotic bodies, which are subsequently taken up by various phagocytic cells or macrophages. This prevents the release of potentially toxic harmful intracellular components to maintain tissue homeostasis.

#### Formation of Death-Inducing Signaling Complex (DISC)

The assembly of the DISC is initiated by the attachment of death receptors like Fas (CD95) to their cognate ligands. The DISC coordinates the death receptors, adaptor proteins, and procaspase-8, which acts as a vital assembly. After engagement with DISC, procaspase-8 is activated by the autocatalytic cleavage to form caspase-8 (Kischkel et al. 1995; Nagata 1994). Activated caspase-8 is generally considered the initiator of caspase in the extrinsic pathway. Therefore, it catalyzes the proteolytic activation of effector caspases like caspase-3, which initiates the execution phase of apoptosis (Boatright et al. 2003).

The death effector domains (DEDs), involved in the interaction between the death receptor and FADD, facilitate the oligomerization of procaspase-8 within the

DISC (Dickens et al. 2012). The conformational changes in procaspase-8 induced by this aggregation promote autocatalytic cleavage to convert it into active forms such as p18 and p10 (Fu et al. 2016).

#### Caspase-8 and Effector Caspase Cascade

Two distinct apoptotic pathways can be elicited by the activated caspase-8. The Type I pathway deals with the direct cleavage and activation of effector caspases, particularly caspase-3 and caspase-7, by the action of caspase-8 for the rapid execution of apoptosis (Kaufmann et al. 2003). On the other hand, the Type II pathway deals with the caspase-8-mediated cleavage of the pro-apoptotic Bcl-2 family proteins Bid into tBid (Li et al. 1998). Then, the translocation of tBid into the mitochondria amplifies the intrinsic apoptosis pathway by activating Bax and Bak. Moreover, additional regulatory mechanisms involve cellular FLICE-like inhibitory proteins (cFLIP), which act as modulators of caspase-8 activity, maintaining the balance between pro-survival and pro-apoptotic signals (Irmler et al. 1997; Kataoka and Tschopp 2004). Understanding the details of caspase-8 activation and its downstream effector caspase cascade is important to address the complexities of the extrinsic apoptotic pathway.

### 4.2.1.3 Crosstalk Between Intrinsic and Extrinsic Pathways

The crosstalk between the intrinsic and extrinsic apoptotic pathways is a highly controlled and complex cellular process with many vital implications for cellular fate. An extremely important mediator in this interconnected process is truncated Bid (tBid), which is generated through Caspase-8-mediated cleavage within the Death-Inducing Signaling Complex (DISC). After entering into the mitochondria, a cascade of events that bridge the extrinsic and intrinsic pathways is arranged by the tBid. The interaction between tBid and anti-apoptotic Bcl-2 family proteins, such as Bcl-2 and Bcl-xL, helps to release the inhibition of anti-apoptotic proteins on pro-apoptotic effectors Bax and Bak (Wei et al. 2001). A conformational change in Bax and Bak is induced by this interaction, resulting in their oligomerization on the mitochondrial outer membrane to initiate the mitochondrial outer membrane permeabilization (MOMP) (Hsu and Youle 1997). The formation of MOMP and the downstream process are the same as mentioned in the intrinsic and extrinsic pathways of apoptosis. This molecular crosstalk ensures a robust and irreversible commitment to programmed cell death.

### 4.2.1.4 Alternative Pathways and Non-caspase-Dependent Apoptosis

Some alternative pathways and non-caspase-dependent apoptosis offer a more detailed perspective on programmed cell death. Other than the classical caspase-dependent apoptotic pathways, non-canonical pathways are also responsible for cellular demise. Studies have revealed that lysosomal membrane permeabilization (LMP) can trigger caspase-independent cell death with the help of cathepsins released from lysosomes to initiate apoptotic cascades (Repnik et al. 2014). Receptor-interacting Protein Kinase 1 (RIPK1) and RIPK3-mediated programmed necrosis or necroptosis. This is distinctly considered an alternative cell death pathway (Galluzzi et al. 2018). These non-canonical pathways of apoptosis often interact with classical caspase-dependent apoptotic signaling and add a layer of complexity for the proper balance between cell survival and cell death.

Lysosomal membrane permeabilization (LMP) is critical in caspase-independent cell death. Various stimuli, such as reactive oxygen species or lysosomotropic agents, can trigger LMP and stimulate the release of lysosomal cathepsins into the cytoplasm for the subsequent activation of apoptotic cascades (Boya and Kroemer 2008; Repnik et al. 2014). This non-canonical pathway adds an additional layer of protection where caspase-dependent pathways may be compromised or inhibited. Moreover, another non-canonical mode of cell death has been carried out to complete programmed necrosis or necroptosis. The activated Receptor-Interacting Protein Kinase 1 (RIPK1) and RIPK3 form the necrosome complex, the key player in Necroptosis (Pasparakis and Vandenabeele 2015). Necroptosis is effective in various pathological conditions like neurodegenerative disorders and inflammatory disorders (Galluzzi et al. 2018).

Instead of operating separately, these non-canonical pathways often interact and modulate the classical apoptotic signaling pathways. This complex network of cell death signaling shows the cellular capacity to act against different natures and intensities of the stressors.

## 4.2.2 Regulation of Apoptosis

### 4.2.2.1 Bcl-2 Family Proteins

The B-cell lymphoma 2 (Bcl-2) family of proteins controls a sophisticated molecular cascade involved in the intrinsic (mitochondrial) and extrinsic pathway of apoptosis. Key molecular events are involved in this pathway, making complex interconnected mechanisms for determining cell fate.

#### Anti-apoptotic Proteins and Their Functions

The cellular integrity is preserved by counteracting Anti-apoptotic proteins against pro-apoptotic proteins. Thus, the anti-apoptotic proteins act as guardians against cell death. Bcl-2, one of this family's prominent members, occupies the outer membrane of mitochondria and prevents the release of cytochrome c, a major event in the intrinsic apoptosis pathway (Adams and Cory 2007). Bcl-2 forms heterodimers with pro-apoptotic proteins such as Bax to neutralize their effect (Oltval et al. 1993). Another essential member of anti-apoptotic proteins is Mcl-1, which plays a vital role in cell survival. It prevents the activation of Bak, a member of the pro-apoptotic family protein (Auner et al. 2010). Additionally, the role of Bcl-xL is an important example of pro-survival machinery. It prevents the opening of the mitochondrial permeability transition pore, a crucial event in apoptosis (Emily et al. 2001). Finally, these anti-apoptotic proteins act as important gatekeepers which prevent the unjustified initiation of programmed cell death. Thus, it ensures cells' survival and proper functioning in various physiological conditions.

### Pro-apoptotic Proteins and Their Functions

The orchestrated cellular demise is promoted by the counteraction of pro-apoptotic proteins against the anti-apoptotic proteins for the completion of apoptosis. Bax, one of the important pro-apoptotic family proteins, operates at the mitochondrial level. After the activation, a conformational change occurs in Bax, and then it translocates to the mitochondrial outer membrane to form oligomers responsible for the membrane's permeabilization. Therefore, apoptogenic factors such as cytochrome c are released into the cytoplasm (Hsu and Youle 1997; Wei et al. 2001). Bak is another major player in pro-apoptotic family proteins. It has functional similarities with Bax and contributes to mitochondrial outer membrane permeabilization (Dewson et al. 2009). Another category of pro-apoptotic regulators is BH3-only proteins such as Bid, Bim, and Puma. These proteins can sense cellular stress and transduce pro-apoptotic signals to activate Bax and Bak or directly involve anti-apoptotic proteins. That is why they function as sentinels (Letai et al. 2002; Puthalakath et al. 2001). The cell's fate is ultimately determined by the interplay between pro-apoptotic and anti-apoptotic proteins.

### Regulation by Bcl-2 Homology 3 (BH3)-Only Proteins

The fine balance between pro-apoptotic and anti-apoptotic proteins is maintained by Bcl-2 Homology 3 (BH3). Only proteins that are a subgroup of the Bcl-2 family proteins. The activation mechanism of Bax and Bak by BH3-only proteins is finely tuned where some of the BH3-only proteins, such as Bim, Bid, and Puma, act as sensitizers or direct activators for the initiation of MOMP (Westphal et al. 2011). These molecular events describe how the Bcl-2 family of proteins combines various signals to determine whether a cell undergoes apoptosis or survives.

### Regulation by BH3-Mimetics and Interactions with Anti-apoptotic Proteins

Recent studies have explained the importance of BH3-mimetics. Synthetic compounds are designed in the laboratory to mimic the BH3 domain of pro-apoptotic proteins to target and neutralize anti-apoptotic proteins (Labi et al. 2006). These BH3 mimetics, such as ABT-737 and ABT-199, target specific Bcl-2 family of proteins including Bcl-2 and Bcl-xL. Therefore, ABT-737 and ABT-199 subsequently bind to Bcl-2 and Bcl-xL, distorting their interaction with pro-apoptotic members and leading to apoptosis (Oltersdorf et al. 2005).

### Non-apoptotic Functions of Bcl-2 Family of Proteins

Apart from their role in apoptosis, the Bcl-2 family of proteins exhibit non-apoptotic functions. The involvement of Bcl-2 family proteins in cellular processes such as DNA repair, autophagy, and cell cycle progression has been found to play a crucial role (Heath-Engel et al. 2012). These non-apoptotic functions highlight the multifaceted roles of Bcl-2 family proteins beyond their classical role in apoptosis, which contributes to making decisions for cellular fate.

#### 4.2.2.2 Bcl-2 Family Dynamics

Several key effectors other than Bax and Bak are involved in regulating apoptosis. One such key effector is the Mitochondrial Apoptosis-inducing Channel, or MAC, which is formed by the oligomerization of Bax and Bak at the outer membrane of mitochondria. The release of apoptogenic factors for further amplifying the apoptotic signal is mediated by the MAC complex, which acts as a channel (Große et al. 2016). Furthermore, PUMA (p53 Up-regulated Modulator of Apoptosis), a typical example of the BH3-only proteins, can initiate apoptosis. Various stress signals, including DNA damage, can activate PUMA, which can neutralize anti-apoptotic Bcl-2 proteins, setting free the pro-apoptotic capability of Bax and Bak (Villunger et al. 2003). Another key player is the multidomain pro-apoptotic protein Noxa, which helps disrupt the anti-apoptotic interactions within the Bcl-2 family proteins. Doing this can contribute to the mitochondrial outer membrane permeabilization and the release of cytochrome c (Chiara et al. 2012). The collaborative interaction of these pro-apoptotic proteins highlights the complex regulatory network that decides between the life and death of cells.

#### 4.2.2.3 Mechanism of Caspase Activation

Caspases are a family of cysteine-dependent aspartate-directed proteases that play a crucial role in apoptosis. They are activated by proximity-induced dimerization and proteolytic cleavage. This proteolytic cleavage helps to activate Initiator caspases like caspase-8 and caspase-9. For example, the death-inducing signaling complex (DISC) formation is induced by the attachment of death ligands to their specific death receptors in the extrinsic apoptosis pathway. This ligand–receptor binding triggers auto-catalytic cleavage and activation of pro-caspase-8. Similarly, the released cytochrome c in cytosol triggers apoptosome formation, which is responsible for the auto-catalytic cleavage and activating pro-caspase-9 in the intrinsic apoptosis pathway. Therefore, initiator caspases activate effector caspases like caspase-3 by proximity-induced dimerization. After the activation, conformational changes occur in effector caspases to cleave cellular substrates (Fuentes-Prior and Salvesen 2004).

#### 4.2.2.4 Role of Inhibitors of Apoptosis Proteins (IAPs)

Cellular fate is determined by the dynamic molecular interaction between pro-apoptotic and anti-apoptotic members of the Bcl-2 family proteins. Any change in this fine balance can lead to either cell survival or programmed cell death (Westphal et al. 2011). The regulation of apoptosis is accomplished by the inhibition of caspases, which are also inhibited by Inhibitors of Apoptosis Proteins (IAPs) such as XIAP and influence the sequential mitochondrial events (Strasser and Vaux 2020). This interconnected regulatory network ensures tight control over the initiation and execution of the intrinsic pathway.

The inhibitors of apoptosis proteins (IAPs) are a family of proteins vital for the regulation of apoptotic process. Several IAPs, such as cIAP1 (cellular inhibitor of apoptosis 1), cIAP2 (cellular inhibitor of apoptosis 2), and XIAP (X-linked inhibitor of apoptosis protein), have anti-apoptotic effects because they block apoptosis.

These IAPs inhibit the activity of the central executioners of apoptosis, known as caspases (Liston et al. 2003). Specifically, XIAP can directly bind to caspases-3, caspase-7, and caspase-9 to inhibit them for the effective suppression of apoptosis at multiple key points in the cascade (Deveraux et al. 1997; Riedl et al. 2001). On the other hand, cIAP1 and cIAP2 regulate the extrinsic pathway of apoptosis by modifying the activity of caspase-8 (Bertrand et al. 2009). Apart from the inhibition of caspase, IAPs also help to activate pro-survival signaling pathways such as the Nuclear Factor-kappa B (NF-κB) pathway for the further enhancement of their ability to promote cell survival (Varfolomeev et al. 2007). Dysregulation of these IAPs can cause various diseases, including cancer. However, their overexpression can lead to the prevention of apoptosis. These highlights have the potential therapeutic significance for targeted clinical interventions.

The interplay between IAPs and ubiquitination processes highlights several crucial aspects of their function. A conserved Baculovirus IAP Repeat (BIR) domain of IAPs is responsible for the inhibition of caspases. However, these proteins have ubiquitin ligase activity, which allows them to ubiquitinate themselves and other proteins involved in apoptotic signaling pathways. The NF-κB pathway is regulated by this ubiquitination activity. The activation of NF-κB can be induced by cIAP1 and cIAP2-mediated ubiquitination of Receptor-Interacting Protein 1 (RIP1). Thus, these IAPs inhibit the proteasomal degradation of NF-κB (Varfolomeev et al. 2007). IAPs play a dual role, acting as both apoptosis inhibitors and pro-survival signaling mediators. Additionally, the dysregulation of IAPs in cancer often leads to enhanced cell survival that ultimately contributes to tumorigenesis and therapeutic resistance.

## 4.3 Clinical Implication

### 4.3.1 Physiological Functions of Apoptosis

#### 4.3.1.1 Embryonic Development and Tissue Remodeling

Embryonic development is a highly regulated physiological process consisting of events that give rise to the different tissues and organs. Several molecular and cellular processes, including cell differentiation, division, migration, and programmed cell death, come together to complete this process at the cellular level. The formation of functional tissues and organs needs spatially and temporally coordinated molecular and cellular processes. Gastrulation is a fundamental process of embryonic development. It helps to set up three primary germ layers: ectoderm, mesoderm, and endoderm. Organogenesis and tissue differentiation start from the process of gastrulation. Signaling pathways like Wnt and BMP initiate gastrulation by inducing changes in cell fate and triggering morphogenetic movements (De Robertis 2006).

Throughout embryonic development, apoptosis helps to shape up the tissues and organs. It is crucial in sculpting and rectifying embryonic structures through programmed cell death. The formation of digits during limb development is a typical example of the physiological role of apoptosis. Apoptosis plays an essential role in the formation of digits by removing interdigital tissue to create well-defined and separated digits (Curtin and Cotter 2003). The precise elimination of cells between developing digits is known as interdigital apoptosis, which is controlled by the activation of caspases, especially caspase-3 (Hamburger and Hamilton 1951; An et al. 2019). Apoptosis is also very crucial for neural development. Apoptosis helps to shape neural structures by eliminating excess neurons to establish functional neural circuits (Helgren et al. 1997). This programmed elimination of cells is very important to establish connections for optimization of neural function during embryogenesis. The role of apoptosis is very crucial in tissue patterning during embryonic development. It removes the unwanted cells to form body cavities and sculpt organs. A precise balance of apoptosis to create internal organs is known as cavitation (Martin 1998). It is found that apoptosis has a significant role in tissue regeneration, especially liver regeneration. The Apoptotic hepatocytes induce compensatory cell regeneration that ensures tissue function reestablishment to maintain organ integrity (Watson et al. 2016).

#### 4.3.1.2 Immune System Regulation

Apoptosis plays a vital role in immune cell development, particularly in the thymus. It is responsible for T cell development, and in bone marrow, it is responsible for the B cell development. The self-reactive T cell undergoes apoptosis to prevent autoimmunity during negative clonal selection in the thymus (Strasser et al. 1991). B cells are eliminated through apoptosis to ensure self-tolerance in bone marrow (Melchers 2015). Apoptosis helps to terminate immune responses to prevent tissue damage and unwanted inflammation. Activated immune cells are eliminated through a process called Activation-Induced Cell Death (AICD) (Green 2019). Virus-infected or altered self-cells are removed by Fas/Fas ligand (FasL)-mediated apoptosis to enhance antitumor and antiviral immune responses (Nagata 2018). Autoreactive lymphocytes are eliminated by apoptosis as it is integral to immune tolerance. Apoptosis contributes to the resolution of the immune response by removing activated immune cells to reduce inflammation (Serhan and Savill 2005).

#### 4.3.1.3 Maintenance of Tissue Homeostasis

Apoptosis ensures a fine balance between cell proliferation and cell death. It prevents the accumulation of cellular debris by eliminating aged, damaged, or redundant cells. Apoptosis helps to shed off the enterocytes to facilitate the continuous renewal of the cell lining in the intestinal epithelium (Potten 1990). Apoptosis can give protection against uncontrolled cell growth and hyperplasia, which can cause cancer. The accumulation of mutations can be prevented through apoptosis by eliminating potentially harmful cells to maintain tissue integrity (Hanahan and Weinberg 2011). Inflammatory cells that are no longer needed or have completed their function are removed by apoptosis to maintain tissue homeostasis (Serhan and Savill 2005).

#### 4.3.1.4 DNA Damage Response (DDR) and Cell Cycle Control

Apoptosis eliminates the cells having irreparable genomic lesions. If the damage is beyond repair capacity, the ATM and ATR kinases can sense the DNA damage and activate downstream effectors for programmed cell death (Jackson and Bartek 2009). The p53, a tumor suppressor protein, plays a key role in coordinating apoptosis against DNA damage. After DNA damage, pro-apoptotic genes, including BAX and PUMA, are transactivated by the activated p53. Therefore, mitochondrial outer membrane permeabilization occurs to initiate apoptosis (Vousden and Prives 2009). Apoptosis acts as a fail-safe mechanism to control the cell cycle for the prevention of damaged cell proliferation. Proteins like p53 and CHK2 regulate the cell cycle to halt its progression for repair. If repair is not successfully working out, damaged cells are eliminated at risk of genomic instability (Shaltiel et al. 2014).

### 4.3.2 Dysregulation of Apoptosis in Disease

The imperative function of dysregulations in apoptotic signaling in preserving the delicate balance between cell survival and cell death that is required to prevent diseases is highlighted by the establishment of the evasion of apoptosis as a prominent hallmark of cancer, diabetes, cardiovascular diseases, autoimmune diseases, and neurological disorders (Fig. 4.4).

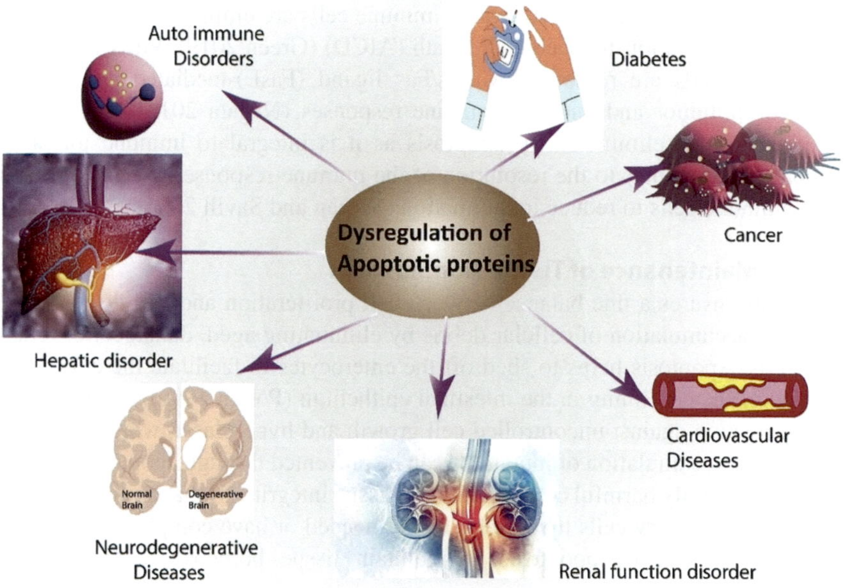


**Fig. 4.4** Dysregulation of apoptosis in disease

#### 4.3.2.1 Cancer

**Resistance to Apoptosis in Cancer Cells**

The dysregulation of Bcl-2 family proteins is frequently observed in cancer cells, enhancing anti-apoptotic signals and fortifying resistance to programmed cell death. Overexpression of anti-apoptotic proteins like Bcl-2 and Bcl-xL effectively suppresses the intrinsic apoptotic pathway, reinforcing cancer cells' resilience against apoptosis (Czabotar et al. 2014). Mutations in the tumor suppressor gene p53, a pivotal apoptosis regulator, further exacerbate this resistance by allowing damaged cells to evade apoptotic mechanisms, thereby fostering genomic instability and cell survival (Vousden and Prives 2009). This survival advantage in apoptosis-resistant cancer cells promotes the emergence of aggressive phenotypes and uncontrolled proliferation (Strasser and Vaux 2020), facilitating their dissemination via bloodstream or lymphatic routes, which enhances metastatic potential (Green and Llambi 2015). Consequently, such apoptosis-resistant cancer cells display insensitivity to conventional treatments like radiotherapy and chemotherapy aimed at inducing programmed cell death, thereby complicating therapeutic interventions (D'arcy 2019).

**Therapeutic Strategies Targeting Apoptosis in Cancer**

Within the field of cancer therapeutics, novel approaches have emerged to combat resistance to apoptosis, which is crucial for restoring the delicate balance favoring programmed cell death. Venetoclax, a BH3 mimetic, disrupts interactions between anti-apoptotic Bcl-2 proteins and their pro-apoptotic counterparts, thereby promoting apoptosis in cancer cells (Westphal et al. 2011). Nutlin-3, an MDM2 inhibitor, stabilizes and activates p53 by obstructing its binding with MDM2, reinstating p53's role as a tumor suppressor and triggering cell death (Vassilev et al. 2004). PAC-1 directly activates downstream caspases, bypassing upstream regulatory mechanisms to induce apoptosis in cancer cells (Lavrik 2010). Immune checkpoint inhibitors like PD-1/PD-L1 or CTLA-4 enhance anti-tumor immune responses, indirectly fostering apoptosis and aiding in eliminating cancer cells (Pardoll 2012). Therapeutic combinations of targeted therapies, chemotherapy, and immunotherapy synergistically target mechanisms of apoptosis resistance to enhance cancer cell death (Galluzzi et al. 2018). Precision medicine approaches tailor treatments based on individual molecular profiles, such as BRAF inhibitors for melanoma with specific genetic mutations, optimizing apoptotic induction in cancer therapy (Flaherty et al. 2010). These diverse strategies underscore the multifaceted battle against apoptosis resistance in cancer, highlighting the evolving landscape of oncological treatments.

#### 4.3.2.2 Neurodegenerative Diseases

**Apoptosis in Neurological Disorders**

In Alzheimer's disease, mitochondrial dysfunction precipitates the intrinsic pathway of apoptosis, exacerbated by oxidative stress stemming from the accumulation of Amyloid-beta (Aβ) peptides within mitochondria, culminating in cytochrome c release and apoptosis initiation (Reddy 2009). Another hallmark, hyperphosphorylated tau, activates caspases, fostering Tau-Induced apoptosis and compromising mitochondrial integrity, thereby exacerbating neurodegeneration and cognitive decline (Gendron and Petrucelli 2009). Similarly, Parkinson's disease manifests mitochondrial dysfunction induced by Alpha-synuclein accumulation in mitochondria, triggering apoptosis in dopaminergic neurons (Martin 2010). Additionally, in Parkinson's, the unfolded protein response (UPR) mediates apoptosis induced by Endoplasmic Reticulum (ER) stress, precipitating dopaminergic neuron degeneration (Hetz and Mollereau 2014). In Huntington's disease, the mutant huntingtin protein catalyzes caspase activation, initiating apoptotic cascades that lead to selective neuronal death in specific brain regions (Hermel et al. 2004). Furthermore, mutant huntingtin's influence on transcriptional dysregulation disrupts cellular homeostasis, further contributing to apoptotic neuronal demise (Ross and Tabrizi 2011). These mechanisms underscore the intricate apoptotic pathways implicated in neurodegenerative diseases, highlighting potential targets for therapeutic interventions.

**Potential Therapeutic Interventions**

Therapeutic strategies targeting amyloid-beta (Aβ) aggregation aim to mitigate apoptosis in Alzheimer's disease. Aducanumab, an anti-Aβ antibody, has potential to alter the disease course (Sevigny et al. 2016). Approaches targeting hyperphosphorylated tau through antibodies or small molecules aim to prevent tau aggregation, with ongoing research assessing their efficacy (Congdon and Sigurdsson 2018). Immunotherapies enhancing alpha-synuclein clearance in Parkinson's disease aim to reduce apoptosis rates, showing promise in preclinical studies (Teil et al. 2020). Mitochondrial-targeted therapies, like coenzyme Q10 supplementation, aim to prevent apoptosis induced by mitochondrial dysfunction, with clinical trials evaluating their effectiveness (Subramaniam and Chesselet 2013). Strategies employing small interfering RNAs (siRNAs) or antisense oligonucleotides (ASOs) aim to reduce mutant huntingtin protein levels in Huntington's disease, with initial clinical trials focusing on safety and efficacy (Tabrizi et al. 2019). Histone deacetylase (HDAC) inhibitors represent another approach to counteract transcriptional dysregulation in Huntington's disease, with multiple clinical trials investigating their potential therapeutic benefits (Wild and Tabrizi 2017). These diverse therapeutic avenues underscore ongoing efforts to modulate apoptotic pathways implicated in neurodegenerative disorders.

#### 4.3.2.3 Autoimmune Disorders

**Apoptosis in Autoimmune Pathogenesis**

Defective apoptosis in autoimmune disorders often perpetuates autoreactive immune cells' survival, fostering self-reactive lymphocyte persistence and driving the initiation and progression of autoimmune diseases (Strasser et al. 1991). Impaired clearance of apoptotic cells can lead to secondary necrosis, releasing immunogenic cellular contents that trigger inflammatory responses, thereby exacerbating the autoimmune process by releasing pro-inflammatory cytokines (Qing et al. 2008). Dysregulated apoptosis also impacts regulatory T cells (Tregs), compromising their suppressive capacity and promoting autoimmunity (Korn et al. 2007). These mechanisms underscore the critical role of apoptosis in maintaining immune tolerance and the pathogenesis of autoimmune conditions.

#### Immunomodulatory Approaches

Autoimmune diseases arise from dysregulated apoptotic pathways involving Fas/FasL interactions, which can be targeted therapeutically to induce apoptosis in autoreactive T cells using anti-Fas agonistic antibodies (Bodmer et al. 2000). Sensitizing autoreactive T cells to apoptosis via Bcl-2 inhibitors with immunomodulatory agents aims to restore immune tolerance and curb autoimmune responses (Strasser 2005). Tumor Necrosis Factor-alpha (TNF-α) plays a pivotal role in autoimmune pathogenesis, and drugs like Infliximab, an anti-TNF-α antibody, mitigate inflammation and apoptosis in diseases such as Rheumatoid arthritis (Phelan et al. 2008). Lower doses of IL-2 promote T regulatory (Treg) cell survival, suppressing their apoptotic rates and enhancing immunosuppressive functions (Koreth et al. 2011). Efferocytosis, crucial for efficient apoptotic cell clearance, prevents secondary necrosis and autoimmune triggers; strategies modulating receptors like MerTK involved in efferocytosis show promise as immunomodulatory approaches (Krysko and Vandenabeele 2010). Dysregulation of the complement system also contributes to autoimmune pathology; immunomodulation targeting complement activation can mitigate excessive inflammation and apoptotic processes (Ricklin et al. 2010).

### 4.3.2.4 Other Diseases Related to Apoptosis

#### Cardiovascular Disorders

Various regulated cell death (RCD) pathways play a significant role in myocardial infarction. Research using mouse models has provided valuable insights. For instance, mice with cardiomyocyte-specific deletions or overexpression of apoptosis-related genes, such as Bak1, Bax, BCL2, BCL-XL, and Bbc3, show different levels of protection against myocardial ischemia-reperfusion injury (Vaseva et al. 2012). These indicate that apoptosis and mitochondrial permeability transition (MPT)-driven necrosis (Whelan et al. 2012) contribute significantly to tissue damage during such events. The role of caspases in myocardial infarction shows mixed results. Broad-spectrum caspase inhibitors (Mersmann et al. 2008) and XIAP-mimicking peptides (Souktani et al. 2009) provide some level of protection. Yet, the deletion of caspases (Casp3, Casp7) does not offer cardioprotection, showing their absence only delays cell death rather than preventing it (Inserte et al. 2016). The contribution of the intrinsic apoptosis pathway to heart failure and cardiomyopathy is still under debate (Vitale et al. 2023). In a mouse model with desmin deletion, overexpression of BCL2 reduces cardiac damage and hypertrophy, leading to improved cardiac function (Weisleder et al. 2004). However, it also increases necrosis due to alternative cell death pathways. Additionally, mice lacking Casp3 are more susceptible to experimental cardiomyopathy, partly due to impaired pro-survival AKT1 signaling (Khalil et al. 2012) and increased RCD-driven inflammation (Rodriguez-Ruiz et al. 2019).

Therapeutic strategies that target apoptosis-related pathways have shown promise in various rodent models of myocardial dysfunction. Caspase inhibition has demonstrated cardioprotective effects in models induced by endotoxin, burn injury, or hypoxia (Vitale et al. 2023). Similarly, inhibition of BAX prevents doxorubicin-induced cardiotoxicity without affecting its anti-cancer properties (Amgalan et al. 2020), while the endothelial cell-specific expression of BCL2A1A improves survival in heart transplantation models (Smyth et al. 2017).

The complex connection is between apoptosis and atherosclerosis (Vitale et al. 2023). For example, deleting Casp3 promotes plaque development (Grootaert et al. 2016), whereas the absence of CAD protects against the disease (Chao et al. 2016). Deleting Mcl1 conditionally in myeloid cells promotes atherosclerosis (Fontaine et al. 2019), whereas inhibiting BCL-XL through genetic or pharmacological means decreases atherosclerosis by reducing platelet numbers (Lee et al. 2021). Additionally, deleting BIM, specifically in macrophages or leukocytes, has a limited impact on plaque formation, particularly during the initial phases of atherosclerosis (Temmerman et al. 2017).

**Renal Dysfunctions Related to Apoptosis**

Studies using mouse models have highlighted the protective effects of certain genetic deletions against kidney injury. For example, deleting Bax in all cells, specifically in kidney cells, reduces acute kidney damage in mice experiencing experimental renal ischemia (Wei et al. 2013). Similar protective effects are observed in Bid knockout mice (Wei et al. 2006) and transgenic mice with kidney-specific expression of BCL-XL (Chien et al. 2007). Additionally, deleting Bax and Bak1 in kidney proximal tubules reduces tubular apoptosis and improves inflammation and fibrosis in a mouse model of renal fibrosis caused by unilateral ureteral obstruction (Mei et al. 2017).

Caspases, especially those associated with apoptosis, have a notable impact on renal conditions, potentially extending beyond intrinsic apoptosis. Deleting Casp3, for instance, mitigates microvascular damage and fibrosis in mice exposed to ischemia-reperfusion injury (Yang et al. 2018), resulting in improved long-term outcomes (Lan et al. 2021). Moreover, the absence of CASP3 improves survival rates in mice with chronic kidney disease stemming from a congenital cystin 1 (Cys1) mutation (Tao et al. 2008). CASP3-deficient mice exhibit elevated CASP7 levels and reduced BCL2 expression, aligning with clinical findings of decreased BCL2 levels in patients with polycystic kidney disease (Duplomb et al. 2017). Broad-spectrum caspase inhibitors have shown promise in diminishing kidney damage and improving renal function across diverse experimental models, encompassing renal ischemia, polycystic kidney disease, glomerulonephritis, lupus nephritis, and diabetic renal disease (Vitale et al. 2023). These favorable outcomes are likely attributed to decreased post-RCD inflammation rather than direct apoptosis inhibition. However, specific targeting of apoptotic caspases is essential to ascertain whether these effects stem from intrinsic apoptosis inhibition or other RCD pathways

(Belavgeni et al. 2020). Certain investigations highlight constraints associated with caspase inhibition. For example, the wide-ranging inhibitor Z-VAD-FMK may exacerbate renal impairment in cisplatin nephrotoxicity models by impeding cyto-protective autophagy (Herzog et al. 2012). Moreover, Z-VAD-FMK's efficacy is limited in osmotic nephrosis and contrast-induced acute kidney injury, potentially due to its inhibition of CASP8 and facilitation of necroptosis (Linkermann et al. 2013). The involvement of BCL-XL in kidney health is very complex. Abrupt decrease of BCL-XL in all mature mouse tissues, except hematopoietic cells, results in significant renal tubular degeneration and fatal anemia because of declined erythropoietin synthesis (Brinkmann et al. 2020). This emphasizes the importance of maintaining a delicate equilibrium when modulating apoptotic pathways to prevent unintended adverse consequences.

#### Hepatic Diseases Related to Apoptosis

Acute liver injuries and chronic liver diseases, whether linked to alcohol or not, often involve apoptosis. Hepatocytes are especially susceptible to this process due to their high expression of BID. BID is a protein that links death receptor (DR) signaling to mitochondrial outer membrane permeabilization (MOMP) through CASP8-dependent cleavage, making it challenging to differentiate between intrinsic and extrinsic apoptotic pathways (Yin et al. 1999). Research in ischemia-reperfusion injury models indicates that liver protection is significantly enhanced by deleting BID or BCL2L11 (BIM), overexpressing BCL2, or utilizing broad-spectrum caspase inhibitors (DuBray Jr et al. 2015). Mice deficient in BIM exhibit resistance to liver damage from viral hepatitis (Lauer et al. 2012). Furthermore, removing the BIM or PUMA genes (but not BOK) mitigates liver injury induced by acetaminophen toxicity (Chen et al. 2019). Pre-treatment with Z-VAD-FMK improves survival rates after hepatectomy (Yoshida et al. 2007).

In cases of alcohol-related liver disease, the absence of BID provides some protection against fibrosis. However, inflammation and steatosis still occur (Roychowdhury et al. 2012). In non-alcoholic steatohepatitis (NASH), reducing BID levels or using antisense oligonucleotides to target BID decreases liver inflammation and fibrosis (Eguchi et al. 2016). However, BID deficiency does not mitigate liver injury and fibrosis caused by bile duct ligation, which is a model for obstructive cholestasis (Devi Nalapareddy et al. 2009).

Overexpressing MCL1, but not BCL2, provides protection against liver damage resulting from bile duct ligation (Mitchell et al. 2011). Deleting XIAP or CASP3 affects liver damage primarily through inflammatory responses rather than apoptosis (Zilu et al. 2019). Furthermore, removing the MCL1 and p53 genes promotes fibrosis and tumor development, while the deletion of Bak1 decreases liver toxicity (Weng et al. 2011). CASP2 plays a role in non-alcoholic steatohepatitis (NASH) by enhancing lipogenesis and steatohepatitis through the activation of SREBP (Kim et al. 2018). Blocking CASP2 can prevent diet-induced steatosis and the progression of NASH (Machado et al. 2016). Broad-spectrum caspase inhibitors such as emricasan and VX-166 have decreased liver injury, inflammation, and fibrosis across various models (Barreyro et al. 2015). Additionally, CASP3-specific inhibitors offer protection against acute liver injury caused by bile duct ligation (Gracia-Sancho et al. 2019).

### 4.3.3 Clinical Diagnostic Approaches

#### 4.3.3.1 Biomarkers of Apoptosis

- Cleaved caspase-3 is a commonly known biomarker of apoptosis. Active caspase-mediated apoptosis can be identified by increased levels of cleaved caspase-3. This biomarker has been used in the diagnosis of cancer and neurodegenerative diseases (Porter and Jänicke 1999).
- As cytochrome c has a significant contribution to the intrinsic apoptotic pathway, cytochrome c in the cytoplasm acts as a biomarker for detecting apoptosis and mitochondrial dysfunction. This biomarker is widely used in cancer research and neurodegenerative disease studies (Liu et al. 1996).
- The Terminal deoxynucleotidyl transferase dUTP Nick End Labeling (TUNEL) assay can detect these apoptotic DNA fragments. TUNEL is a well-known clinical diagnostics approach used in cancer detection to identify apoptotic cells within tissues (Gavrieli et al. 1992).
- A perfect balance between anti-apoptotic like Bcl-2 and pro-apoptotic like Bax members of the Bcl-2 family plays a vital role in proper apoptosis functioning. The ratio of Bax/Bcl-2 acts as an indicator of apoptotic sensitivity. It is also used to diagnose cancer and evaluate the response to therapy (Reed 1997).
- Annexin V can bind to phosphatidylserine, which is externalized to the non-cytosolic surface of the plasma membrane during apoptotic conditions. It is combined with propidium iodide (PI) to discriminate necrotic cells, early apoptotic and late apoptotic cells using flow cytometry for apoptosis quantification (Fig. 4.5) (Vermes et al. 1995).

#### 4.3.3.2 Imaging Techniques for Apoptotic Cells

- TUNEL Assay: Apoptotic cells have fragmented DNA, which can be visualized by Fluorescence microscopy coupled Terminal deoxynucleotidyl transferase dUTP Nick End Labelling (TUNEL) assay. Fluorescent labels help visualize fragmented DNA to precisely detect apoptotic cells in tissues (Gavrieli et al. 1992).
- Live-Cell Imaging: An early apoptotic event, Phosphatidylserine externalization on the non-cytosolic phase of the plasma membrane, can be monitored in real time using Annexin V staining with fluorophores. This live-cell imaging technique can detect the progression of apoptosis in living cells (Vermes et al. 1995).
- Mitochondrial Imaging: Confocal microscopy coupled with mitochondrial-specific dyes is used for mitochondrial imaging to visualize mitochondrial morphology and dynamics during apoptosis (Kuznetsov et al. 2010).
- Magnetic Resonance Imaging (MRI): Non-invasive detection of apoptotic cells can be done in vivo using Magnetic Resonance Imaging (MRI) with Annexin

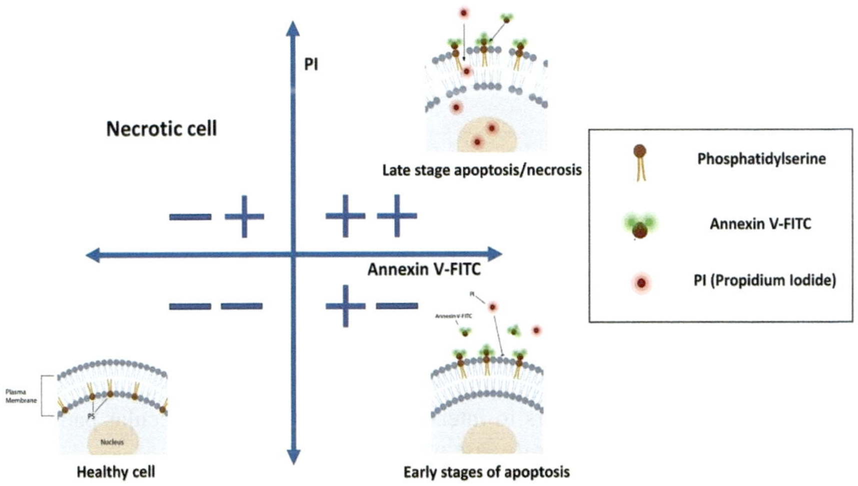


**Fig. 4.5** Flow cytometric analysis of apoptotic cells

V-functionalized contrast agents. This technique is widely used in clinical diagnostics (Zeng et al. 2015).

- Positron Emission Tomography (PET): Apoptotic cell density can be quantified in vivo using Positron Emission Tomography (PET) imaging coupled with radiolabeled Annexin V. One of the well-known Annexin V with PET tracers is 18F, which is used for sensitive and quantitative imaging of apoptosis (Ghanem-Zoubi et al. 2023).
- Multiphoton Microscopy: One of the most advanced imaging techniques is Multiphoton microscopy, which uses non-linear optical processes for imaging biological structures. The advantage of using this technique is reduced phototoxicity and enhanced imaging depth. These features are useful for real-time visualization of apoptotic in tissues (Zipfel et al. 2003).
- Fluorescence Lifetime Imaging Microscopy (FLIM): The fluorescence decay rate of fluorophores is measured by an advanced fluorescence microscopy technique known as FLIM. This technique provides information about protein–protein interactions and cellular metabolism during apoptosis (Becker 2012).

#### 4.3.3.3 Flow Cytometry and Apoptosis Detection

##### Flow Cytometry

- Flow cytometry is a high-throughput quantitative analytical technique widely used to detect apoptotic cells based on different cellular parameters. Phosphatidylserine exposure is a crucial hallmark of apoptotic cells. Annexin-V staining coupled with propidium iodide (PI) helps to distinguish different apoptotic stages (Fig. 4.5). Annexin V-positive and Propidium Iodide (PI)-negative result is shown by early apoptotic cells, while Annexin V-positive and Propidium Iodide (PI)-positive result is shown by late apoptotic factors (Vermes et al. 1995).

Mitochondrial membrane potential ($\Delta\Psi m$) is also measured by flow cytometry using specialized dyes such as JC-1 or TMRE. These dyes help to assess mitochondrial depolarization, which is an indicator of apoptosis (Poot et al. 1996).

## 4.4 Conclusion

The intricate molecular mechanism of apoptosis is pivotal in both physiological processes and therapeutic interventions. It is a fundamental process regulating cellular life and death, maintaining cellular homeostasis through a delicate balance between pro-apoptotic and anti-apoptotic factors. However, any slight disruption in this balance can lead to the pathogenesis of various diseases, such as neurodegenerative disorders, cancer, and autoimmune diseases. Therapeutic strategies aim to target specific apoptotic factors to intervene in dysregulated molecular pathways, offering promising avenues for disease management.

Current advancements in clinical diagnostics are enhancing our understanding and management of apoptosis-related diseases. Advanced imaging techniques and the identification of biomarkers are crucial in accurately assessing disease states and guiding therapeutic decisions. Despite these advancements, much remains to be explored within the complex molecular pathways of apoptosis. Future research directions should focus on developing novel therapeutic strategies and refining diagnostic approaches to achieve better clinical outcomes.

This chapter summarizes the foundational concepts of apoptosis, providing insights into its mechanisms and clinical implications. By highlighting various therapeutic and diagnostic approaches, it aims to underscore the importance of ongoing research efforts in unraveling the complexities of apoptosis. Continuing exploration of apoptosis-related pathways promises to open new avenues for improved treatment strategies and personalized medicine approaches tailored to individual patient needs.

## References


Adams JM, Cory S (2007) The Bcl-2 apoptotic switch in cancer development and therapy. Oncogene 26(9):1324–1337

Alberts B, Johnson A, Lewis J, Morgan D, Raff M, Roberts K, Walter P (2014) Molecular biology of the cell, 6th edn. Garland Science, New York

Amgalan D, Garner TP, Pekson R, Jia XF, Yanamandala M, Paulino V, Kitsis RN (2020) A small-molecule allosteric inhibitor of BAX protects against doxorubicin-induced cardiomyopathy. Nat Cancer 1(3):315–328

An W, Liu M, Cao S (2019) Apoptotic pathway as the therapeutic target for anticancer traditional Chinese medicines. Front Pharmacol 10:464828

Ashkenazi A (2002) Targeting death and decoy receptors of the tumour-necrosis factor superfamily. Nat Rev Cancer 2(6):420–430

Ashkenazi A, Dixit VM (1998) Death receptors: signaling and modulation. Science 281(5381):1305–1308

Auner HW, Beham-Schmid C, Dillon N, Sabbattini P (2010) The life span of short-lived plasma cells is partly determined by a block on activation of apoptotic caspases acting in combination with endoplasmic reticulum stress. Blood 116(18):3445–3455
Barreyro FJ, Holod S, Finocchietto PV, Camino AM, Aquino JB, Avagnina A, Gores GJ (2015) The pan-caspase inhibitor Emricasan (IDN-6556) decreases liver injury and fibrosis in a murine model of non-alcoholic steatohepatitis. Liver Int 35(3):953–966
Becker W (2012) Fluorescence lifetime imaging–techniques and applications. J Microsc 247(2):119–136
Belavgeni A, Meyer C, Stumpf J, Hugo C, Linkermann A (2020) Ferroptosis and necroptosis in the kidney. Cell Chem Biol 27(4):448–462
Bertrand MJ, Doiron K, Labbé K, Korneluk RG, Barker PA, Saleh M (2009) Cellular inhibitors of apoptosis cIAP1 and cIAP2 are required for innate immunity signaling by the pattern recognition receptors NOD1 and NOD2. Immunity 30(6):789–801
Boatright KM, Renatus M, Scott FL, Sperandio S, Shin H, Pedersen IM, Salvesen GS (2003) A unified model for apical caspase activation. Mol Cell 11(2):529–541
Bodmer JL, Holler N, Reynard S, Vinciguerra P, Schneider P, Juo P, Tschopp J (2000) TRAIL receptor-2 signals apoptosis through FADD and caspase-8. Nat Cell Biol 2(4):241–243
Boya P, Kroemer G (2008) Lysosomal membrane permeabilization in cell death. Oncogene 27(50):6434–6451
Brinkmann K, Waring P, Glaser SP, Wimmer V, Cottle DL, Tham MS, Strasser A (2020) BCL-XL exerts a protective role against anemia caused by radiation-induced kidney damage. EMBO J 39(24):e105561
Chao ML, Guo J, Cheng WL, Zhu XY, She ZG, Huang Z et al (2016) Loss of caspase-activated DNase protects against atherosclerosis in apolipoprotein E–deficient mice. J Am Heart Assoc 5(12):e004362
Chen L, Willis SN, Wei A, Smith BJ, Fletcher JI, Hinds MG, Huang DC (2005) Differential targeting of prosurvival Bcl-2 proteins by their BH3-only ligands allows complementary apoptotic function. Mol Cell 17(3):393–403
Chen D, Ni HM, Wang L, Ma X, Yu J, Ding WX, Zhang L (2019) p53 up-regulated modulator of apoptosis induction mediates acetaminophen-induced necrosis and liver injury in mice. Hepatology 69(5):2164–2179
Chiara F, Gambalunga A, Sciacovelli M, Nicolli A, Ronconi L, Fregona D, Trevisan A (2012) Chemotherapeutic induction of mitochondrial oxidative stress activates GSK-3α/β and Bax, leading to permeability transition pore opening and tumor cell death. Cell Death Dis 3(12):e444–e444
Chien CT, Shyue SK, Lai MK (2007) Bcl-xL augmentation potentially reduces ischemia/reperfusion induced proximal and distal tubular apoptosis and autophagy. Transplantation 84(9):1183–1190
Chipuk JE, McStay GP, Bharti A, Kuwana T, Clarke CJ, Siskind LJ, Green DR (2012) Sphingolipid metabolism cooperates with BAK and BAX to promote the mitochondrial pathway of apoptosis. Cell 148(5):988–1000
Coleman ML, Sahai EA, Yeo M, Bosch M, Dewar A, Olson MF (2001) Membrane blebbing during apoptosis results from caspase-mediated activation of ROCK I. Nat Cell Biol 3(4):339–345
Congdon EE, Sigurdsson EM (2018) Tau-targeting therapies for Alzheimer disease. Nat Rev Neurol 14(7):399–415
Curtin JF, Cotter TG (2003) Apoptosis: historical perspectives. Essays Biochem 39:1–10
Czabotar PE, Lessene G, Strasser A, Adams JM (2014) Control of apoptosis by the BCL-2 protein family: implications for physiology and therapy. Nat Rev Mol Cell Biol 15(1):49–63
D'Amelio M, Tino E, Cecconi F (2008) The apoptosome: emerging insights and new potential targets for drug design. Pharm Res 25:740–751
D'arcy MS (2019) Cell death: a review of the major forms of apoptosis, necrosis and autophagy. Cell Biol Int 43(6):582–592
Danial NN, Korsmeyer SJ (2004) Cell death: critical control points. Cell 116(2):205–219

De Robertis EM (2006) Spemann's organizer and self-regulation in amphibian embryos. Nat Rev Mol Cell Biol 7(4):296–302
Deveraux QL, Takahashi R, Salvesen GS, Reed JC (1997) X-linked IAP is a direct inhibitor of cell-death proteases. Nature 388(6639):300–304
Devi Nalapareddy P, Schüngel S, Hong JY, Manns MP, Jaeschke H, Vogel A (2009) The BH3-only protein bid does not mediate death-receptor-induced liver injury in obstructive cholestasis. Am J Pathol 175(3):1077–1085
Dewson G, Kratina T, Czabotar P, Day CL, Adams JM, Kluck RM (2009) Bak activation for apoptosis involves oligomerization of dimers via their α6 helices. Mol Cell 36(4):696–703
Dickens LS, Boyd RS, Jukes-Jones R, Hughes MA, Robinson GL, Fairall L, MacFarlane M (2012) A death effector domain chain DISC model reveals a crucial role for caspase-8 chain assembly in mediating apoptotic cell death. Mol Cell 47(2):291–305
DuBray BJ Jr, Conzen KD, Upadhya GA, Gunter KL, Jia J, Knolhoff BL et al (2015) BH3-only proteins contribute to steatotic liver ischemia–reperfusion injury. J Surg Res 194(2):653–658
Duplomb L, Droin N, Bouchot O, Thauvin-Robinet C, Bruel AL, Thevenon J, Faivre L (2017) A constitutive BCL2 down-regulation aggravates the phenotype of PKD1-mutant-induced polycystic kidney disease. Hum Mol Genet 26(23):4680–4688
Eguchi A, Du Jeu XDM, Johnson CD, Nektaria A, Feldstein AE (2016) Liver Bid suppression for treatment of fibrosis associated with non-alcoholic steatohepatitis. J Hepatol 64(3):699–707
Elmore S (2007) Apoptosis: a review of programmed cell death. Toxicol Pathol 35(4):495–516
Emily HYC, Wei MC, Weiler S, Flavell RA, Mak TW, Lindsten T, Korsmeyer SJ (2001) BCL-2, BCL-XL sequester BH3 domain-only molecules preventing BAX-and BAK-mediated mitochondrial apoptosis. Mol Cell 8(3):705–711
Flaherty KT, Puzanov I, Kim KB, Ribas A, McArthur GA, Sosman JA et al (2010) Inhibition of mutated, activated BRAF in metastatic melanoma. N Engl J Med 363(9):809–819
Fontaine MA, Westra MM, Bot I, Jin H, Franssen AJ, Bot M, Biessen EA (2019) Low human and murine Mcl-1 expression leads to a pro-apoptotic plaque phenotype enriched in giant-cells. Sci Rep 9(1):14547
Fu TM, Li Y, Lu A, Li Z, Vajjhala PR, Cruz AC, Wu H (2016) Cryo-EM structure of caspase-8 tandem DED filament reveals assembly and regulation mechanisms of the death-inducing signaling complex. Mol Cell 64(2):236–250
Fuentes-Prior P, Salvesen GS (2004) The protein structures that shape caspase activity, specificity, activation and inhibition. Biochem J 384(2):201–232
Galluzzi L, Vitale I, Aaronson SA, Abrams JM, Adam D, Agostinis P, Turk B (2018) Molecular mechanisms of cell death: recommendations of the Nomenclature Committee on Cell Death 2018. Cell Death Differ 25(3):486–541
Gavrieli Y, Sherman Y, Ben-Sasson SA (1992) Identification of programmed cell death in situ via specific labeling of nuclear DNA fragmentation. J Cell Biol 119(3):493–501
Gendron TF, Petrucelli L (2009) The role of tau in neurodegeneration. Mol Neurodegener 4:1–19
Ghanem-Zoubi N, Abu-Elhija J, Kagna O, Mustafa-Hellou M, Qasum M, Haber D, Keidar Z (2023) Predictors of infectious foci on FDG PET/CT in Staphylococcus aureus bacteremia. Sci Rep 13(1):14063
Gracia-Sancho J, Manicardi N, Ortega-Ribera M, Maeso-Díaz R, Guixé-Muntet S, Fernández-Iglesias A, Bosch J (2019) Emricasan ameliorates portal hypertension and liver fibrosis in cirrhotic rats through a hepatocyte-mediated paracrine mechanism. Hepatol Commun 3(7):987–1000
Green DR (2019) The coming decade of cell death research: five riddles. Cell 177(5):1094–1107
Green DR, Llambi F (2015) Cell death signaling. Cold Spring Harb Perspect Biol 7(12):a006080
Grootaert MO, Schrijvers DM, Hermans M, Van Hoof VO, De Meyer GR, Martinet W (2016) Caspase-3 deletion promotes necrosis in atherosclerotic plaques of ApoE knockout mice. Oxidative Med Cell Longev 2016:3087469
Große L, Wurm CA, Brüser C, Neumann D, Jans DC, Jakobs S (2016) Bax assembles into large ring-like structures remodeling the mitochondrial outer membrane in apoptosis. EMBO J 35(4):402–413

Hamburger V, Hamilton HL (1951) A series of normal stages in the development of the chick embryo. J Morphol 88(1):49–92
Hanahan D, Weinberg RA (2000) The hallmarks of cancer. Cell 100(1):57–70
Hanahan D, Weinberg RA (2011) Hallmarks of cancer: the next generation. Cell 144(5):646–674
Heath-Engel HM, Wang B, Shore GC (2012) Bcl2 at the endoplasmic reticulum protects against a Bax/Bak-independent paraptosis-like cell death pathway initiated via p20Bap31. Biochim Biophys Acta (BBA) Mol Cell Res 1823(2):335–347
Helgren ME, Cliffer KD, Torrento K, Cavnor C, Curtis R, DiStefano PS, Lindsay RM (1997) Neurotrophin-3 administration attenuates deficits of pyridoxine-induced large-fiber sensory neuropathy. J Neurosci 17(1):372–382
Hermel E, Gafni J, Propp SS, Leavitt BR, Wellington CL, Young JE, Ellerby LM (2004) Specific caspase interactions and amplification are involved in selective neuronal vulnerability in Huntington's disease. Cell Death Differ 11(4):424–438
Herzog C, Yang C, Holmes A, Kaushal GP (2012) zVAD-fmk prevents cisplatin-induced cleavage of autophagy proteins but impairs autophagic flux and worsens renal function. Am J Physiol Renal Physiol 303(8):F1239–F1250
Hetz C, Mollereau B (2014) Disturbance of endoplasmic reticulum proteostasis in neurodegenerative diseases. Nat Rev Neurosci 15(4):233–249
Hockenbery DM, Oltvai ZN, Yin XM, Milliman CL, Korsmeyer SJ (1993) Bcl-2 functions in an antioxidant pathway to prevent apoptosis. Cell 75(2):241–251
Hsu YT, Youle RJ (1997) Nonionic detergents induce dimerization among members of the Bcl-2 family. J Biol Chem 272(21):13829–13834
Inserte J, Cardona M, Poncelas-Nozal M, Hernando V, Vilardosa Ú, Aluja D, Garcia-Dorado D (2016) Studies on the role of apoptosis after transient myocardial ischemia: genetic deletion of the executioner caspases-3 and-7 does not limit infarct size and ventricular remodeling. Basic Res Cardiol 111:1–10
Irmler M, Thome M, Hahne M, Schneider P, Hofmann K, Steiner V, Tschopp J (1997) Inhibition of death receptor signals by cellular FLIP. Nature 388(6638):190–195
Jackson SP, Bartek J (2009) The DNA-damage response in human biology and disease. Nature 461(7267):1071–1078
Jassem W, Heaton ND (2004) The role of mitochondria in ischemia/reperfusion injury in organ transplantation. Kidney Int 66(2):514–517
Kataoka T, Tschopp J (2004) N-terminal fragment of c-FLIP (L) processed by caspase 8 specifically interacts with TRAF2 and induces activation of the NF-κBsignaling pathway. Mol Cell Biol 24:2627
Kaufmann T, Schlipf S, Sanz J, Neubert K, Stein R, Borner C (2003) Characterization of the signal that directs Bcl-xL, but not Bcl-2, to the mitochondrial outer membrane. J Cell Biol 160(1):53–64
Kerr JF, Wyllie AH, Currie AR (1972) Apoptosis: a basic biological phenomenon with wideranging implications in tissue kinetics. Br J Cancer 26(4):239–257
Khalil H, Peltzer N, Walicki J, Yang JY, Dubuis G, Gardiol N, Widmann C (2012) Caspase-3 protects stressed organs against cell death. Mol Cell Biol 32(22):4523–4533
Kim JY, Garcia-Carbonell R, Yamachika S, Zhao P, Dhar D, Loomba R, Karin M (2018) ER stress drives lipogenesis and steatohepatitis via caspase-2 activation of S1P. Cell 175(1):133–145
Kischkel FC, Hellbardt S, Behrmann I, Germer M, Pawlita M, Krammer PH, Peter ME (1995) Cytotoxicity-dependent APO-1 (Fas/CD95)-associated proteins form a death-inducing signaling complex (DISC) with the receptor. EMBO J 14(22):5579–5588
Koreth J, Matsuoka KI, Kim HT, McDonough SM, Bindra B, Alyea EP III, Soiffer RJ (2011) Interleukin-2 and regulatory T cells in graft-versus-host disease. N Engl J Med 365(22):2055–2066
Korn T, Anderson AC, Bettelli E, Oukka M (2007) The dynamics of effector T cells and Foxp3+ regulatory T cells in the promotion and regulation of autoimmune encephalomyelitis. J Neuroimmunol 191(1–2):51–60

Korsmeyer SJ (2007) Pro-apoptotic cascade activates BID, which oligomerizes BAK or BAX into pores that result in the release of cytochrome c. Cell Death Differ 11:66

Krysko DV, Vandenabeele P (2010) Clearance of dead cells: mechanisms, immune responses and implication in the development of diseases. Apoptosis 15:995–997

Kuznetsov AV, Hermann M, Troppmair J, Margreiter R, Hengster P (2010) Complex patterns of mitochondrial dynamics in human pancreatic cells revealed by fluorescent confocal imaging. J Cell Mol Med 14(1–2):417–425

Labi V, Erlacher M, Kiessling S, Villunger A (2006) BH3-only proteins in cell death initiation, malignant disease and anticancer therapy. Cell Death Differ 13(8):1325–1338

Lan S, Yang B, Migneault F, Turgeon J, Bourgault M, Dieudé M, Hébert MJ (2021) Caspase-3-dependent peritubular capillary dysfunction is pivotal for the transition from acute to chronic kidney disease after acute ischemia-reperfusion injury. Am J Physiol Renal Physiol 321(3):F335–F351

Lauer C, Brunner T, Corazza N (2012) The proapoptotic Bcl-2 family member Bim plays a central role during the development of virus-induced hepatitis. J Immunol 188(2):916–922

Lavrik IN (2010) Systems biology of apoptosis signaling networks. Curr Opin Biotechnol 21(4):551–555

Lazebnik YA, Kaufmann SH, Desnoyers S, Poirier GG, Earnshaw WC (1994) Cleavage of poly (ADP-ribose) polymerase by a proteinase with properties like ICE. Nature 371(6495):346–347

Lee MK, Kraakman MJ, Dragoljevic D, Hanssen NM, Flynn MC, Al-Sharea A, Murphy AJ (2021) Apoptotic ablation of platelets reduces atherosclerosis in mice with diabetes. Arterioscler Thromb Vasc Biol 41(3):1167–1178

Letai A, Bassik MC, Walensky LD, Sorcinelli MD, Weiler S, Korsmeyer SJ (2002) Distinct BH3 domains either sensitize or activate mitochondrial apoptosis, serving as prototype cancer therapeutics. Cancer Cell 2(3):183–192

Li P, Nijhawan D, Budihardjo I, Srinivasula SM, Ahmad M, Alnemri ES, Wang X (1997) Cytochrome c and dATP-dependent formation of Apaf-1/caspase-9 complex initiates an apoptotic protease cascade. Cell 91(4):479–489

Li H, Zhu H, Xu CJ, Yuan J (1998) Cleavage of BID by caspase 8 mediates the mitochondrial damage in the Fas pathway of apoptosis. Cell 94(4):491–501

Linkermann A, Heller JO, Prókai Á, Weinberg JM, De Zen F, Himmerkus N, Krautwald S (2013) The RIP1-kinase inhibitor necrostatin-1 prevents osmotic nephrosis and contrast-induced AKI in mice. J Am Soc Nephrol 24(10):1545–1557

Liston P, Fong WG, Korneluk RG (2003) The inhibitors of apoptosis: there is more to life than Bcl2. Oncogene 22(53):8568–8580

Liu X, Kim CN, Yang J, Jemmerson R, Wang X (1996) Induction of apoptotic program in cell-free extracts: requirement for dATP and cytochrome c. Cell 86(1):147–157

Luo X, Budihardjo I, Zou H, Slaughter C, Wang X (1998) Bid, a Bcl2 interacting protein, mediates cytochrome c release from mitochondria in response to activation of cell surface death receptors. Cell 94(4):481–490

Machado MV, Michelotti GA, Jewell ML, Pereira TA, Xie G, Premont RT, Diehl AM (2016) Caspase-2 promotes obesity, the metabolic syndrome and nonalcoholic fatty liver disease. Cell Death Dis 7(2):e2096–e2096

Martin GR (1998) The roles of FGFs in the early development of vertebrate limbs. Genes Dev 12(11):1571–1586

Martin LJ (2010) Mitochondrial and cell death mechanisms in neurodegenerative diseases. Pharmaceuticals 3(4):839–915

Mattson MP, Kroemer G (2003) Mitochondria in cell death: novel targets for neuroprotection and cardioprotection. Trends Mol Med 9(5):196–205

Mei S, Li L, Wei Q, Hao J, Su Y, Mei C, Dong Z (2017) Double knockout of Bax and Bak from kidney proximal tubules reduces unilateral urethral obstruction associated apoptosis and renal interstitial fibrosis. Sci Rep 7(1):44892

Melchers F (2015) Checkpoints that control B cell development. J Clin Invest 125(6):2203–2210

Mersmann J, Zacharowski PA, Schmitz I, Zacharowski K (2008) Caspase inhibitor zVAD. Fmk reduces infarct size after myocardial ischaemia and reperfusion in rats but not in mice. Resuscitation 79(3):468–474
Mitchell C, Mahrouf-Yorgov M, Mayeuf A, Robin MA, Mansouri A, Fromenty B, Gilgenkrantz H (2011) Overexpression of Bcl-2 in hepatocytes protects against injury but does not attenuate fibrosis in a mouse model of chronic cholestatic liver disease. Lab Investig 91(2):273–282
Nagata S (1994) Fas and Fas ligand: a death factor and its receptor. Adv Immunol 57:129–144
Nagata S (2018) Apoptosis and clearance of apoptotic cells. Annu Rev Immunol 36:489–517
Oltersdorf T, Elmore SW, Shoemaker AR, Armstrong RC, Augeri DJ, Belli BA et al (2005) An inhibitor of Bcl-2 family proteins induces regression of solid tumours. Nature 435(7042):677–681
Oltval ZN, Milliman CL, Korsmeyer SJ (1993) Bcl-2 heterodimerizes in vivo with a conserved homolog, Bax, that accelerates programed cell death. Cell 74(4):609–619
Pang Y, Bai XC, Yan C, Hao Q, Chen Z, Wang JW et al (2015) Structure of the apoptosome: mechanistic insights into activation of an initiator caspase from drosophila. Genes Dev 29(3):277–287
Pardoll DM (2012) The blockade of immune checkpoints in cancer immunotherapy. Nat Rev Cancer 12(4):252–264
Pasparakis M, Vandenabeele P (2015) Necroptosis and its role in inflammation. Nature 517(7534):311–320
Phelan JD, Orekov T, Finkelman FD (2008) Cutting edge: mechanism of enhancement of in vivo cytokine effects by anti-cytokine monoclonal antibodies. J Immunol 180(1):44–48
Poot M, Zhang YZ, Krämer JA, Wells KS, Jones LJ, Hanzel DK et al (1996) Analysis of mitochondrial morphology and function with novel fixable fluorescent stains. J Histochem Cytochem 44(12):1363–1372
Porter AG, Jänicke RU (1999) Emerging roles of caspase-3 in apoptosis. Cell Death Differ 6(2):99–104
Potten CS (1990) A comprehensive study of the radiobiological response of the murine (BDF1) small intestine. Int J Radiat Biol 58(6):925–973
Puthalakath H, Villunger A, O'Reilly LA, Beaumont JG, Coultas L, Cheney RE et al (2001) Bmf: a proapoptotic BH3-only protein regulated by interaction with the myosin V actin motor complex, activated by anoikis. Science 293(5536):1829–1832
Qing X, Pitashny M, Thomas DB, Barrat FJ, Hogarth MP, Putterman C (2008) Pathogenic anti-DNA antibodies modulate gene expression in mesangial cells: involvement of HMGB1 in anti-DNA antibody-induced renal injury. Immunol Lett 121(1):61–73
Reddy PH (2009) Amyloid beta, mitochondrial structural and functional dynamics in Alzheimer's disease. Exp Neurol 218(2):286–292
Reed JC (1997) Double identity for proteins of the Bcl-2 family. Nature 387(6635):773–776
Repnik U, Česen MH, Turk B (2014) Lysosomal membrane permeabilization in cell death: concepts and challenges. Mitochondrion 19:49–57
Ricklin D, Hajishengallis G, Yang K, Lambris JD (2010) Complement: a key system for immune surveillance and homeostasis. Nat Immunol 11(9):785–797
Riedl SJ, Renatus M, Schwarzenbacher R, Zhou Q, Sun C, Fesik SW, Salvesen GS (2001) Structural basis for the inhibition of caspase-3 by XIAP. Cell 104(5):791–800
Rodriguez-Ruiz ME, Buqué A, Hensler M, Chen J, Bloy N, Petroni G, Galluzzi L (2019) Apoptotic caspases inhibit abscopal responses to radiation and identify a new prognostic biomarker for breast cancer patients. Onco Targets Ther 8(11):e1655964
Ross CA, Tabrizi SJ (2011) Huntington's disease: from molecular pathogenesis to clinical treatment. Lancet Neurol 10(1):83–98
Roychowdhury S, Chiang DJ, Mandal P, McMullen MR, Liu X, Cohen JI, Nagy LE (2012) Inhibition of apoptosis protects mice from ethanol-mediated acceleration of early markers of CC l4-induced fibrosis but not steatosis or inflammation. Alcohol Clin Exp Res 36(7):1139–1147
Schittenhelm MM, Kaiser M, Győrffy B, Kampa-Schittenhelm KM (2024) Evaluation of apoptosis stimulating protein of TP53-1 (ASPP1/PPP1R13B) to predict therapy resistance and overall survival in acute myeloid leukemia (AML). Cell Death Dis 15(1):25

Schneider-Brachert W, Tchikov V, Neumeyer J, Jakob M, Winoto-Morbach S, Held-Feindt J, Schütze S (2004) Compartmentalization of TNF receptor 1 signaling: internalized TNF receptosomes as death signaling vesicles. Immunity 21(3):415–428

Serhan CN, Savill J (2005) Resolution of inflammation: the beginning programs the end. Nat Immunol 6(12):1191–1197

Sevigny J, Chiao P, Bussière T, Weinreb PH, Williams L, Maier M, Sandrock A (2016) The antibody aducanumab reduces Aβ plaques in Alzheimer's disease. Nature 537(7618):50–56

Shaltiel IA, Aprelia M, Saurin AT, Chowdhury D, Kops GJ, Voest EE, Medema RH (2014) Distinct phosphatases antagonize the p53 response in different phases of the cell cycle. Proc Natl Acad Sci 111(20):7313–7318

Smyth LA, Meader L, Xiao F, Woodward M, Brady HJM, Lechler R, Lombardi G (2017) Constitutive expression of the anti-apoptotic Bcl-2 family member A1 in murine endothelial cells leads to transplant tolerance. Clin Exp Immunol 188(2):219–225

Souktani R, Pons S, Guegan C, Bouhidel O, Bruneval P, Zini R, Ghaleh B (2009) Cardioprotection against myocardial infarction with PTD-BIR3/RING, a XIAP mimicking protein. J Mol Cell Cardiol 46(5):713–718

Strasser A (2005) The role of BH3-only proteins in the immune system. Nat Rev Immunol 5(3):189–200

Strasser A, Vaux DL (2020) Cell death in the origin and treatment of cancer. Mol Cell 78(6):1045–1054

Strasser A, Whittingham S, Vaux DL, Bath ML, Adams JM, Cory S, Harris AW (1991) Enforced BCL2 expression in B-lymphoid cells prolongs antibody responses and elicits autoimmune disease. Proc Natl Acad Sci 88(19):8661–8665

Subramaniam SR, Chesselet MF (2013) Mitochondrial dysfunction and oxidative stress in Parkinson's disease. Prog Neurobiol 106:17–32

Tabrizi SJ, Ghosh R, Leavitt BR (2019) Huntingtin lowering strategies for disease modification in Huntington's disease. Neuron 101(5):801–819

Takada E, Okubo K, Yano Y, Iida K, Someda M, Hirasawa A, Matsuzaki K (2020) Molecular mechanism of apoptosis by amyloid β-protein fibrils formed on neuronal cells. ACS Chem Neurosci 11(5):796–805

Tao Y, Zafar I, Kim J, Schrier RW, Edelstein CL (2008) Caspase-3 gene deletion prolongs survival in polycystic kidney disease. J Am Soc Nephrol 19(4):749–755

Teil M, Arotcarena ML, Faggiani E, Laferriere F, Bezard E, Dehay B (2020) Targeting α-synuclein for PD therapeutics: a pursuit on all fronts. Biomol Ther 10(3):391

Temmerman L, Westra MM, Bot I, van Vlijmen BJ, Van Bree N, Bot M et al (2017) Leukocyte Bim deficiency does not impact atherogenesis in ldlr−/− mice, despite a pronounced induction of autoimmune inflammation. Sci Rep 7(1):3086

Twig G, Elorza A, Molina AJ, Mohamed H, Wikstrom JD, Walzer G, Shirihai OS (2008) Fission and selective fusion govern mitochondrial segregation and elimination by autophagy. EMBO J 27(2):433–446

Varfolomeev E, Blankenship JW, Wayson SM, Fedorova AV, Kayagaki N, Garg P et al (2007) IAP antagonists induce autoubiquitination of c-IAPs, NF-κB activation, and TNFα-dependent apoptosis. Cell 131(4):669–681

Vaseva AV, Marchenko ND, Ji K, Tsirka SE, Holzmann S, Moll UM (2012) p53 opens the mitochondrial permeability transition pore to trigger necrosis. Cell 149(7):1536–1548

Vassilev LT, Vu BT, Graves B, Carvajal D, Podlaski F, Filipovic Z, Liu EA (2004) In vivo activation of the p53 pathway by small-molecule antagonists of MDM2. Science 303(5659):844–848

Vermes I, Haanen C, Steffens-Nakken H, Reutellingsperger C (1995) A novel assay for apoptosis flow cytometric detection of phosphatidylserine expression on early apoptotic cells using fluorescein labelled annexin V. J Immunol Methods 184(1):39–51

Villunger A, Michalak EM, Coultas L, Mullauer F, Böck G, Ausserlechner MJ, Strasser A (2003) p53-and drug-induced apoptotic responses mediated by BH3-only proteins puma and noxa. Science 302(5647):1036–1038

Vitale I, Pietrocola F, Guilbaud E, Aaronson SA, Abrams JM, Adam D, Pasparakis M (2023) Apoptotic cell death in disease—current understanding of the NCCD 2023. Cell Death Differ 30(5):1097–1154
Vousden KH, Prives C (2009) Blinded by the light: the growing complexity of p53. Cell 137(3):413–431
Wakatsuki S, Tokunaga S, Shibata M, Araki T (2017) GSK3B-mediated phosphorylation of MCL1 regulates axonal autophagy to promote Wallerian degeneration. J Cell Biol 216(2):477–493
Wang D, Wan X (2023) Progress in the study of molecular mechanisms of cell pyroptosis in tumor therapy. Int Immunopharmacol 118:110143
Watson EC, Koenig MN, Grant ZL, Whitehead L, Trounson E, Dawson G, Coultas L (2016) Apoptosis regulates endothelial cell number and capillary vessel diameter but not vessel regression during retinal angiogenesis. Development 143(16):2973–2982
Wei MC, Zong WX, Cheng EHY, Lindsten T, Panoutsakopoulou V, Ross AJ, Korsmeyer SJ (2001) Proapoptotic BAX and BAK: a requisite gateway to mitochondrial dysfunction and death. Science 292(5517):727–730
Wei Q, Yin XM, Wang MH, Dong Z (2006) Bid deficiency ameliorates ischemic renal failure and delays animal death in C57BL/6 mice. Am J Physiol Renal Physiol 290(1):F35–F42
Wei Q, Dong G, Chen JK, Ramesh G, Dong Z (2013) Bax and Bak have critical roles in ischemic acute kidney injury in global and proximal tubule–specific knockout mouse models. Kidney Int 84(1):138–148
Weisleder N, Taffet GE, Capetanaki Y (2004) Bcl-2 overexpression corrects mitochondrial defects and ameliorates inherited desmin null cardiomyopathy. Proc Natl Acad Sci 101(3):769–774
Weng SY, Yang CY, Li CC, Sun TP, Tung SY, Yen JJY, Yang-Yen HF (2011) Synergism between p53 and Mcl-1 in protecting from hepatic injury, fibrosis and cancer. J Hepatol 54(4):685–694
Westphal D, Dewson G, Czabotar PE, Kluck RM (2011) Molecular biology of Bax and Bak activation and action. Biochim Biophys Acta (BBA) Mol Cell Res 1813(4):521–531
Whelan RS, Konstantinidis K, Wei AC, Chen Y, Reyna DE, Jha S et al (2012) Bax regulates primary necrosis through mitochondrial dynamics. Proc Natl Acad Sci 109(17):6566–6571
Wild EJ, Tabrizi SJ (2017) Therapies targeting DNA and RNA in Huntington's disease. Lancet Neurol 16(10):837–847
Yang B, Lan S, Dieudé M, Sabo-Vatasescu JP, Karakeussian-Rimbaud A, Turgeon J, Hébert MJ (2018) Caspase-3 is a pivotal regulator of microvascular rarefaction and renal fibrosis after ischemia-reperfusion injury. J Am Soc Nephrol 29(7):1900–1916
Yin XM, Wang K, Gross A, Zhao Y, Zinkel S, Klocke B, Korsmeyer SJ (1999) Bid-deficient mice are resistant to Fas-induced hepatocellular apoptosis. Nature 400(6747):886–891
Yoshida N, Iwata H, Yamada T, Sekino T, Matsuo H, Shirahashi K, Takemura H (2007) Improvement of the survival rate after rat massive hepatectomy due to the reduction of apoptosis by caspase inhibitor. J Gastroenterol Hepatol 22(11):2015–2021
Youle RJ, Van Der Bliek AM (2012) Mitochondrial fission, fusion, and stress. Science 337(6098):1062–1065
Yu C, Minemoto Y, Zhang J, Liu J, Tang F, Bui TN, Lin A (2004) JNK suppresses apoptosis via phosphorylation of the proapoptotic Bcl-2 family protein BAD. Mol Cell 13(3):329–340
Yuan J, Yankner BA (2000) Apoptosis in the nervous system. Nature 407(6805):802–809
Yuan Y, Zheng Y, Zhang X, Chen Y, Wu X, Wu J, Chen Z (2017) BNIP3L/NIX-mediated mitophagy protects against ischemic brain injury independent of PARK2. Autophagy 13(10):1754–1766
Zeng W, Wang X, Xu P, Liu G, Eden HS, Chen X (2015) Molecular imaging of apoptosis: from micro to macro. Theranostics 5(6):559
Zilu S, Qian H, Haibin W, Chenxu G, Deshuai L, Qiang L, Minxuan X (2019) Effects of XIAP on high fat diet-induced hepatic steatosis: a mechanism involving NLRP3 inflammasome and oxidative stress. Aging (Albany NY) 11(24):12177
Zipfel WR, Williams RM, Christie R, Nikitin AY, Hyman BT, Webb WW (2003) Live tissue intrinsic emission microscopy using multiphoton-excited native fluorescence and second harmonic generation. Proc Natl Acad Sci 100(12):7075–7080

# Apoptosis and Mitochondria: Cell Death Mechanisms

5

Chirantan Majumder, Anirban Manna, Satyajit Halder, and Kuladip Jana

**Abstract**

Mitochondrion, the gift to the eukaryotic cells through endosymbiosis of a bacterial progenitor, not only the wallet of power, but also the cellular calcium pool, mediates non-shivering thermogenesis and cell death. Apoptosis is a self-destructive mechanism, essential for maintaining different cellular mechanisms like sculpting tissues for organ development, tissue homeostasis, and removal of unwanted cells. Here, we review and discuss about the mechanisms of release of different apoptotic factors from mitochondria into cytosol that carries out apoptosis.

**Keywords**

Apoptosis · Mitochondria · Cytochrome c · Bax · Bcl-2 · Bcl-xl · Mitochondrial fission

## 5.1 Introduction

The most mighty of nature's law is this: that out of death, she brings life. (Harman Melville)

Life and death is not only a very dynamic matter of philosophy, but also a highly challenging thing for the mighty "unknown driving force" of the universe to maintain the balance of life-form, and the most interesting and intriguing fact about this matter is an in-built single cell that is not an exception. As nothing is more

C. Majumder · A. Manna · S. Halder · K. Jana (✉)
Division of Molecular Medicine, Bose Institute, Unified Academic Campus, Kolkata, West Bengal, India
e-mail: kuladip@jcbose.ac.in





K. Jana (ed.), *Apoptosis and Human Health: Understanding Mechanistic and Therapeutic Potential*, https://doi.org/10.1007/978-981-97-7905-5_5

fascinating than nature's canvas, she didn't fail to surprise the living world, even with single-cell organisms. Single-cell organisms try to survive by relatively faster propagation. On the contrary, multi-cellular organisms use self-destructive mechanisms to clear the infected or unwanted cell units and avoid hostile situations to normal healthy cells. Cell death is generally characterized by chromatin condensation, blebs formation in the cell membrane, cell shrinkage, and formation of lobular structures in the nuclear membrane (Kerr et al. 1972). There are many self-destructive mechanisms, both programmed and non-programmed. The programmed cell mechanisms include type-I programmed cell death (PCD) or apoptosis and type-II programmed cell death (PCD) or autophagy (Bursch et al. 2000), and non-programmed include necrosis (Yan et al. 2020).

Cysteine-aspartate proteases, also known as **caspases,** are the master regulators of apoptosis (Turk and Stoka 2007; McIlwain et al. 2013). Caspases cleave some cellular proteins that activate them. Then, those activated proteins destroy several cellular components essential for cellular homeostasis and survival, resulting in cell death (Green 2022). Inhibition of protease activity by different caspase inhibitors, like Z-VAD-FMK, significantly protects against cell death (Sakthivel et al. 2022). So, caspase-mediated programmed cell death is tightly regulated (Kumar 2007). There are two major pathways of apoptosis: the extrinsic and the intrinsic pathways, also known as the mitochondrial pathways (Fulda and Debatin 2006). The mitochondrion is the site where both pro-apoptotic and anti-apoptotic mediators interact. At the same time, it is also the site where various apoptotic components are generated, like apoptotic mediators such as cytochrome c, Smac (second mitochondria-derived activator of apoptosis), and omi. Cytochrome c is one of the major elements of **apoptosome** or **wheel of death** once released into the cytosol. Apoptosome activates initiator caspase, caspase-9, which further activates executioner caspase, caspase-3, causing apoptosis. Smac and omi, on the other hand, inhibit inhibitors of apoptosis protein (IAPs) that, in normal conditions, inhibit caspases. Thus, by relieving the inhibitory effects of IAPs, SMAC, and omi induce apoptosis (Green and Evan 2002; Yang et al. 2003; Kalkavan and Green 2017). Studies of mitochondrial ultra-structural details give deeper insights into the high degree of regulation of apoptosis by mitochondria by developing a spatial arrangement between apoptotic mediators from their substrate (Wang and Youle 2009; Duranova et al. 2020).

Mitochondria produce various apoptotic mediators, but they remain inactive until released into the cytosol, where they interact with their substrate (Wang and Youle 2009). So, the release of mitochondrial apoptotic factors to the cytosol is one of the key events of apoptosis, which is a complex and tightly regulated phenomenon. B-cell lymphoma (BCL-2) family of proteins, mitochondrial outer membrane permeabilization (MOMP), formation of permeability transition pore complex (PTPC), and mitochondrial fission regulate the release of such mitochondrial apoptotic effectors (Crompton 1999; Zorov et al. 2000; Wang 2001; Wang and Youle 2009).

## 5.2 Mitochondrial Involvement in Cell Death

### 5.2.1 Mitochondrial Components of Death

Mitochondria are the site where different types of apoptotic factors are generated and stored. Mitochondrial membranes act as the spatial barrier for such apoptotic factors, become open or leaky by some specific cues, and protect the cells from unwanted death. Here, we will discuss the major mitochondrial apoptotic factors.

#### 5.2.1.1 Mitochondrial Cytochrome c

Cytochrome c (Cyt c) is one of the components of mitochondrial electron transport and the key component of the intrinsic pathway of apoptosis. Cyt c is the 15 kDa protein (Roucou et al. 2001) that remains in inter-membrane space (IMS). Under normal physiological conditions, cyt c neutralizes reactive oxygen species (ROS). Still, in apoptotic conditions, cyt c, along with co-factor p66shc, produces ROS and oxidizes cardiolipin, a lipoprotein that holds cyt c in normal conditions, leading to making way for cyt c release into cytosol (Hüttemann et al. 2011).

Once a cell is exposed to apoptotic stimuli, cyt c is released into the cytosol. In the cytosol, cyt c binds with apoptotic protein activating factor-1(Apaf-1) in an ATP-dependent manner and produces apoptosome (Wang 2001), leading to activation of pro-caspase9 to active caspase-9. Activated caspase-9 then activates the executioner caspase, caspase-3, causes apoptosis (Green 2000; Wang 2001).

#### 5.2.1.2 Inhibitor of Apoptosis Proteins (IAPs) Inhibitors (IAPIs)

It is believed that tumor cells acquire resistance by over-expression of IAPs. They are structurally related proteins and were identified first in baculovirus. Blocking apoptosis by mammalian IAPs is mediated through binding and inhibiting caspases or caspase-independent processes. Eight human IAPs have been identified: c-IAP1, c-IAP2, NAIP, Survivin, XIAP, Bruce, ILP-2, and Livin. One of the main features of IAPs is that they have one or more conserved repeats of 70 amino-acid residues-domain, called baculovirus IAP/inhibitory repeats (BIR), located at the N-terminus. Except for NIAP and survivin, human IAPs also contain another conserved domain, called the RING finger domain, located at the C-terminus of the protein. IAPs inhibit apoptosis by either inhibiting caspases or regulating signal transduction pathways, or exerting E3 ligase activity (Nachmias et al. 2004).

Survivin, XIAP, c-IAP-1, and c-IAP-2 are the IAPs that directly inhibit caspases −3, 7, and 9. Compared to other human IAP activities, XIAP is best characterized as containing 3 BIR domains: BIR1, BIR2, and BIR3. BIR3 directly binds to the area of caspase 9, exposed after cleavage (Takahashi et al. 1998). BIR1 and BIR2, on the other hand, inhibit caspase 3 and 7 by inhibiting substrate access (Riedl et al. 2001). IAP, Livin with only one BIR domain, can also inhibit caspase 3, 7, and 9 (Sharma et al. 2006).

Proteasome-mediated clearance of cellular proteins is one of the major pathways for maintaining cellular homeostasis (Chondrogianni and Gonos 2010). It is essential for eliminating damaged and misfolded proteins, cell-cycle regulation, and maintenance of cellular transcription factors (Jesenberger and Jentsch 2002). The c-terminal RING-finger domain of IAPs acts as E3 ubiquitin ligase (Yang et al. 2000). In response to different apoptotic cues, the RING-finger domain of XIAP targets caspase 3 by ubiquitinylation and protects cells from unwanted caspase-induced apoptotic stress. In the presence of apoptotic signals, a second mitochondria-derived activator of caspases (Smac) is released from the mitochondria along with cyt c, which can promote caspase activation by IAP inhibition. Smac/DIABLO interacts with BIR2 and BIR3 domains, inhibits IAPs, and activates caspases, leading to the activation of apoptosis (Nachmias et al. 2004). A four-amino acid containing a highly conserved motif at the N-terminus of Smac mediates this binding (Wu et al. 2000).

#### 5.2.1.3 Mitochondrial Caspases

Different tissues like the brain, kidney, thymus, and heart contain pro-caspases −2, −3, and −9, localized in mitochondria and cytosol. But in cardiomyocytes and neurons, procaspase-9 is predominantly present in mitochondria, and upon specific apoptotic signals, it is released into the cytosol to induce apoptosis (Roucou et al. 2001).

#### 5.2.1.4 Apoptosis-Inducing Factors (AIF)

AIF is an FAD-containing NADH-dependent oxidoreductase located in mitochondrial inter-membrane space, and its enzymatic activity is still unknown. The AIF gene, also known as AIFM and PCDC8, is located in humans' Xq25–26 region. Translation of AIF mRNA produces 67 kDa precursor, composed of residues 1–613 in humans (Sevrioukova 2011) (Fig. 5.1). Human AIF is composed of three major units: N-terminal mitochondrial leading sequence (MLS), two nuclear leading sequences (NLS), and the NAD-binding and FAD-binding motifs (Susin et al. 1999). After the translation, the precursor is imported to mitochondria in a non-native form, indicating that the folding of AIF in the cytosol is somehow prevented from blocking the events of unwanted apoptosis (Susin et al. 1999).

AIF does not have any nuclease activity but promotes DNA disruption by the removal of chromatin-associated proteins or by recruiting proteases or nucleases and forming a DNA degradation complex called degradosomes (Susin et al. 1999, 2000).

## 5.3 Molecular Mechanisms of Release of Mitochondrial Death Mediators

Different apoptogenic factors are localized in the intermembrane space (IMS) of mitochondria, including mitochondrial caspases, caspase-2, 3, and 9 (specific tissue sites), SMAC/DIABLO, AIF, and cytochrome c. So, the outer mitochondrial membrane plays a key role as the barrier to those apoptotic factors' release into cytosol and induction of apoptosis. Two major hypotheses have been described about the mechanism (Roucou et al. 2001).

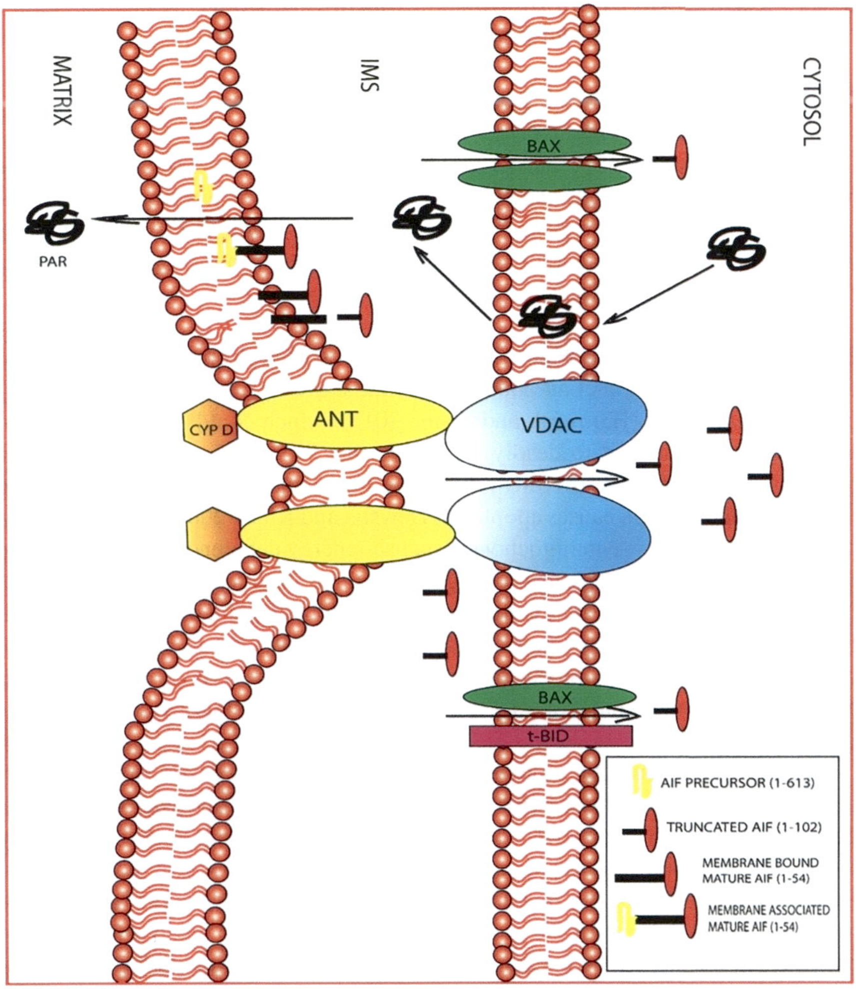


**Fig. 5.1** Schematic diagram of cleavage and release of apoptogenic AIF from mitochondria into cytosol. Local or cytoplasmic proteases mediated the proteolysis of the membrane tether AIF into mature ones, entering the intermembrane space through the permeabilization of the outer membrane. Detached AIF can translocate to the cytoplasm with the involvement of the PTP complex (composed of VDAC, ANT, and cyclophilin D) or through pores formed by the proapoptotic Bcl-2 family. AIF can be translocated into the nucleus by PAR. ANT, adenine nucleotide translocase; PAR, poly (ADP-ribose); CYP D, cyclophilin D; PTP, permeability transition pore; VDAC, voltage-dependent anion channel. (Figure was created by Adobe Illustrator 2021)

## 5.3.1 Formation of Conductance Channels at the Outer Mitochondrial Membrane (OMM)

### 5.3.1.1 Permeability Transition Pore Formation: *The Non-Swelling Model*

Experimental results show that using low concentrations of Bax can induce the release of cytochrome c into the cytosol without any matrix swelling, and the process could be inhibited by cyclosporin A (CsA). The dearth of matrix-swelling is characterized by the formation of PTP and the release of mitochondrial apoptotic factors into the cytosol (Roucou et al. 2001).

#### Bax Channels: One of the Major Guardians of the Mitochondrial Gate

Bax is cytosolic (Yethon et al. 2003), one of the BCL-2 proapoptotic family members along with Bak (Hardwick and Soane 2013), which could be converted from harmless bax monomers to deadly oligomers that form leaky pores on the outer mitochondrial membrane (MOMP), resulting in the release of different apoptotic factors into the cytosol, causes apoptosis (Dewson and Kluck 2009). The release has two effects: lack of cyt c in mitochondria causes energy deficiency and activation of downstream proteolytic cascade, enough for cellular destruction (Liu et al. 1996). Under the same experimental conditions, Bax channels could act in two different forms: type A, attributed by small, homogeneous, well-regulated, and voltage-gated, whereas Type B is a large, noisy, and voltage-independent. Type B channels could be converted into a population of Type A channels, irreversibly by applying La3+ (Lin et al. 2011).

Like other pro-apoptotic protein members, Bax and Bak contain BCL-2 homology domains (BHD). Three BH-domains are BH-1, BH-2, and BH-3. The BH-1 domain is located between BH-2 and BH-3. The BH-2 domain is located near the C-terminal transmembrane domain, whereas the BH-3 domain is near the N-terminal BH4 domain. Both Bax and Bak are composed of nine α-helices. The hydrophobic α-5 is located at the core of the protein, surrounded by amphipathic α-helices. The α-9 has the transmembrane domain and helps tether the protein with the outer mitochondrial membrane (OMM). The α-2 helix is at the hydrophobic BH-3 domain, where it is hidden very nicely in the inactive form until activated. When activated, this hydrophobic domain is exposed to form activated homo and heterodimeric units. A hydrophobic groove is present in Bax and Bak, composed of a C-terminus of α-2 to an N-terminus of α-5 and residues of α-8 (Fig. 5.2a). This groove is believed to be the site of interaction with the proapoptotic BH-3 domain, through which they are regulated. When inactive, this hydrophobic groove of Bax remains in the transmembrane domain. Bax has a rear-hydrophobic pocket composed of α-1/6 helices, masked by α-1/2 loop.

It is now believed after several studies that under proper apoptotic cues, the BH-3-only proteins bind to the α-1/6 rear hydrophobic pocket, which causes a conformational change in the α-9 helix, resulting in its displacement from the hydrophobic pocket followed by binding of this transmembrane domain into the OMM, leads to the formation of bax-PTPs (Fig. 5.2b).

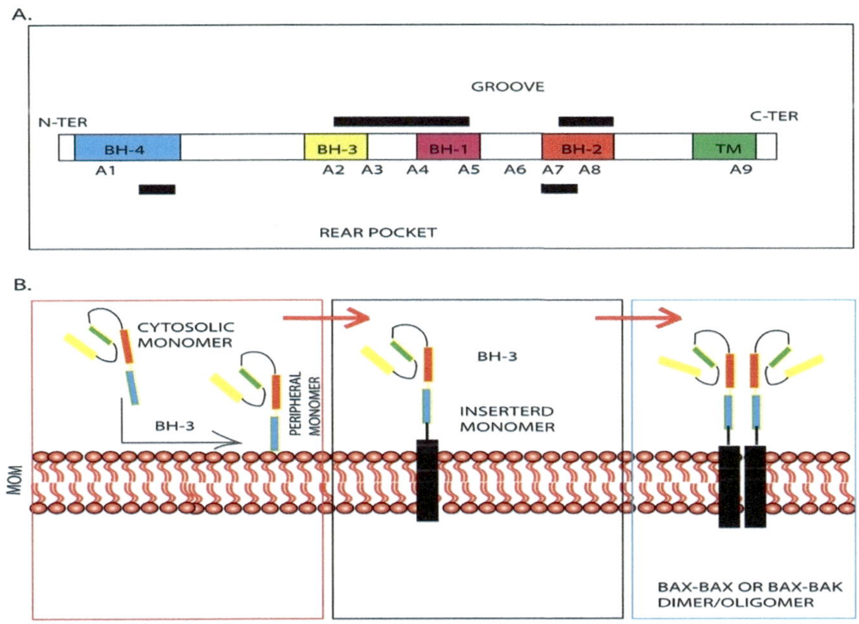


**Fig. 5.2** The members of the Bcl-2 family regulate Bax and Bak transitions. (**a**) Schematic representation of nine α-helices containing Bax/Bak protein (A represents α in the diagram). (**b**) Schematic representation of BH-3 dependent activation and oligomerization of Bax and Bak. (Figure was created by Adobe Illustrator 2021)

The non-swelling model also includes forming Bax/Bak and Bax/t-Bid heterodimeric channels at OMM (Westphal et al. 2011; Westphal et al. 2014).

## 5.3.2 Rupture of Mitochondrial Membrane

### 5.3.2.1 Formation of Permeability Transition Pore (PTP): *Mitochondrial Swelling Model*

The molecular attributes of the PTP structure are not well defined. But in general, it is composed of 3 major components; voltage-dependent anion channels (VDAC) at the outer mitochondrial membrane (OMM), cyclophilin D in the mitochondrial matrix, and adenine nucleotide transporter (ANT) in the inner mitochondrial membrane (IMM). PTP is activated by several cytosolic factors, like increased levels of cytosolic calcium (ca $^{2+}$), reactive oxygen species (ROS), and mitochondrial depolarization. On the contrary, PTP opening is inhibited by magnesium ion ($Mg^{2+}$), acidic pH of the mitochondrial matrix, cyclosporine, and mitochondrial hyperpolarization. Activation and opening of PTP are characterized by mitochondrial depolarization and matrix swelling because of increased solute concentration. Mitochondrial matrix swelling leads to rupture of OMM because the surface area of OMM is less than IMM. Induction of PTP causes the release of different mitochondrial apoptotic factors like cyt c, Smac/DIABLO, omi, etc., leading to apoptosis (Roucou et al. 2001).

#### 5.3.2.2 VDAC Closure Model

The closure of VDAC is the result of mitochondrial hyperpolarization. Closed VDAC leads to failure of mitochondria to exchange ATP/ADP with cytosol and reduced activity of F0-F1-ATPase in the IMM. The decreased activity of F0 causes increased mitochondrial membrane potential followed by swelling of the mitochondrial matrix rupture of OMM and release of mitochondrial apoptotic factors into the cytosol, resulting in apoptosis (Roucou, Antonsson).

#### 5.3.2.3 Mitochondrial Fission

Mitochondrial fission is a cellular phenomenon that is a result of the amalgamation of two distinct and different mechanisms: inhibition of mitochondrial fusion proteins like mitofusins (Mfn-1 and -2) and activation and recruitment of mitochondrial fission proteins like Dynamin-related protein 1 (Drp1) through phosphorylation (at Ser616 by cyclin-dependent kinase-1/cyclin B) (Roe and Qi 2018) at the mitochondrion-endoplasmic reticulum (ER) junction. DRP-1 is recruited on the OMM, where it interacts with OMM localized adaptor proteins: human fission factor-1 (Fis-1), mitochondrial fission factor (MFF), and mitochondrial dynamics protein of 49 and 51 kDa (MiD49/51) (Susin et al. 2000; Yoon et al. 2003; Otera et al. 2010; Palmer et al. 2011; Losón et al. 2013). The pro-apoptotic protein PUMA is believed the major player in the recruitment of DRP-1 on OMM (Wang et al. 2009). PUMA removes the Bax and Bad from the cytosolic sequestering protein and helps recruit DRP-1 from the cytosol to OMM (Ren et al. 2010). Once on OMM, the P-DRP-1, with several receptors, makes an oligomeric ring encircling the mitochondrion, and by using the energy from GTP-hydrolysis, it constricts the mitochondrial membrane until the lipid bilayer gets destabilized to ruptured from each other (Mears et al. 2011). DRP-1 then gets turned to a monomeric state, translocated to the cytosol, and keeps the multiple mitochondria behind. As rupture of both OMM and IMM occurs in the process of fission, it induces the release of multiple apoptotic factors, including cytochrome c, into the cytosol, resulting in apoptosis (Hall et al. 2014) (Fig. 5.3).

## 5.4 The Mitochondrial Pathway of Apoptosis (Intrinsic Pathway)

Mitochondrion, the energy source of cells, converts nutrients into energy and stores them in the form of ATP, essential for cellular homeostasis and maintenance of cellular entity in all terms (Casanova et al. 2023). Besides ATP production, mitochondria are the master regulators of different cellular functions, such as cellular calcium pool, radical generation, and cell death or apoptosis (Godoy et al. 2021). The mitochondrial pathway of apoptosis is highly regulated by the permeability of OMM and IMM. OMM allows solutes up to 5 KDa and includes ETC substrates and products,

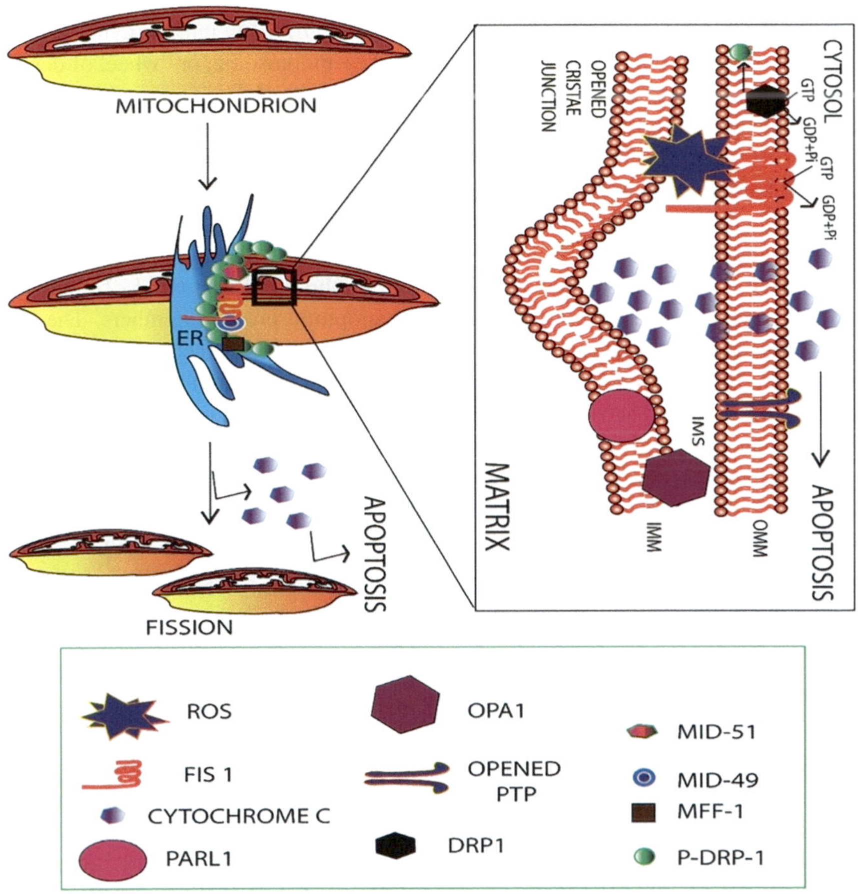


**Fig. 5.3** Role of mitochondrial fission proteins in changing mitochondrial morphology through fission. Fission proteins like Drp1 interact with Fis1, MFF, and MiD49/51 to induce mitochondrial fission. BAK and BAX binding to MFN2 on mitochondrial OMM stabilizing Drp1 binding leads to mitochondrial fragmentation and the release of cytochrome c, inducing apoptosis. (Figure was created Adobe Illustrator 2021)

whereas the IMM is highly impermeable and makes decisions of apoptosis (Redza-Dutordoir and Averill-Bates 2016). Activation of the intrinsic pathway of apoptosis needs, formation of permeability transition pores (PTP), composed of cyclophilin D, present in the matrix, voltage-dependent anion channel (VDAC) in OMM and adenine nucleotide transporter (ANT) in IMM (Halestrap 2009; Grimm 2012; Bernardi 2013). The permeability of IMM could be increased by allowing solutes under 1.5 KDa, under specific conditions (e.g., stress), allowing protons to move into the matrix, resulting in alteration of OXPHOS (Halestrap 2009), leading to swelling of the mitochondrial matrix. As the surface area of OMM is lesser than the IMM, it leads to rupture of OMM due to high matrix pressure and resulting outer mitochondrial membrane permeabilization (MOMP), causing the release of different pro-apoptotic factors like cyt c, endo G, SMAC/DIABLO into the cytosol. Once in the cytosol, cyt c binds with Apaf-1 and makes apoptosome, or "wheel of death," that recruits pro-caspase9 and converts it into caspase9, which in turn causes proteolytic cleavage of pro-caspase3 to active caspase3, leads to cell death. Smac/DIABLO is an IAPI that, when in the cytosol, inhibits inhibitors of apoptosis proteins, which are involved in the inhibition of caspases in normal conditions, leading to apoptosis. AIF, on the other hand, when released into cytosol from mitochondria, causes DNA damage and leads to cell death (Orrenius et al. 2015).

MOMP is highly controlled by the B-cell lymphoma protein (BCL-2) family of proteins, which includes both pro- and anti-apoptotic protein members. The pro-apoptotic factors are Bax, Bak, and Bok. The anti-apoptotic factors of the BCL-2 family are Bcl-XL, Bcl-2, Bcl-W, A1, and Mcl-1. Under normal physiological conditions, the balanced activity between these two types of proteins maintains homeostasis. Under stressful conditions, the balance is lost, and the cytosolic pro-apoptotic proteins like Bak, and Bax, start incorporating into the OMM and making homo- or heterodimeric channels, leading to MOMP, causing cell death (Leibowitz and Yu 2010; Kaufmann et al. 2012; Glory et al. 2014; Luna-Vargas and Chipuk 2016) (Fig. 5.4).

## 5.5 The Extrinsic Pathway of Apoptosis Can Communicate with the Intrinsic Pathway

The extrinsic or death receptor pathway of apoptosis is governed by the tumor necrosis factor receptor (TNF-R) superfamily. The members of this family include Death receptor-1 (DR-1) (also known as TNF-R1, CD120a, or P55), DR-2 (Fas, CD-95, or Apo-1), DR-3 (Apo-3, TRAMP, or LARD), DR-4 (TNF-related apoptosis-inducing ligand (TRAIL)-R1 or Apo-2), DR-5 (TRAIL R-2), and DR-6 (TNF receptor superfamily member 21). The common attribute of all these receptors is the presence of the death domain (DD) (Mahmood and Shukla 2010). The binding of TNF-$\alpha$ with the TNF R1 or DR-1 makes it active by trimerization and causes recruitment and activation of TNFR-associated adaptor protein with death domain (TRADD) to the exposed cytosolic segment of TNFR1 (Mahmood and Shukla 2010). TNF-receptor associated factor-2 (Traf-2) and RIP-1is then activated by TRADD, which further activates NF-K$\beta$ (Hayden and Ghosh 2014), which in turn up-regulates inhibitors of apoptosis proteins (IAPs) (Lee and Collins 2001). The FADD-TRADD associative activity can directly recruit and activate caspase 8, which leads to direct activation of executioner caspases like caspase 3, 6, and 7, causing apoptosis or by t-BID-mediated indirect apoptosis (Thomas et al. 2017). Cell survival or apoptosis through TNFR1 activation is regulated by a regulatory protein called FLICE inhibitory protein (FLIP) (Krueger et al. 2001).

The binding of Fas L with Fas receptor (DR2) makes it active by trimerization and can recruit and activate caspase 8, which could further activate central executioner caspases, caspase 3, −6, and 7, and mediate apoptosis (Kaufmann et al.

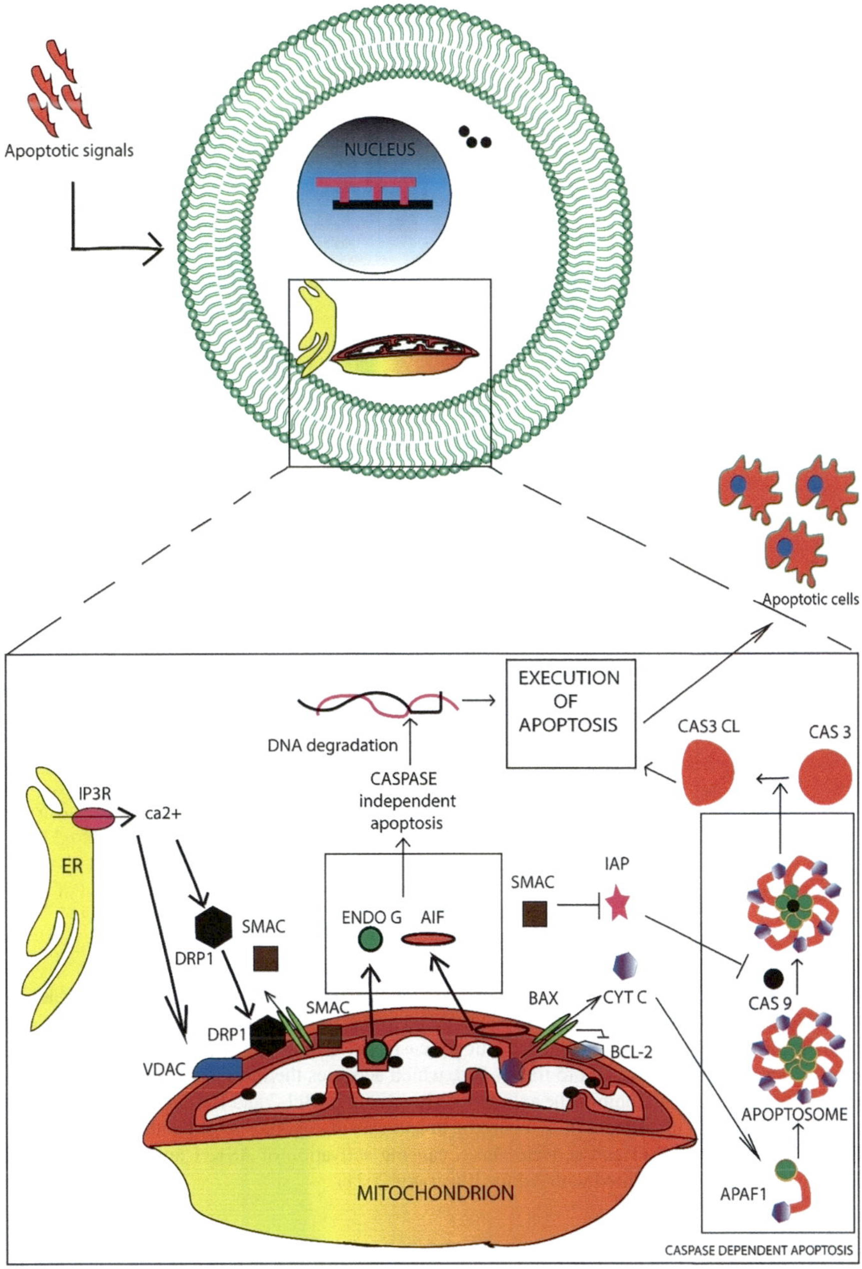


**Fig. 5.4** Apoptotic cues induce the accumulation of BH3-only proteins, causing BAX/BAK oligomerization, MOMP, and release of apoptotic factors of intermembrane space into the cytosol. Cytochrome *c* leads to apoptosome formation, which leads to activation of executioner caspases and induces apoptosis. Smac neutralizes inhibitors of apoptosis proteins (IAP), the caspase inhibitor. Apoptosis-inducing factors (AIF) induce caspase-independent apoptosis by DNA degradation. (Figure was created with Adobe Illustrator 2021)

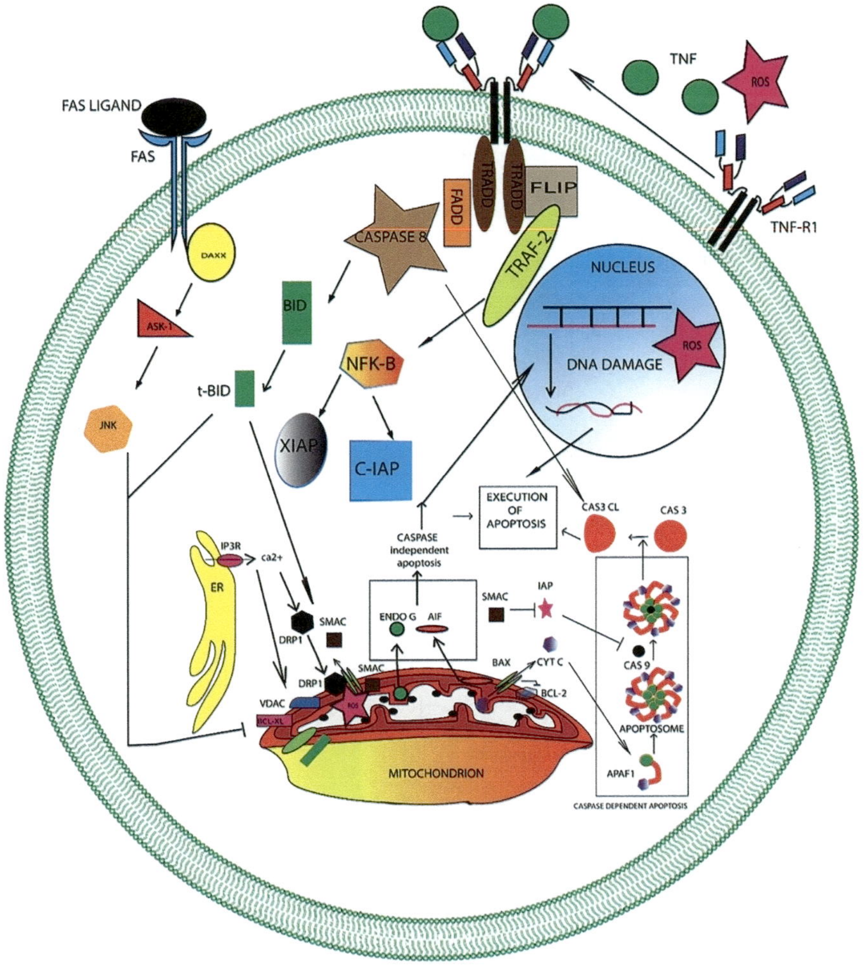


**Fig. 5.5** Extrinsic pathways can communicate with intrinsic pathways. Activation of TNF-R1 by TNF induces recruitment of adaptor protein FADD and procaspase-8, causing activation of caspases-8, which can directly activate executioner caspases-3, 6, and result in apoptosis. Activated caspase-8 could also cleave Bid to form t-Bid, which activates the intrinsic pathway of apoptosis through cross-talk. t-Bid inhibits the anti-apoptotic activity of Bcl-2 and Bcl-XL and activates Bax and Bak, leading to the release of cytochrome c, and causing apoptosis. Activation of Fas by Fas ligand (Fas L) can recruit and activate Daxx, causing activation of ASK-1 and JNK, leading to apoptosis. (Figure was created with Adobe Illustrator 2021)

2012). Activated Fas receptor recruits and activates Daxx, followed by ASK-1 and JNK activation, which inhibits anti-apoptotic, Bcl-2, and Bcl-Xl and executes apoptosis (Yang et al. 1997; Matsuzawa et al. 2002). (Fig. 5.5).

## 5.6 Conclusion

Even after breathtaking progress in the understanding of the role of mitochondria in cell death, many questions are still unanswered. The biochemistry regarding the release of mitochondrial apoptogenic factors and the relation between the release of apoptogenic factors and loss of mitochondrial functions like matrix alkalinization and uncoupling of OXPHOS seems most elusive. Even after fantastic advancements have been made in the last few years, some open questions are still remaining. In particular, how the apoptotic signals are transduced and integrated into mitochondria? Are BH-3 only proteins play as the master regulators of signal transduction to mitochondria? To answer these questions, more advanced studies are required.

## References

Bernardi P (2013) The mitochondrial permeability transition pore: a mystery solved? Front Physiol 4:95. https://doi.org/10.3389/fphys.2013.00095

Bursch W et al (2000) Programmed cell death (PCD): apoptosis, autophagic PCD, or others? Ann N Y Acad Sci 926(1):1–12. https://doi.org/10.1111/j.1749-6632.2000.tb05594.x

Casanova A et al (2023) Mitochondria: it is all about energy. Front Physiol 14:1114231. https://doi.org/10.3389/fphys.2023.1114231

Chondrogianni N, Gonos ES (2010) Proteasome function determines cellular homeostasis and the rate of aging. Adv Exp Med Biol 694:38–46. https://doi.org/10.1007/978-1-4419-7002-2_4

Crompton M (1999) The mitochondrial permeability transition pore and its role in cell death. Biochem J 341(Pt 2):233–249

Dewson G, Kluck RM (2009) Mechanisms by which Bak and Bax permeabilise mitochondria during apoptosis. J Cell Sci 122(16):2801–2808. https://doi.org/10.1242/jcs.038166

Duranova H et al (2020) Mitochondria: a worthwhile object for ultrastructural qualitative characterization and quantification of cells at physiological and pathophysiological states using conventional transmission electron microscopy. Acta Histochem 122(8):151646. https://doi.org/10.1016/j.acthis.2020.151646

Fulda S, Debatin K-M (2006) Extrinsic versus intrinsic apoptosis pathways in anticancer chemotherapy. Oncogene 25(34):4798–4811. https://doi.org/10.1038/sj.onc.1209608

Glory A, Bettaieb A, Averill-Bates DA (2014) Mild thermotolerance induced at 40 °C protects cells against hyperthermia-induced pro-apoptotic changes in Bcl-2 family proteins. Int J Hyperthermia 30(7):502–512. https://doi.org/10.3109/02656736.2014.968641

Godoy JA et al (2021) Mitostasis, calcium and free radicals in health, aging and neurodegeneration. Biomol Ther 11(7):1012. https://doi.org/10.3390/biom11071012

Green DR (2000) Apoptotic pathways: paper wraps stone blunts scissors. Cell 102(1):1–4. https://doi.org/10.1016/s0092-8674(00)00003-9

Green DR (2022) Caspases and their substrates. Cold Spring Harb Perspect Biol 14(3):a041012. https://doi.org/10.1101/cshperspect.a041012

Green DR, Evan GI (2002) A matter of life and death. Cancer Cell 1(1):19–30. https://doi.org/10.1016/S1535-6108(02)00024-7

Grimm S (2012) The ER-mitochondria interface: the social network of cell death'. Biochim Biophys Acta 1823(2):327–334. https://doi.org/10.1016/j.bbamcr.2011.11.018

Halestrap AP (2009) What is the mitochondrial permeability transition pore? J Mol Cell Cardiol 46(6):821–831. https://doi.org/10.1016/j.yjmcc.2009.02.021

Hall AR et al (2014) Mitochondrial fusion and fission proteins: novel therapeutic targets for combating cardiovascular disease. Br J Pharmacol 171(8):1890–1906. https://doi.org/10.1111/bph.12516

Hardwick JM, Soane L (2013) Multiple functions of BCL-2 family proteins. Cold Spring Harb Perspect Biol 5(2):a008722. https://doi.org/10.1101/cshperspect.a008722
Hayden MS, Ghosh S (2014) Regulation of NF-κB by TNF family cytokines. Semin Immunol 26(3):253–266. https://doi.org/10.1016/j.smim.2014.05.004
Hüttemann M et al (2011) The multiple functions of cytochrome c and their regulation in life and death decisions of the mammalian cell: from respiration to apoptosis. Mitochondrion 11(3):369–381. https://doi.org/10.1016/j.mito.2011.01.010
Jesenberger V, Jentsch S (2002) Deadly encounter: ubiquitin meets apoptosis. Nat Rev Mol Cell Biol 3(2):112–121. https://doi.org/10.1038/nrm731
Kalkavan H, Green D (2017) MOMP, cell suicide as a BCL-2 family business. Cell Death Differ 25:46. https://doi.org/10.1038/cdd.2017.179
Kaufmann T, Strasser A, Jost PJ (2012) Fas death receptor signalling: roles of bid and XIAP. Cell Death Differ 19(1):42–50. https://doi.org/10.1038/cdd.2011.121
Kerr JFR, Wyllie AH, Currie AR (1972) Apoptosis: a basic biological phenomenon with wide-ranging implications in tissue kinetics. Br J Cancer 26(4):239–257
Krueger A et al (2001) FLICE-inhibitory proteins: regulators of death receptor-mediated apoptosis. Mol Cell Biol 21(24):8247–8254. https://doi.org/10.1128/MCB.21.24.8247-8254.2001
Kumar S (2007) Caspase function in programmed cell death. Cell Death Differ 14(1):32–43. https://doi.org/10.1038/sj.cdd.4402060
Lee RT, Collins T (2001) Nuclear factor-κB and cell survival. Circ Res 88(3):262–264. https://doi.org/10.1161/01.RES.88.3.262
Leibowitz B, Yu J (2010) Mitochondrial signaling in cell death via the Bcl-2 family. Cancer Biol Ther 9(6):417–422
Lin SH et al (2011) Bax forms two types of channels, one of which is voltage-gated. Biophys J 101(9):2163–2169. https://doi.org/10.1016/j.bpj.2011.09.041
Liu X et al (1996) Induction of apoptotic program in cell-free extracts: requirement for dATP and cytochrome c. Cell 86(1):147–157. https://doi.org/10.1016/s0092-8674(00)80085-9
Losón OC et al (2013) Fis1, Mff, MiD49, and MiD51 mediate Drp1 recruitment in mitochondrial fission. Mol Biol Cell 24(5):659–667. https://doi.org/10.1091/mbc.E12-10-0721
Luna-Vargas MPA, Chipuk JE (2016) Physiological and pharmacological control of BAK, BAX, and beyond. Trends Cell Biol 26(12):906–917. https://doi.org/10.1016/j.tcb.2016.07.002
Mahmood Z, Shukla Y (2010) Death receptors: targets for cancer therapy. Exp Cell Res 316(6):887–899. https://doi.org/10.1016/j.yexcr.2009.12.011
Matsuzawa A et al (2002) Physiological roles of ASK1-mediated signal transduction in oxidative stress- and endoplasmic reticulum stress-induced apoptosis: advanced findings from ASK1 knockout mice. Antioxid Redox Signal 4:415–425. https://doi.org/10.1089/15230860260196218
McIlwain DR, Berger T, Mak TW (2013) Caspase functions in cell death and disease. Cold Spring Harb Perspect Biol 5(4):a008656. https://doi.org/10.1101/cshperspect.a008656
Mears JA et al (2011) Conformational changes in Dnm1 support a contractile mechanism for mitochondrial fission. Nat Struct Mol Biol 18(1):20–26. https://doi.org/10.1038/nsmb.1949
Nachmias B, Ashhab Y, Benyehuda D (2004) The inhibitor of apoptosis protein family (IAPs): an emerging therapeutic target in cancer. Semin Cancer Biol 14(4):231–243. https://doi.org/10.1016/j.semcancer.2004.04.002
Orrenius S, Gogvadze V, Zhivotovsky B (2015) Calcium and mitochondria in the regulation of cell death. Biochem Biophys Res Commun 460(1):72–81. https://doi.org/10.1016/j.bbrc.2015.01.137
Otera H et al (2010) Mff is an essential factor for mitochondrial recruitment of Drp1 during mitochondrial fission in mammalian cells. J Cell Biol 191(6):1141–1158. https://doi.org/10.1083/jcb.201007152
Palmer CS et al (2011) MiD49 and MiD51, new components of the mitochondrial fission machinery. EMBO Rep 12(6):565–573. https://doi.org/10.1038/embor.2011.54

Redza-Dutordoir M, Averill-Bates DA (2016) Activation of apoptosis signalling pathways by reactive oxygen species. Biochim Biophys Acta (BBA) Mol Cell Res 1863(12):2977–2992. https://doi.org/10.1016/j.bbamcr.2016.09.012

Ren D et al (2010) BID, BIM, and PUMA are essential for activation of the BAX- and BAK-dependent cell death program. Science (New York, N.Y.) 330(6009):1390–1393. https://doi.org/10.1126/science.1190217

Riedl SJ et al (2001) Structural basis for the inhibition of caspase-3 by XIAP. Cell 104(5):791–800. https://doi.org/10.1016/s0092-8674(01)00274-4

Roe AJ, Qi X (2018) Drp1 phosphorylation by MAPK1 causes mitochondrial dysfunction in cell culture model of Huntington's disease. Biochem Biophys Res Commun 496(2):706–711. https://doi.org/10.1016/j.bbrc.2018.01.114

Roucou X, Antonsson B, Martinou J-C (2001) Involvement of mitochondria in apoptosis. Cardiol Clin 19(1):45–55. https://doi.org/10.1016/S0733-8651(05)70194-6

Sakthivel D, Bolívar BE, Bouchier-Hayes L (2022) Cellular autophagy, an unbidden effect of caspase inhibition by zVAD-fmk. FEBS J 289(11):3097–3100. https://doi.org/10.1111/febs.16346

Sevrioukova IF (2011) Apoptosis-inducing factor: structure, function, and redox regulation. Antioxid Redox Signal 14(12):2545–2579. https://doi.org/10.1089/ars.2010.3445

Sharma SK, Straub C, Zawel L (2006) Development of peptidomimetics targeting IAPs. Int J Pept Res Ther 12(1):21–32. https://doi.org/10.1007/s10989-005-9003-2

Susin SA et al (1999) Molecular characterization of mitochondrial apoptosis-inducing factor. Nature 397(6718):441–446. https://doi.org/10.1038/17135

Susin SA et al (2000) Two distinct pathways leading to nuclear apoptosis. J Exp Med 192(4):571–580. https://doi.org/10.1084/jem.192.4.571

Takahashi R et al (1998) A single BIR domain of XIAP sufficient for inhibiting caspases. J Biol Chem 273(14):7787–7790. https://doi.org/10.1074/jbc.273.14.7787

Thomas CN et al (2017) Caspases in retinal ganglion cell death and axon regeneration. Cell Death Discov 3(1):1–13. https://doi.org/10.1038/cddiscovery.2017.32

Turk B, Stoka V (2007) Protease signalling in cell death: caspases versus cysteine cathepsins. FEBS Lett 581(15):2761–2767. https://doi.org/10.1016/j.febslet.2007.05.038

Wang X (2001) The expanding role of mitochondria in apoptosis. Genes Dev 15(22):2922–2933

Wang C, Youle RJ (2009) The role of mitochondria in apoptosis. Annu Rev Genet 43:95–118. https://doi.org/10.1146/annurev-genet-102108-134850

Wang J-X, Li Q, Li P-F (2009) Apoptosis repressor with caspase recruitment domain contributes to chemotherapy resistance by abolishing mitochondrial fission mediated by dynamin-related protein-1. Cancer Res 69(2):492–500. https://doi.org/10.1158/0008-5472.CAN-08-2962

Westphal D et al (2011) Molecular biology of Bax and Bak activation and action. Biochim Biophys Acta 1813(4):521–531. https://doi.org/10.1016/j.bbamcr.2010.12.019

Westphal D, Kluck RM, Dewson G (2014) Building blocks of the apoptotic pore: how Bax and Bak are activated and oligomerize during apoptosis. Cell Death Differ 21(2):196–205. https://doi.org/10.1038/cdd.2013.139

Wu G et al (2000) Structural basis of IAP recognition by Smac/DIABLO. Nature 408(6815):1008–1012. https://doi.org/10.1038/35050012

Yan G, Elbadawi M, Efferth T (2020) Multiple cell death modalities and their key features (review). World Acad Sci J 2(2):39–48. https://doi.org/10.3892/wasj.2020.40

Yang X et al (1997) Daxx, a novel Fas-binding protein that activates JNK and apoptosis. Cell 89(7):1067–1076

Yang Y et al (2000) Ubiquitin protein ligase activity of IAPs and their degradation in proteasomes in response to apoptotic stimuli. Science (New York, N.Y.) 288(5467):874–877. https://doi.org/10.1126/science.288.5467.874

Yang Q-H et al (2003) Omi/HtrA2 catalytic cleavage of inhibitor of apoptosis (IAP) irreversibly inactivates IAPs and facilitates caspase activity in apoptosis. Genes Dev 17(12):1487–1496. https://doi.org/10.1101/gad.1097903

Yethon JA et al (2003) Interaction with a membrane surface triggers a reversible conformational change in Bax normally associated with induction of apoptosis. J Biol Chem 278(49):48935–48941. https://doi.org/10.1074/jbc.M306289200
Yoon Y et al (2003) The mitochondrial protein hFis1 regulates mitochondrial fission in mammalian cells through an interaction with the dynamin-like protein DLP1. Mol Cell Biol 23(15):5409. https://doi.org/10.1128/MCB.23.15.5409-5420.2003
Zorov DB et al (2000) Reactive oxygen species (Ros-induced) Ros release. J Exp Med 192(7):1001–1014

# Apoptosis: ER Stress and Disease Pathology

# 6

Satyajit Halder, Shraman Jana, Chirantan Majumder, and Kuladip Jana

**Abstract**

Multicellular creatures depend on the carefully controlled process of apoptosis, often known as programmed cell death. Through this technique, organisms can get rid of cells that are redundant or even hazardous, including ones that contain mutations or are infested with viruses. Apoptotic cells experience a series of morphological and metabolic alterations as part of a cascade of self-destruction. Cellular homeostasis requires effective control over the programmed cell death process. A crucial quality-control organelle for protein homeostasis, or "proteostasis," is the endoplasmic reticulum (ER). This encompasses multiple processes of protein quality control, including autophagy, protein chaperones, and ER-associated degradation. ER stress is brought on by a build-up of misfolded or unfolded proteins that disrupt proteostasis. An adaptive unfolded protein response (UPR) is triggered to maintain proteostasis. The activation of inositol-requiring enzyme 1 (IRE1), activating transcription factor 6 (ATF6), and protein kinase R-like ER kinase (PERK) are all involved in this reaction. Transcription and protein processing are just two of the many epigenetic factors that are impacted by ER stress. Elevated levels of cellular stress can cause harm to lipids, proteins, nucleic acids, membranes, and organelles, which may initiate processes leading to cell death, including apoptosis. Reactive oxygen species (erROS) formed from the endoplasmic reticulum (ER) are also essential for controlling major apoptosis pathways and cell signaling. The ER, death receptors, and mitochondria con-

S. Halder · C. Majumder · K. Jana (✉)
Division of Molecular Medicine, Bose Institute, Unified Academic Campus, Kolkata, West Bengal, India
e-mail: kuladip@jcbose.ac.in

S. Jana
IISER Kolkata, Campus Rd, Mohanpur, Haringhata Farm, West Bengal, India


K. Jana (ed.), *Apoptosis and Human Health: Understanding Mechanistic and Therapeutic Potential*, https://doi.org/10.1007/978-981-97-7905-5_6

trol these pathways. With an emphasis on metabolic, inflammatory, and neoplastic disorders, this book chapter explores the current understanding of ER stress's role in the principal apoptotic pathways associated with various human diseases. It draws attention to the complex interactions and feedback loops that exist between these signaling pathways and how they relate to apoptosis, underscoring the need for more study to understand these relationships fully.

**Keywords**

Apoptosis · Endoplasmic reticulum (ER) stress · erROS · Unfolded protein response (UPR) · Mitochondria · diseases

## 6.1 Introduction

Proteins are essential macromolecules that require precise three-dimensional structures to function effectively. The endoplasmic reticulum (ER) is responsible for the folding of secretory and membrane proteins, a process prone to errors. In the ER, nascent proteins are assisted by molecular chaperones and folding enzymes, collectively known as ER chaperones. Properly folded proteins are transported to the Golgi apparatus, while misfolded proteins are retained in the ER, retro-translocated to the cytoplasm via the ER-associated degradation (ERAD) machinery, and subsequently degraded by the proteasome. To manage the load of protein folding, the ER expresses ER chaperones and ERAD components.

When cells produce an excess of secretory proteins, surpassing the ER's folding capacity and ERAD machinery, unfolded proteins accumulate in the ER. These misfolded proteins expose normally hidden hydrophobic amino acid residues, leading to the formation of protein aggregates. Such aggregates can trigger apoptotic cell death and are implicated in various conformational diseases, including cancer, diabetes mellitus, and neurodegenerative disorders. To counteract this stress, eukaryotic cells activate a suite of defense mechanisms known as the ER stress response or unfolded protein response (UPR) (Harding et al. 2002; Mori 2000; Patil and Walter 2001; Schröder and Kaufman 2005).

The mammalian ER stress response comprises four main mechanisms. First, there is a reduction in protein synthesis to prevent further accumulation of misfolded proteins. Second, the transcriptional activation of ER chaperone genes enhances the cell's folding capacity. Third, the transcriptional induction of ERAD component genes enhances the ERAD machinery's capabilities. Finally, severely damaged cells are eliminated through apoptosis, thereby protecting the organism's survival. The homeostasis of protein folding in the ER is closely linked to the ER's redox state.

When ER stress is severe or prolonged, or if the UPR is defective, pro-apoptotic signaling pathways are activated. Both reducing and oxidizing agents can disrupt protein folding, leading to ER stress. Disulfide bond formation in the ER lumen is particularly sensitive to changes in redox balance. During oxidative protein folding, thiol groups on the cysteines of substrate peptides are oxidized, forming disulfide bonds and releasing hydrogen peroxide ($H_2O_2$) as a by-product. Dysregulated disulfide bond formation and breakdown in a stressed ER can contribute to the accumulation of reactive oxygen species (ROS) and oxidative stress (Malhotra and Kaufman 2007; Cao and Kaufman 2012; Ushioda et al. 2008; Dorner et al. 1990).

In this chapter, we will explore the fundamentals of the unfolded protein response (UPR) regulated by ER stress and oxidative stress within the cell and its role in initiating apoptotic cell death. Understanding these pathways provides insights into the molecular mechanisms underlying numerous conformational disorders and can inform the development of therapeutic strategies.

## 6.2 Mechanisms of the ER Stress Response

The mammalian ER stress response comprises four key mechanisms: translational attenuation, increased expression of ER chaperones, expression of ERAD components, and activation of apoptosis. These regulatory mechanisms ensure cellular homeostasis during ER stress (Fig. 6.1).

### 6.2.1 PERK Pathway

The PERK pathway is one of the key sensors of the unfolded protein response (UPR) within the endoplasmic reticulum (ER), along with IRE1 and ATF6. These sensors trigger downstream signaling mechanisms to restore ER balance by either reducing the accumulation of unfolded proteins or enhancing the ER's capacity to manage them. The detection of misfolded proteins in the ER is mediated by the luminal domain of PERK, while its cytoplasmic domain is responsible for its kinase activity (Harding et al. 1999). In a non-stressed state, PERK's activation is inhibited by its interaction with the ER chaperone BiP, which binds to its luminal domain. During ER stress, BiP dissociates from PERK, leading to PERK activation through processes of oligomerization and trans-phosphorylation (Bertolotti et al. 2000). Activated PERK then phosphorylates eukaryotic initiation factor 2 alpha (eIF2α) at serine 51. This modification reduces the overall rate of protein synthesis, thereby decreasing the influx of newly synthesized proteins into the ER (Boyce et al. 2005).

Interestingly, the phosphorylation of eIF2α specifically enhances the translation of certain mRNAs, such as ATF4, which feature upstream open reading frames (uORFs) in their 5′ untranslated regions (UTRs). ATF4 is a transcription factor that regulates the expression of genes involved in amino acid metabolism, oxidative stress responses, and autophagy. In situations of prolonged or intense ER stress, the transcription factor CHOP is also activated. CHOP promotes apoptosis by downregulating genes that inhibit cell death and upregulating genes that promote apoptosis, thereby leading the cell toward programmed death when the stress is overwhelming.

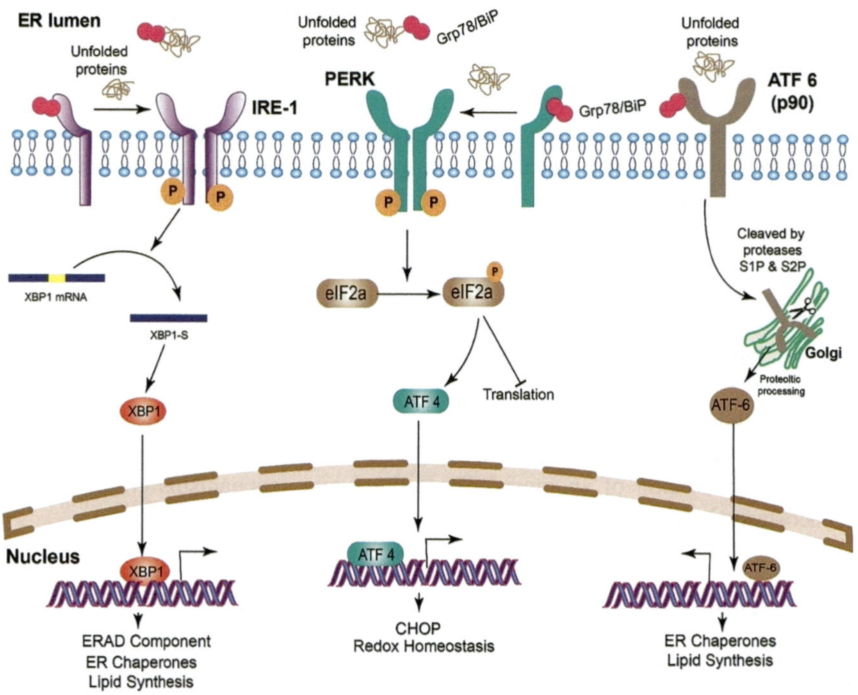


**Fig. 6.1** Depicts the pathways through which mammalian cells manage ER stress. These pathways—PERK, ATF6, and IRE1—coordinate the cellular response to the accumulation of unfolded proteins in the ER. The PERK pathway involves a transmembrane kinase known as PERK, which phosphorylates eIF2α. This phosphorylation reduces global protein synthesis while enhancing the translation of ATF4. ATF4, in turn, activates the transcription of various genes essential for coping with ER stress, including CHOP. The ATF6 pathway centers around ATF6, a transmembrane transcription factor. In response to ER stress, ATF6 is transported to the Golgi apparatus, where it is cleaved by site-1 protease (S1P) and site-2 protease (S2P). This cleavage releases the cytosolic domain of ATF6, which then enters the nucleus to stimulate the transcription of genes that code for ER chaperones, thereby improving the protein folding capacity of the cell. The third pathway, mediated by IRE1, features transmembrane RNase activity. IRE1 splices XBP1 mRNA to produce the active form of XBP1 (pXBP1(S)). This transcription factor then promotes the expression of genes involved in ER-associated degradation (ERAD), which helps remove misfolded proteins from the ER lumen

### 6.2.2 ATF6 Pathway

ATF6 is a type II transmembrane protein situated on the endoplasmic reticulum (ER) membrane, responsible for detecting unfolded proteins via its luminal domain. Its cytoplasmic portion includes a DNA-binding domain, a transcriptional activation domain, and a basic leucine zipper (bZIP) motif. Normally, the ER chaperone BiP binds to ATF6's luminal domain, preventing its movement to the Golgi apparatus. During periods of ER stress, BiP dissociates from ATF6, allowing ATF6 to relocate to the Golgi. In the Golgi, ATF6 undergoes cleavage by site-1 protease

(S1P) and site-2 protease (S2P). This cleavage produces the cytoplasmic fragment of ATF6, known as pATF6(N), which then moves to the nucleus. In the nucleus, pATF6(N) binds to the ER stress response element (ERSE), leading to the transcription of genes that code for various ER chaperones, such as BiP, GRP94, and calreticulin. The S2P cleavage occurs through a process called regulated intramembrane proteolysis (RIP), a mechanism also involved in processing sterol regulatory element-binding proteins (SREBPs), which are crucial for regulating cholesterol metabolism (Yoshida et al. 2001; Shen et al. 2002; Ye et al. 2000; Mori 2005).

### 6.2.3 IRE1 Pathway

IRE1 is a membrane protein located on the endoplasmic reticulum (ER) with a cytoplasmic domain that includes both kinase and RNase activities. Its luminal domain is responsible for detecting unfolded proteins within the ER, similar to PERK. There are two gene variants of IRE1: IRE1α and IRE1β. Under normal ER conditions, the ER chaperone BiP binds to IRE1, inhibiting its activation. However, in response to ER stress, BiP dissociates from IRE1, leading to its activation through dimerization and subsequent trans-phosphorylation. Once activated, IRE1α employs a unique splicing mechanism to convert XBP1 pre-mRNA into its functional form. XBP1 acts as a crucial transcription factor that regulates the expression of genes involved in various aspects of the unfolded protein response (UPR), including ER protein folding, intracellular trafficking, ER-associated degradation (ERAD), phospholipid synthesis, and the expansion of the ER membrane (Yoshida et al. 1998; Kaneko and Nomura 2003). Additionally, IRE1's RNase domain plays a role in regulated IRE1-dependent decay of mRNA (RIDD), a process that targets and degrades specific mRNAs localized to the ER during stress conditions (Sriburi et al. 2004).

## 6.3 ER Stress and Apoptosis

The buildup of unfolded proteins in the endoplasmic reticulum (ER) can dramatically limit cell survival. If the PERK, ATF6, and IRE1 pathways fail to address ER stress adequately, cells may resort to apoptosis as a final measure to safeguard the organism (Kadowaki et al. 2004; Kim et al. 2006). Several processes have been implicated in ER stress-induced apoptosis, frequently necessitating the simultaneous activation of several death pathways, although this feature remains under dispute (Fig. 6.2).

One important mechanism implicated in ER stress-induced apoptosis is the CHOP pathway. During ER stress, the ATF6 and PERK pathways lead to the synthesis of CHOP-GADD153, a key transcription factor (Ma et al. 2002). Cells lacking in CHOP display decreased rates of programmed cell death under ER stress, underscoring CHOP's crucial function in mediating apoptosis.

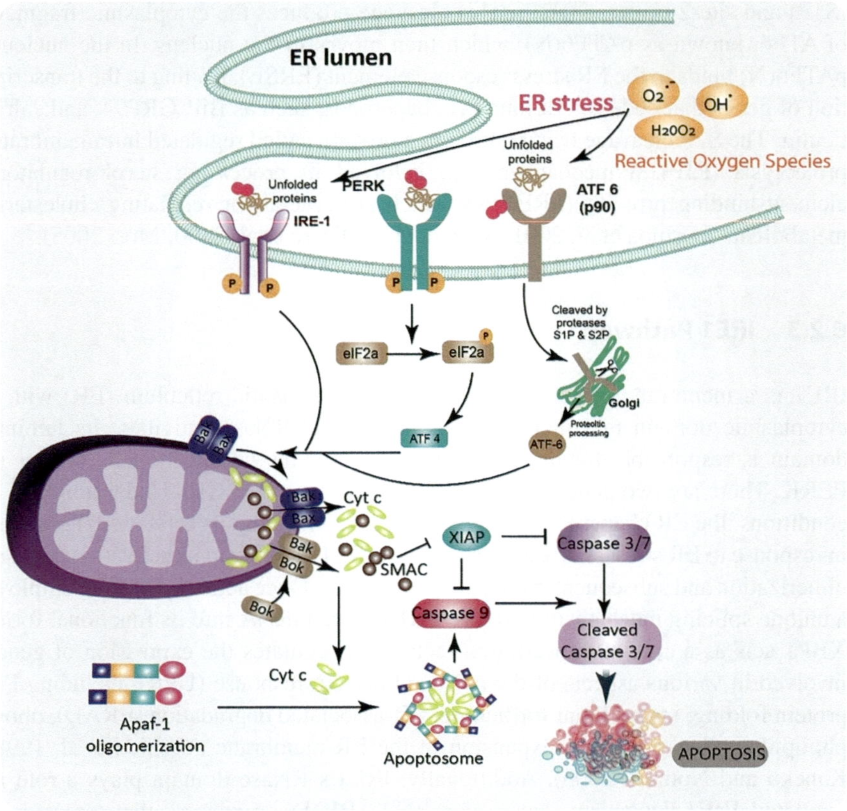


**Fig. 6.2** ER stress-mediated apoptotic cell death

CHOP regulates numerous genes connected to apoptosis, including GADD34, ERO1, DR5 (death receptor 5), and carbonic anhydrase VI (Marciniak et al. 2004; Novoa et al. 2001; Sok et al. 1999; Wang et al. 1998). ERO1 encodes an oxidase that provides a hyper-oxidizing environment in the ER (Marciniak et al. 2004). GADD34, interacting with protein phosphatase 2C, contributes to dephosphorylating eIF2α, therefore boosting the synthesis of ER client proteins (Novoa et al. 2001). DR5, a cell surface receptor, can stimulate caspase activation (Yamaguchi and Wang 2004). Carbonic anhydrase VI impacts cellular pH and many processes (Sok et al. 1999; Wang et al. 1998).

The signaling mechanisms linking CHOP to apoptosis are not fully explored, although the IRE1-TRAF2-ASK1 pathway provides another essential apoptotic route. IRE1's cytoplasmic domain binds to TRAF2 (tumor necrosis factor receptor-associated factor 2), which then connects with Jun kinase (JNK) and stress-activated protein kinase (SAPK) (Urano et al. 2000). This complex, combining TRAF2 and ASK1 (apoptosis signal-regulating kinase 1), phosphorylates and activates JNK (Nishitoh et al. 2002). TRAF2 also interacts with caspase-12, regulating its activation. Additionally, IRE1-TRAF2 activation leads to the expression of the transcriptional repressor ATF3, further contributing to apoptosis (Zhang et al. 2001).

The IRE1-TRAF2-ASK1 pathway is crucial for controlling apoptosis in response to ER stress. Several caspases, including caspases 2, 3, 4, 7, 9, and 12, are critical participants in ER stress-induced cell death (Morishima et al. 2002). Caspase-12, which connects with the ER membrane, becomes activated during ER stress, potentially through calpain (Nakagawa et al. 2000). This activation leads to the downstream activation of caspase-9 and subsequently caspase-3, terminating in cell death (Nakagawa et al. 2000). Research has revealed that caspase-12 mutant animals are resistant to ER stress-induced apoptosis but still respond to other death stimuli, underlining caspase-12's particular role in ER stress (Nakagawa et al. 2000; Saleh et al. 2006).

The significance of caspase-12 in diseases such as Alzheimer's disease, polyglutamine disorders, ischemia, and viral infections underlines its importance in pathological settings (Nakagawa et al. 2000; Saleh et al. 2006). However, the functional significance of human caspase-12, which exhibits many variants, remains contested (Saleh et al. 2006), suggesting that other, yet undiscovered, caspases may potentially contribute to ER stress-induced apoptosis in humans.

Members of the Bcl2 family are also critical in controlling ER stress-induced apoptosis. These proteins govern the recruitment of pro-apoptotic members to the ER membrane, enabling caspase-12 activation, whereas anti-apoptotic members prevent this process. Under ER stress, CHOP suppresses the anti-apoptotic protein Bcl2, leading to increased oxidative damage and apoptosis. Conversely, Bcl2 overexpression can prevent caspase-12 activation and apoptosis during ER stress.

Among the pro-apoptotic factors, BAD (Bcl2 antagonist of cell death) is activated and dephosphorylated in response to ER stress (Zha et al. 1996). Additionally, Bax (Bcl2-associated X protein) and Bak (Bcl-2 homologous antagonist killer), situated in the ER and mitochondrial membranes, oligomerize and activate caspase-12 during ER stress (Scorrano et al. 2003; Wei et al. 2001). Notably, Bax and Bak interact with IRE1α during ER stress, altering its function (Hetz et al. 2006). Bax and Bak are necessary for most apoptotic forms (Wei et al. 2001).

Furthermore, BH3-only members such as PUMA (p53 upregulated modulator of apoptosis) and NOXA (neutrophil NADPH oxidase factor) are increased by p53 under ER stress (Nakano and Vousden 2001; Oda et al. 2000). Cells missing PUMA and NOXA demonstrate resistance to ER stress-induced apoptosis, confirming their role in inducing cell death under such conditions.

In summary, ER stress-induced apoptosis includes a complex interaction of numerous pathways and regulators, including the IRE1-TRAF2-ASK1 pathway, caspases, and Bcl2 family members, showing the diverse character of cellular responses to ER stress. Additionally, the involvement of inhibitor of apoptosis (IAP) family members, such as HIAP2 (human inhibitor of apoptosis 2), IAP-2, and XIAP (X-linked inhibitor of apoptosis), underscores their role in modulating apoptosis during ER stress (Hu et al. 2004; Zhang et al. 2000; Wootz et al. 2006). NRF1 and NRF2, transcription factors implicated in oxidative stress responses, also contribute to ER stress-specific apoptosis (Cullinan et al. 2003; Wang and Chan 2006). During myogenesis, ATF6 further participates in apoptosis by controlling BiP and CHOP levels, with inhibitors such as AEBSF influencing this process (Nakanishi et al. 2005).

## 6.4 ER Stress Mediated Cell Death and Human Diseases

The endoplasmic reticulum (ER) serves a key role in manufacturing and folding proteins, ensuring they attain the correct three-dimensional shape. When the ER gets overburdened with unfolded or misfolded proteins, it generates an ER stress response aimed at reducing this stress. This reaction involves increasing the synthesis of molecular chaperones and components of the ER-associated degradation (ERAD) pathway, which work together to appropriately fold or remove defective proteins (Behrends et al. 2006).

Smaller protein aggregates, coupled with reactive oxygen species (ROS), present a greater risk under ER stress compared to larger aggregates. These smaller aggregates can interact with transcription factors like CREB and TATA, affecting the ubiquitin-proteasome system. Conversely, bigger aggregates and molecular chaperones such as the TCP1-ring complex (TRiC) and Heat Shock Protein 70 (HSP70) provide protection by limiting the creation of harmful aggregates and boosting cellular survival during stress. Moreover, interactions between estrogen receptors and the tumor suppressor Breast Cancer 1 (BRCA1) alter Nrf2, a major regulator of cellular redox balance, metabolism, and antioxidant defenses. This underscores the need to regulate ROS levels to decrease oxidative stress and maintain cellular health. Failure in the ER stress response can lead to conformational disorders, commonly known as "folding diseases," which are often related with ER stress-induced protein misfolding and aggregation (Behrends et al. 2006; Wang and Di 2014; Zhang et al. 2002).

Pancreatic β cells, responsible for manufacturing insulin, are especially susceptible to ER stress due to their intensive protein synthesis, which constitutes approximately half of the proteins in these cells. Additionally, their high energy requirements and low quantities of antioxidant enzymes make β cells sensitive to oxidative stress. Dysfunction or death of β cells, driven by ER and oxidative stress, plays a crucial role in both type I and type II diabetes mellitus.

In type 1 diabetes mellitus (IDDM), the PERK pathway, which responds to ER stress, has been intensively researched due to its relevance in β cells that undergo ER stress from excessive insulin secretion. Mutations in PERK lead to Wolcott–Rallison syndrome, an early-onset variant of IDDM, and PERK-deficient animals exhibit symptoms comparable to IDDM. Moreover, mice lacking CHOP show a delayed onset of diabetes, equivalent to those with mutations at the eIF2α phosphorylation site (Ser51Ala), demonstrating a vital role for ER stress in β-cell failure (Fonseca et al. 2005).

Beyond PERK, other ER stress mechanisms are also implicated in IDDM. In the Akita diabetes mouse model, elevated expression of BiP, GRP94, ORP150, and HRD1, coupled with activation of ATF6 and XBP1, suggests the participation of numerous ER stress pathways (Nozaki et al. 2004). Deficiency in p58IPK, a modulator of the ER stress response, also leads to IDDM-like symptoms (Ladiges et al. 2005). Furthermore, the ER stress-induced activation of the WFS1 gene, associated with Wolfram syndrome (juvenile diabetes), implies a larger role of ER stress in β-cell failure (Fonseca et al. 2005).

In type 2 diabetes mellitus (NIDDM), defined by insulin resistance rather than decreased insulin production, obesity—a prominent risk factor—induces ER stress and leads to insulin resistance. Mice lacking in XBP1 have greater ER stress and poorer insulin tolerance, suggesting that XBP1 plays a protective role in sustaining β-cell function under stress conditions (Ozcan et al. 2004). Additionally, a polymorphism in ATF6 has been connected with NIDDM in Pima Indians, underlining the hereditary propensity to ER stress-related processes in diabetes (Thameem et al. 2006).

ER stress-mediated by ROS is also involved in liver illnesses such as chronic viral hepatitis, alcoholic hepatic steatosis, and non-alcoholic hepatic steatosis. Alcohol metabolism affects sulfur amino acid metabolism, resulting in homocysteine buildup and consequent ER stress. This stress stimulates ATF6 and IRE1 pathways, contributing to liver damage and potentially hepatocellular cancer (Kaplowitz and Ji 2006; Arai et al. 2006; Shuda et al. 2003). Additionally, lower glutathione levels in acute hepatitis aggravate protein aggregation and ER stress, further compromising liver function (Galligan et al. 2012).

In renal disorders, ER stress contributes to nephrotoxicity and renal fibrosis. For instance, drug-induced renal tubular injury from acetaminophen stimulates the PERK pathway, raises CHOP expression, and promotes cell death (Lorz et al. 2004; Naidoo 2009). Age-related renal fibrosis is also increased by ER stress, implicating it in chronic kidney disorders (Adam et al. 2012).

ER stress generated by substances like homocysteine stimulates macrophage apoptosis, which contributes to plaque development in atherosclerosis. This entails the activation of ER stress indicators and apoptotic pathways in macrophages within atherosclerotic plaques (Lawrence de Koning et al. 2003; Werstuck et al. 2001). Moreover, ER stress-induced cholesterol synthesis through SREBP activation further enhances macrophage death, aggravating atherosclerosis progression (Feng et al. 2003).

ER stress also plays a role in inflammation associated with illnesses such as inflammatory bowel disease (IBD) and autoimmune ailments. In these disorders, the accumulation of misfolded proteins promotes the ER stress response, which contributes to inflammation and immunological dysregulation (Endo et al. 2006).

Finally, ER stress contributes to heart disorders, including cardiac hypertrophy and cardiomyopathy. In these situations, apoptosis of cardiac myocytes triggered by ER stress affects heart function and promotes disease progression (Okada et al. 2004; Hamada et al. 2004).

## 6.5 Conclusion

Recent research has illuminated the intricate interplay between apoptosis and endoplasmic reticulum (ER) stress, revealing them as closely connected biological processes that regulate various signaling pathways within cells. Apoptosis refers to programmed cell death, which plays a crucial role in tissue homeostasis, development, and immunity. On the other hand, ER stress occurs when there is an accumulation of unfolded or misfolded proteins in the ER lumen, triggering the unfolded protein response (UPR) to restore ER function or induce apoptosis if the stress is severe and prolonged. Studies have shown that ER stress can induce apoptosis, and conversely, apoptosis can exacerbate ER stress by altering ER homeostasis and protein folding capacity. This bidirectional relationship is significant as it impacts a wide range of human pathologies, including neurodegenerative diseases, metabolic disorders, inflammatory conditions, and cancer, as well as influencing normal physiological processes. Targeting the signaling pathways activated by ER stress and apoptosis holds promise for developing new therapeutic strategies. Future research directions should focus on gaining a comprehensive understanding of the mechanisms underlying their interactions in both physiological and pathological contexts. This includes exploring how specific conditions, such as aging or disease states, influence the crosstalk between these stress responses. To achieve this, advanced experimental techniques such as real-time in vitro and in vivo imaging systems are crucial. These technologies allow researchers to observe and analyze the dynamics of ER stress and apoptosis responses at the cellular and organismal levels with high temporal and spatial resolution. Additionally, innovative conditional genetic animal models can be developed to dissect the roles of specific genes and pathways involved in mediating these stress responses. Moreover, further mechanistic and physiological studies using newly developed high-throughput technologies are needed to elucidate the precise molecular mechanisms underlying the interactions between ER stress and apoptosis. This knowledge will not only deepen our understanding of disease mechanisms but also pave the way for the development of more targeted and effective therapeutic interventions aimed at mitigating the impact of ER stress-related diseases.

In summary, ongoing and future research efforts focused on unraveling the complex interplay between ER stress and apoptosis are poised to provide critical insights that could revolutionize the diagnosis, treatment, and prevention of diseases associated with these cellular processes.

## References

Adam J, Bollée G, Fougeray S, Noël LH, Antignac C, Knebelmann B et al (2012) Endoplasmic reticulum stress in UMOD-related kidney disease: a human pathologic study. Am J Kidney Dis 59(1):117–121

Arai M, Kondoh N, Imazeki N, Hada A, Hatsuse K, Kimura F et al (2006) Transformation-associated gene regulation by ATF6 alpha during hepatocarcinogenesis. FEBS Lett 580(1):184–190

Behrends C, Langer CA, Boteva R, Böttcher UM, Stemp MJ, Schaffar G et al (2006) Chaperonin TRiC promotes the assembly of polyQ expansion proteins into nontoxic oligomers. Mol Cell 23(6):887–897

Bertolotti A, Zhang Y, Hendershot LM, Harding HP, Ron D (2000) Dynamic interaction of BiP and ER stress transducers in the unfolded-protein response. Nat Cell Biol 2(6):326–332

Boyce M, Bryant KF, Jousse C, Long K, Harding HP, Scheuner D, Kaufman RJ, Ma D, Coen DM, Ron D, Yuan J (2005) A selective inhibitor of eIF2alpha dephosphorylation protects cells from ER stress. Science 307(5711):935–939

Cao SS, Kaufman RJ (2012) Unfolded protein response. Curr Biol 22(16):R622–R626

Cullinan SB, Zhang D, Hannink M, Arvisais E, Diehl JA (2003) Nrf2 is a direct PERK substrate and effector of PERK-dependent cell survival. Mol Cell Biol 23(20):7198–7209

Dorner AJ, Wasley LC, Raney P, Haugejorden S, Green M, Kaufman RJ (1990) The stress response in Chinese hamster ovary cells. Regulation of ERp72 and protein disulfide isomerase expression and secretion. J Biol Chem 265(35):22029–22034

Endo M, Mori M, Akira S, Gotoh T (2006) C/EBP homologous protein (CHOP) is crucial for the induction of caspase-11 and the pathogenesis of lipopolysaccharide-induced inflammation. J Immunol 176(10):6245–6253

Feng B, Yao PM, Li Y, Devlin CM, Zhang D, Harding HP et al (2003) The endoplasmic reticulum is the site of cholesterol-induced cytotoxicity in macrophages. Nat Cell Biol 5(9):781–792

Fonseca SG, Fukuma M, Lipson KL, Nguyen LX, Allen JR, Oka Y et al (2005) WFS1 is a novel component of the unfolded protein response and maintains homeostasis of the endoplasmic reticulum in pancreatic beta-cells. J Biol Chem 280(47):39609–39615

Galligan JJ, Smathers RL, Shearn CT, Fritz KS, Backos DS, Jiang H et al (2012) Oxidative stress and the ER stress response in a murine model for early-stage alcoholic liver disease. J Toxicol 2012:Article ID 207594

Hamada H, Suzuki M, Yuasa S, Mimura N, Shinozuka N, Takada Y et al (2004) Dilated cardiomyopathy caused by aberrant endoplasmic reticulum quality control in mutant KDEL receptor transgenic mice. Mol Cell Biol 24(18):8007–8017

Harding HP, Zhang Y, Ron D (1999) Protein translation and folding are coupled by an endoplasmic-reticulum-resident kinase. Nature 397(6716):271–274

Harding HP, Calfon M, Urano F, Novoa I, Ron D (2002) Transcriptional and translational control in the mammalian unfolded protein response. Annu Rev Cell Dev Biol 18:575–599

Hetz C, Bernasconi P, Fisher J, Lee AH, Bassik MC, Antonsson B, Brandt GS, Iwakoshi NN, Schinzel A, Glimcher LH, Korsmeyer SJ (2006) Proapoptotic BAX and BAK modulate the unfolded protein response by a direct interaction with IRE1alpha. Science 312(5773):572–576

Hu P, Han Z, Couvillon AD, Exton JH (2004) Critical role of endogenous Akt/IAPs and MEK1/ERK pathways in counteracting endoplasmic reticulum stress induced cell death. J Biol Chem 279(47):49420–49429

Kadowaki H, Nishitoh H, Ichijo H (2004) Survival and apoptosis signals in ER stress: the role of protein kinases. J Chem Neuroanat 28(1–2):93–100

Kaneko M, Nomura Y (2003) ER signaling in unfolded protein response. Life Sci 74(2–3):199–205

Kaplowitz N, Ji C (2006) Unfolding new mechanisms of alcoholic liver disease in the endoplasmic reticulum. J Gastroenterol Hepatol 21(Suppl. S3):S7–S9

Kim R, Emi M, Tanabe K, Murakami S (2006) Role of the unfolded protein response in cell death. Apoptosis 11(1):5–13

Lawrence de Koning AB, Werstuck GH, Zhou J, Austin RC (2003) Hyperhomocysteinemia and its role in the development of atherosclerosis. Clin Biochem 36(6):431–441

Lorz C, Justo P, Sanz A, Subirá D, Egido J, Ortiz A (2004) Paracetamol-induced renal tubular injury: a role for ER stress. J Am Soc Nephrol 15(2):380–389

Ma Y, Brewer JW, Diehl JA, Hendershot LM (2002) Two distinct stress signaling pathways converge upon the CHOP promoter during the mammalian unfolded protein response. J Mol Biol 318(5):1351–1365

Malhotra JD, Kaufman RJ (2007) Endoplasmic reticulum stress and oxidative stress: a vicious cycle or a double-edged sword? Antioxid Redox Signal 9(12):2277–2293

Marciniak SJ, Yun CY, Oyadomari S, Novoa I, Zhang Y, Jungreis R, Nagata K, Harding HP, Ron D (2004) CHOP induces death by promoting protein synthesis and oxidation in the stressed endoplasmic reticulum. Genes Dev 18(24):3066–3077

Mori K (2000) Tripartite management of unfolded proteins in the endoplasmic reticulum. Cell 101(5):451–454

Mori K (2005) Endoplasmic reticulum stress signaling transmitted by ATF6 mediates apoptosis during muscle development. J Cell Biol 169(4):555–560

Morishima N, Nakanishi K, Takenouchi H, Shibata T, Yasuhiko Y (2002) An endoplasmic reticulum stress-specific caspase cascade in apoptosis. Cytochrome c-independent activation of caspase-9 by caspase-12. J Biol Chem 277(37):34287–34294

Naidoo N (2009) The endoplasmic reticulum stress response and aging. Rev Neurosci 20(1):23–37

Nakagawa T, Zhu H, Morishima N, Li E, Xu J, Yankner BA, Yuan J (2000) Caspase-12 mediates endoplasmic-reticulum-specific apoptosis and cytotoxicity by amyloid-beta. Nature 403(6765):98–103

Nakanishi K, Sudo T, Morishima N (2005) Endoplasmic reticulum stress signaling transmitted by ATF6 mediates apoptosis during muscle development. J Cell Biol 169(4):555–560

Nakano K, Vousden KH (2001) PUMA, a novel proapoptotic gene, is induced by p53. Mol Cell 7(3):683–694

Nishitoh H, Matsuzawa A, Tobiume K, Saegusa K, Takeda K, Inoue K, Hori S, Kakizuka A, Ichijo H (2002) ASK1 is essential for endoplasmic reticulum stress-induced neuronal cell death triggered by expanded polyglutamine repeats. Genes Dev 16(11):1345–1355

Novoa I, Zeng H, Harding HP, Ron D (2001) Feedback inhibition of the unfolded protein response by GADD34-mediated dephosphorylation of eIF2alpha. J Cell Biol 153(5):1011–1022

Nozaki JI, Kubota H, Yoshida H, Naitoh M, Goji J, Yoshinaga T et al (2004) The endoplasmic reticulum stress response is stimulated through the continuous activation of transcription factors ATF6 and XBP1 in Ins2+/Akita pancreatic beta cells. Genes Cells 9(3):261–270

Oda E, Ohki R, Murasawa H, Nemoto J, Shibue T, Yamashita T, Tokino T, Taniguchi T, Tanaka N (2000) Noxa, a BH3-only member of the Bcl-2 family and candidate mediator of p53-induced apoptosis. Science 288(5468):1053–1058

Okada K, Minamino T, Tsukamoto Y, Liao Y, Tsukamoto O, Takashima S et al (2004) Prolonged endoplasmic reticulum stress in hypertrophic and failing heart after aortic constriction: possible contribution of endoplasmic reticulum stress to cardiac myocyte apoptosis. Circulation 110(6):705–712

Ozcan U, Cao Q, Yilmaz E, Lee AH, Iwakoshi NN, Ozdelen E et al (2004) Endoplasmic reticulum stress links obesity, insulin action, and type 2 diabetes. Science 306(5695):457–461

Patil C, Walter P (2001) Intracellular signaling from the endoplasmic reticulum to the nucleus: the unfolded protein response in yeast and mammals. Curr Opin Cell Biol 13(3):349–355

Saleh M, Vaillancourt JP, Graham RK, Huyck M, Srinivasula SM, Alnemri ES, Steinberg MH, Nolan V, Baldwin CT, Hotchkiss RS, Buchman TG, Zehnbauer BA, Hayden MR, Farrer LA, Roy S, Nicholson DW (2006) Differential modulation of endoplasmic reticulum stress-induced apoptosis by caspase-2 and caspase-12. J Biol Chem 281(10):7260–7270

Schröder M, Kaufman RJ (2005) ER stress and the unfolded protein response. Mutat Res 569(1–2):29–63

Scorrano L, Oakes SA, Opferman JT, Cheng EH, Sorcinelli MD, Pozzan T, Korsmeyer SJ (2003) BAX and BAK regulation of endoplasmic reticulum $Ca^{2+}$: a control point for apoptosis. Science 300(5616):135–139

Shen J, Chen X, Hendershot L, Prywes R (2002) ER stress regulation of ATF6 localization by dissociation of BiP/GRP78 binding and unmasking of Golgi localization signals. Dev Cell 3(1):99–111

Shuda M, Kondoh N, Imazeki N, Tanaka K, Okada T, Mori K et al (2003) Activation of the ATF6, XBP1 and grp78 genes in human hepatocellular carcinoma: a possible involvement of the ER stress pathway in hepatocarcinogenesis. J Hepatol 38(5):605–614

Sok J, Wang XZ, Batchvarova N, Kuroda M, Harding H, Ron D (1999) CHOP-dependent stress-inducible expression of a novel form of carbonic anhydrase VI. Mol Cell Biol 19(1):495–504

Sriburi R, Jackowski S, Mori K, Brewer JW (2004) XBP1: a link between the unfolded protein response, lipid biosynthesis, and biogenesis of the endoplasmic reticulum. J Cell Biol 167(1):35–41
Thameem F, Farook VS, Bogardus C, Prochazka M (2006) Association of amino acid variants in the activating transcription factor 6 gene (ATF6) on 1q21-q23 with type 2 diabetes in Pima Indians. Diabetes 55(3):839–842
Urano F, Wang X, Bertolotti A, Zhang Y, Chung P, Harding HP, Ron D (2000) Coupling of stress in the ER to activation of JNK protein kinases by transmembrane protein kinase IRE1. Science 287(5453):664–666
Ushioda R, Hoseki J, Araki K, Jansen G, Thomas DY, Nagata K (2008) ERdj5 is required as a disulfide reductase for degradation of misfolded proteins in the ER. Science 321(5888):569–572
Wang W, Chan JY (2006) Nrf1 is targeted to the ER membrane by an N-terminal transmembrane domain: inhibition of nuclear translocation and transacting function. J Biol Chem 281(28):19676–19687
Wang L, Di LJ (2014) BRCA1 and estrogen/estrogen receptor in breast cancer: where they interact? Int J Biol Sci 10(5):566–575
Wang XZ, Kuroda M, Sok J, Batchvarova N, Kimmel R, Chung P, Zinszner H, Ron D (1998) Identification of novel stress-induced genes downstream of chop. EMBO J 17(13):3619–3630
Wei MC, Zong WX, Cheng EH, Lindsten T, Panoutsakopoulou V, Ross AJ, Roth KA, MacGregor GR, Thompson CB, Korsmeyer SJ (2001) Proapoptotic BAX and BAK: a requisite gateway to mitochondrial dysfunction and death. Science 292(5517):727–730
Werstuck GH, Lentz SR, Dayal S, Hossain GS, Sood SK, Shi YY et al (2001) Homocysteine-induced endoplasmic reticulum stress causes dysregulation of the cholesterol and triglyceride biosynthetic pathways. J Clin Invest 107(10):1263–1273
Wootz H, Hansson I, Korhonen L, Lindholm D (2006) XIAP decreases caspase-12 cleavage and calpain activity in spinal cord of ALS transgenic mice. Exp Cell Res 312(10):1890–1898
Yamaguchi H, Wang HG (2004) CHOP is involved in endoplasmic reticulum stress-induced apoptosis by enhancing DR5 expression in human carcinoma cells. J Biol Chem 279(44):45495–45502
Ye J, Rawson RB, Komuro R, Chen X, Davé UP, Prywes R, Brown MS, Goldstein JL (2000) ER stress induces cleavage of membrane-bound ATF6 by the same proteases that process SREBPs. Mol Cell 6(6):1355–1364
Yoshida H, Haze K, Yanagi H, Yura T, Mori K (1998) Identification of the cis-acting endoplasmic reticulum stress response element responsible for transcriptional induction of mammalian glucose-regulated proteins. Involvement of basic leucine zipper transcription factors. J Biol Chem 273(50):33741–33749
Yoshida H, Matsui T, Yamamoto A, Okada T, Mori K (2001) XBP1 mRNA is induced by ATF6 and spliced by IRE1 in response to ER stress to produce a highly active transcription factor. Cell 107(7):881–891
Zha J, Harada H, Yang E, Jockel J, Korsmeyer SJ (1996) Serine phosphorylation of death agonist BAD in response to survival factor results in binding to 14-3-3 not BCL-X(L). Cell 87(4):619–628
Zhang K, Wong HN, Song B, Miller CN, Scheuner D, Kaufman RJ (2000) The unfolded protein response sensor IRE1alpha is required at 2 distinct steps in B cell lymphopoiesis. J Clin Invest 115(2):268–281
Zhang C, Kawauchi J, Adachi MT, Hashimoto Y, Oshiro S, Aso T, Kitajima S (2001) Activation of JNK and transcriptional repressor ATF3/LRF1 through the IRE1/TRAF2 pathway is implicated in human vascular endothelial cell death by homocysteine. Biochem Biophys Res Commun 289(3):718–724
Zhang P, McGrath B, Li S, Frank A, Zambito F, Reinert J et al (2002) The PERK eukaryotic initiation factor 2 alpha kinase is required for the development of the skeletal system, postnatal growth, and the function and viability of the pancreas. Mol Cell Biol 22(11):3864–3874

# 7 Apoptosis: Oxidative Stress and Antioxidants

Satyajit Halder, Shraman Jana, Anirban Manna, and Kuladip Jana

**Abstract**

Reactive oxygen species (ROS) are fundamental to apoptosis, the programmed cell death essential for multicellular organisms' survival and proper functioning. ROS production within cells is meticulously balanced by various antioxidant defenses. At physiological levels, ROS are vital for regulating cell cycle progression, proliferation, differentiation, migration, and apoptosis. Additionally, they play crucial roles in maintaining redox homeostasis, bolstering immune function, and activating diverse cellular signaling pathways. However, excessive ROS can cause significant damage to proteins, nucleic acids, lipids, membranes, and organelles, thereby triggering cell death mechanisms like apoptosis. Apoptosis consists of three distinct phases: induction, effector, and execution, each significantly influenced by ROS and oxidative stress. Mitochondrial dysfunction is a pivotal event in apoptosis, leading to the release of apoptotic factors such as cytochrome C, apoptosis-inducing factor (AIF), and apoptotic protease-activating factor-1 (APAF-1), which activate caspases and result in cell death. This process is tightly regulated and is linked to various chronic diseases, including cancer, neurodegenerative disorders, and cardiovascular diseases. ROS-induced DNA damage, including mutations and gene rearrangements, can either prompt apoptosis or contribute to oncogenesis by affecting proto-oncogenes and tumor suppressor genes. Understanding the balance between ROS production and apoptosis

S. Halder · A. Manna · K. Jana (✉)
Division of Molecular Medicine, Bose Institute, Unified Academic Campus, Kolkata, West Bengal, India
e-mail: kuladip@jcbose.ac.in

S. Jana
IISER Kolkata, Campus Rd, Mohanpur, Haringhata Farm, West Bengal, India





K. Jana (ed.), *Apoptosis and Human Health: Understanding Mechanistic and Therapeutic Potential*, https://doi.org/10.1007/978-981-97-7905-5_7

regulation is critical for maintaining cellular homeostasis. ROS are central to cell signaling and regulate major apoptosis pathways mediated by mitochondria, death receptors, and the endoplasmic reticulum (ER). This review explores the role of ROS in these apoptosis pathways and examines the complex interactions between different ROS-mediated signaling mechanisms, highlighting areas of ongoing research in this field.

**Keywords**

Apoptosis · ROS · mitochondrial ROS · ER stress · diseases · antioxidant therapy

## 7.1 Introduction

Apoptosis, also known as programmed cell death (PCD), is vital for developing and maintaining homeostasis in multicellular organisms. This mechanism eliminates damaged, infected, or potentially cancerous cells. Nevertheless, an imbalance in apoptotic cell death, whether too little or too much, can result in negative biological outcomes. Reactive oxygen species (ROS) are reactive molecules crucial for cellular signaling and redox regulation in biological systems. These species include free radicals such as superoxide anion (O2•−) and hydroxyl radical (OH•), as well as non-radical molecules like hydrogen peroxide ($H_2O_2$). The balance between ROS production and antioxidant defenses, comprising enzymatic (e.g., superoxide dismutase, catalase, glutathione peroxidase) and non-enzymatic (e.g., vitamins C and E, glutathione) scavengers, is critical for maintaining cellular homeostasis (Halliwell 2011).

Physiologically, ROS serve as signaling molecules in diverse processes such as cell proliferation, differentiation, migration, and programmed cell death (apoptosis) (Covarrubias et al. 2008). They activate intricate signaling pathways, including PI3K/Akt, MAPK, Nrf2/Keap1, NF-κB, and p53, which modulate cellular responses ranging from survival to apoptosis and autophagy (Bae et al. 2011). However, dysregulation of ROS levels leads to oxidative stress, where excessive ROS production overwhelms antioxidant defenses, resulting in cellular damage to proteins, lipids, nucleic acids, and organelles like mitochondria (Kehrer and Klotz 2015) (Fig. 7.1).

Endogenous sources of ROS include mitochondrial respiration, enzymatic reactions, and NADPH oxidases, while exogenous factors such as environmental pollutants, UV radiation, and certain drugs also contribute to ROS generation (Brieger et al. 2012; Dickinson and Chang 2011). Superoxide reacts with nitric oxide (NO•) to form peroxynitrite (ONOO−), and hydrogen peroxide can generate hydroxyl radicals (OH•) via the Fenton reaction in the presence of metal ions, highlighting the complex chemistry of ROS in biological systems (West and Marnett 2006; Pallepati and Averill-Bates 2011). Cellular detoxification mechanisms involve enzymes like catalase and glutathione peroxidases, which mitigate the harmful effects of ROS by converting them into less reactive species (Flusberg and Sorger 2015) (Fig. 7.1).

Living organisms are constantly exposed to environmental changes and toxic components like pesticides, heavy metals, and radiation, which disrupt cellular homeostasis and lead to stress. This stress can cause cellular damage, disease, or

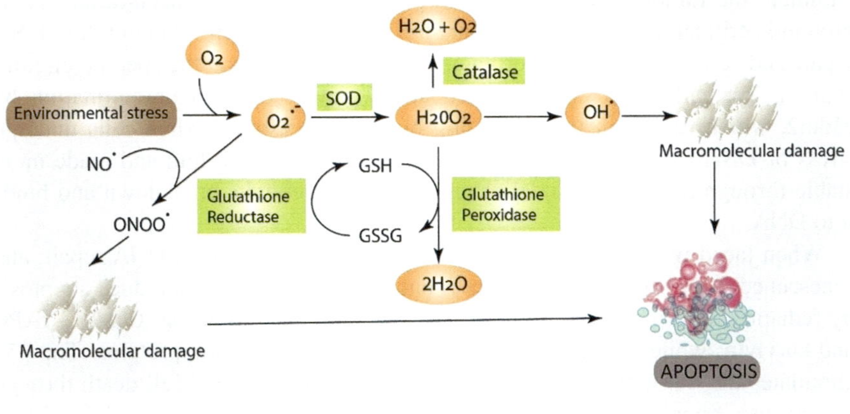


**Fig. 7.1** Environmental stress induces the production of reactive oxygen species (ROS), leading to cellular damage and apoptosis

death, contributing to conditions such as cancer, diabetes, chronic lung diseases, and neurodegenerative disorders (Franco et al. 2009; Gibson 2013). Environmental pollutants generate reactive oxygen species (ROS) through processes like redox cycling, catalyzed by enzymes such as NADPH cytochrome P450 reductases. High levels of ROS, including hydroxyl radicals and peroxynitrite, severely damage macromolecules, resulting in apoptosis or necrosis, while moderate levels may induce autophagy, a survival mechanism that can also lead to cell death. Understanding ROS-induced damage and balancing cell survival and death pathways is crucial for mitigating these health impacts. Future research efforts should focus on elucidating the specific roles of ROS in health and disease contexts, exploring antioxidant therapies, and developing interventions to restore redox balance and mitigate oxidative damage in biological systems (Halliwell 2012).

Toxic compounds, known as xenobiotics, generate free radicals through a one-electron reduction process catalyzed by cytochrome P450 reductases. These radicals react rapidly with oxygen to form superoxide (O2•−), which can then react with nitric oxide (NO•) to produce peroxynitrite (ONOO−) or undergo dismutation to generate hydrogen peroxide ($H_2O_2$), catalyzed by superoxide dismutase (SOD). Hydrogen peroxide can either be detoxified by antioxidants such as catalase and glutathione peroxidase (GSH/GPx) or participate in the Fenton reaction, catalyzed by metal ions, to produce hydroxyl radicals (OH•). Both hydroxyl radicals and peroxynitrite cause severe damage to cellular proteins, lipids, and nucleic acids, ultimately leading to apoptosis

## 7.2 Apoptosis and ROS

Lower concentrations of reactive oxygen species (ROS), such as hydrogen peroxide ($H_2O_2$), are linked to cell survival responses. In contrast, greater concentrations can initiate cell death processes, such as apoptosis. Reactive oxygen species (ROS)

stimulate the tumor suppressor protein p53 activity at low concentrations. This protein is critical in regulating cellular stress responses by either facilitating DNA repair and cell survival or triggering apoptosis, depending on the specific circumstances. Typically, p53 is maintained at low levels through ubiquitination by Mdm2, which results in its destruction by the proteasome. When cells undergo stress or DNA damage, the protein p53 is liberated from Mdm2 and made more stable through post-translational changes. This prevents its breakdown and binds it to DNA.

When faced with modest stress, p53 triggers cell cycle arrest, DNA repair, and senescence. Nevertheless, when there is enough damage, p53 can induce apoptosis by reducing the expression of pro-survival proteins such as Bcl-2, Bcl-XL, IAPs, and survivin, while increasing the expression of pro-apoptotic proteins. The p53 stimulates the transcription of genes that promote programmed cell death through the intrinsic apoptosis pathway, including Bax, Bid, Puma, Noxa, and Apaf-1. It also triggers the production of extrinsic pro-apoptotic proteins such as Fas, FasL, DR-4, and DR-5. In addition, p53, located in the cytosol, can move to the mitochondria and directly interact with anti-apoptotic proteins (Bcl-2, Bcl-XL, Mcl-1) and pro-apoptotic proteins (Bax, Bak). This interaction results in the permeabilization of the outer membrane of the mitochondria, leading to the release of factors that induce apoptosis. This process was described by Leu et al. in 2004. The presence of p53 in the cytosol can directly trigger the activation of Bax by inducing changes in its structure. This process increases mitochondrial outer membrane permeabilization (MOMP) and apoptosis (Chipuk et al. 2004).

Apoptosis induced by hydrogen peroxide ($H_2O_2$) has been linked to the upregulation of proteins such as p53, Puma, Noxa, and Bax, along with phosphorylation of p53 at Ser15 and Ser46 across various cell types, including rat neural AF5 cells, glioma, colon cancer, and human cervical carcinoma HeLa cells (Chipuk and Green 2008; Chipuk et al. 2005). In HeLa cells, p53 acts as an upstream regulator, facilitating the upregulation of Puma and the translocation of Bax to the mitochondria during $H_2O_2$-induced apoptosis (Chipuk et al. 2005).

The redox regulation of caspase activity involves post-translational modifications of their catalytic cysteine residues (Denisov et al. 2013). The catalytic cysteines in most caspases are prone to oxidation, while procaspase-9, procaspase-3, and caspase-3 can undergo S-glutathiolation (Denisov et al. 2013). Although oxidants like $H_2O_2$ can activate caspases, they can also irreversibly inhibit their enzymatic activity (Denisov et al. 2013). This inhibitory effect is noted in caspases-3, −8, and − 9. Additionally, cellular glutathione (GSH) can modulate caspase-3 activity through S-glutathiolation, reducing its susceptibility to proteolytic cleavage and leading to apoptosis resistance (Denisov et al. 2013). However, this process's precise mechanisms and consequences in apoptosis regulation remain unclear. Specifically, it is still unknown which upstream apoptotic components are targets for S-glutathiolation and which cysteine residues are particularly susceptible (Denisov et al. 2013).

### 7.2.1 Mitochondrial Pathway of Apoptosis and ROS

The activation of the mitochondrial pathway of apoptosis is triggered by different cellular stresses, including damage to mitochondrial DNA (mtDNA), deprivation of growth factors, exposure to heat shock, lack of oxygen, endoplasmic reticulum stress, and developmental signals (Chipuk and Green 2008; Denisov et al. 2013). ROS are strongly linked to this system, as mitochondria are the principal source of intracellular ROS, created through electron leakage from the respiratory electron transport chain. These ROS can damage adjacent structures, including mtDNA, affecting the transcription of proteins essential in the electron transport chain. This leads to increased ROS formation, loss of mitochondrial membrane potential, and decreased ATP synthesis, ending in death.

ROS such as hydrogen peroxide ($H_2O_2$) and superoxide anion can cause cytochrome c release from mitochondria, initiating apoptosis via the mitochondrial pathway (Denisov et al. 2013). Key components of the mitochondrial permeability transition pore (MPTP), including voltage-dependent anion channel (VDAC), adenine nucleotide translocase (ANT), and cyclophilin D, are targets for ROS-induced oxidative alterations that increase MPTP opening (Denisov et al. 2013). $H_2O_2$ causes early mitochondrial membrane hyperpolarization, followed by membrane potential collapse, translocation of pro-apoptotic proteins Bax and Bad to the mitochondria, and cytochrome c release. This loss disturbs the electron transport system, further boosting ROS generation.

In HeLa cells, p53 operates as an upstream regulator of $H_2O_2$-induced apoptosis via the mitochondrial route. $H_2O_2$ induces both caspase-dependent and caspase-independent apoptosis (Chipuk et al. 2005). The caspase-dependent mechanism involves loss of mitochondrial membrane permeability, cytochrome c release, and caspase-9-mediated activation of caspase-3. In contrast, the caspase-independent process requires the release of apoptosis-inducing factor (AIF) from mitochondria and its transport to the nucleus (Denisov et al. 2013). Additionally, oxidative stress-induced activation of c-Jun N-terminal kinase (JNK) can phosphorylate and inactivate anti-apoptotic proteins such as Bcl-2 and Bcl-XL while activating pro-apoptotic Bcl-2 family members.

ROS also targets cardiolipin, a phospholipid that attaches cytochrome c to the inner mitochondrial membrane (IMM). Oxidation of cardiolipin by ROS lowers its binding affinity for cytochrome c, increasing free cytochrome c levels that can trigger the apoptotic cascade. Oxidative stress and mitochondrial calcium ($Ca^{2+}$) buildup can both trigger mitochondrial permeability transition (MPT), leading to IMM pore opening. Factors like ATP depletion, oxidative stress, low pH, and high inorganic phosphate levels also facilitate pore opening, producing osmotic swelling, outer mitochondrial membrane (OMM) rupture, and release of pro-apoptotic proteins such as cytochrome c. The degree of pore opening appears to rely on the $Ca^{2+}$ concentration, with larger levels encouraging necrosis and lower levels promoting apoptosis. Moreover, ROS production can exacerbate $Ca^{2+}$-induced cellular damage by reducing the threshold $Ca^{2+}$ concentration required for MPT (Denisov et al. 2013).

Disruption of cellular glutathione (GSH) redox state through GSH oxidation or outflow contributes to oxidant-mediated apoptosis. GSH is key in shielding cells against apoptosis triggered by diverse stimuli, highlighting its central involvement in cellular defense systems against oxidative stress (Denisov et al. 2013).

### 7.2.2 Death Receptor Pathway and ROS

Links between ROS and extrinsic-induced apoptosis also exist. The cytokine TNF-α is critical in complex signaling networks that govern cell survival or death. TNF-α can also promote caspase-independent cell death by necrosis (necroptosis) involving ROS generated from mitochondria and non-mitochondrial sources.

ROS and TNF-α interact in a positive feedback loop. At low levels, ROS regulate the TNF-R1 signaling pathway, whereas greater doses can lead to DNA damage and apoptosis. The role of mitochondrial ROS on NF-κB activation remains contested. However, ROS can activate NF-κB through TNF-R1 signaling, boosting the upregulation of antioxidants like catalase and MnSOD (Circu and Aw 2010). These antioxidants inhibit TNF-induced apoptosis by neutralizing mitochondrial-derived ROS. Additionally, NF-κB increases transcription of anti-apoptotic genes like IAP and Bcl-XL, which suppress TNF-R1-mediated apoptosis by preventing caspase-8 activation (Schieber and Chandel 2014). However, TNF-induced ROS can suppress NF-κB activation, reducing NF-κB-mediated survival signaling and inducing apoptosis. In contrast, mitochondrial ROS seem to increase TNF-mediated NF-κB activation by suppressing phosphatases that regulate kinases mediating NF-κB signaling. This inhibition leads to higher IκB phosphorylation, degradation, and subsequent NF-κB activation.

The methods by which TNF stimulation promotes mitochondrial ROS production remain unknown. Upon TNF-R1 activation, caspase-8 binds to ROS modulator-1 (ROMO-1) in the OMM (Filomeni et al. 2015). ROMO1 sequesters Bcl-XL, causing mitochondrial membrane potential loss and enhanced ROS production. ROS then activates ASK-1, generally coupled to decreased thioredoxin under basal conditions. Oxidation of thioredoxin by ROS releases ASK-1, allowing it to activate downstream JNK and p38 MAPK, hence increasing TNF-induced apoptosis. ASK-1 can also be inactivated by ROS-regulated protein phosphatase-5. ROS can additionally inhibit JNK-inactivating phosphatases, leading to JNK activation, cytochrome c release, caspase-3 activation, and death (Filomeni et al. 2015).

ROS such as $H_2O_2$ can activate Fas-mediated apoptosis. In HeLa cells, $H_2O_2$ upregulates FasL, increases FADD translocation to the plasma membrane, and activates caspase-8. Activated caspase-8 subsequently processes caspase-2, leading to Bid cleavage into t-Bid and increased mitochondrial-induced apoptosis. Moreover, oxidative stress initiates death receptor-mediated apoptosis in various cell types, including upregulation of FasL and Fas in murine intestinal epithelial cells (Gogvadze et al. 2006) and HUVECs (Circu and Aw 2010), as well as activation of caspase-8 and caspase-3 in HL-60 cells (Schieber and Chandel 2014).

During stress conditions, death-associated protein 6 (Daxx) interacts with Fas via its DD, activating ASK-1 and promoting JNK. Active JNK phosphorylates Bcl-2, promoting apoptosis via the mitochondrial route. The Nox family proteins serve as substantial generators of ROS (O2 • −) within cells (Filomeni et al. 2015), situated specifically in subcellular compartments like lipid rafts, caveolae, endosomes, and the nucleus. Death ligand-receptor interaction for Fas and TNF-α stimulates lipid raft formation, Nox recruitment, and ROS generation, producing lipid raft-derived redox signaling platforms that activate death receptors and trigger apoptosis. Compared to standard ligand/receptor-induced apoptotic signaling, the physiological importance of ROS-dependent receptor-mediated apoptosis remains incompletely understood.

ROS are implicated in controlling TRAIL-mediated apoptosis (Filomeni et al. 2015), although the underlying processes are unknown. ROS can upregulate DR gene expression, such as TRAIL, a process blocked by antioxidants like catalase and N-acetylcysteine. Furthermore, enhanced ROS production sensitizes cells to TRAIL-mediated apoptosis.

ROS can activate p53 Bcl-2 proteins. ROS induces cardiolipin oxidation, leading to the release of cytochrome c (Cyt c) into the cytosol. Additionally, ROS causes mitochondrial membrane depolarization and/or opening of Bax/Bak channels on the outer mitochondrial membrane (OMM), facilitating the release of cytochrome c into the cytosol. In the cytosol, cytochrome c forms the apoptosome complex with Apaf-1 and procaspase-9, activating caspase-9. Caspase-9 subsequently activates effector caspases such as caspase-3, which cleavages cellular proteins and cell apoptosis (Fig. 7.2).

### 7.2.3 ER Pathway of Apoptosis and ROS

Maintaining the correct redox state is essential in the endoplasmic reticulum (ER) lumen because an oxidative environment is necessary for producing disulfide bonds and protein folding. In contrast to the more reducing cytosol, the ER lumen's oxidative milieu is characterized by a high ratio of oxidized to reduced glutathione (GSSG/GSH). Chaperone proteins, most notably protein disulfide isomerase (PDI), control the production of disulfide bonds in proteins.

The most common chaperone in the ER, PDI, is essential for the disulfide bond synthesis in developing proteins. It undergoes a two-electron reduction at its active site and catalyzes the oxidation of cysteine residues within polypeptide substrates (Appenzeller-Herzog and Ellgaard 2008). To return PDI to its oxidized form and produce reactive oxygen species (ROS), the ER membrane-associated protein endoplasmic reticulum oxidoreductin-1 (ERO1) is necessary.

Protein thiol groups on PDI undergo an electron transfer dependent on flavin adenine dinucleotide (FAD) to molecular oxygen ($O_2$) and ERO1. Hydrogen peroxide ($H_2O_2$) is produced in this reaction when $O_2$ reoxidizes the reduced form of FAD (FADH2), which acts as an electron acceptor in the process (Gross et al. 2006). The

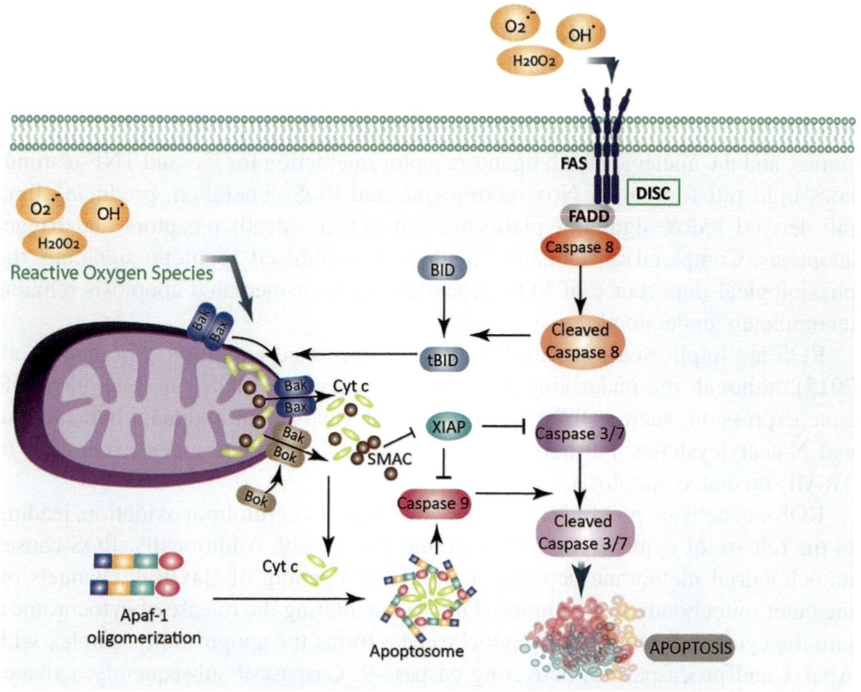


**Fig. 7.2** The activation of the mitochondrial (intrinsic) pathway of apoptosis by reactive oxygen species (ROS) occurs both exogenously and endogenously

nature of the $H_2O_2$ produced in the ER during oxidative protein folding is yet unknown. It may be harmful or serve as a signaling molecule.

Excessive generation of $H_2O_2$ through ERO1 can be caused by elevated glutathione (GSH) levels in the ER. Increased GSH causes larger amounts of oxidized glutathione (GSSG) when ERO1 is active because GSH is a substrate for PDI-catalyzed oxidation (Harding et al. 2003). The redox equilibrium in the ER lumen is upset by this increase in $H_2O_2$ and GSSG, which exacerbates the ER stress (Frand and Kaiser 1998).

Furthermore, redox control affects the activities of both PDI and ERO1, suggesting that they perform best within a particular redox range. Since chronic ER stress is linked to several illnesses, including diabetes and neurodegenerative disorders, disruptions in this balance may have a substantial impact on cellular survival and function.

The need for ERO1-mediated disulfide production is increased when endoplasmic reticulum (ER) proteostasis is disrupted. Hydrogen peroxide ($H_2O_2$) created from misfolded proteins in the ER is produced more frequently due to their ineffective cycles of misfolded disulfide link creation and reduction. This triggers the unfolded protein response (UPR) signaling pathway and adds to ER stress. Reactive oxygen species (ROS) produced by ER stress have unclear exact causes and characteristics. Furthermore, eliminating misfolded proteins' erroneously coupled disulfide linkages might deplete glutathione (GSH), raising ROS levels. According to estimates, the ER's disulfide bond production process during oxidative protein folding generates about 25% of all cellular ROS.

Compared to proteins with fewer disulfide links, those rich in disulfide bonds have a major role in the formation of ROS in the ER. ER stress-mediated cell death can be brought on by the ROS created as byproducts of protein oxidation in the ER. Glutathione peroxidases 7 and 8 and peroxiredoxin 4 are examples of antioxidants found in the ER that detoxify $H_2O_2$ and are crucial for preserving ER homeostasis because they stop oxidative changes and misfolding of newly generated proteins, which can lead to ER stress.

A change in oxidative phosphorylation brought on by releasing calcium ($Ca^{2+}$) from the ER or energy depletion may cause increased ROS generation during protein misfolding. Antioxidants such as butylated hydroxyanisole (BHA) might reduce the effects of excessive ROS by causing protein misfolding, which in turn triggers UPR activation and the induction of C/EBP homologous protein (CHOP). Additionally, high ROS can cause erroneous disulfide bond formation and inactivate the thiol-disulfide exchange between PDI and ERO1, impairing protein folding.

While severe or sustained oxidative stress and protein misfolding can initiate apoptotic signaling pathways, modest oxidative stress that activates the UPR is an adaptive response to preserve cell survival and function. According to Malhotra and Kaufman, persistent ER stress triggers the activation of inositol-requiring enzyme 1 alpha (IRE1α), which in turn triggers the activation of p38 mitogen-activated protein kinase (p38MAPK) and apoptosis signal-regulating kinase 1 (ASK1). These processes further activate CHOP and contribute to the formation of ROS. Acute ROS production from CHOP activation promotes an overoxidizing environment, increases PDI activity, and sets off a vicious cycle that results in the formation of more disulfide bonds and reoxidation of PDI coupled with the buildup of misfolded proteins.

According to Oakes and Papa (2015), ERO1 can also activate the inositol trisphosphate receptor (IP3R), which results in significant $Ca^{2+}$ transport from the ER to the mitochondria and mitochondrial-mediated cell death. ASK1, whose activity is directly impacted by the oxidative state, regulates c-Jun N-terminal kinase (JNK), which controls ER-stress-mediated apoptosis. ASK1 conjugation with tumor necrosis factor receptor-associated factor 2 (TRAF2) and subsequent JNK activation require thioredoxin-mediated ASK1 oxidation (Hetz and Mollereau 2014).

Apart from eukaryotic initiation factor 2 alpha (eIF2α), nuclear factor erythroid 2-related factor 2 (Nrf2) can also be phosphorylated by protein kinase RNA-like endoplasmic reticulum kinase (PERK). Through the induction of antioxidant genes like heme oxygenase 1 (HO-1), this phosphorylation causes the Nrf2-Kelch-like ECH-associated protein 1 (Keap1) complex to dissociate, promoting the expression of genes with antioxidant response elements (AREs) and preventing oxidative stress. The essential enzyme for synthesizing GSH, glutamate-cysteine ligase (GCL), is also expressed more when Nrf2 is present.

ROS can increase $Ca^{2+}$ leakage from the ER lumen under ER stress. The ER and mitochondria have intimate interactions through specialized compartments called mitochondria-associated membranes (MAMs), which regulate the movement of ions and small molecules between these organelles. Because mitochondria are close to the ER, they absorb large amounts of $Ca^{2+}$ that are released from the ER. By supplying NADH for mitochondrial respiration and ATP synthesis, elevated mitochondrial $Ca^{2+}$ levels activate $Ca^{2+}$-sensitive matrix dehydrogenases, boosting metabolic activity. Through complexes I and III, increased mitochondrial respiration increases the formation of ROS. Furthermore, the opening of the mitochondrial permeability transition pore (MPTP) caused by mitochondrial $Ca^{2+}$ overload can release ATP, GSH, and cytochrome c, which raises ROS levels even further (Giorgi et al. 2018).

The pro-apoptotic BH3-only Bcl-2 protein Bim can also be expressed in response to ROS. Bim is phosphorylated by JNK, which releases it from inhibitory dynein complexes and activates the Bax and Bak proteins. Bax-dependent apoptosis results from signal transmission from the ER to mitochondria made possible by $Ca^{2+}$ release from the ER via Bax-Bak channels.

The generation of ROS in the mitochondria and ER stress are functionally related. According to Fu et al., the ATF-4/CHOP pro-apoptotic pathway increases the formation of ROS in the mitochondria through its transcriptional target, ERO1. ERO1 levels are critical for the production of ROS by mitochondria and are highest in MAMs. ERO1 facilitates $Ca^{2+}$ efflux from the ER-mediated by IP3R, which is absorbed by mitochondria through the MAM. Thus, in a CHOP-dependent manner, ER stress causes oxidative stress and damages mitochondrial function, ultimately resulting in cell death (Grootjans et al. 2016).

Different cell types have their UPR activated upon exposure to $H_2O_2$. For example, brief exposure of HeLa cells to $H_2O_2$ causes cleavage of activating transcription factor 6 (ATF6) and enhances the expression of ER stress sensors p-PERK, p-eIF2α, and p-IRE1α (Zhang and Kaufman, 2006). Exposure to $H_2O_2$ causes GRP78/BiP, p-PERK, and p-eIF2α expression in human oral keratinocytes and oral cancer cells. $H_2O_2$ raises the expression of BiP in mesenchymal stem cells (Chen et al. 2008). Extended periods of exposure to $H_2O_2$ cause apoptosis mediated by the ER. The ER proteases calpain, caspase-4, caspase-12, and caspase-7 exhibit enhanced enzymatic activity in HeLa cells when CHOP expression rises concurrently with the breakdown of calpain inhibitor calpastatin. The activation of ER-mediated apoptosis by $H_2O_2$ is confirmed by the reduction of this activation by calcium chelators and inhibitors of calpain and caspase-7.

Under modest pro-oxidant stress, the adaptive survival response is driven by UPR; in severe circumstances, however, ER-mediated apoptosis takes over (Fig. 7.3).

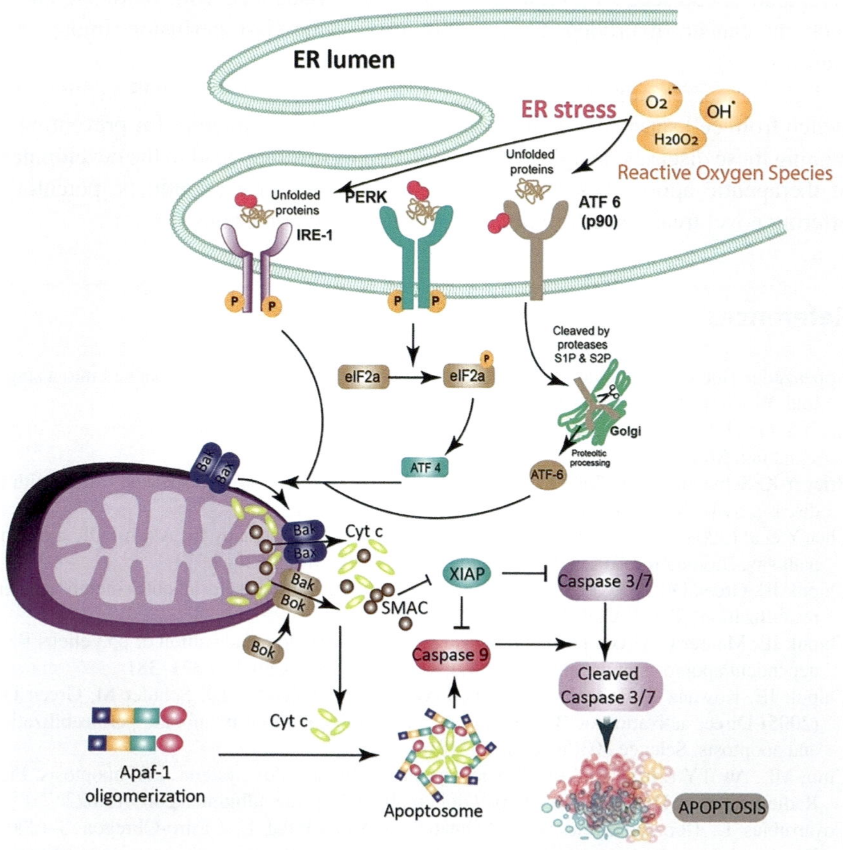


**Fig. 7.3** ROS activates the ER stress response, leading to apoptosis

## 7.3 Conclusion

Reactive oxygen species (ROS) are pivotal in cell signaling and regulation, particularly in apoptosis pathways involving mitochondria, death receptors, and the endoplasmic reticulum (ER). Additionally, ROS plays a significant role in other regulated pathways of cell survival and cell death, including autophagy and necroptosis. The complex interactions and crosstalk between these signaling pathways, influenced by ROS, remain an active research area. Understanding these interactions is critical, as deregulation can lead to pathological conditions associated with oxidative stress, such as cancer, neurodegenerative diseases, ischemia-reperfusion injury, and diabetes.

Future research aimed at elucidating how stress sensors like ROS facilitate the switch from cell survival to cell death could inform new strategies for preventing or treating these diseases. Insights into these mechanisms may lead to the development of therapeutic approaches that modulate ROS levels or their effects, potentially offering novel treatments for oxidative stress-related conditions.

## References

Appenzeller-Herzog C, Ellgaard L (2008) The human PDI family: versatility packed into a single fold. Biochim Biophys Acta 1783(4):535–548

Bae YS, Oh H, Rhee SG, Yoo YD (2011) Regulation of reactive oxygen species generation in cell signaling. Mol Cell 32(6):491–509

Brieger K, Schiavone S, Miller FJ Jr, Krause KH (2012) Reactive oxygen species: from health to disease. Swiss Med Wkly 142:13659–13664

Chen Y et al (2008) $H_2O_2$ induces mesenchymal stem cell senescence by the Akt/mTOR signaling pathway. Biomed Rep 7(4):200–206. https://doi.org/10.3892/br.2017.974

Chipuk JE, Green DR (2008) How do BCL-2 proteins induce mitochondrial outer membrane permeabilization? Trends Cell Biol 18(4):157–164

Chipuk JE, Maurer U, Green DR, Schuler M (2004) Pharmacologic activation of p53 elicits Bax-dependent apoptosis in the absence of transcription. Cancer Cell 4(5):371–381

Chipuk JE, Kuwana T, Bouchier-Hayes L, Droin NM, Newmeyer DD, Schuler M, Green DR (2005) Direct activation of Bax by p53 mediates mitochondrial membrane permeabilization and apoptosis. Science 303(5660):1010–1014

Circu ML, Aw TY (2010) Reactive oxygen species, cellular redox systems, and apoptosis. Free Radic Biol Med 48(6):749–762. https://doi.org/10.1016/j.freeradbiomed.2009.12.022

Covarrubias L, Hernandez-Garcia D, Schnabel D, Salas-Vidal E, Castro-Obregon S (2008) Function of reactive oxygen species during animal development: passive or active? Dev Biol 320(1):1–11

Denisov AY, Sprules T, Moldoveanu T, Batey RA, Gehring K, Leber B (2013) Solution structure of Bcl-2 in complex with the BH3 peptide of Beclin-1: insights into the mechanism of autophagy regulation. Cell Death Dis 4(4):e581

Dickinson BC, Chang CJ (2011) Chemistry and biology of reactive oxygen species in signaling or stress responses. Nat Chem Biol 7(7):504–511

Filomeni G, De Zio D, Cecconi F (2015) Oxidative stress and autophagy: the clash between damage and metabolic needs. Cell Death Differ 22(3):377–388

Flusberg DA, Sorger PK (2015) Surviving apoptosis: life-death signaling in single cells. Trends Cell Biol 25(7):446–458

Franco R, Sanchez-Olea R, Reyes-Reyes EM, Panayiotidis MI (2009) Environmental toxicity, oxidative stress and apoptosis: menage a trois. Mutat Res 674(1–2):3–22

Frand AR, Kaiser CA (1998) The ERO1 gene of yeast is required for oxidation of protein dithiols in the endoplasmic reticulum. Mol Cell 1(2):161–170

Gibson SB (2013) Investigating the role of reactive oxygen species in regulating autophagy. Methods Enzymol 528:217–235

Giorgi C et al (2018) Mitochondrial calcium homeostasis and signaling in cell death. Front Oncol 8:264. https://doi.org/10.3389/fonc.2018.00264

Gogvadze V, Orrenius S, Zhivotovsky B (2006) Mitochondria in apoptosis. In: Hockenbery DM (ed) Bcl-2 protein family: essential regulators of cell death. Springer, pp 99–124

Grootjans J et al (2016) Regulated necrosis: the expanding network of non-apoptotic cell death pathways. Nat Rev Mol Cell Biol 17(2):117–131. https://doi.org/10.1038/nrm.2015.9

Gross E, Sevier CS, Heldman N, Vitu E, Bentzur M, Kaiser CA, Thorpe C, Fass D (2006) Generating disulfides enzymatically: reaction products and electron acceptors of the endoplasmic reticulum thiol oxidase Ero1p. Proc Natl Acad Sci 103(2):299–304

Halliwell B (2011) Free radicals and antioxidants—quo vadis? Trends Pharmacol Sci 32(3):125–130

Halliwell B (2012) Free radicals and antioxidants: updating a personal view. Nutr Rev 70(5):257–265

Harding HP, Calfon M, Urano F, Novoa I, Ron D (2003) Transcriptional and translational control in the mammalian unfolded protein response. Annu Rev Cell Dev Biol 18(1):575–599

Hetz C, Mollereau B (2014) Disturbance of endoplasmic reticulum proteostasis in neurodegenerative diseases. Nat Rev Neurosci 15(4):233–249. https://doi.org/10.1038/nrn3689

Kehrer JP, Klotz L-O (2015) Free radicals and related reactive species as mediators of tissue injury and disease: implications for health. Crit Rev Toxicol 45(9):765–798

Oakes SA, Papa FR (2015) The role of endoplasmic reticulum stress in human pathology. Annual Review of Pathology: Mechanisms of Disease 10:173–194. https://doi.org/10.1146/annurev-pathol-012513-104649

Pallepati P, Averill-Bates DA (2011) reactive oxygen species, cell death signaling and apoptosis in embryonic kidney cells during hyperosmotic stress. Biochem Biophys Res Commun 409(3):448–453. https://doi.org/10.1016/j.bbrc.2011.05.010

Schieber M, Chandel NS (2014) ROS function in redox signaling and oxidative stress. Curr Biol 24(10):R453–R462

West JD, Marnett LJ (2006) Endogenous reactive intermediates as modulators of cell signaling and cell death. Chem Res Toxicol 19(2):173–194. https://doi.org/10.1021/tx050321u

Zhang K, Kaufman RJ (2006) The unfolded protein response: a stress signaling pathway critical for health and disease. Neurology 66(2 Suppl 1):S102–S109. https://doi.org/10.1212/01.wnl.0000192306.98198.ec

# Caspase Independent Apoptosis: New Insights and Current Perspectives

8

Sushweta Mahalanobish, Noyel Ghosh, and Parames C. Sil


**Abstract**

Apoptosis is an evolutionarily conserved cell death modality that regularly maintains a harmonious homeostasis between cellular genesis and demise. Caspases are a unique class of cysteine proteases that exist as inactive zymogens within the cells and follow a sequential cascade of catalytic activation for onset of apoptosis induction. Although the importance of caspases in apoptosis has received huge attention, considerable focus on caspase-independent mammalian cell death needs to be elucidated. The phenomenon of caspase-independent cell death (CICD) in mammals narrates a process of "determination" between cell survival and death upstream of caspase activation or without reliance on a caspase trigger, primarily in case of permeabilization of the outer mitochondrial membrane. Often, cells undergoing CICD exhibit distinctive features, such as the formation of large cytoplasmic vacuoles, accumulation of autophagosomes, peripheral condensation of the nuclei, etc. However, unlike apoptosis where cells exhibit a relatively uniform and consistent set of characteristics, cells undergoing CICD can manifest a wide range of variable traits depending on varied factors such as the initial stimulus and the type of cells involved. This chapter aims to explore the molecular mechanisms of CICD and its potential roles in the onset of various diseases from a pathophysiological perspective.



**Keywords**

Apoptosis · Caspase · Cell cycle · Mitochondria



S. Mahalanobish · N. Ghosh · P. C. Sil (✉)
Division of Molecular Medicine, Bose Institute, Calcutta, West Bengal, India
e-mail: parames@jcbose.ac.in







K. Jana (ed.), *Apoptosis and Human Health: Understanding Mechanistic and Therapeutic Potential*, https://doi.org/10.1007/978-981-97-7905-5_8

## Abbreviations

| | |
|---|---|
| ANT | Adenine nucleotide translocase |
| AIF | Apoptosis-inducing factor |
| CRP | C-reactive protein |
| HaCaT | Human keratinocytes |
| IMS | Intermembrane space |
| IMCE | Immorto-Min colonic epithelial cells |
| LDH | Lactate dehydrogenase |
| MI | Myocardial infarction |
| PLA2 | Phospholipase A2 |
| ROS | Reactive oxygen species |
| RIP-1 | Receptor interacting protein |
| TNF | Tumor necrosis factor |

## 8.1 Introduction

Apoptosis is a highly regulated mechanism of cell death that governs various physiological processes in multicellular organisms (Hale et al. 1996). Additionally, it plays a role in regulating pathological conditions such as stroke, sepsis, myocardial infarction (MI), Alzheimer's, Huntington's, and Parkinson's disease. Over the last decade, extensive research has illuminated apoptosis's involvement in numerous diseases. Consequently, inhibiting apoptosis has emerged as a potential therapeutic approach to mitigate pathological progression. Typically, therapeutic strategies target caspases, a family of cysteine protease enzymes highly conserved from invertebrates to humans. However, evidence suggests that caspase-independent pathways also significantly contribute to cell death induction. Caspase-independent cell death (CICD) refers to instances where apoptosis fails to occur due to the absence of caspases. Nonetheless, CICD shares common features with apoptotic cell death, including mitochondrial outer membrane permeabilization (MOMP), which is crucial for both forms of death. Notably, typical caspase-dependent features like externalization of phosphatidylserine in the outer leaflet of the plasma membrane and extensive chromatin condensation are notably absent during CICD. Instead, cells undergoing CICD exhibit formation of large cytoplasmic vacuoles, peripheral nuclear condensation, and accumulation of autophagosomes. Furthermore, CICD displays slower kinetics compared to apoptosis (Tait and Green 2008). Caspase-independent programmed cell death (CICD) was initially identified in fission yeast. In Schizosaccharomyces pombe, the yeast genome lacks any caspase genes (Jürgensmeier et al. 1997). From an evolutionary perspective, CICD appears to have evolved earlier compared to caspase-dependent classical apoptosis pathways. Various intracellular factors such as IFN-gamma, TNF-alpha, Fas/APO-1, nitric oxide (NO), molecular iodine ($I_2$), and CD47 are implicated in initiating this CI-PCD event (Okuno et al. 1998; Mateo et al. 2002; Johansson et al. 2004; Shrivastava et al. 2006). This chapter delves into the molecular mechanisms of CICD and explores their potential significance in devising effective therapeutic strategies for various diseases.

## 8.2 Mechanisms of Caspase-Independent Cell Death (CICD)

The molecular mechanism of death receptor-mediated apoptotic cascade is extensively understood. Essentially, it involves the interaction between ligands and their corresponding death receptors, leading to dimerization of intracellular adaptor molecules, subsequently activating caspase-8 and its downstream apoptotic effectors (Lavrik et al. 2005). Intriguingly, several studies have investigated instances of cell death occurring in the absence of caspases following death receptor interaction (Fig. 8.1) (Jäättelä and Tschopp 2003). In the absence of caspases, cell death may occur either via necroptosis (dependent on ligand and death receptor interaction) or through mitochondrial outer membrane permeabilization (MOMP)-induced caspase-independent cell death (CICD) (Degterev et al. 2005).

### 8.2.1 Necroptosis

#### 8.2.1.1 Death Receptor Mediated Necroptosis

Receptor-interacting protein AQ2(RIP-1), an intracellular adaptor molecule, is also involved in regulating death receptor-induced necroptosis. This kinase protein is responsible for recruiting the IκB kinase complex to the TNF receptor to activate NFκB. Loss of RIP-1 or lack of RIP-1 kinase activity inhibits the onset of necroptosis in the presence of TNF or CD95 (Holler et al. 2000; Degterev et al. 2008). Besides its role in NFκB activation, RIP-1 kinase regulates necroptosis (Kelliher et al. 1998). RIP-1 disrupts the interaction between adenine nucleotide translocase (ANT) and cyclophilin D, thereby triggering mitochondrial dysfunction (Temkin et al. 2006). Interestingly, RIP-1-induced necroptosis has been suggested in a separate study (Yu et al. 2004). Low levels of caspase-8 are essential to maintain cell viability. RNA interference application or zVAD-FMK treatment causes the downregulation of caspase-8 expression and induces necroptosis.

#### 8.2.1.2 Ligand Dependent Necroptosis

Tumor necrosis factor (TNF) plays a central role in inducing necroptosis. It has been observed that the use of the caspase inhibitor z-VAD-FMK in a mouse model can trigger rapid cell death within 4 h (Teng et al. 2005). Phospholipase A2 (PLA2) activation by TNF stimulation is responsible for the necroptotic event in the presence of caspase inhibitors. Another caspase inhibitor, CrmA, also triggers necroptosis following TNF treatment (Vercammen et al. 1998). Interestingly, Fas-mediated apoptosis and necroptosis share some common molecular events. During caspase-mediated apoptosis, FADD acts as an adaptor molecule linking death receptor oligomerization to caspase-8 activation. Again, FADD is necessary for necroptosis after Fas activation, as the loss of FADD expression hinders both Fas-induced

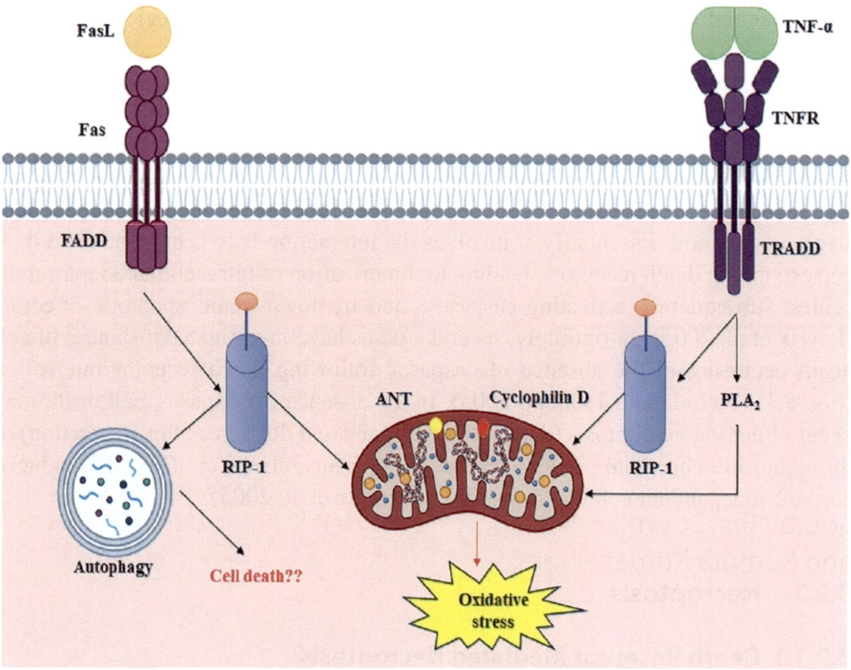


**Fig. 8.1**

apoptosis and necroptosis events (Holler et al. 2000). Moreover, FADD oligomerization can itself trigger necroptosis (Kawahara et al. 1998). However, the loss of FADD does not impede TNF-induced necroptosis events (Holler et al. 2000).

### 8.2.2 Mitochondria and CICD

Necroptotic events mediated by death receptors are somewhat restricted to specific cell types. However, mitochondrial outer membrane permeabilization (MOMP) inducing caspase-independent cell death (CICD) occurs universally across a wide variety of cell types. The overexpression of Bax can induce CICD in the presence of caspase inhibitors, along with the resulting MOMP, which leads to the loss of mitochondrial membrane potential (Xiang et al. 1996). MOMP can initiate CICD through various mechanisms, primarily by interfering with mitochondrial function and/or disrupting the mitochondrial proteome system (Fig. 8.2). Mitochondria serve various functional roles such as acetyl CoA and NADH production, buffering intracellular calcium levels, and being the sole powerhouse for ATP generation. During apoptosis, caspase-3-mediated cleavage of the p75 subunit of mitochondrial respiratory chain complex I disrupts the electron transport chain and inhibits ATP production. However, blocking the activity of caspase-mediated p75 cleavage does not rescue mitochondria from losing their membrane potential and causing CICD. This

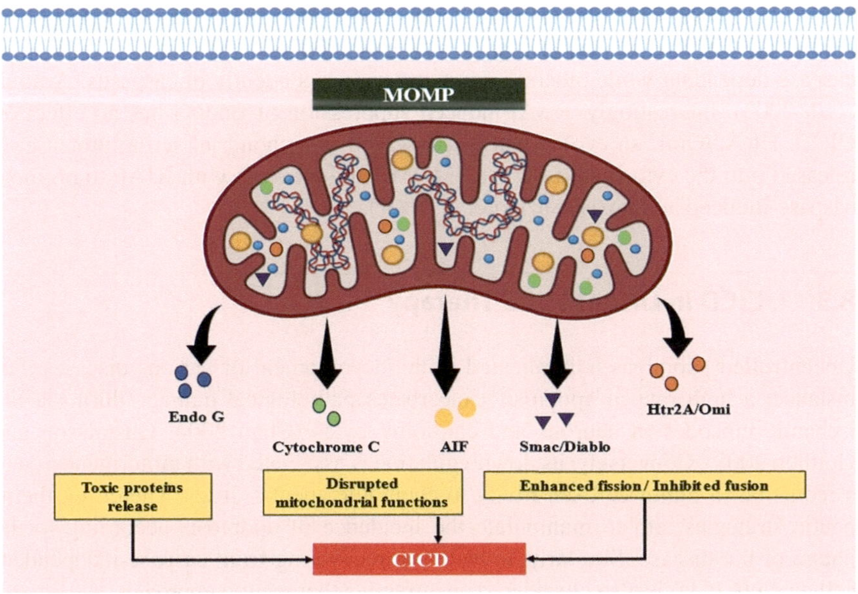


**Fig. 8.2** Role of mitochondria in caspase-independent cell death

is due to the gradual release of respiratory chain components following MOMP, resulting in failed ATP generation that ultimately leads to CICD (Ricci et al. 2004). In transformed cells, disrupted ATP production due to MOMP cannot initiate CICD as glycolysis is the primary system for ATP generation in these cell types (Colell et al. 2007). Additionally, the maintenance of mitochondrial membrane potential ($\Delta\psi m$) is essential for mitochondrial protein import, and disruption of $\Delta\psi m$ disturbs mitochondrial homeostasis (Gasser et al. 1982).

The dramatic fragmentation of the mitochondrial network also contributes to MOMP, where increased rates of mitochondrial fission are associated with MOMP and resulting CICD (Arnoult 2007). When MOMP occurs, cytochrome C and various other proteins in the mitochondrial intermembrane space (IMS) such as HtrA2/Omi, Smac/Diablo, Endonuclease G (Endo G), and apoptosis-inducing factor (AIF) are released (Du et al. 2000; Susin et al. 1999; Suzuki et al. 2001; Verhagen et al. 2000). AIF is slowly released from mitochondria and translocated to the nucleus in the absence of caspases, but its release is accelerated in the presence of caspases, inducing chromatin condensation and triggering CICD (Susin et al. 1999). AIF also regulates aerobic respiration, as demonstrated in Harlequin mice, which exhibit reduced AIF levels along with disrupted mitochondrial respiration and resulting neuronal and retinal degeneration. The lack of AIF expression is associated with defective electron transport chain complex I/III activity, indicating that mitochondrial release of AIF following MOMP perturbs mitochondrial homeostasis and promotes CICD (Klein et al. 2002). Endonuclease G (EndoG), another mitochondrial endonuclease, is involved in CICD by translocating to the nucleus following MOMP and causing DNA degradation (Li et al. 2001). However, the exact role of EndoG in mediating CICD remains unclear, with some studies suggesting its release is caspase-dependent while others show it occurs independently of caspases (Arnoult et al. 2003). Interestingly, RNAi-induced suppression of EndoG has no effect on CICD. HtrA2/Omi, an evolutionarily conserved mitochondrial serine protease, is released into the cytoplasm following MOMP and interacts with XIAP to promote caspase-induced apoptosis (Suzuki et al. 2001).

## 8.3 CICD in Disease and Therapy

Uncontrolled apoptosis is implicated in the development of various diseases. For instance, a high rate of apoptosis exacerbates pathological damage during acute ischemic episodes in neurons and cardiomyocytes (Chan 2004; Gustafsson and Gottlieb 2008). Conversely, its downregulation is associated with tumorigenesis and a reduction in tumor cell sensitivity to chemotherapeutic drugs. Therefore, therapeutic strategies aim to manipulate the incidence of apoptosis according to the nature of the disease. Similarly, inducing or recovering from caspase-independent cell death (CICD) is also considered an attractive therapeutic approach.

Ischemia, occurring during events like stroke or cardiac infarction, leads to MOMP-dependent apoptotic cell death. Several studies have shown that the administration of caspase inhibitors yields significant clinical outcomes in murine stroke models, suggesting that cells undergoing apoptosis can effectively recover from CICD-inducing conditions (Balsam et al. 2005; Braun et al. 2007). Manipulating metabolic pathways or restoring autophagy may offer better outcomes. However, in stroke models with Apaf-1 or caspase-9 null mice, there is no evidence that caspase inhibition protects cells from CICD-inducing conditions. Due to CICD, dying neurons may release survival factors that promote the survival of surrounding tissue. Additionally, inhibiting inflammatory caspase-1 may also have a protective effect.

In cancer, neoplastic cells evade apoptotic events at multiple stages to facilitate their metastatic process (Green and Evan 2002). Tumor cells erect apoptotic barriers at different levels, including signals induced by oncogenes, detachment from the extracellular matrix, and signaling driven by hypoxia. They acquire machinery to inhibit apoptosis, with the upregulation of anti-apoptotic Bcl-2 expression being a well-studied effect to inhibit MOMP (Chipuk et al. 2008). Evidence-based studies also indicate that inhibiting caspase function downstream of MOMP is another survival strategy for most cancerous cells, possibly indicating cells escaping from CICD to promote tumorigenesis. Downregulation of Apaf-1 expression has been observed in several tumor cell lines, associated with promoter methylation in metastatic melanoma and poor prognosis (Fujimoto et al. 2004). Tumor cells also suppress apoptosome-dependent caspase activity through various mechanisms. For example, the oncogenic Bcr-Abl kinase inhibits apoptosome formation by blocking caspase-9/Apaf-1 interactions. Intracellular ions or nucleotides can modulate apoptosome function, where the function of Apaf-1 or caspase-9 may remain unaltered even when compromised (Zlobec et al. 2007). Furthermore, various tumors overexpress the endogenous caspase inhibitor XIAP, which promotes tumorigenesis and inhibits chemosensitivity of tumor cells. XIAP is involved in TGF-β and NFκB signaling pathways to promote tumorigenesis (Tamm et al. 2004). Since tumor cells block the intrinsic apoptosis pathway and recover from CICD, blocking cellular recovery with autophagy or glycolysis inhibitors like chloroquine or 2-deoxyglucose could be effective strategies to combat them.

## 8.4 Role of Natural Compounds in CICD

Various natural compounds exhibit antitumorigenic natures by modulating CICD pathways (Bhadra 2022). Vitamin D and its analogues, CB 1093 EB, and 1089 administration in T47D and MCF7 (breast cancer cell lines) induce CICD by stimulating cathepsin D release and inhibiting Bcl-2 expression without involving caspases or p53 (Mathiasen et al. 1999). Administration of these compounds ranging from 1 to 100 nM concertation in the presence of Caspase inhibitors did not suppress the growth. Another example is staurosporine, Streptomyces staurosporeus derived alkaloid, which triggers the release of cytochrome c through cathepsin D in human fibroblasts. In T-lymphocytes, cathepsin D induced Bax activity helps to relocate AIF and stimulates apoptosis (Bidère et al. 2003; Johansson et al. 2003). Staurosporine exposure in T lymphocytes induces the onset of early caspase-independent phase such as peripheral chromatin condensation and cell shrinkage. At this stage, release of AIF into IMS and relocation of Bax into the OMM happens independently of Bid. Effect of different cathepsins to control the Bax activity and AIF release was thoroughly investigated. Translocation of lysosomal cathepsins to the cytosol induces Bax activation and resulting in release of mitochondrial AIF that mediates early apoptosis. Thrombospondin-1, a functional glycoprotein, can induce the onset of CICD in NB4 cells (promyelocytic leukemia), monocytes, and monocyte-derived dendritic cells by interacting with thrombospondin-1 membrane receptors αvβ3 and CD47 (Saumet et al. 2005). Although thrombospondin-1 causes mitochondrial membrane depolarization, the release of endonuclease G, AIF, cytochrome c, which are involved in non-caspase apoptotic pathways, does not appear in thrombospondin-1-mediated CICD.

Selenite, can promote CICD in Hep-2 and HeLa (cervical cancer cell lines). This signaling event involves stress induced p53 and p38 activation, Bax accumulation, AIF, and Smac/DIABLO release without involving caspase. Moreover, use of Z-VAD-FMK (caspase inhibitor) in these cell lines induces cell death. Time and dose-dependent application of Selenite suppresses synthesis of DNA and induces DNA damage, histone H2A.X phosphorylation, and activation of p53 and p38 signaling cascade. Eventually, p53- and p38-dependent signaling cascade promotes Bax accumulation, alters mitochondrial dynamics, and release of AIF and Smac/DIABLO, to trigger CICD (Rudolf et al. 2008).

Administration of Lipoic acid in HL-60 (leukemia cells) shows a strong antitumor activity by inducing CICD. Time and dose-dependent application of Lipoic acid suppresses cellular proliferation by inhibiting cell cycle progression. Lipoic acid reduces the expression of anti-apoptotic protein Bcl-2, upregulates Bax expression and promotes the translocation of mitochondrial AIF and cytochrome c to the nucleus without involving caspases (Selvakumar and Hsieh 2008).

N-acetylcysteine and glutathione are the example of Thiol-reducing antioxidants which can suppress the onset of apoptosis. Application of Berberine (natural isoquinoline alkaloid) can inhibit the growth of BxPC-3 cells (human pancreatic cancer cells) by a caspase-independent pathway (Pinto-Garcia et al. 2010). When berberine is applied in mouse immorto-Min colonic epithelial cells (IMCE), it triggers caspase-independent cell death by producing ROS and therefore releases the cathepsin B and stimulates PARP-dependent AIF translocation. Berberine inhibits the colon colony formation, and promotes the release of lactate dehydrogenase (LDH) in IMCE cells (Wang et al. 2012).

Different types of natural flavonoids like myricetin, quercetin have shown pronounced caspase inhibitory activity. They can inhibit the function of caspase 1, 3, and 7 and stimulate the apoptosis in MDA-MB-231cell (epithelial human breast cancer cell) by turning on non-classical apoptotic cascade. Therefore, they can be utilized as potent caspase-specific inhibitors (White et al. 2012). Although how these natural flavonoids provide a safeguard against cancer is not fully elucidated, it may be assumed that these compounds can suppress the NF-kB-signaling pathway as NF-kB can effectively inhibit the onset of apoptosis and speed up the proliferation and metastasis of cancer cells. Matrine, Sophora subprostrata's root derived alkaloid, can stimulate both caspase-dependent and caspase-independent apoptosis simultaneously in HepG2 cells by activating Bid-mediated AIF nuclear translocation cascade (Zhou et al. 2014). Application of Bid specific siRNA, Bid inhibitor BI-6C9, or ROS scavenger Tiron can inhibit the nuclear translocation of AIF and prevent the cell death.

Tomatine, glycoalkaloid present in tomato plant, can trigger caspase-free cell death in both in vivo and in vitro condition. In CT-26 cells, (Balb/c mouse derived colorectal cell line), tomatine significantly downregulated the expression of survivin (apoptosis inhibitor) and promotes the nuclear translocation of mitochondrial AIF to nucleus. However, it is not able to induce the proteolytic cleavage to make active caspase-8, -9, and -3 in CT-26 cells (Kim et al. 2015).

Resveratrol, a stilbenoid, can promote the onset of CICD by altering mitochondrial membrane potential, reducing Bcl-2 expression, upregulating nitric oxide and ROS production, and suppressing NF-kB activity (Pozo-Guisado et al. 2005). Interestingly, oxyresveratrol, a hydroxyl-substituted stilbene with potent tyrosinase inhibitor function, can suppress cell proliferation by inducing sub-G1 cell cycle arrest and induce apoptosis in CICD pathway by altering mitochondrial membrane potential and AIF nuclear translocation in MDA-MB-231 cells (triple negative breast cancer cells) (Sunil kumar et al. 2020). Molecular docking-based investigation revealed that interaction of oxyresveratrol to the S1 site of caspase-3 can inhibit the function of this executioner caspase. Using QVD-OPH, a pan caspase inhibitor, the cell viability and morphology of MDA-MB-231 shows similar effect when compared to oxyresveratrol treatment. Moreover, the immunoblot study indicates complete absence of the cleaved caspase during apoptosis which also confirms the occurrence of CICD.

Curcumin, a phenolic derivative of Curcuma longa, can promote apoptosis in immortalized human keratinocytes (HaCaT) cells (Song et al. 2021). Treating HaCaT cells with 24-μM curcumin dose causes cell cycle arrest at G2/M phase and induces 33.95% apoptotic cell death. HaCaT cells exhibited specific apoptotic morphology along with proper nuclear matrix-intermediate filament arrangement. Using Mass spectrometry, 16 differentially expressed nuclear matrix proteins were identified that contain AIF and caspase 3. Immunofluorescence assays revealed the translocation of mitochondrial AIF to the nucleus and the AIF expression level is elevated inside the nucleus. Interestingly, silencing of AIF or use of Z-VAD-FMK (caspase inhibitor) before curcumin treatment makes HaCaT cells nonresponsive to apoptosis. Curcumin can also cause depolarization of mitochondrial membrane, resulting in release of mitochondrial cytochrome c to the cytoplasm, and upregulated Bax/Bcl-2 ratio in HaCaT cells which ultimately triggers the caspase-3 activation. All together, these observations indicate the role of curcumin to induce caspase-dependent as well as caspase-independent cell death in HaCaT cells (Song et al. 2021).

*Clostridium difficile* toxin B, virulent factor released from *Clostridium difficile*, can induce caspase-dependent in HeLa cells where promote caspase independent mode of apoptosis in MCF-7 cell. Thiosulfinates, derived from *Allium tuberosum* L. upregulates the AIF and Bax/ Bcl2 ratio in PC3 cell line (human prostate cancer cell line) to induce both CDCD and CICD (Qa'Dan et al. 2002).

*Chelidonine majus* L. has wide range of isoquinoline alkaloids like chelidonine, sanguinarine, chelerythrine, coptisine, protopine, and phenolic acids and flavonoid which can participate in CICD as well as Caspase Dependent Cell Death (CDCD) when applied on Human Glioblastoma Cell Line T98G. Chelidonine exhibits time and dose dependent effect to induce apoptosis. It causes the activation of Caspase-9 and 3 but when pan-caspase inhibitor was applied, chelidonine induced apoptosis can be blocked partially. Interestingly, Chelidonine participates in CICD by promoting mitochondrial AIF translocation to nucleus. It also induces cell cycle arrest in G2/M phase and thereby block mitotic onset. In T98G cells, chelidonine treatment induces the activation of cyclin-dependent kinase, CDK1, which in turn prolongs G2/M, degrades Mcl-1 (an apoptotic inhibitor), promotes mitochondrial instability to induce apoptosis (Lee et al. 2019).

The rhizome part of Zingiber officinale triggers caspase-independent paraptosis by inducing DNA damage, ER stress, nuclear translocation of AIF in A549 cells (non-small lung cancer cells) and MDA-MB-231 cells (Nedungadi et al. 2021). Figure 8.3 shows the structure of important natural compounds involved in CICD.

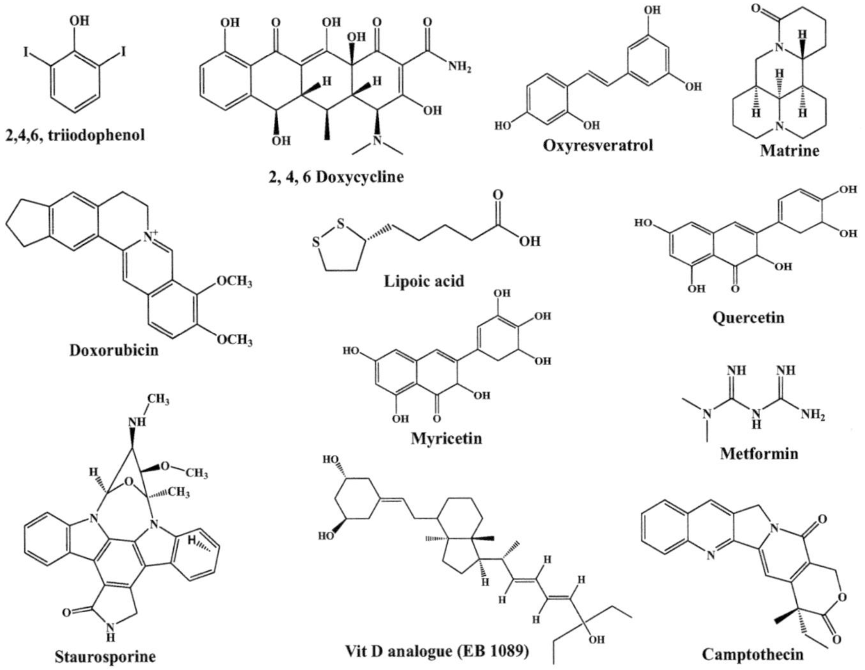


**Fig. 8.3** Natural compounds involved in caspase-independent cell death

## 8.5 Conclusions and Perspective

Apoptosis is recognized as a pivotal regulator in numerous pathological conditions, and modulating its activity holds great promise as a therapeutic strategy. Although inhibiting caspase activity has shown to mitigate adverse outcomes during ischemia in animal models, its effectiveness in clinical settings appears limited. Importantly, implementing caspase inhibition strategies is challenging due to the activation of caspase-independent cell death (CICD), which occurs through the release of AIF, EndoG, and Omi/HtrA2 from mitochondria to the cytosol in the absence of caspase. Therefore, a comprehensive understanding of the interplay between caspases and caspase-independent pathways is imperative to devise more effective therapeutic approaches.

## References

Arnoult D (2007) Mitochondrial fragmentation in apoptosis. Trends Cell Biol 17(1):6–12
Arnoult D, Gaume B, Karbowski M, Sharpe JC, Cecconi F, Youle RJ (2003) Mitochondrial release of AIF and EndoG requires caspase activation downstream of Bax/Bak-mediated permeabilization. EMBO J 22(17):4385–4399

Balsam LB, Kofidis T, Robbins RC (2005) Caspase-3 inhibition preserves myocardial geometry and long-term function after infarction. J Surg Res 124(2):194–200
Bhadra K (2022) A mini review on molecules inducing caspase-independent cell death: a new route to cancer therapy. Molecules 27(19):6401
Bidère N, Lorenzo HK, Carmona S, Laforge M, Harper F, Dumont C, Senik A (2003) Cathepsin D triggers Bax activation, resulting in selective apoptosis-inducing factor (AIF) relocation in T lymphocytes entering the early commitment phase to apoptosis. J Biol Chem 278(33):31401–31411
Braun JS, Prass K, Dirnagl U, Meisel A, Meisel C (2007) Protection from brain damage and bacterial infection in murine stroke by the novel caspase-inhibitor Q-VD-OPH. Exp Neurol 206(2):183–191
Chan PH (2004) Mitochondria and neuronal death/survival signaling pathways in cerebral ischemia. Neurochem Res 29:1943–1949
Chipuk JE, Fisher JC, Dillon CP, Kriwacki RW, Kuwana T, Green DR (2008) Mechanism of apoptosis induction by inhibition of the anti-apoptotic BCL-2 proteins. Proc Natl Acad Sci 105(51):20327–20332
Colell A, Ricci JE, Tait S, Milasta S, Maurer U, Bouchier-Hayes L, Fitzgerald P, Guio-Carrion A, Waterhouse NJ, Li CW, Mari B (2007) GAPDH and autophagy preserve survival after apoptotic cytochrome c release in the absence of caspase activation. Cell 129(5):983–997
Degterev A, Hitomi J, Germscheid M, Ch'en IL, Korkina O, Teng X, Abbott D, Cuny GD, Yuan C, Wagner G, Hedrick SM (2008) Identification of RIP1 kinase as a specific cellular target of necrostatins. Nat Chem Biol 4(5):313–321
Degterev A, Huang Z, Boyce M, Li Y, Jagtap P, Mizushima N, Cuny GD, Mitchison TJ, Moskowitz MA, Yuan J (2005) Chemical inhibitor of nonapoptotic cell death with therapeutic potential for ischemic brain injury. Nat Chem Biol 1(2):112–119
Du C, Fang M, Li Y, Li L, Wang X (2000) Smac, a mitochondrial protein that promotes cytochrome c–dependent caspase activation by eliminating IAP inhibition. Cell 102(1):33–42
Fujimoto A, Takeuchi H, Taback B, Hsueh EC, Elashoff D, Morton DL, Hoon DS (2004) Allelic imbalance of 12q22–23 associated with APAF-1 locus correlates with poor disease outcome in cutaneous melanoma. Cancer Res 64(6):2245–2250
Gasser SM, Ohashi A, Daum G, Böhni PC, Gibson J, Reid GA, Yonetani T, Schatz G (1982) Imported mitochondrial proteins cytochrome b2 and cytochrome c1 are processed in two steps. Proc Natl Acad Sci 79(2):267–271
Green DR, Evan GI (2002) A matter of life and death. Cancer Cell 1(1):19–30
Gustafsson ÅB, Gottlieb RA (2008) Heart mitochondria: gates of life and death. Cardiovasc Res 77(2):334–343
Hale AJ, Smith CA, Sutherland LC, Stoneman VE, Longthorne VL, Culhane AC, Williams GT (1996) Apoptosis: molecular regulation of cell death. Eur J Biochem 236(1):1–26
Holler N, Zaru R, Micheau O, Thome M, Attinger A, Valitutti S, Bodmer JL, Schneider P, Seed B, Tschopp J (2000) Fas triggers an alternative, caspase-8–independent cell death pathway using the kinase RIP as effector molecule. Nat Immunol 1(6):489–495
Jäättelä M, Tschopp J (2003) Caspase-independent cell death in T lymphocytes. Nat Immunol 4(5):416–423
Johansson U, Higginbottom K, Londei M (2004) CD47 ligation induces a rapid caspase-independent apoptosis-like cell death in human monocytes and dendritic cells. Scand J Immunol 59(1):40–49
Johansson AC, Steen H, Öllinger K, Roberg K (2003) Cathepsin D mediates cytochrome c release and caspase activation in human fibroblast apoptosis induced by staurosporine. Cell Death Differ 10(11):1253–1259
Jürgensmeier JM, Krajewski S, Armstrong RC, Wilson GM, Oltersdorf T, Fritz LC, Reed JC, Ottilie S (1997) Bax-and Bak-induced cell death in the fission yeast Schizosaccharomyces pombe. Mol Biol Cell 8(2):325–339
Kawahara A, Ohsawa Y, Matsumura H, Uchiyama Y, Nagata S (1998) Caspase-independent cell killing by Fas-associated protein with death domain. J Cell Biol 143(5):1353–1360

Kelliher MA, Grimm S, Ishida Y, Kuo F, Stanger BZ, Leder P (1998) The death domain kinase RIP mediates the TNF-induced NF-kappaB signal. Immunity 8(3):297–303

Kim SP, Nam SH, Friedman M (2015) The tomato glycoalkaloid α-tomatine induces caspase-independent cell death in mouse colon cancer CT-26 cells and transplanted tumors in mice. J Agric Food Chem 63(4):1142–1150

Klein JA, Longo-Guess CM, Rossmann MP, Seburn KL, Hurd RE, Frankel WN, Bronson RT, Ackerman SL (2002) The harlequin mouse mutation downregulates apoptosis-inducing factor. Nature 419(6905):367–374

Lavrik IN, Golks A, Krammer PH (2005) Caspases: pharmacological manipulation of cell death. J Clin Invest 115(10):2665–2672

Lee YK, Lee KW, Kim M, Lee Y, Yoo J, Hwangbo C, Park KH, Kim KD (2019) Chelidonine induces caspase-dependent and caspase-independent cell death through G 2/M arrest in the T98G human glioblastoma cell line. Evidence-Based Complementary and Alternative Medicine

Li LY, Luo X, Wang X (2001) Endonuclease G is an apoptotic DNase when released from mitochondria. Nature 412(6842):95–99

Mateo V, Brown EJ, Biron G, Rubio M, Fischer A, Deist FL, Sarfati M (2002) Mechanisms of CD47-induced caspase-independent cell death in normal and leukemic cells: link between phosphatidylserine exposure and cytoskeleton organization. Blood J Am Soc Hematol 100(8):2882–2890

Mathiasen IS, Lademann U, Jäättelä M (1999) Apoptosis induced by vitamin D compounds in breast cancer cells is inhibited by Bcl-2 but does not involve known caspases or p53. Cancer Res 59(19):4848–4856

Nedungadi D, Binoy A, Vinod V, Vanuopadath M, Nair SS, Nair BG, Mishra N (2021) Ginger extract activates caspase independent paraptosis in cancer cells via ER stress, mitochondrial dysfunction, AIF translocation and DNA damage. Nutr Cancer 73(1):147–159

Okuno SI, Shimizu S, Ito T, Nomura M, Hamada E, Tsujimoto Y, Matsuda H (1998) Bcl-2 prevents caspase-independent cell death. J Biol Chem 273(51):34272–34277

Pinto-Garcia L, Efferth T, Torres A, Hoheisel JD, Youns M (2010) Berberine inhibits cell growth and mediates caspase-independent cell death in human pancreatic cancer cells. Planta Med 76(11):1155–1161

Pozo-Guisado E, Merino JM, Mulero-Navarro S, Lorenzo-Benayas MJ, Centeno F, Alvarez-Barrientos A, Salguero PMF (2005) Resveratrol-induced apoptosis in MCF-7 human breast cancer cells involves a caspase-independent mechanism with downregulation of Bcl-2 and NF-κB. Int J Cancer 115(1):74–84

Qa'Dan M, Ramsey M, Daniel J, Spyres LM, Safiejko-Mroczka B, Ortiz-Leduc W, Ballard JD (2002) Clostridium difficile toxin B activates dual caspase-dependent and caspase-independent apoptosis in intoxicated cells. Cell Microbiol 4(7):425–434

Ricci JE, Muñoz-Pinedo C, Fitzgerald P, Bailly-Maitre B, Perkins GA, Yadava N, Scheffler IE, Ellisman MH, Green DR (2004) Disruption of mitochondrial function during apoptosis is mediated by caspase cleavage of the p75 subunit of complex I of the electron transport chain. Cell 117(6):773–786

Rudolf E, Rudolf K, Červinka M (2008) Selenium activates p53 and p38 pathways and induces caspase-independent cell death in cervical cancer cells. Cell Biol Toxicol 24:123–141

Saumet A, Slimane MB, Lanotte M, Lawler J, Dubernard V (2005) Type 3 repeat/C-terminal domain of thrombospondin-1 triggers caspase-independent cell death through CD47/αvβ3 in promyelocytic leukemia NB4 cells. Blood 106(2):658–667

Selvakumar E, Hsieh TC (2008) Regulation of cell cycle transition and induction of apoptosis in HL-60 leukemia cells by Lipoic acid: role in cancer prevention and therapy. J Hematol Oncol 1(1):1–8

Shrivastava A, Tiwari M, Sinha RA, Kumar A, Balapure AK, Bajpai VK, Sharma R, Mitra K, Tandon A, Godbole MM (2006) Molecular iodine induces caspase-independent apoptosis in human breast carcinoma cells involving the mitochondria-mediated pathway. J Biol Chem 281(28):19762–19771

Song W, Ren YJ, Liu LL, Zhao YY, Li QF, Yang HB (2021) Curcumin induced the cell death of immortalized human keratinocytes (HaCaT) through caspase-independent and caspase-dependent pathways. Food Funct 12(18):8669–8680

Sunil kumar D, Drishya G, Chandrasekharan A, Shaji SK, Bose C, Jossart J, Perry JJP, Mishra N, Kumar GB, Nair BG (2020) Oxyresveratrol drives caspase-independent apoptosis-like cell death in MDA-MB-231 breast cancer cells through the induction of ROS. Biochem Pharmacol 173:113724

Susin SA, Lorenzo HK, Zamzami N, Marzo I, Snow BE, Brothers GM, Mangion J, Jacotot E, Costantini P, Loeffler M, Larochette N (1999) Molecular characterization of mitochondrial apoptosis-inducing factor. Nature 397(6718):441–446

Suzuki Y, Imai Y, Nakayama H, Takahashi K, Takio K, Takahashi R (2001) A serine protease, HtrA2, is released from the mitochondria and interacts with XIAP, inducing cell death. Mol Cell 8(3):613–621

Tait SW, Green DR (2008) Caspase-independent cell death: leaving the set without the final cut. Oncogene 27(50):6452–6461

Tamm I, Richter S, Scholz F, Schmelz K, Oltersdorf D, Karawajew L, Schoch C, Haferlach T, Ludwig WD, Wuchter C (2004) XIAP expression correlates with monocytic differentiation in adult de novo AML: impact on prognosis. Hematol J 5(6):489–495

Temkin V, Huang Q, Liu H, Osada H, Pope RM (2006) Inhibition of ADP/ATP exchange in receptor-interacting protein-mediated necrosis. Mol Cell Biol 26(6):2215–2225

Teng X, Degterev A, Jagtap P, Xing X, Choi S, Denu R, Yuan J, Cuny GD (2005) Structure–activity relationship study of novel necroptosis inhibitors. Bioorg Med Chem Lett 15(22):5039–5044

Vercammen D, Brouckaert G, Denecker G, Van de Craen M, Declercq W, Fiers W, Vandenabeele P (1998) Dual signaling of the Fas receptor: initiation of both apoptotic and necrotic cell death pathways. J Exp Med 188(5):919–930

Verhagen AM, Ekert PG, Pakusch M, Silke J, Connolly LM, Reid GE, Moritz RL, Simpson RJ, Vaux DL (2000) Identification of DIABLO, a mammalian protein that promotes apoptosis by binding to and antagonizing IAP proteins. Cell 102(1):43–53

Wang L, Liu L, Shi Y, Cao H, Chaturvedi R, Calcutt MW, Hu T, Ren X, Wilson KT, Polk DB, Yan F (2012) Berberine induces caspase-independent cell death in colon tumor cells through activation of apoptosis-inducing factor. PLoS One 7(5):e36418

White JB, Beckford J, Yadegarynia S, Ngo N, Lialiutska T, d'Alarcao M (2012) Some natural flavonoids are competitive inhibitors of caspase-1, -3, and-7 despite their cellular toxicity. Food Chem 131(4):1453–1459

Xiang J, Chao DT, Korsmeyer SJ (1996) BAX-induced cell death may not require interleukin 1β-converting enzyme-like proteases. Proc Natl Acad Sci 93(25):14559–14563

Yu L, Alva A, Su H, Dutt P, Freundt E, Welsh S, Baehrecke EH, Lenardo MJ (2004) Regulation of an ATG7-beclin 1 program of autophagic cell death by caspase-8. Science 304(5676):1500–1502

Zhou H, Xu M, Gao Y, Deng Z, Cao H, Zhang W, Wang Q, Zhang B, Song G, Zhan Y, Hu T (2014) Matrine induces caspase-independent program cell death in hepatocellular carcinoma through bid-mediated nuclear translocation of apoptosis inducing factor. Mol Cancer 13(1):1–11

Zlobec I, Minoo P, Baker K, Haegert D, Khetani K, Tornillo L, Terracciano L, Jass JR, Lugli A (2007) Loss of APAF-1 expression is associated with tumour progression and adverse prognosis in colorectal cancer. Eur J Cancer 43(6):1101–1107

# 9 Role of Apoptosis in Cancer: War of the Worlds, Therapeutic Targets and Strategies

Aiindrila Dhara, Ramandeep Kaur, Rajarshi Chattopadhyay, Souhadri Das, Shuvam Pal, and Nirmalya Sen

**Abstract**

Evading apoptosis is a pioneer hallmark of all cancers, and cancer therapeutics worldwide mainly focus on activating programmed cell death mechanisms. Various cellular mechanisms involving loss of tumour suppressors, activation of oncogenic growth factors, and epigenetic silencing act as primary anti-apoptotic signals for promoting cancer. The emergence of additional cancer hallmarks in the last decade that bypasses or dampen apoptosis signals, like metabolic reprogramming, immune evasion, phenotypic plasticity, microbiota interactions and epigenetic reprogramming, indicates an arsenal of strategies deployed by this multifaced disease against cell death. Combinatorial therapies are being engaged in clinics to synergise the cell death potentials of these agents. However, with the rise of therapy-resistant cancers and severe side effects of chemotherapy drugs, new strategies are required to treat the disease. The chapter will cover various aspects of cell death and cancer, emphasising apoptotic evasion, mechanism failures, therapy resistance, transcriptional rewiring, and modern combinatorial therapeutic strategies.

A. Dhara
Molecular Oncology Laboratory, Department of Biological Sciences, Bose Institute, Kolkata, India

University of Cambridge, Cambridge, UK

Masaryk University, Brno, Czech Republic

R. Kaur · S. Das · S. Pal · N. Sen (✉)
Molecular Oncology Laboratory, Department of Biological Sciences, Bose Institute, Kolkata, India

R. Chattopadhyay
Department of Zoology, Ramakrishna Mission Vidyamandira, Belur Math, India


K. Jana (ed.), *Apoptosis and Human Health: Understanding Mechanistic and Therapeutic Potential*, https://doi.org/10.1007/978-981-97-7905-5_9

**Keywords**

Apoptosis · Cancer · Oncogenes · Resistance · Cell death

## 9.1 Introduction: Alien Inside the Body

Metazoans maintain homeostasis in cell numbers via various life and death signals required for normal growth and development. For example, signals for division lead to the overall growth of the organism and maintenance of normal physiology, while cell death results in the removal of unwanted, damaged, or diseased cells. Among various pathways of cell death identified till date, apoptosis remains the most widely studied. The term 'apoptosis' was coined by Kerr and colleagues to describe the unique morphology of cell death, which works in contrast to mitosis but complements it in regulating cell populations (Kerr et al. 1972). In today's world, apoptosis can be defined as a genetically orchestrated energy-dependent cascade of biochemical and molecular events resulting in the death of cells. Apoptosis is physiologically essential for vital processes like tissue homeostasis, developmental remodelling, immune system functions, and wound healing (Elmore 2007; Rai et al. 2005; Singh et al. 2019; Suzanne and Steller 2013). Malfunction of apoptosis results in several human diseases and pathological conditions. For example, low apoptosis or impaired clearance of apoptotic cells results in autoimmune diseases, infectious diseases, and neoplastic conditions, while excessive apoptosis causes neuronal disorders, cardiovascular diseases, and inflammatory conditions (Favaloro et al. 2012; Perniok et al. 1998; Thompson 1995; Xu et al. 2019) (Table 9.1).

Presently, apoptosis is categorised under regulated cell death (RCD), which involves the caspase family of proteases as per the Nomenclature Committee on Cell Death (NCCD) (Vitale et al. 2023). On the contrary, the hallmarks of cancer involve evasion of cell death mechanisms that act as natural barriers to tumour formation, indicating a warlike relationship between cancer and apoptosis(Hanahan 2022). Apoptosis is often triggered by various stress signals that a cancer cell experiences during oncogenesis or anticancer therapy. These signals are mostly conveyed between apoptotic regulators through counterbalancing pro- and anti-apoptotic BCL-2 family members that work as regulatory proteins (Adams and Cory 2007). The BCL-2 family proteins connect the two arms of the apoptotic machinery used for processing signals, one receiving external signals (extrinsic) and another sensing and processing intracellular signals (intrinsic), culminating in activating executioner caspases. Nonetheless, various cancers have evolved to nullify signals that trigger apoptotic machinery and can be categorised into (a) impaired death receptor signalling (extrinsic), (b) disrupted balance of pro-apoptotic and anti-apoptotic proteins (intrinsic), (c) reduced caspase activity (intrinsic) (Wong 2011).

Given the fact that cancer evolves from healthy cells of our body undergoing genetic and environmental insults, processes like apoptosis are nature's way to alienate and eliminate such diseases. To date, cancer has been the most studied disease with respect to apoptosis malfunction and apoptosis-inducing agents as compared to others (Fig. 9.1). Moreover, therapy-resistant cancers are the new threat

**Table 9.1** Lists several diseases associated with impaired or hyperactivated apoptosis mechanisms

| SL. No. | Disease | Mechanism of failure | Possible treatment | Reference PMID |
|---|---|---|---|---|
| 1 | Alzheimer's Disease | Hyperphosphorylation of Tau/(Aβ) Amyloid Overproduction leading to neuritic plaques which activates caspases | Anti-Tau/Anti-Amyloid Drugs (Tau clearance Immunotherapy) | 33302541, 28138104, 32165850 |
| 2 | Huntington's Disease | Duplication of CAG repeats in the Huntingtin gene results in neurodegeneration in the striatum and cortex with apoptotic features | New treatments aim to reduce amounts of mutant Huntingtin protein | 18588526, 36176885 |
| 3 | Parkinson's Disease | Loss of dopaminergic neurons via apoptosis due to intracellular aggregation of α-synuclein. | Dopamine-based medications | 30702835, 32044947 |
| 4 | Amyotrophic Lateral Sclerosis | Characterised by degeneration of the large pyramidal neurons in the motor cortex and spinal tracts and subsequent death | RNAi; Administration of Neuroprotective agents | 11724913, 34372914, 36121612 |
| 5 | Retinitis Pigmentosa | Death of Rods and cone cells associated with oxidative stress. | Transplantation, Gene Therapy, Cytokine Therapy, Hyperbaric Oxygen Therapy | 28962928, 25345673 |
| 6 | Myasthenia Gravis | Autoimmune antibodies against anti-acetylcholine receptor (AChR) with defective apoptosis | Administration of Immunosuppressants and thymectomy in selected patients | 27408701, 36537972 |
| 7 | Systemic Lupus Erythematosus | Loss of self-tolerance accompanied by autoreactive T and B cell activation that causes tissue damage and the generation of pathogenic autoantibodies due to defective apoptosis | Reducing glucocorticoid consumption, avoiding flare-ups, and improving quality of life | 23131610, 37054414 |
| 8 | Inflammatory Intestinal Disease | A decrease in the consortium of bacteria with anti-inflammatory capabilities and an increase of bacteria with inflammatory abilities due to defective apoptosis | Pharmacotherapy or surgical resection (if necessary) | 30611442, 34988090 |

(continued)

**Table 9.1** (continued)

| SL. No. | Disease | Mechanism of failure | Possible treatment | Reference PMID |
|---|---|---|---|---|
| 9 | Pulmonary Inflammation | Bacteria flourish following their direct infiltration into the airspaces by bacterial byproducts that harm host phagocytes (neutrophils and macrophages) or airway epithelial cells, change the normal flora of the respiratory tract, and induces epithelial injury. | Adjunct anti-inflammatory treatment with corticosteroids | 19956071, 31559145 |
| 10 | Aplastic Anaemia | Haematopoietic stem and progenitor cells (HSPC) are destroyed by autoreactive T lymphocytes in a genetically vulnerable host. | Immunosuppressive agents or bone marrow transplantation (in extreme cases) | 30047412, 6116472 |
| 11 | Myelodysplastic Syndrome | Varying degrees of cytopenias, dysplasia in one or more myeloid cell lineages, and ineffective haematopoiesis can be hold accountable as the major mechanism of failure. The two epigenetic mechanisms that have been discovered to be disrupted in MDS are DNA methylation and histone modification. | Enhancing lifestyle and reducing erythrocyte and platelet transfusions | 25548754, 36396662 |
| 12 | Lymphocytopenia | T cell destruction through antibody-dependent cytotoxicity (ADCC) or complement-dependent cytotoxicity (CDC) | Allogeneic haematopoietic stem cell transplantation, interleukin 2, interferon gamma, interleukin 7, and other factors | 32865518, 27931306 |
| 13 | G6PD Deficiency | Mutation in X-linked gene G6PD are inherited. RBCs with a G6PD deficit are much more susceptible to oxidative damage and haemolysis. | Packed red blood cell transfusions or hyperbaric oxygen therapy. | 32702756, 29262208 |
| 14 | Myocardial Infarction | Thrombotic occlusion of a coronary vessel resulting from the rupture of a vulnerable plaque | Thrombolytic drugs, Lipid-lowering treatment | 26426469, 30725761 |

(continued)

**Table 9.1** (continued)

| SL. No. | Disease | Mechanism of failure | Possible treatment | Reference PMID |
|---|---|---|---|---|
| 15 | Ischaemic Renal Damage | An abrupt, temporary interruption of blood supply to the specific organ. When exposed to hypoxia and reperfusion, IRD is typically accompanied by a strong inflammatory and oxidative stress response that disrupts organ function. | Antioxidants, nitric oxide and nitric oxide synthase inhibitors, PPAR agonists | 26060833, 18038125 |
| 16 | Polycystic Kidney | Fluid-filled cysts begin to form and expand uncontrollably in the kidney, and as they spread, renal function is eventually lost. | Reducing cAMP levels, or nephrectomy | 34566674, 26380198, 30422529 |

clinicians perceive against which various drugs are failing (Housman et al. 2014; Vasan et al. 2019). Hence, understanding the various aspects of apoptosis evasion by cancers will aid in developing new therapeutics against refractory cancers (Jan and Chaudhry 2019; Shahar and Larisch 2020). The chapter will focus on various aspects of cell death and its relation to cancers, emphasising apoptotic evasion, mechanism failures, therapy resistance, transcriptional rewiring and modern combinatorial therapeutic strategies.

## 9.2 Mechanisms of Apoptosis: The Allies

The mechanism of apoptosis consists of two core pathways that work via activation of caspases but differ in their mode of initiation (Fig. 9.2). These two pathways are extrinsic (external Death Receptor mediated pathway) and the intrinsic (mitochondrial mediated pathway) pathways (Jan and Chaudhry 2019). Extrinsic pathway is initiated through binding of extracellular ligands such as TNF, Fas-L, and TRAIL (death ligands) to their specific proapoptotic receptors located on cell membrane. Binding of ligands leads to oligomerisation of receptors which helps recruit adaptor proteins such as Fas-associated death domain (FADD) and TRADD (TNFR1-associated death domain protein). Finally, DISC (death-inducing signalling complex) is formed which activates a series of downstream factors including Caspase-8,10 (initiator caspase). On activation, Caspase-8,10 directly activates executioner caspases like 3, 6, and 7 resulting in apoptosis. Alternately, Caspase 8,10 promotes BID cleavage and its subsequent translocation into mitochondria, which results in Cytochrome c release, a prerequisite for apoptosis via intrinsic pathway (Billen et al. 2008). Interestingly, Caspase 8,10-mediated BID cleavage is

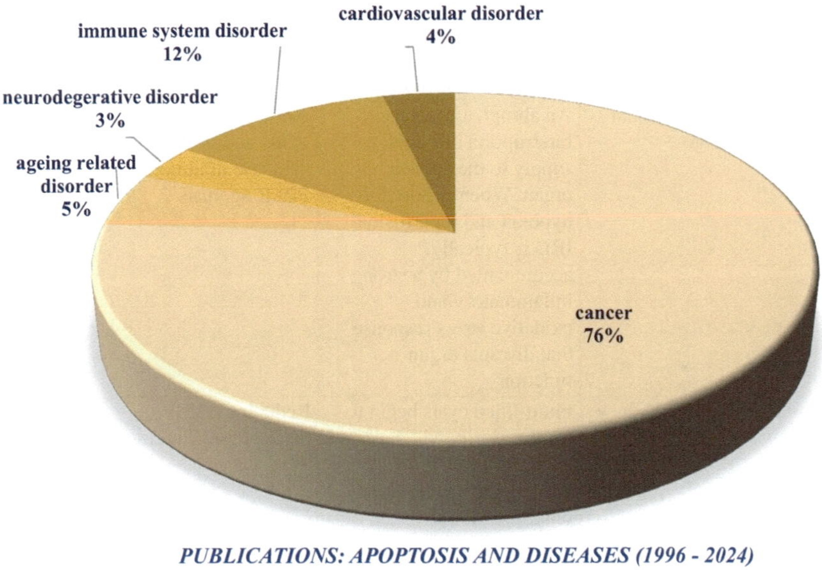


**Fig. 9.1** Chart depicting the percentage of diseases associated with apoptosis malfunction and apoptosis-related drug discovery. Publication numbers were curated from PubMed (https://pubmed.ncbi.nlm.nih.gov/) with keywords apoptosis and indicated diseases from 1996 to 2024 and plotted against the total number of publications

a crucial link between the extrinsic and intrinsic apoptotic pathways and often acts to amplify death signals through mitochondrial events (Ozoren and El-Deiry 2002). The intrinsic pathway is initiated through mitochondrial outer membrane permeabilisation (MOMP) triggered by various extra and intra-cellular stresses, including oxidative stress, DNA damage, irradiation, and treatment with cytotoxic drugs (Kalkavan and Green 2018). MOMP results in Cytochrome c release and apoptosome formation, which irreversibly mark the cells for apoptosis by activating a series of caspases (Li et al. 1997; Liu et al. 1996). Counterbalancing pro- and anti-apoptotic BCL-2 family members are known to regulate various intrinsic pathway events like MOMP and Cytochrome c release (Adams and Cory 2007). For example, pro-apoptotic members like BAX (BCL-2-associated X protein), BAK (BCL-2 antagonist killer 1), BIM, BID, and PUMA positively regulate Cytochrome c release via mitochondrial permeabilisation. In contrast, the anti-apoptotic members such as BCL-2, BCL-XL, and MCL1 inhibit MOMP (Bleicken et al. 2014; Kharbanda et al. 1997; Kim et al. 2009; Nijhawan et al. 2003; Yang et al. 1997). Fine-tuned regulation of BAX and BAK (pore formers) is achieved via several activator and sensitiser proteins of the BCL-2 members containing BH3-only domain (Letai et al. 2002; Wang et al. 1996). Nonetheless, activators-like BIM, BID, and PUMA directly bind and activate proapoptotic BAX and BAK. At the same time, sensitisers like BAD, NOXA, and BIK inhibit anti-apoptotic proteins that prevent BAX or BAK-mediated apoptosis (Wang et al. 1996; Nakano and Vousden 2001; O'Connor et al. 1998;

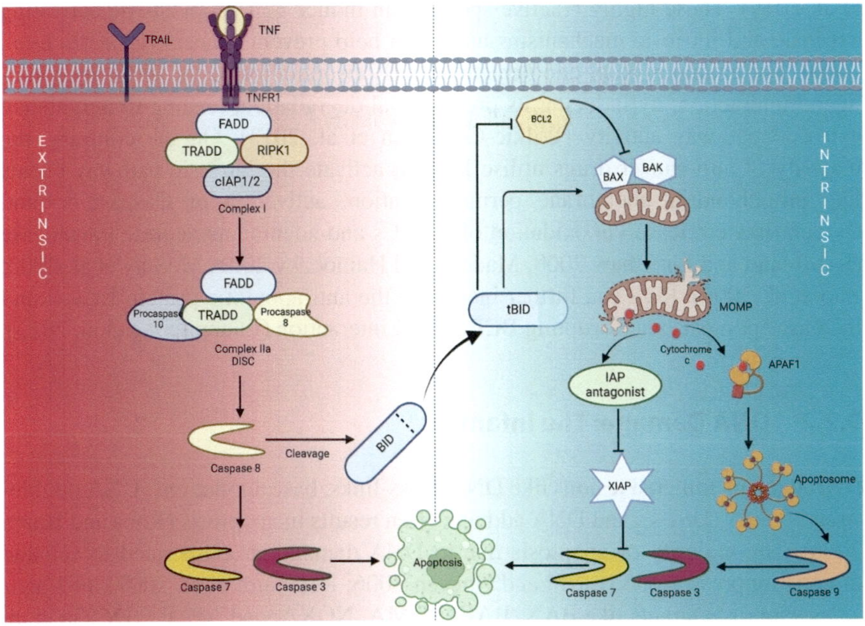


**Fig. 9.2** Core apoptotic pathways indicating intrinsic (mitochondrial pathway) and extrinsic (cell death receptors) mechanisms

Youle and Strasser 2008; Zha et al. 1996). Upon release, Cytochrome c interacts with Apaf1,which oligomerises into a heptamericapoptosome that recruits procaspase-9, the first initiator caspase of the intrinsic pathway. Cleaved Caspase 9 further activates executioner Caspases 3, 6, and 7 for apoptosis, thus unifying both pathways. Apoptosis inhibitory proteins such as XIAP and c-IAP1 can block caspase 3 activation and subsequent apoptosis while small peptides like Smac/DIABLO can promotes apoptosis by blocking IAPs, indicating regulation downstream of mitochondria (Deveraux and Reed 1999; van Loo et al. 2002).

While caspase acts as an executioner for the apoptosis pathways, the upstream events leading to their activation widely vary from genetic mutations, DNA damage, osmotic stress, growth factor withdrawal, loss of cell–matrix interaction, oxidative stress, hypoxia, deregulated cell cycle, and even mitochondrial injury. We have discussed a few events that often act as allies for inducing apoptosis in cells.

### 9.2.1 Reactive Oxygen Species: The Cavalry

Reactive oxygen species (ROS) constitute a group of reactive oxygen or nitrogen intermediates that act as signal transducers for various biological processes. While ROS are often generated during normal metabolism, default antioxidant functions help in ROS clearance within cells to avoid ROS-mediated cell death (Snezhkina et al. 2019). These highly reactive species can induce cell apoptosis by activating extrinsic and intrinsic mechanisms and often help prevent cancers (Chandra et al. 2000). For example, ROS can deplete c-FLIP, an inhibitor of the death receptor–ligand interaction, by inducing its proteasomal degradation, leading to heightened extrinsic pathway activity (Wilkie-Grantham et al. 2013). On the contrary, the majority of anti-cancer drugs utilise ROS to activate the intrinsic pathway events like mitochondrial membrane permeabilisation, activation of Caspase 9, and Cytochrome c release via oxidation of VDACs and adenine nucleotide translocase (Kaufmann and Earnshaw 2000; Madesh and Hajnoczky 2001; McStay et al. 2002; Zuo et al. 2009). ROS can further inactivate the anti-apoptotic protein BCL-2 and increase apoptosis via disrupting BCL2-ERK interaction (Luanpitpong et al. 2013).

### 9.2.2 DNA Damage: The Infantry

DNA damage inflicted lesions like DNA cross-links, base alkylations, DNA double-strand breaks (DSBs), and DNA adducts often results in apoptosis (Roos and Kaina 2006). The majority of apoptosis due to DNA damage is consecrated to activate tumour suppressor p53 (Lavin and Gueven 2006; Reinhardt et al. 2007). p53 activates apoptosis targets like BAX, BAK, PUMA, NOXA, and APAF1 (De Zio et al. 2013; Wei et al. 2006). Further, p53 can induce cell death by directly activating BAK oligomerisation and mitochondrial membrane permeabilisation (Marchenko and Moll 2014). P53-independent apoptosis is essential for tumours where p53 is mutated or absent. Alternate tumour suppressors like p73 and NF-kB (nuclear factor-kB) often result in p53-independent cell deaths (Bauer et al. 1998; Melino et al. 2004). Caspase 2 and 9 have been reported in p53-independent apoptosis in cancer cells treated with cisplatin, etoposide, or UV radiation (Lassus et al. 2002). Platinum-based drugs likely result in sustained JNK and p38 kinase activation, leading to AP-1 and FASL-mediated apoptosis (Brozovic et al. 2004). DNA damage sensors like PARP1 have been shown to activate apoptosis in both BRCA mutant and silent tumours (Weaver and Yang 2013). All cancer initiation starts with unresolved DNA lesions and faulty repair pathways; hence DNA damage response is a significant determinant between apoptosis and cancer.

### 9.2.3 Cell Cycle Checkpoints: The Border Police

Cell division is strictly controlled by various cell cycle checkpoints, and their deregulation is one of the hallmarks of cancers (Hanahan and Weinberg 2011). Checkpoints activated in cells as a response to DNA damage and chromosomal aberrations undergo cell cycle arrest and are often hurled towards apoptosis when cells fail to repair such damage to the genome. Checkpoints are regulated by crucial protein families like retinoblastoma (RB), E2F, CDK inhibitory subunits (CKIs mainly Cip/Kip), and cyclin-dependent kinases (CDKs) (Pucci et al. 2000). While tumour suppressor, RB remains hypo-phosphorylated by the p53-p21-RB activation axis, thereby preventing unwanted cell division; the RB-E2F complex acts as the gatekeeper of the G1/S checkpoint by altering the transcription of cell cycle regulatory genes (Engeland 2022). DNA damage sensor kinases like ATM and ATR initiate phosphorylation events, resulting in the activation/deactivation of the above checkpoint regulatory proteins (Matsuoka et al. 1998). DSB-activated ATM during G1 phase not only phosphorylates checkpoint kinase 2 (CHK2) which subsequently inhibits CDC25A, CyclinA/Cdk2, CyclinE/CDK2 complexes and blocks S phase progression, but also backhandedly activates p53-p21 axis leading to further inhibition of CyclinA/CyclinE/CDK2 complexes, activation of DNA repair proteins and apoptosis (Banin et al. 1998; Falck et al. 2001; Gu et al. 1993). DNA damage during the S-phase often results in ATR activation, which results in simultaneous inhibition of S-phase (by proteasomal degradation of CDC25A and deactivation of CyclinA/Cyclin E/CDK2 complexes) and activation of G2/M checkpoint (by activating checkpoint kinase 1; CHK1, WEE1, and inhibition of CDC25A, CyclinB/CDK1 complex), thereby preventing progress of cells with damaged genome through cell cycle (Cimprich and Cortez 2008; Nigg 2001; Xiao et al. 2003).

## 9.3 Bypass Mechanism in Cancer: The Axis

To defeat the dedicated crew of programmed cell death, cancer utilises loopholes within cells to achieve malignancy. Cancer weaponises cellular mechanisms like gene amplification, chimeric gene fusion, noncoding RNA expression, and altered protein functions for circumventing cellular safeguards (Fig. 9.3). Here, we discuss such bypass mechanisms of cancers.

### 9.3.1 Militia at Gene Level (DNA amplification/ Transcriptional Turnovers)

Pro-survival gene BCL-2 discovered in B Cell Follicular Lymphoma was shown to form fusion gene t (14;18) via translocation resulting in constitutive expression and cancer cell survival (Fukuhara and Rowley 1978). Another fusion occurring between BACH2-BCL2L1 due to translocation t(Favaloro et al. 2012; Kalkavan and Green 2018) is observed in leukaemias (Turkmen et al. 2011). Recent DNA level studies by Beroukhim et al. with more than 3000 patient samples encompassing almost 26 types of cancer show copy number amplification of anti-apoptotic genes like BCL-2 and BCL-xl, indicating a causal relationship between various malignancies and DNA amplification (Beroukhim et al. 2010). Amplifications of the *BCL2* family genes, with a significant gain in copy number, are often associated with poor prognosis in various tumours of lung, bone, and breast (Kendall et al. 2007; Smith et al. 2006; Tonon et al. 2005). On the other hand, proapoptotic genes BK, *Bax, PUMA*, and BOK are affected by frame shift mutations and deep deletions in various cancers (Beroukhim et al. 2010; Ionov et al. 2000).

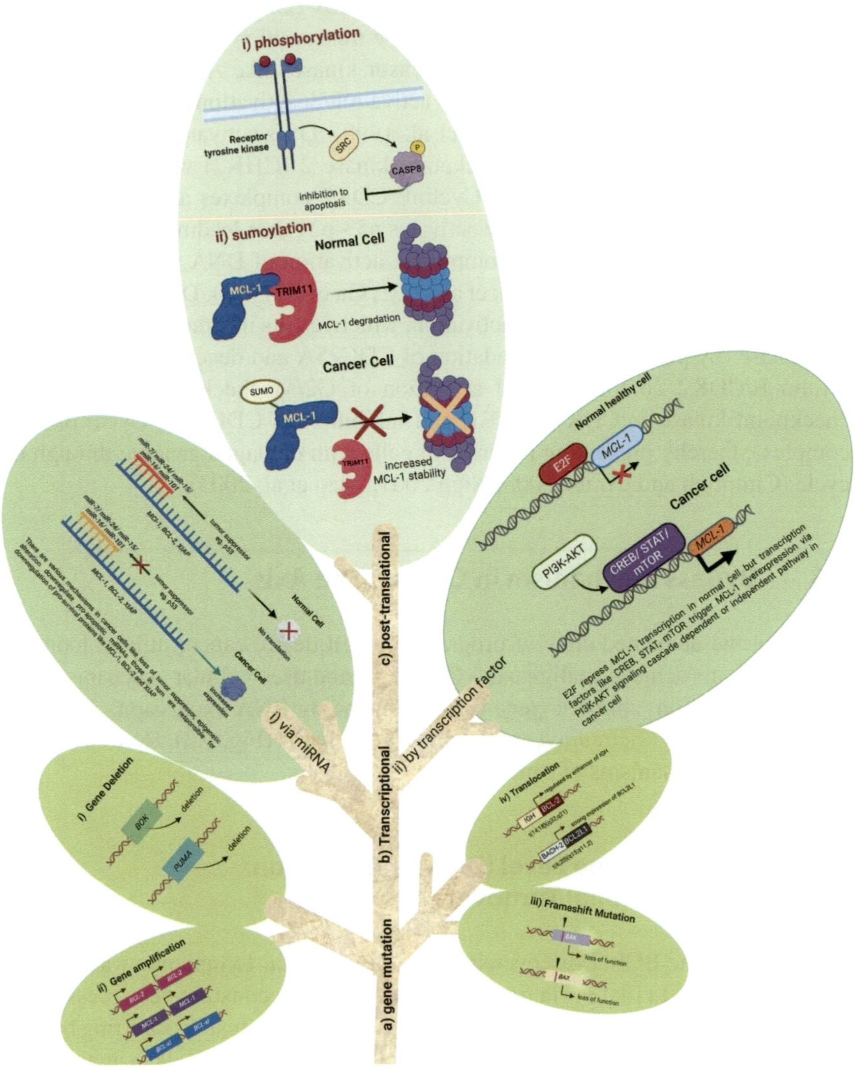


**Fig. 9.3** Anti-apoptotic mechanisms at the level of DNA, RNA, and proteins harnessed by cancer cells for their growth

Following the central dogma hierarchy, deregulation in transcription affects the expression of the pro and anti-apoptotic genes, resulting in oncogenesis. For example, *MCL-1* is known to be transcriptionally repressed by cell cycle regulators like E2F, which binds directly to the *MCL-1* promoter in healthy cells (Croxton et al. 2002). Again, transcription factors response element-binding protein (CREB) and signal transducer and activator of transcription 3 (STAT3), via phosphatidylinositol 3-kinase (PI3K-AKT) signalling has been shown to promote lung cancer (Kendall et al. 2007; Huang et al. 2000; Wang et al. 1999). Malignant tumours lose regulation of the cell death program due to such malfunction by the signalling cascade.

Apart from transcription factors, microRNAs' downregulation can help cancers skip apoptotic signals. *miR-15,miR-16,* and *miR-101* act as tumour suppressor by inhibiting *MCL-1* transcription and are downregulated in chronic lymphocytic leukaemia and hepatocellular carcinoma (Cimmino et al. 2005; Su et al. 2009). Another pro-survival gene, *XIAP*, is inhibited by *miR-7, miR-24*. Cancers of the cervix and lung adapt to downregulate these miRNAs resulting in *XIAP* elevation and caspase inactivation (Liu et al. 2013; Xie et al. 2013). Most Chronic Lymphocytic Leukaemia have downregulated miR-15 and miR-16, which act as tumour suppressors by downregulating BCL-2 expression (Cimmino et al. 2005). *miR-101* is diminished in hepatocellular carcinoma, which results in elevated *MCL-1* expression and apoptosis evasion (Su et al. 2009).

### 9.3.2 Terrorising Translation

Cancer can hijack the post-translational machinery, resulting in altered protein function or turnover for tumour suppressors, oncogenes, and apoptotic regulators. Deregulated post-translational modifications (PTMs) affect apoptosis by directly inhibiting the pro-apoptotic proteins or stabilising pro-survival factors. For example, HER2-amplified breast cancers resistant to kinase inhibitor lapatinib have been shown to skip apoptosis by altering the stability of both pro and anti-apoptotic protein targets of MOMP(Kurokawa et al. 2013). MCL-1 is also deregulated via phosphorylation-induced ubiquitination in cancers. In normal healthy cells, MCL-1 is subjected to phosphorylation by JNK, p38 MAPK, CK2, and GSK3, followed by degradation via ubiquitination system (Tong et al. 2018). However, this system gets corrupted in cancer as high ERK and AKT kinases inhibit GSK3, thereby blocking the degradation of MCL-1 and resulting in enhanced pro-survival signals. Additionally, FBW7, a substrate-binding partner of E3 ubiquitin ligase responsible for the degradation of MCL-1, is mutated or downregulated in various cancers, resulting in increased oncogenesis (Sancho et al. 2022).

Parallelly, increased expression of deubiquitinase enzymes that can remove ubiquitin from oncogenic proteins and prevent their degradation has been observed in various cancers. Ubiquitin-specific peptidase 9 X-linked (USP9X), a deubiquitinating enzyme stabilising MCL-1, is upregulated in several myelomas (Schwickart et al. 2010). PTMs can also decrease the half-life of pro-apoptotic proteins (Balmanno and Cook 2009). ERK-mediated phosphorylation of pro-apoptotic BIM at S69 makes it susceptible to degradation via SCFb-TrCP, c-CBL, and TRIM2 ubiquitin ligases, as shown in a transgenic mouse model (Akiyama et al. 2003; Dehan et al. 2009; Hubner et al. 2008; Luciano et al. 2003). The ubiquitin E3 ligase complex ElonginB/C–Cullin2–CIS has also been shown to degrade BIM protein (Zhang et al. 2008). Another pro-apoptotic protein, BAX, which is one of the final players relaying apoptotic signals, is subject to degradation by E3 ligase IBR domain-containing 2 (IBRDC2) or RNF144B, on the mitochondrial outer membrane. This short half-life of BAX has been associated with poor prognosis in chronic lymphocytic lymphoma and prostate cancer (Fernald and Kurokawa 2013).

## 9.4 Other Cell Death Types in the Prevention of Cancers: Brothers at Arm

Besides apoptosis, necrosis and autophagy-dependent cell deaths are the major cell death processes linked to various physiological and pathological aspects, including cancers (Fig. 9.4) (Green and Llambi 2015). Autophagy is a catabolic process mainly triggered by nutrient starvation, metabolic stress, and bioenergetic failures (Kroemer et al. 2010). Although the motive of autophagy remains survival, removing critically stressed cells via autophagic death has been observed as a caspase-independent death process (Shen et al. 2012). Autophagic genes like *BECLIN 1*, *UVRAG*, and *BIF-1* are required to form phagophores and are often depleted in breast, prostate, ovary, cervical, and liver cancers. Knockout mice of other autophagy regulators like ATG5, ATG7, and ATG9 show increased proliferation, metastasis, and oncogenesis, indicating tumour suppressor roles of autophagy (Yun and Lee 2018). In contrast, RAS-dependent cancers like lung, colon, and pancreatic cancers that are under high stress often use autophagy for survival benefits (Guo et al. 2011; Liu and Ryan 2012; Rabinowitz and White 2010). Necrosis or necrotic cell death is often considered a signal-independent/passive death pathway characterised by loss of cell membrane integrity and cytoplasmic leakage when cells undergo extensive damage (Yuan and Kroemer 2010). However, signalling through RIP and MLKL kinases often results in the formation of RIPoptosome complexes, which leads to rapid necrotic cell death called necroptosis (Degterev et al. 2005; Tenev et al. 2011). Similar to autophagy, the role of necroptosis in cancer is debatable. While necrosis can initiate a severe immune response for curbing cancer growth by activating inflammatory pathways, many cancers hijack such inflammatory signals to activate immunosuppression, metastasis, and tumorigenesis (Gong et al. 2019).

Apart from major types of cell death, namely apoptosis, autophagic, and necrotic cell death, various other forms of cell death have been discovered lately (Tang et al. 2019). These have now become an integral part of regulated cell death (RCD) and are found to be elemental in cancer prevention (Koren and Fuchs 2021). We have tabulated and illustrated various alternate death pathways concerning their mechanisms, potential role in cancer prevention, and therapeutics (Table 9.2, Fig. 9.5).

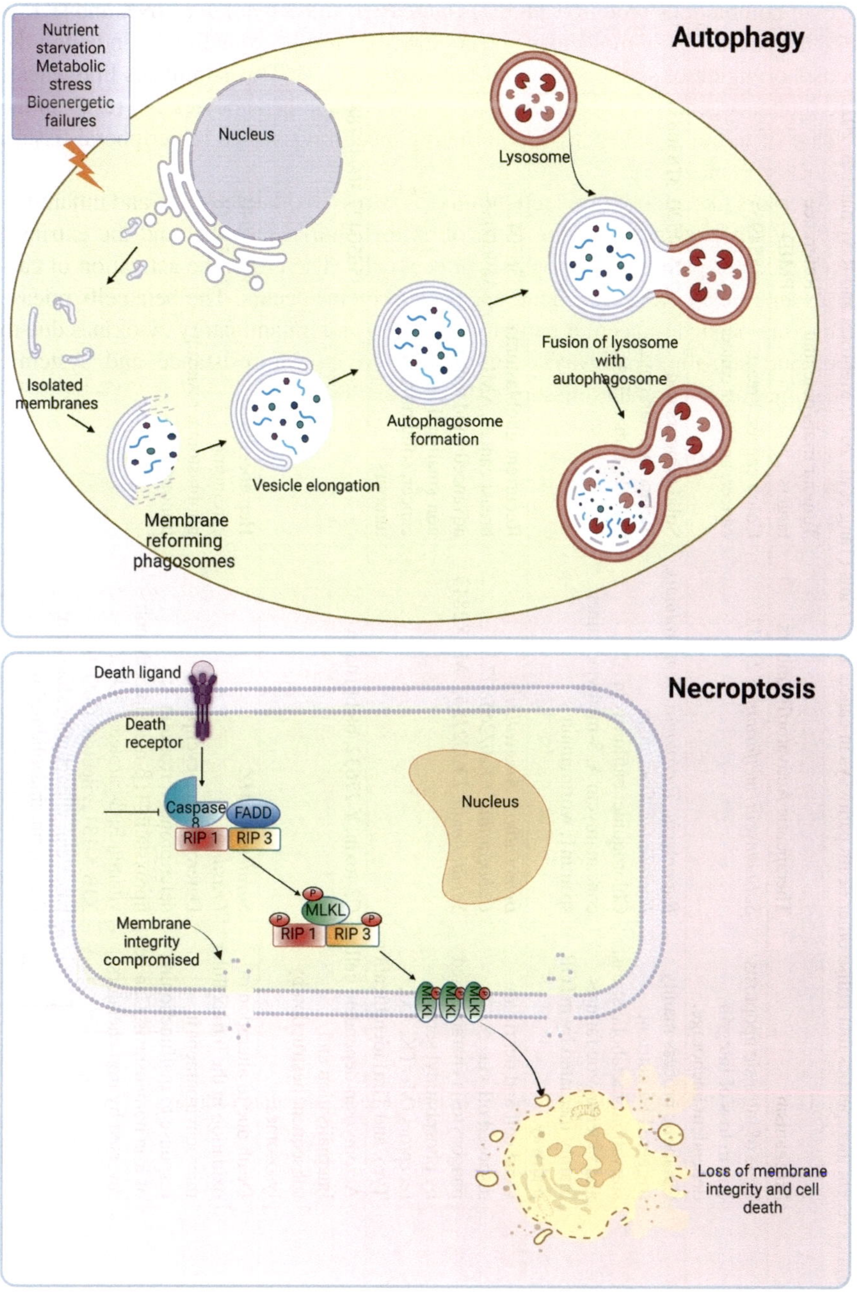


**Fig. 9.4** Molecular mechanisms of autophagy and necroptosis, which represent major death pathways alongside apoptosis

**Table 9.2** Compiles recent therapeutics that utilise other death pathways in cancer drug trials (Tang et al. 2019; Mehrotra and Ravichandran 2022)

| RCD Type | Mechanism | Therapeutic Activator/Inhibitors | Tumour trials/potential targets | Reference PMID |
|---|---|---|---|---|
| Anoikis | Type of intrinsic apoptosis due to loss of integrin-dependent anchorage. | *Salinomycin, metformin*, Y27632. | Colon cancer, breast cancer, and liver cancer | 33854965 |
| Autophagy | Catabolic process mainly triggered by nutrient starvation, metabolic stress, and bioenergetic failures mainly mediated by mTOR pathway | *Rapamycin, Tacrolimus, Temsirolimus, Everolimus*, Dactolisib Chloroquine; bafilomycin A1; concanamycin A, 3-methyladenine; spautin1; wortmannin | Solid tumours and melanoma, Sarcoma, Glioblastoma, PDAC | 24574520, 31836831 |
| Efferocytosis | Dead cell and debris are disposed off via phagocytosis characterised by phosphatidyl serine receptors (MerTK, Axl, Tyro, and Tim) identification | *Bemcentinib, Amuvatinib, Cabozantini, PF-07265807, Sabatolimab, LY3321367*. MRX2843 | Recurrent glioblastoma, breast cancer, AML advanced solid tumours, non-small cell lung cancer, solid refractory tumours | 35650427, 36960055 |
| Entosis | Actomyosin-dependent cell internalisation and subsequent destruction by lysosomes | C3-toxin, Y-27632, blebbistatin | | 35650427,30948788 |
| Ferroptosis | Death due to alteration of oxidation in the intracellular microenvironment in response to malfunctioning of iron homeostasis. Can be reversed by iron chelators | *Sorafenib, TMZ, Erastin,* Deferoxamine; cyclipirox, deferiprone; ferrostatin-1; liproxstatin-1; β-mercaptoethanol; vitamin E; β-carotene; NAC; XJB-5-131; zileuton; CoQ10; baicalein; vildagliptin; alogliptin; linagliptin; thiazolidinedione; rosiglitazone; selenium | Hepatocellular carcinoma, glioblastoma, resistant tumours | 36581640, 35151318 |

| | | | | |
|---|---|---|---|---|
| Mitotic catastrophe | Tumour suppressive death mechanism to control cells with deregulated mitosis | *Taxanes and Vinca alkaloids, Alisertib, Ispinesib , GSK923295, AZD4877, B12536, Griseofulvin, AZD7762* | Pancreatic cancer, glioblastoma, head and neck cancer, urothelial cancer, Breast cancer, solid tumours, refractory cancers, Leukaemias | 26491220 |
| Necroptosis | Death pathway is characterised by cell membrane integrity loss and depends on RIPoptosome formation via RIP/MLKL kinases. | *Staurosporine, Neoalbaconol, Sorafenib, 5'FU, Smacmimetics* Necrostatin-1; GSK872; HS-1371; necrosulfonamide | Leukaemia, nasopharyngeal carcinoma, multiple myeloma; Hodgkin's lymphoma, colorectal cancer, pancreatic cancer | 31122251, 30948788 |
| NETotic cell death | Characterised by NADPH oxidase-driven ROS production and histone citrullination, which results in NET release | Tetrahydroisoquinolines; cl-amidine; lactoferrin; DNase, Reparixin, GSK484/DNase, Cl-amidine | Breast cancer, pancreatic tumours, Lung cancer, Sivelestad | 34261496 |
| Parthanatos | Death characterised by PARP1-hyperactivation and the AIFM1-MIF pathway resulting in DNA degradation | *ZINC253504760, Sepantronium bromide (YM155)*, BYK204165; AG-14361; iniparib, Talazoparib, Lapatinib, Olaparib. | Leukaemia, esophageal cancer, advanced breast cancer, colon cancer | 35806303 |

(continued)

**Table 9.2** (continued)

| RCD Type | Mechanism | Therapeutic Activator/Inhibitors | Tumour trials/potential targets | Reference PMID |
|---|---|---|---|---|
| Pyroptosis | Activation of CASP1, CASP3, and GSDMD; GSDMD cleavage; GSDMD-N-mediated pore formation; driven by GSDMD after cleavage of this protein by CASP1 and CASP11 in response to PAMPs and DAMPs | *Polyphyllin VI (PPVI), 5'FU, Paclitaxel, Lobaplatin, Famotidine,* Ac-YVAD-cmk; VX765; wedelolactone; Ac-FLTD-CMK; MCC950; isoliquiritigenin; glybenclamide; CY-09; oridonin | Lung cancer, Glioma, colorectal cancer, esophageal cancer, gastric cancer, breast cancer | 35970871 |
| Lysosome-dependent cell death | Lysosomal membrane permeabilisation; release of lysosomal hydrolytic enzymes; lysosomal iron-induced oxidative injury mediated by releasing hydrolytic enzymes (cathepsins) or iron upon LMP | *Chloroquine*, *Mefloquine, Siramesine* CA-074Me; deferoxamine; NAC | Burkitt's lymphoma, CML, AML, CLL | 30071644 |
| Alkaliptosis | Intracellular alkalinisation; activation of NF-κB; caspase-independent | *JTC801* , NAC, N-acetyl alanine acid; IMD0354; CAY10657; SC514 | Pancreatic cancer, leukaemia | 31467365 |
| Oxeiptosis | ROS-dependent; activation of KEAP1; NFE2L2-independent; caspase-independent; lack of nuclear translocation of AIFM1. Oxygen radical-induced form of cell death driven by the activation of the KEAP1-PGAM5-AIFM1 pathway | *Sanguinarine (SNG), Alantolactone (ALT)* NAC | Colorectal cancer, Ovarian cancer | 36914635 |

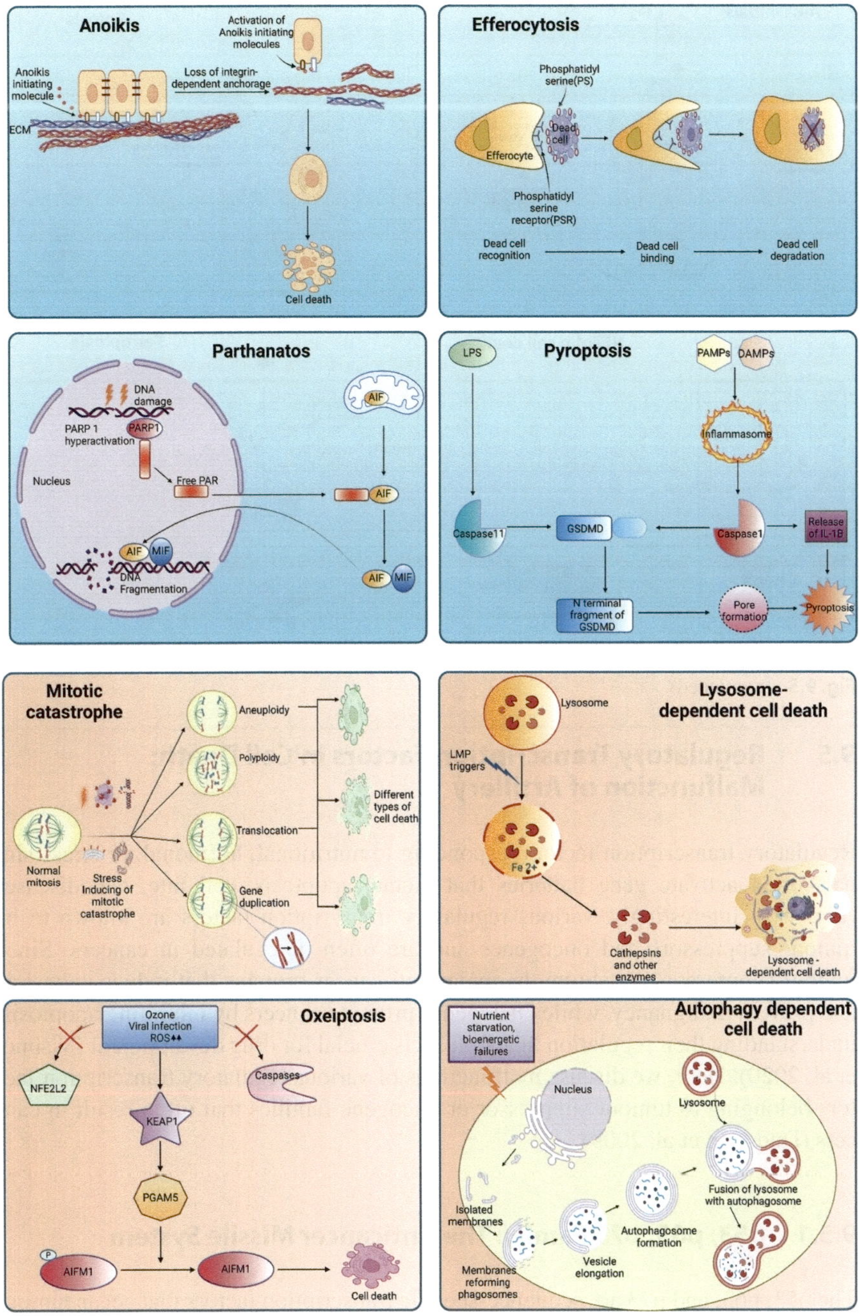


**Fig. 9.5** Various cell death pathways with an overview of molecular mechanisms preventing cancer

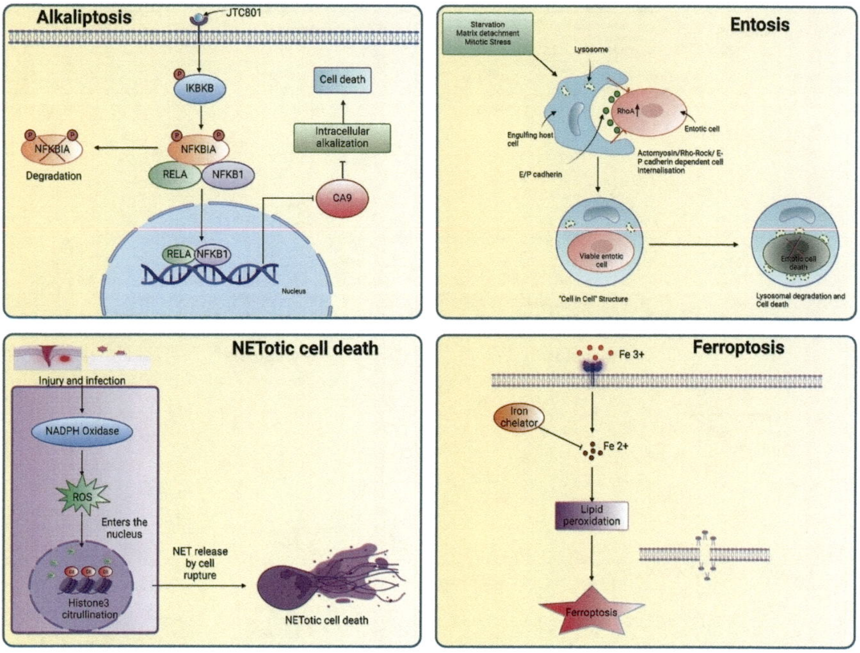


**Fig. 9.5** (continued)

## 9.5 Regulatory Transcription Factors in Cell Death: Malfunction of Artillery

Regulatory transcription factors responding to nutritional, hormonal, or stress signals often activate gene batteries that dictate apoptosis, cell fate, and diseases (Fig. 9.6). Interestingly, various regulatory transcription factors are known to be tumour suppressors and oncogenes and are often deregulated in cancers. Since tumour suppressors constitute the major anti-cancer proteins that induce apoptosis and prevent malignancy, while Oncogenes promote cancers by inhibiting apoptosis, understanding their regulation in apoptosis is crucial for drug development (Vishnoi et al. 2020). Here, we discuss malfunctions of various regulatory transcription factors belonging to tumour suppressor or oncogene families that often result in cancers (Dudgeon et al. 2009).

### 9.5.1 p53, p63, p73 Family: The Anticancer Missile System

The p53, p63, and p73 are sequence-specific transcription factors that act as tumour suppressors. These proteins are induced in response to various genotoxic stresses and, in turn, transcribe various regulators of cell cycle arrest and apoptosis-like *p21, GADD45, 14-3-3 Sigma, BAX, BID, NOXA, PUMA, PIDD, SLUG, PIG3, SCOTIN,*

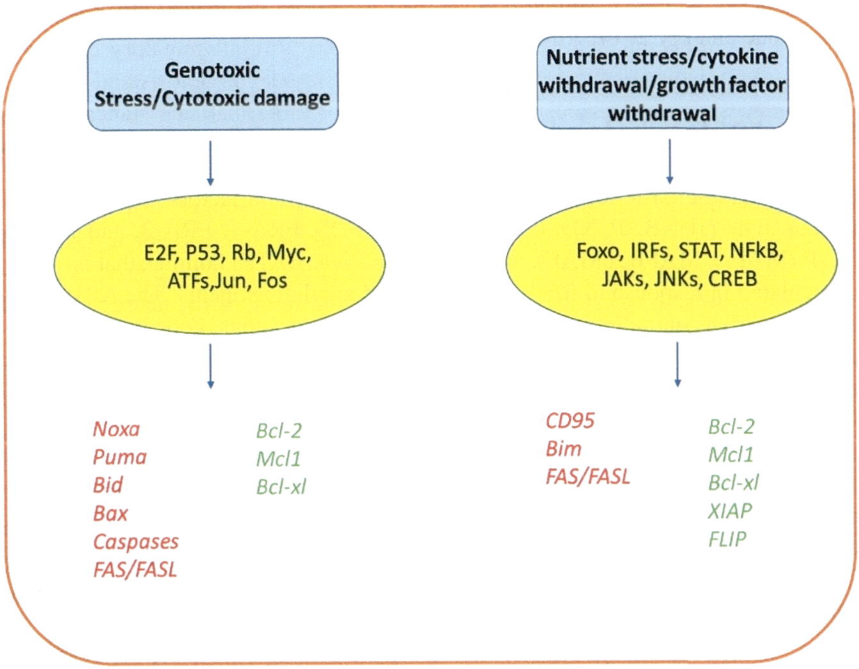


**Fig. 9.6** Transcriptional network activation in response to various stress and growth factor withdrawal regulates pro and anti-apoptotic genes

*FasL, REDD1*, etc. The p53 family of proteins can sense the degree of damage and maintain the balance between cell cycle arrest and apoptosis. In case of irreparable damage to cells, they activate the apoptosis regulators and prevent damaged cells from proliferation. The p53 family can also indirectly activate Caspase 1, 6, and 8, resulting in apoptosis.

Although p53 is known to activate both extrinsic (Fas, DR4, and DR5) and intrinsic (BAX, NOXA, PUMA, BID, P53AIP1) regulators of apoptosis, mutant p53 versions are often reported in various cancers (Chen et al. 2022; de Andrade et al. 2021). P53 mutants are primarily inactive due to loss of transactivation or DNA binding functions and contribute to carcinogenesis by mostly competing with the wild-type version(Sigal and Rotter 2000). Hotspot mutations in codons 175, 245, 248, 273, and 282 are the most commonly reported mutant p53 versions with ‘gain in function’ roles in various oncogenic processes like metastasis, stemness, and drug resistance(Chiang et al. 2021; Escoll et al. 2017; Heinlein et al. 2008; Lang et al. 2004; Sampath et al. 2001; Varley et al. 1999; Yang et al. 2022). Similar to p53 mutant versions, mutant p63 (TAp63 alpha, TAp63 Beta, TAp63 Gamma, Delta Np63 Alpha, Delta Np63 Beta, and Delta Np63 Gamma) and mutant p73 (Delta Np73, ex2p73 and ex2/3p73) also compete with their wild-type homologs and impairs their anti-cancer functions(Melino et al. 2003; Murray-Zmijewski et al.

2006; Yang et al. 1998). Mutant p63 and p73 have been implicated in various cancers (Inoue and Fry 2014; Xu et al. 2023).

### 9.5.2 AP1 Family: Landmine Valley

Class of dimeric basic leucine zipper (bZIP) transcription factors that include the famous JUN (JUNB, JUND, c-JUN), FOS (c-FOS, FRA-1, FRA-2, and FOSB), MAF (MAFA, MAFB, cMAF), and ATF (ATF2/3/4, BATF) families that are often upregulated in response to irradiations, UV and alkylating agents. The AP1 family members often contradict each other during cell fate decisions. For example, c-JUN can positively regulate proliferation via the repression of tumour suppressors and activation of CyclinD1 transcription, while JUNB antagonistically activates tumour suppressors and repress Cyclin D (Shaulian and Karin 2001).

Although AP1 proteins have various roles in development, apoptosis, autophagy, and physiology, malfunctional AP1 members are often associated with various cancers of the lung, breast, skin, brain, GI tract, and bone(Ameyar et al. 2003; Caputto et al. 2014; Gil et al. 2012; Jiang et al. 2020; Jochum et al. 2001; Kappelmann et al. 2014; Trop-Steinberg and Azar 2017). AP1 transcription factor Jun plays a significant role in developing leukaemias and lymphomas (Blonska et al. 2015; Fan et al. 2017; Ott et al. 2007; Szremska et al. 2003). AP1 family members c-JUN, JUNB promote proliferation in ALK-positive ALCL by inducting PI3K/AKT signalling in these leukaemias (Atsaves et al. 2015). Owing to their property of heterodimerisation, AP1 members have been implicated in various oncogenic hallmarks like JUNB in metastasis(Hyakusoku et al. 2016), c-Jun in cell proliferation, FRA-1/c-JUN in angiogenesis(Shaulian and Karin 2001), and JUN/FOS in tumour-promoting inflammatory signals(Ji et al. 2019).

### 9.5.3 NFkB, FOXO, IRF, and STAT: Horseman of Apocalypse

These family of transcription factors acts as sensors of growth factor signals and are activated in response to withdrawal of growth factors, cytokines, or hormones. NFkB family consists of dimeric transcription factors that are widely known for their anti-apoptotic, tumour-initiating, and immune-evasive roles(Burstein and Duckett 2003; Wang et al. 2014; Xia et al. 2018). NFkB, by controlling immune responses, often influences tumour development and promotes various cancers(Ben-Neriah and Karin 2011). NFkB's persistent expression in breast, colon, and lymphatic cancers is associated with proliferation, invasion, and metastasis (Sau et al. 2016; Sun 2017; Walther et al. 2015; Xia et al. 2014). NFkB can transcriptionally regulate proliferative genes like CyclinD and antiapoptotic factors like *BCL2, BCL-xl, BIRC5 (*Survivin*),* and *cIAPs* (Chen et al. 2000; Cui et al. 2017; Hinz et al. 1999; James et al. 2006). NFkB can also deactivate extrinsic pathways via its transcriptional target *DCR1,c-FLIP*, which antagonises TRAIL, FAS, or TNF-mediated apoptosis (Bernard et al. 2001; Micheau et al. 2001).

The forkhead box O transcription factors or FOXO family consists of four members, namely, FOXO1,3,4,6. It is negatively regulated via phosphorylation, which results in its cytoplasmic localisation and inhibition of transcriptional function(Greer and Brunet 2005). Various apoptosis regulators like *BIM, FasL, IGFBP1, TRAIL, and PUMA* are transcriptional targets of FOXO family members (Dijkers et al. 2000; Modur et al. 2002; You et al. 2021). The PI3K-AKT growth axis signals dictate FOXO phosphorylation and inactivation (Brunet et al. 1999). During nutrient deprivation, FOXO gets dephosphorylated and relocated into the nucleus, initiating apoptosis by activating the above target genes, indicating their role as tumour suppressors (Accili and Arden 2004). The deactivation of FOXO members is implicated in cancers of the prostate, breast, ovaries, and brain(Fu and Tindall 2008). FOXO can form gene fusions with various other proteins like *PAX3, PAX7* where they can act as oncogenes, for example, *PAX3-FOXO1* in Rhabdomyosarcomas, *PAX-FOXO3* and *FOXO4* fusions in AMLs (Fu and Tindall 2008).

IRFs or interferon regulatory factors are helix turn helix family of transcription factors that recognise interferon-stimulated response elements at promoters of interferon-inducible genes and play roles in immunity and cancers (Alsamman and El-Masry 2018). *Irf2* has been reported to promote cell transformation and enhance tumorigenicity in nude mice, while *Irf1* expression reversed the phenotype(Harada et al. 1993). IRF1 has been shown to reduce proliferation in a p21-dependent manner in various cell types, indicating its tumour-suppressive roles (Coccia et al. 1999; Passioura et al. 2005). IRF1 can activate various caspases and death receptors like Caspase 1, Caspase 8, and TRAIL, under interferon-gamma stimulation that results in apoptosis of ovarian, Ewing sarcoma, medulloblastoma, and breast cancer cells(Fulda and Debatin 2002; Kim et al. 2002). Deletions of IRF1 have been reported in various cancers, including leukaemia, breast, GI tract, and liver cancers. Oncogenic IRF2 has been associated with tumour size, metastasis, and poor survival in pancreatic cancer(Cui et al. 2012). Other IRF families like *IRF3, IRF4,* and *IRF7*'s overexpression have been implicated in various cancers of the lungs, lymph nodes, blood, and nervous system(Takaoka et al. 2008).

IRF1 is known to activate STAT1, a major player in downregulation of cell growth (Nguyen et al. 1997). STAT family belongs to latent cytoplasmic proteins, which, upon cytokine stimulation, are phosphorylated by JAK (Janus Protein Kinase), resulting in their nuclear localisation and activation of downstream target genes (Tolomeo and Cascio 2021). STAT family members STAT1 and STAT3 play controversial roles in promoting or inhibiting apoptosis, as various pro-apoptotic (*FAS, FasL, Caspase 1/8*) and antiapoptotic (*BCL-XL, BCL2,MCL-1, BIRC5*) genes are their targets (Oritani et al. 1999; Stephanou et al. 2000; Stephanou et al. 2001). Oncogenic STAT3 can also activate *MMP2, MMP3, MMP7,* and *VEGF,* thereby regulating angiogenesis and metastasis in various cancers (Kamran et al. 2013). STATs are involved in ovarian, lung, breast, melanoma, and prostate cancer progression (Ebersbach et al. 2021; Loh et al. 2019; Yang et al. 2020).

### 9.5.4 MYC and E2F: The Evil Duo

Oncogene *MYC* belongs to bHLH and Leucine zipper domain-containing transcription factors and works in conjunction with its heterodimer partner MAX, thus regulating transcription through E box elements (Blackwell et al. 1990; Blackwood and Eisenman 1991). *MYC* is one of the most frequently deregulated oncogenes owing to translocation, copy number amplification, and enhanced protein stability in >50% of all cancers (Dalla-Favera et al. 1982; Sears et al. 2000; Zack et al. 2013). Thus, *MYC* expression and function is strictly regulated in normal cells via various mechanisms like (a) post-transcriptionally repressed by micro RNAs like *miR-24, let-7,* (b) post-translationally protein stability is altered via kinases, ubiquitine ligases, and other MYC interactors, (c) via activation of apoptotic and cell cycle checkpoint regulators like p53, ARF, BIM, and PTEN under various genotoxic and nutritional stress(Stine et al. 2015). In cancers, hyper-activated MYC transcriptionally induces several cell cycle genes like *ODC, CUL1, CDK4, CDc25A,* and *ID2,* eventually activating cyclin E/CDK2 complexes, resulting in enhanced proliferation (Matsumura et al. 2003). Backhandedly, MYC can also transcriptionally repress cell cycle blockers like *p15 (CDKN2B), p21 (CDKN1A),* and *P27 (CDKN1B),* resulting in enhanced oncogenesis (Garcia-Gutierrez et al. 2019). Additionally, MYC can deregulate biochemical pathways like nucleotide biosynthesis, energy metabolism, protein synthesis, and genome maintenance, resulting in transformation and tumour progression (Stine et al. 2015; Dang 2012; Wolpaw and Dang 2018). MYC-driven cancers include myeloma, lymphoma, colon carcinoma, leukaemia, brain cancers, breast carcinoma, and prostate and liver cancer (Dang 2012; Gabay et al. 2014). Surprisingly, MYC-driven cancers are prone to the phenomena of 'oncogenic addition' and are considered for *MYC*-targeted therapies(Gabay et al. 2014).

Another critical cell cycle regulatory factor, the Early region 2 binding factor (*E2F*) transcription factor, plays a major role in conjunction with MYC in driving various oncogenic processes like proliferation, metastasis, stemness maintenance, angiogenesis, and chemoresistance (Aguda et al. 2008; Johnson and Schneider-Broussard 1998; Nakajima et al. 2023). E2F oncogenic activity is known to be upregulated in almost all cancers due to inactivation of the CDK-RB-E2F axis, which is regulated by pocket factors p107, p130, Rb, or CDK deregulation(Johnson and Schneider-Broussard 1998; Di Fiore et al. 2013; Kent and Leone 2019). MYC has been reported as the upstream transcription factor for the *E2F* family, and recently, sequential activation of *E2F* via MYC-driven pathways has been implicated in CDK inhibitor therapy-resistant breast cancer (Kim et al. 2023; Leone et al. 2001). All E2F member's protein stability depends upon the activity of APC/CCDC20 or CCNF ubiquitin ligases, which are well-known tumour suppressors that control the activation of E2F oncogenes(Budhavarapu et al. 2012; Clijsters et al. 2019; Peart et al. 2010). Apart from the deregulated cell cycle, E2F plays significant roles in the deregulation of other cellular processes like chromosomal instability, genomic duplication, DNA damage response, and replicative stress, which results in aggressive tumours (Johnson and Schneider-Broussard 1998; Kent and Leone 2019).

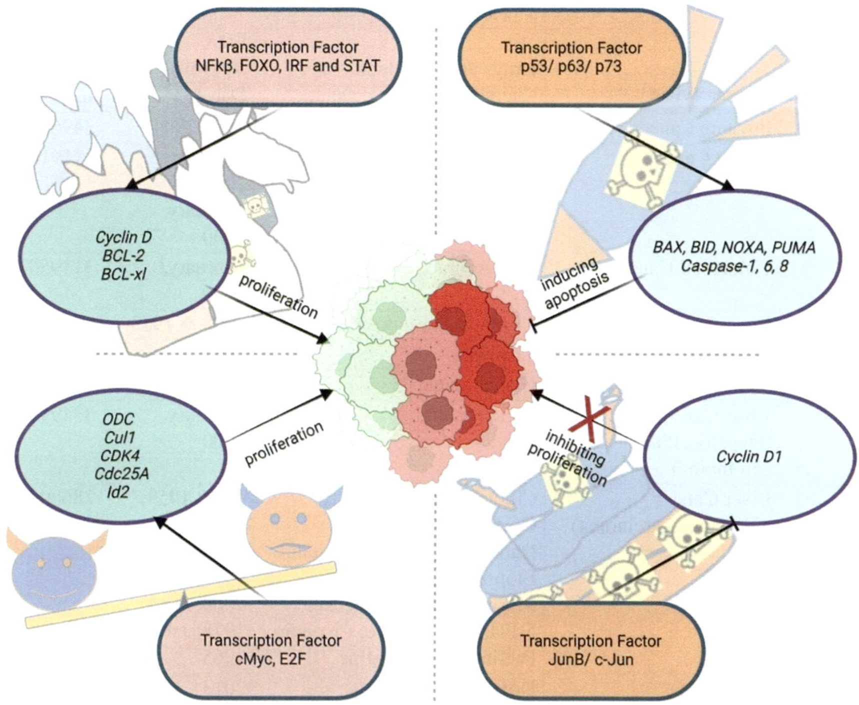


**Fig. 9.7** Interplay of master regulatory transcription factors belonging to tumour suppressors or oncogenes that control cell fate and cancer progression by altering apoptosis and cell proliferation functions

The four key transcription factor networks described above comprises key nodes regulating cell fate, cancer progression, and prevention. Other key nodes that are interconnected with these major players might help us understand the fine tuning of these network and provide drug targets for cancers. Together, these master regulatory transcription factors dictate the outcomes of oncogenesis (Fig. 9.7).

## 9.6 Conclusion: Truce of Uncertainty

With plethora of tumour suppressors deactivated and oncogenes hyperactivated during oncogenesis, cancer treatment requires a precise understanding of molecular mechanisms for drug development. Unfortunately, chemotherapy drugs used clinically often become obsolete due to the development of therapy resistance cancers (Table 9.3).

Modern-day interventions involve multi-modal approaches like combinatorial therapies, precision medicine, immune therapies, and alternate medicines. Here, we discuss a few mechanisms of therapy resistance and recent developments in combinatorial therapies for cancers.

**Table 9.3** Depicting FDA-approved drugs and resistance in cancers

| SL. No. | Cancer Types | Drugs against which Chemo-resistance developed | FDA Approval Date | Reference ID |
|---|---|---|---|---|
| 1 | Prostate Cancer | Docetaxel | May, 2004 | 34965937 |
| 2 | Breast Cancer | Doxorubicin & 5-Fuorouracil | February, 2013 & February, 2000 | 31191817 |
| 3 | Colorectal Cancer | 5-Fuorouracil | February, 2000 | 31395078 |
| 4 | Lung Cancer | Cisplatin & Etoposide | 1978 & 1983 | 36932427 |
| 5 | Ovarian Cancer | Cisplatin | 1978 | 35042862 |
| 6 | Pancreatic Cancer | Gemcitabine | April, 2011 | 31514451 |
| 7 | Liver Cancer (Hepatocellular carcinoma) | 5-Fuorouracil | February, 2000 | 33499874 |
| 8 | Liver Cancer (Cholangiocarcinoma) | Chlorambucil, Cyclophosphamide, Daunorubicin, Doxorubicin, Etoposide, Irinotecan, Mitoxantrone, Topotecan, 5-FU, Gemcitabine, Cisplatin, Paclitaxel, Vinblastine | 1957; 1959; 1979; 2013; 1983; 1998; 1987; 1998; 2000; 2011; 1978; 1992; 1965 | 28600147 |
| 9 | Lymphoma | Doxorubicin | 2013 | 35488020 |
| 10 | Leukaemia | Cytarabine | 1999 | 34836965 |
| 11 | Cervical Cancer | Cisplatin | 1978 | 27354763 |
| 12 | Sarcoma (Bone Cancer) | Doxorubicin | 2013 | 35406568 |
| 13 | Melanoma (Skin Cancer) | Doxorubicin & Cisplatin | 2013 & 1978 | 28382191 |
| 14 | Oral Cancer | Cisplatin & 5-FU | 1978 & 2000 | 33768115 |

### 9.6.1 Drug Resistance: Trojan Horse of the Enemy

Resistance in cancers; either extrinsic/acquired or intrinsic evolve due to various mechanisms including (a) altered drug export/import, (b) cell death inhibition (apoptosis suppression), (c) altering in the drug metabolism and detoxification, (d) epigenetic changes, (e) overexpression of pro-survival/anti-apoptotic genes, (f) enhances DNA-repair and target gene amplification (Fig. 9.8). Apart from these, various factors like tumour heterogeneity, microenvironment, immune system, and microRNAs also play essential roles in drug resistance (Mansoori et al. 2017). Membrane-bound exporters belonging to the ABC (ATP Binding Cassette) family of drug exporters play a significant role in chemoresistance (Emran et al. 2022). For example, mutations of *ABCB1/MDR1* and their overexpression are involved in acquired resistance to colchicine, vincristine, taxol, irinotecan via efflux mechanisms in many cancers. Another transporter protein, ABCC2,

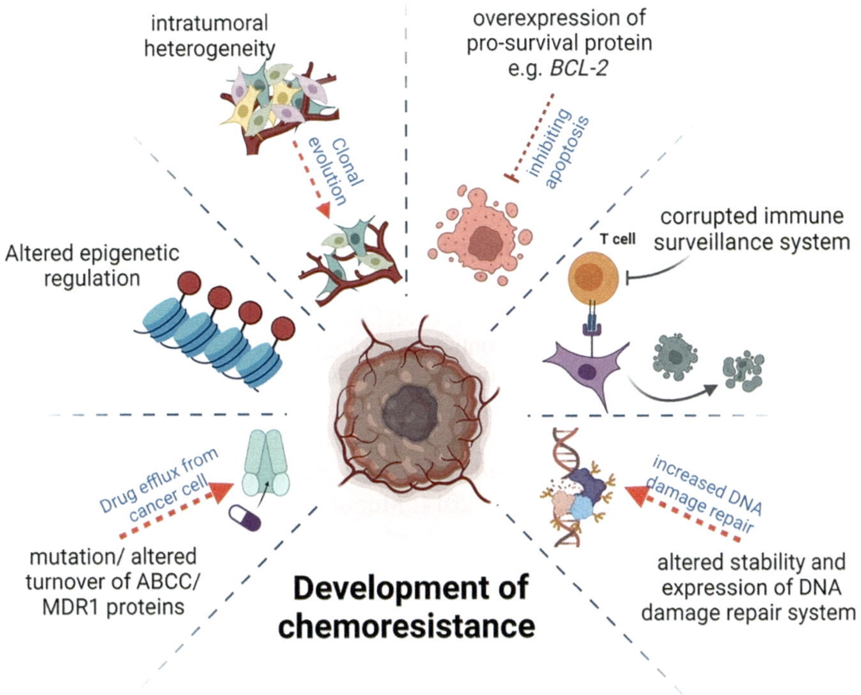


**Fig. 9.8** Various mechanisms of chemotherapy resistance utilised by tumours to overcome therapy-induced apoptosis or death

overexpressed in esophageal cancers tends to provide resistance towards cisplatin. Again, ABCC5 overexpression promotes resistance to thiopurine analogs, pacilitaxels in head and neck cancer while ABCG2 expression results in resistance towards doxorubicin analogs. Parallelly, overexpression of anti-apoptotic regulators like BCL-2 and growth stimulators like EGFR, E2F, and MYC promote chemoresistance in various cancers (Emran et al. 2022). Various microRNAs like miR*34a, 217, 320a, 96, 23b* are known to cause resistance against taxel groups, purine analogs, platinum-based drugs, adriamycin, doxorubicin, in various cancers, including prostate cancer, cervical cancer, gastric cancer, breast cancer, and ovarian cancer(Bukowski et al. 2020). Moreover, DNA repair proteins like ERCC1 increase repair function and resistance towards platinum-based agents (Bukowski et al. 2020; Rocha et al. 2018). DNA repair kinases *ATM, ATR, and PRKDC are often amplified in various cancers and result* in acquired chemoresistance due to heightened DNA repair functions (Bukowski et al. 2020). Xenobiotic metabolism regulators belonging to cytochrome (CYP) families like CYP1A6, CYP1A2, and CYP1B1 are often unregulated in specific cancers, resulting in modification and detoxification of mitoxantrone, flutamide, docetaxel, and paclitaxel leading to chemoresistance (Pathania et al. 2018). Altered expression of xenobiotic class II enzymes, including glutathione-S-transferases (*GST*s), gamma-glutamyltransferases (*GT*s), uridinediphospho-glucuronosyltransferases (*UGT*s) also increase multi drug resistance in various cancers (Pathania et al. 2018; Li et al. 2017). Thus, clear understanding of resistance mechanisms, tumour heterogeneity, and immune signal disruption is necessary for developing clinical intervention towards refractory cancers.

### 9.6.2 Recent Clinical Advances for Refractory Cancers; Smart Bombs and Magic Bullets

Personalised cancer therapy using omics technologies for effective drug solutions had been encouraged and proved effective in the western world (Krzyszczyk et al. 2018; Nimmesgern et al. 2017). However, lack of early diagnosis, variability among races, access to sequencing technologies, and poor healthcare system in developing countries had increased the burden of both refractory cancers and drug discovery(Adeniji et al. 2021; Mills 2014; Morash et al. 2018). Standard chemotherapy agents which are currently being used as interventions for cancer fall under the class of (a) alkylating agents, (b) antimetabolites, (c) mixed class. Owing to the failure of classical drugs in case of advanced drug-resistant disease, the concept of 'smart bombs' in chemotherapy are gaining importance through clinical trials for various tumours (Anand et al. 2023; Smith et al. 2021). Smart bombs concept includes the usage of standard chemotherapeutics in combination with modulators/inhibitors of immune checkpoints (CTLA4, PD-1/PD-L1), tumour-specific receptor antibodies (antibody–drug conjugates against HER2, EGFR), epigenetic modifiers (DNMTs, HDACs), and kinase inhibitors which yield better survival in patients (Feng and De Carvalho 2022; Han et al. 2020; Hughes et al. 2016; Modi et al. 2020; Wu et al. 2021; Zhang et al. 2023). For example, targeted therapies using CAR-T cells (Chimeric antigen receptor) alongside immune checkpoint blockers (anti-CTLA4/ anti-PD1 antibodies like ipilimumab, Nivolumab, and Pembrolizumab) and chemo showed encouraging results against various refractory and immunologically 'cold' tumours (Grosser et al. 2019). Similarly, a recent clinical trial for EnHerTu, a HER2-specific combination of Trastuzumab and Deruxtecan, decreased tumour incidence in previously treated Her2-positive breast cancer patients (Modi et al. 2020). Other antibody–drug conjugates against EGFR receptors showed promising results against glioblastoma, metastatic solid tumours, and NSCLC (Yu et al. 2022). Epigenetic modifiers like DNMT, EZH2, HDAC inhibitors (AzaC, Decitabine, Entinostat) are being used with regular replication blockers for refractory cancers of ovaries, prostate, Lungs and Hodgkin lymphoma (NCT00477386, NCT03709550, NCT00866333, NCT00086827).

Efficient drug targeting and delivery using radiotracers/radionuclides are being reutilised as 'magic bullets' against refractory cancers. Repurposing historical radiotracers/radionuclides with tumour-specific receptor targeting have been approved by FDA for refractory prostate cancer ($^{68}$Ga-PSMA-11/$^{177}$Lu-PSMA-617), neuroendocrine tumours ($^{68}$Ga-DOTA-TATE/$^{177}$Lu-DOTA-TATE), and several

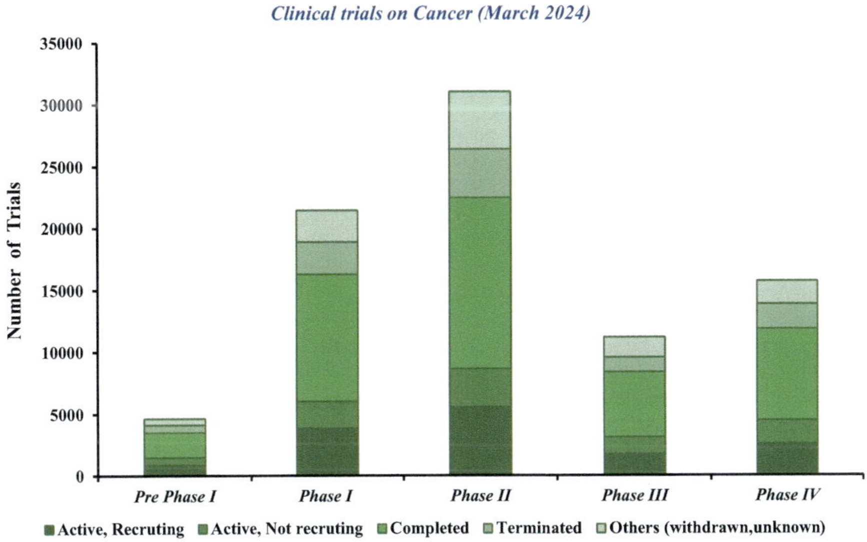


**Fig. 9.9** Chart depicting clinical trials on cancers based on phase, recruitment, and completion status till March 2024. The number of trials is based on data from https://clinicaltrials.gov/ with disease type as 'cancer'

haematological and solid tumours are currently under the clinical trials of these magic bullets (Caers et al. 2022; Echavidre et al. 2023; Fallah et al. 2023; Hennrich and Kopka 2019). Apart from well-established tumour-specific receptors like PSMA (Prostate-specific membrane antigen) and SSTR2 (Somatostatin Receptor 2), various others like Fibroblast Activation Protein Inhibitor (FAPI; NCT05410821 and NCT05400967), Gastrin-Releasing Peptide Receptor (GRPR), Human Epidermal Growth Factor Receptor 2 (HER2), Mucin-16 (MUC16), and Mucin-5AC (MUC5AC) are being investigated for effective theranostics against various refractory tumours (Echavidre et al. 2023). Another approach to combine FDA-approved radiotracers like DOTA-TATE, $^{177}$Lu-PSMA with immunotherapy modules like pembrolizumab, Nivolumab (anti-PD-L1 antibodies) are showing promising results under clinical trials (NCT03658447, NCT03805594, and NCT05150236) for many refractory cancers(Echavidre et al. 2023).

Killing cancer cells through the activation of apoptosis and other forms of regulated cell death has been the backbone of cancer research. With WHO and GLOBOCAN 2020 estimates of 10 million deaths worldwide, the fight against cancer has become a challenge. In the search for a cancer cure, thousands of clinical trials at various phases have been registered, completed, and terminated as per the https://clinicaltrials.gov/ (Fig. 9.9). Current understanding states that early detection, diagnosis, and intervention can be perceived as the cure for many cancers (Crosby et al. 2022; Ding and Chen 2022; Loud and Murphy 2017). Modern medicine, lifestyle improvements, and improved screening for various cancers will undoubtedly lower the burden of disease in future.

## References


Accili D, Arden KC (2004) FoxOs at the crossroads of cellular metabolism, differentiation, and transformation. Cell. 117(4):421–426

Adams JM, Cory S (2007) The Bcl-2 apoptotic switch in cancer development and therapy. Oncogene. 26(9):1324–1337

Adeniji AA, Dulal S, Martin MG (2021) Personalized medicine in oncology in the developing world: barriers and concepts to improve Status Quo. World J Oncol. 12(2–3):50–60

Aguda BD, Kim Y, Piper-Hunter MG, Friedman A, Marsh CB (2008) MicroRNA regulation of a cancer network: consequences of the feedback loops involving miR-17-92, E2F, and Myc. Proc Natl Acad Sci U S A. 105(50):19678–19683

Akiyama T, Bouillet P, Miyazaki T, Kadono Y, Chikuda H, Chung UI et al (2003) Regulation of osteoclast apoptosis by ubiquitylation of proapoptotic BH3-only Bcl-2 family member Bim. EMBO J. 22(24):6653–6664

Alsamman K, El-Masry OS (2018) Interferon regulatory factor 1 inactivation in human cancer. Biosci Rep. 38(3):BSR20171672

Ameyar M, Wisniewska M, Weitzman JB (2003) A role for AP-1 in apoptosis: the case for and against. Biochimie. 85(8):747–752

Anand U, Dey A, Chandel AKS, Sanyal R, Mishra A, Pandey DK et al (2023) Cancer chemotherapy and beyond: current status, drug candidates, associated risks and progress in targeted therapeutics. Genes Dis. 10(4):1367–1401

de Andrade KC, Khincha PP, Hatton JN, Frone MN, Wegman-Ostrosky T, Mai PL et al (2021) Cancer incidence, patterns, and genotype-phenotype associations in individuals with pathogenic or likely pathogenic germline TP53 variants: an observational cohort study. Lancet Oncol. 22(12):1787–1798

Atsaves V, Zhang R, Ruder D, Pan Y, Leventaki V, Rassidakis GZ et al (2015) Constitutive control of AKT1 gene expression by JUNB/CJUN in ALK+ anaplastic large-cell lymphoma: a novel crosstalk mechanism. Leukemia. 29(11):2162–2172

Balmanno K, Cook SJ (2009) Tumour cell survival signalling by the ERK1/2 pathway. Cell Death Differ. 16(3):368–377

Banin S, Moyal L, Shieh S, Taya Y, Anderson CW, Chessa L et al (1998) Enhanced phosphorylation of p53 by ATM in response to DNA damage. Science. 281(5383):1674–1677

Bauer MK, Vogt M, Los M, Siegel J, Wesselborg S, Schulze-Osthoff K (1998) Role of reactive oxygen intermediates in activation-induced CD95 (APO-1/Fas) ligand expression. J Biol Chem. 273(14):8048–8055

Ben-Neriah Y, Karin M (2011) Inflammation meets cancer, with NF-kappaB as the matchmaker. Nat Immunol. 12(8):715–723

Bernard D, Quatannens B, Vandenbunder B, Abbadie C (2001) Rel/NF-kappaB transcription factors protect against tumor necrosis factor (TNF)-related apoptosis-inducing ligand (TRAIL)-induced apoptosis by up-regulating the TRAIL decoy receptor DcR1. J Biol Chem. 276(29):27322–27328

Beroukhim R, Mermel CH, Porter D, Wei G, Raychaudhuri S, Donovan J et al (2010) The landscape of somatic copy-number alteration across human cancers. Nature. 463(7283):899–905

Billen LP, Shamas-Din A, Andrews DW (2008) Bid: a Bax-like BH3 protein. Oncogene. 27(Suppl 1):93–104

Blackwell TK, Kretzner L, Blackwood EM, Eisenman RN, Weintraub H (1990) Sequence-specific DNA binding by the c-Myc protein. Science. 250(4984):1149–1151

Blackwood EM, Eisenman RN (1991) Max: a helix-loop-helix zipper protein that forms a sequence-specific DNA-binding complex with Myc. Science. 251(4998):1211–1217

Bleicken S, Jeschke G, Stegmueller C, Salvador-Gallego R, Garcia-Saez AJ, Bordignon E (2014) Structural model of active Bax at the membrane. Mol Cell. 56(4):496–505

Blonska M, Zhu Y, Chuang HH, You MJ, Kunkalla K, Vega F et al (2015) Jun-regulated genes promote interaction of diffuse large B-cell lymphoma with the microenvironment. Blood. 125(6):981–991
Brozovic A, Fritz G, Christmann M, Zisowsky J, Jaehde U, Osmak M et al (2004) Long-term activation of SAPK/JNK, p38 kinase and fas-L expression by cisplatin is attenuated in human carcinoma cells that acquired drug resistance. Int J Cancer. 112(6):974–985
Brunet A, Bonni A, Zigmond MJ, Lin MZ, Juo P, Hu LS et al (1999) Akt promotes cell survival by phosphorylating and inhibiting a Forkhead transcription factor. Cell. 96(6):857–868
Budhavarapu VN, White ED, Mahanic CS, Chen L, Lin FT, Lin WC (2012) Regulation of E2F1 by APC/C Cdh1 via K11 linkage-specific ubiquitin chain formation. Cell Cycle. 11(10):2030–2038
Bukowski K, Kciuk M, Kontek R (2020) Mechanisms of multidrug resistance in cancer chemotherapy. Int J Mol Sci. 21(9):3233
Burstein E, Duckett CS (2003) Dying for NF-kappaB? Control of cell death by transcriptional regulation of the apoptotic machinery. Curr Opin Cell Biol. 15(6):732–737
Caers J, Duray E, Vrancken L, Marcion G, Bocuzzi V, De Veirman K et al (2022) Radiotheranostic agents in hematological malignancies. Front Immunol. 13:911080
Caputto BL, Cardozo Gizzi AM, Gil GA (2014) c-Fos: an AP-1 transcription factor with an additional cytoplasmic, non-genomic lipid synthesis activation capacity. Biochim Biophys Acta. 1841(9):1241–1246
Chandra J, Samali A, Orrenius S (2000) Triggering and modulation of apoptosis by oxidative stress. Free Radic Biol Med. 29(3–4):323–333
Chen C, Edelstein LC, Gelinas C (2000) The Rel/NF-kappaB family directly activates expression of the apoptosis inhibitor Bcl-x(L). Mol Cell Biol. 20(8):2687–2695
Chen X, Zhang T, Su W, Dou Z, Zhao D, Jin X et al (2022) Mutant p53 in cancer: from molecular mechanism to therapeutic modulation. Cell Death Dis. 13(11):974
Chiang YT, Chien YC, Lin YH, Wu HH, Lee DF, Yu YL (2021) The function of the Mutant p53-R175H in cancer. Cancers (Basel). 13(16):4088
Cimmino A, Calin GA, Fabbri M, Iorio MV, Ferracin M, Shimizu M et al (2005) miR-15 and miR-16 induce apoptosis by targeting BCL2. Proc Natl Acad Sci U S A. 102(39):13944–13949
Cimprich KA, Cortez D (2008) ATR: an essential regulator of genome integrity. Nat Rev Mol Cell Biol. 9(8):616–627
Clijsters L, Hoencamp C, Calis JJA, Marzio A, Handgraaf SM, Cuitino MC et al (2019) Cyclin F controls cell-cycle transcriptional outputs by directing the degradation of the three activator E2Fs. Mol Cell. 74(6):1264–77 e7
Coccia EM, Del Russo N, Stellacci E, Orsatti R, Benedetti E, Marziali G et al (1999) Activation and repression of the 2-5A synthetase and p21 gene promoters by IRF-1 and IRF-2. Oncogene. 18(12):2129–2137
Crosby D, Bhatia S, Brindle KM, Coussens LM, Dive C, Emberton M et al (2022) Early detection of cancer. Science. 375(6586):eaay9040
Croxton R, Ma Y, Song L, Haura EB, Cress WD (2002) Direct repression of the Mcl-1 promoter by E2F1. Oncogene. 21(9):1359–1369
Cui L, Deng Y, Rong Y, Lou W, Mao Z, Feng Y et al (2012) IRF-2 is over-expressed in pancreatic cancer and promotes the growth of pancreatic cancer cells. Tumour Biol. 33(1):247–255
Cui X, Shen D, Kong C, Zhang Z, Zeng Y, Lin X et al (2017) NF-kappaB suppresses apoptosis and promotes bladder cancer cell proliferation by upregulating survivin expression in vitro and in vivo. Sci Rep. 7:40723
Dalla-Favera R, Bregni M, Erikson J, Patterson D, Gallo RC, Croce CM (1982) Human c-myc onc gene is located on the region of chromosome 8 that is translocated in Burkitt lymphoma cells. Proc Natl Acad Sci U S A. 79(24):7824–7827
Dang CV (2012) MYC on the path to cancer. Cell. 149(1):22–35
De Zio D, Cianfanelli V, Cecconi F (2013) New insights into the link between DNA damage and apoptosis. Antioxid Redox Signal. 19(6):559–571

Degterev A, Huang Z, Boyce M, Li Y, Jagtap P, Mizushima N et al (2005) Chemical inhibitor of nonapoptotic cell death with therapeutic potential for ischemic brain injury. Nat Chem Biol. 1(2):112–119

Dehan E, Bassermann F, Guardavaccaro D, Vasiliver-Shamis G, Cohen M, Lowes KN et al (2009) betaTrCP- and Rsk1/2-mediated degradation of BimEL inhibits apoptosis. Mol Cell. 33(1):109–116

Deveraux QL, Reed JC (1999) IAP family proteins--suppressors of apoptosis. Genes Dev. 13(3):239–252

Di Fiore R, D'Anneo A, Tesoriere G, Vento R (2013) RB1 in cancer: different mechanisms of RB1 inactivation and alterations of pRb pathway in tumorigenesis. J Cell Physiol. 228(8):1676–1687

Dijkers PF, Medema RH, Lammers JW, Koenderman L, Coffer PJ (2000) Expression of the pro-apoptotic Bcl-2 family member Bim is regulated by the forkhead transcription factor FKHR-L1. Curr Biol. 10(19):1201–1204

Ding X, Chen L (2022) Analysis of the role of continuous early intervention in improving the quality of life of breast cancer patients. Comput Intell Neurosci. 2022:3054587

Dudgeon C, Qiu W, Sun Q, Zhang L, Yu J (2009) Transcriptional regulation of apoptosis. In: Dong Z, Yin X-M (eds) Essentials of apoptosis: a guide for basic and clinical research. Humana Press, Totowa, pp 239–260

Ebersbach C, Beier AK, Thomas C, Erb HHH (2021) Impact of STAT proteins in tumor progress and therapy resistance in advanced and metastasized prostate cancer. Cancers (Basel). 13(19):4854

Echavidre W, Fagret D, Faraggi M, Picco V, Montemagno C (2023) Recent pre-clinical advancements in nuclear medicine: pioneering the path to a limitless future. Cancers (Basel). 15(19):4839

Elmore S (2007) Apoptosis: a review of programmed cell death. Toxicol Pathol. 35(4):495–516

Emran TB, Shahriar A, Mahmud AR, Rahman T, Abir MH, Siddiquee MF et al (2022) Multidrug resistance in cancer: understanding molecular mechanisms, immunoprevention and therapeutic approaches. Front Oncol. 12:891652

Engeland K (2022) Cell cycle regulation: p53-p21-RB signaling. Cell Death Differ. 29(5):946–960

Escoll M, Gargini R, Cuadrado A, Anton IM, Wandosell F (2017) Mutant p53 oncogenic functions in cancer stem cells are regulated by WIP through YAP/TAZ. Oncogene. 36(25):3515–3527

Falck J, Mailand N, Syljuasen RG, Bartek J, Lukas J (2001) The ATM-Chk2-Cdc25A checkpoint pathway guards against radioresistant DNA synthesis. Nature. 410(6830):842–847

Fallah J, Agrawal S, Gittleman H, Fiero MH, Subramaniam S, John C et al (2023) FDA approval summary: Lutetium Lu 177 Vipivotide Tetraxetan for patients with metastatic castration-resistant prostate cancer. Clin Cancer Res. 29(9):1651–1657

Fan F, Bashari MH, Morelli E, Tonon G, Malvestiti S, Vallet S et al (2017) The AP-1 transcription factor JunB is essential for multiple myeloma cell proliferation and drug resistance in the bone marrow microenvironment. Leukemia. 31(7):1570–1581

Favaloro B, Allocati N, Graziano V, Di Ilio C, De Laurenzi V (2012) Role of apoptosis in disease. Aging (Albany NY). 4(5):330–349

Feng S, De Carvalho DD (2022) Clinical advances in targeting epigenetics for cancer therapy. FEBS J. 289(5):1214–1239

Fernald K, Kurokawa M (2013) Evading apoptosis in cancer. Trends Cell Biol. 23(12):620–633

Fu Z, Tindall DJ (2008) FOXOs, cancer and regulation of apoptosis. Oncogene. 27(16):2312–2319

Fukuhara S, Rowley JD (1978) Chromosome 14 translocations in non-Burkitt lymphomas. Int J Cancer. 22(1):14–21

Fulda S, Debatin KM (2002) IFNgamma sensitizes for apoptosis by upregulating caspase-8 expression through the Stat1 pathway. Oncogene. 21(15):2295–2308

Gabay M, Li Y, Felsher DW (2014) MYC activation is a hallmark of cancer initiation and maintenance. Cold Spring Harb Perspect Med. 4(6):a014241

Garcia-Gutierrez L, Delgado MD, Leon J (2019) MYC Oncogene contributions to release of cell cycle brakes. Genes (Basel). 10(3):244

Gil GA, Silvestre DC, Tomasini N, Bussolino DF, Caputto BL (2012) Controlling cytoplasmic c-Fos controls tumor growth in the peripheral and central nervous system. Neurochem Res. 37(6):1364–1371

Gong Y, Fan Z, Luo G, Yang C, Huang Q, Fan K et al (2019) The role of necroptosis in cancer biology and therapy. Mol Cancer. 18(1):100

Green DR, Llambi F. Cell death signaling. Cold Spring Harb Perspect Biol. 2015;7(12).

Greer EL, Brunet A (2005) FOXO transcription factors at the interface between longevity and tumor suppression. Oncogene. 24(50):7410–7425

Grosser R, Cherkassky L, Chintala N, Adusumilli PS (2019) Combination immunotherapy with CAR T cells and checkpoint blockade for the treatment of solid tumors. Cancer Cell. 36(5):471–482

Gu Y, Turck CW, Morgan DO (1993) Inhibition of CDK2 activity in vivo by an associated 20K regulatory subunit. Nature. 366(6456):707–710

Guo JY, Chen HY, Mathew R, Fan J, Strohecker AM, Karsli-Uzunbas G et al (2011) Activated Ras requires autophagy to maintain oxidative metabolism and tumorigenesis. Genes Dev. 25(5):460–470

Han J, Xu Y, Zhou Y, Yang A, Cui J, Chen P et al (2020) The effect of TKI therapy and chemotherapy treatment delivery sequence on total progression-free survival in patients with advanced EGFR-mutated NSCLC. Oncol Lett. 20(1):391–400

Hanahan D (2022) Hallmarks of cancer: new dimensions. Cancer Discov. 12(1):31–46

Hanahan D, Weinberg RA (2011) Hallmarks of cancer: the next generation. Cell. 144(5):646–674

Harada H, Kitagawa M, Tanaka N, Yamamoto H, Harada K, Ishihara M et al (1993) Anti-oncogenic and oncogenic potentials of interferon regulatory factors-1 and -2. Science. 259(5097):971–974

Heinlein C, Krepulat F, Lohler J, Speidel D, Deppert W, Tolstonog GV (2008) Mutant p53(R270H) gain of function phenotype in a mouse model for oncogene-induced mammary carcinogenesis. Int J Cancer. 122(8):1701–1709

Hennrich U, Kopka K (2019) Lutathera((R)): the first FDA- and EMA-approved radiopharmaceutical for peptide receptor radionuclide therapy. Pharmaceuticals (Basel). 12(3):114

Hinz M, Krappmann D, Eichten A, Heder A, Scheidereit C, Strauss M (1999) NF-kappaB function in growth control: regulation of cyclin D1 expression and G0/G1-to-S-phase transition. Mol Cell Biol. 19(4):2690–2698

Housman G, Byler S, Heerboth S, Lapinska K, Longacre M, Snyder N et al (2014) Drug resistance in cancer: an overview. Cancers (Basel). 6(3):1769–1792

Huang HM, Huang CJ, Yen JJ (2000) Mcl-1 is a common target of stem cell factor and interleukin-5 for apoptosis prevention activity via MEK/MAPK and PI-3K/Akt pathways. Blood. 96(5):1764–1771

Hubner A, Barrett T, Flavell RA, Davis RJ (2008) Multisite phosphorylation regulates Bim stability and apoptotic activity. Mol Cell. 30(4):415–425

Hughes PE, Caenepeel S, Wu LC (2016) Targeted therapy and checkpoint immunotherapy combinations for the treatment of cancer. Trends Immunol. 37(7):462–476

Hyakusoku H, Sano D, Takahashi H, Hatano T, Isono Y, Shimada S et al (2016) JunB promotes cell invasion, migration and distant metastasis of head and neck squamous cell carcinoma. J Exp Clin Cancer Res. 35:6

Inoue K, Fry EA (2014) Alterations of p63 and p73 in human cancers. Subcell Biochem. 85:17–40

Ionov Y, Yamamoto H, Krajewski S, Reed JC, Perucho M (2000) Mutational inactivation of the proapoptotic gene BAX confers selective advantage during tumor clonal evolution. Proc Natl Acad Sci U S A. 97(20):10872–10877

James MA, Lee JH, Klingelhutz AJ (2006) Human papillomavirus type 16 E6 activates NF-kappaB, induces cIAP-2 expression, and protects against apoptosis in a PDZ binding motif-dependent manner. J Virol. 80(11):5301–5307

Jan R, Chaudhry GE (2019) Understanding apoptosis and apoptotic pathways targeted cancer therapeutics. Adv Pharm Bull. 9(2):205–218

Ji Z, He L, Regev A, Struhl K (2019) Inflammatory regulatory network mediated by the joint action of NF-kB, STAT3, and AP-1 factors is involved in many human cancers. Proc Natl Acad Sci U S A. 116(19):9453–9462

Jiang X, Xie H, Dou Y, Yuan J, Zeng D, Xiao S (2020) Expression and function of FRA1 protein in tumors. Mol Biol Rep. 47(1):737–752

Jochum W, Passegue E, Wagner EF (2001) AP-1 in mouse development and tumorigenesis. Oncogene. 20(19):2401–2412

Johnson DG, Schneider-Broussard R (1998) Role of E2F in cell cycle control and cancer. Front Biosci. 3:d447–d448

Kalkavan H, Green DR (2018) MOMP, cell suicide as a BCL-2 family business. Cell Death Differ. 25(1):46–55

Kamran MZ, Patil P, Gude RP (2013) Role of STAT3 in cancer metastasis and translational advances. Biomed Res Int. 2013:421821

Kappelmann M, Bosserhoff A, Kuphal S (2014) AP-1/c-Jun transcription factors: regulation and function in malignant melanoma. Eur J Cell Biol. 93(1–2):76–81

Kaufmann SH, Earnshaw WC (2000) Induction of apoptosis by cancer chemotherapy. Exp Cell Res. 256(1):42–49

Kendall J, Liu Q, Bakleh A, Krasnitz A, Nguyen KC, Lakshmi B et al (2007) Oncogenic cooperation and coamplification of developmental transcription factor genes in lung cancer. Proc Natl Acad Sci U S A. 104(42):16663–16668

Kent LN, Leone G (2019) The broken cycle: E2F dysfunction in cancer. Nat Rev Cancer. 19(6):326–338

Kerr JF, Wyllie AH, Currie AR (1972) Apoptosis: a basic biological phenomenon with wide-ranging implications in tissue kinetics. Br J Cancer. 26(4):239–257

Kharbanda S, Pandey P, Schofield L, Israels S, Roncinske R, Yoshida K et al (1997) Role for Bcl-xL as an inhibitor of cytosolic Cytochrome c accumulation in DNA damage-induced apoptosis. Proc Natl Acad Sci U S A. 94(13):6939–6942

Kim S, Armand J, Safonov A, Zhang M, Soni RK, Schwartz G et al (2023) Sequential activation of E2F via Rb degradation and c-Myc drives resistance to CDK4/6 inhibitors in breast cancer. Cell Rep. 42(11):113198

Kim EJ, Lee JM, Namkoong SE, Um SJ, Park JS (2002) Interferon regulatory factor-1 mediates interferon-gamma-induced apoptosis in ovarian carcinoma cells. J Cell Biochem. 85(2):369–380

Kim H, Tu HC, Ren D, Takeuchi O, Jeffers JR, Zambetti GP et al (2009) Stepwise activation of BAX and BAK by tBID, BIM, and PUMA initiates mitochondrial apoptosis. Mol Cell. 36(3):487–499

Koren E, Fuchs Y (2021) Modes of Regulated Cell Death in Cancer. Cancer Discov. 11(2):245–265

Kroemer G, Marino G, Levine B (2010) Autophagy and the integrated stress response. Mol Cell. 40(2):280–293

Krzyszczyk P, Acevedo A, Davidoff EJ, Timmins LM, Marrero-Berrios I, Patel M et al (2018) The growing role of precision and personalized medicine for cancer treatment. Technology (Singap World Sci). 6(3–4):79–100

Kurokawa M, Kim J, Geradts J, Matsuura K, Liu L, Ran X et al (2013) A network of substrates of the E3 ubiquitin ligases MDM2 and HUWE1 control apoptosis independently of p53. Sci Signal. 6(274):ra32

Lang GA, Iwakuma T, Suh YA, Liu G, Rao VA, Parant JM et al (2004) Gain of function of a p53 hot spot mutation in a mouse model of Li-Fraumeni syndrome. Cell. 119(6):861–872

Lassus P, Opitz-Araya X, Lazebnik Y (2002) Requirement for caspase-2 in stress-induced apoptosis before mitochondrial permeabilization. Science. 297(5585):1352–1354

Lavin MF, Gueven N (2006) The complexity of p53 stabilization and activation. Cell Death Differ. 13(6):941–950

Leone G, Sears R, Huang E, Rempel R, Nuckolls F, Park CH et al (2001) Myc requires distinct E2F activities to induce S phase and apoptosis. Mol Cell. 8(1):105–113

Letai A, Bassik MC, Walensky LD, Sorcinelli MD, Weiler S, Korsmeyer SJ (2002) Distinct BH3 domains either sensitize or activate mitochondrial apoptosis, serving as prototype cancer therapeutics. Cancer Cell. 2(3):183–192

Li P, Nijhawan D, Budihardjo I, Srinivasula SM, Ahmad M, Alnemri ES et al (1997) Cytochrome c and dATP-dependent formation of Apaf-1/caspase-9 complex initiates an apoptotic protease cascade. Cell. 91(4):479–489

Li Y, Steppi A, Zhou Y, Mao F, Miller PC, He MM et al (2017) Tumoral expression of drug and xenobiotic metabolizing enzymes in breast cancer patients of different ethnicities with implications to personalized medicine. Sci Rep. 7(1):4747

Liu X, Kim CN, Yang J, Jemmerson R, Wang X (1996) Induction of apoptotic program in cell-free extracts: requirement for dATP and Cytochrome c. Cell. 86(1):147–157

Liu EY, Ryan KM (2012) Autophagy and cancer—issues we need to digest. J Cell Sci. 125(Pt 10):2349–2358

Liu S, Zhang P, Chen Z, Liu M, Li X, Tang H (2013) MicroRNA-7 downregulates XIAP expression to suppress cell growth and promote apoptosis in cervical cancer cells. FEBS Lett. 587(14):2247–2253

Loh CY, Arya A, Naema AF, Wong WF, Sethi G, Looi CY (2019) Signal Transducer and Activator of Transcription (STATs) proteins in cancer and inflammation: functions and therapeutic implication. Front Oncol. 9:48

van Loo G, Saelens X, van Gurp M, MacFarlane M, Martin SJ, Vandenabeele P (2002) The role of mitochondrial factors in apoptosis: a Russian roulette with more than one bullet. Cell Death Differ. 9(10):1031–1042

Loud JT, Murphy J (2017) Cancer screening and early detection in the 21(st) century. Semin Oncol Nurs. 33(2):121–128

Luanpitpong S, Chanvorachote P, Stehlik C, Tse W, Callery PS, Wang L et al (2013) Regulation of apoptosis by Bcl-2 cysteine oxidation in human lung epithelial cells. Mol Biol Cell. 24(6):858–869

Luciano F, Jacquel A, Colosetti P, Herrant M, Cagnol S, Pages G et al (2003) Phosphorylation of Bim-EL by Erk1/2 on serine 69 promotes its degradation via the proteasome pathway and regulates its proapoptotic function. Oncogene. 22(43):6785–6793

Madesh M, Hajnoczky G (2001) VDAC-dependent permeabilization of the outer mitochondrial membrane by superoxide induces rapid and massive Cytochrome c release. J Cell Biol. 155(6):1003–1015

Mansoori B, Mohammadi A, Davudian S, Shirjang S, Baradaran B (2017) The different mechanisms of cancer drug resistance: a brief review. Adv Pharm Bull. 7(3):339–348

Marchenko ND, Moll UM (2014) Mitochondrial death functions of p53. Mol Cell Oncol. 1(2):e955995

Matsumura I, Tanaka H, Kanakura Y (2003) E2F1 and c-Myc in cell growth and death. Cell Cycle. 2(4):333–338

Matsuoka S, Huang M, Elledge SJ (1998) Linkage of ATM to cell cycle regulation by the Chk2 protein kinase. Science. 282(5395):1893–1897

McStay GP, Clarke SJ, Halestrap AP (2002) Role of critical thiol groups on the matrix surface of the adenine nucleotide translocase in the mechanism of the mitochondrial permeability transition pore. Biochem J. 367(Pt 2):541–548

Mehrotra P, Ravichandran KS (2022) Drugging the efferocytosis process: concepts and opportunities. Nat Rev Drug Discov. 21(8):601–620

Melino G, Bernassola F, Ranalli M, Yee K, Zong WX, Corazzari M et al (2004) p73 Induces apoptosis via PUMA transactivation and Bax mitochondrial translocation. J Biol Chem. 279(9):8076–8083

Melino G, Lu X, Gasco M, Crook T, Knight RA (2003) Functional regulation of p73 and p63: development and cancer. Trends Biochem Sci. 28(12):663–670

Micheau O, Lens S, Gaide O, Alevizopoulos K, Tschopp J (2001) NF-kappaB signals induce the expression of c-FLIP. Mol Cell Biol. 21(16):5299–5305

Mills A (2014) Health care systems in low- and middle-income countries. N Engl J Med. 370(6):552–557
Modi S, Saura C, Yamashita T, Park YH, Kim SB, Tamura K et al (2020) Trastuzumab Deruxtecan in previously treated HER2-positive breast cancer. N Engl J Med. 382(7):610–621
Modur V, Nagarajan R, Evers BM, Milbrandt J (2002) FOXO proteins regulate tumor necrosis factor-related apoptosis inducing ligand expression. Implications for PTEN mutation in prostate cancer. J Biol Chem. 277(49):47928–47937
Morash M, Mitchell H, Beltran H, Elemento O, Pathak J (2018) The role of next-generation sequencing in precision medicine: a review of outcomes in oncology. J Pers Med. 8(3):30
Murray-Zmijewski F, Lane DP, Bourdon JC (2006) p53/p63/p73 isoforms: an orchestra of isoforms to harmonise cell differentiation and response to stress. Cell Death Differ. 13(6):962–972
Nakajima R, Zhao L, Zhou Y, Shirasawa M, Uchida A, Murakawa H et al (2023) Deregulated E2F activity as a cancer-cell specific therapeutic tool. Genes (Basel). 14(2):393
Nakano K, Vousden KH (2001) PUMA, a novel proapoptotic gene, is induced by p53. Mol Cell. 7(3):683–694
Nguyen H, Lin R, Hiscott J (1997) Activation of multiple growth regulatory genes following inducible expression of IRF-1 or IRF/RelA fusion proteins. Oncogene. 15(12):1425–1435
Nigg EA (2001) Mitotic kinases as regulators of cell division and its checkpoints. Nat Rev Mol Cell Biol. 2(1):21–32
Nijhawan D, Fang M, Traer E, Zhong Q, Gao W, Du F et al (2003) Elimination of Mcl-1 is required for the initiation of apoptosis following ultraviolet irradiation. Genes Dev. 17(12):1475–1486
Nimmesgern E, Norstedt I, Draghia-Akli R (2017) Enabling personalized medicine in Europe by the European Commission's funding activities. Per Med. 14(4):355–365
O'Connor L, Strasser A, O'Reilly LA, Hausmann G, Adams JM, Cory S et al (1998) Bim: a novel member of the Bcl-2 family that promotes apoptosis. EMBO J. 17(2):384–395
Oritani K, Tomiyama Y, Kincade PW, Aoyama K, Yokota T, Matsumura I et al (1999) Both Stat3-activation and Stat3-independent BCL2 downregulation are important for interleukin-6-induced apoptosis of 1A9-M cells. Blood. 93(4):1346–1354
Ott RG, Simma O, Kollmann K, Weisz E, Zebedin EM, Schorpp-Kistner M et al (2007) JunB is a gatekeeper for B-lymphoid leukemia. Oncogene. 26(33):4863–4871
Ozoren N, El-Deiry WS (2002) Defining characteristics of Types I and II apoptotic cells in response to TRAIL. Neoplasia. 4(6):551–557
Passioura T, Dolnikov A, Shen S, Symonds G (2005) N-ras-induced growth suppression of myeloid cells is mediated by IRF-1. Cancer Res. 65(3):797–804
Pathania S, Bhatia R, Baldi A, Singh R, Rawal RK (2018) Drug metabolizing enzymes and their inhibitors' role in cancer resistance. Biomed Pharmacother. 105:53–65
Peart MJ, Poyurovsky MV, Kass EM, Urist M, Verschuren EW, Summers MK et al (2010) APC/C(Cdc20) targets E2F1 for degradation in prometaphase. Cell Cycle. 9(19):3956–3964
Perniok A, Wedekind F, Herrmann M, Specker C, Schneider M (1998) High levels of circulating early apoptic peripheral blood mononuclear cells in systemic lupus erythematosus. Lupus. 7(2):113–118
Pucci B, Kasten M, Giordano A (2000) Cell cycle and apoptosis. Neoplasia. 2(4):291–299
Rabinowitz JD, White E (2010) Autophagy and metabolism. Science. 330(6009):1344–1348
Rai NK, Tripathi K, Sharma D, Shukla VK (2005) Apoptosis: a basic physiologic process in wound healing. Int J Low Extrem Wounds. 4(3):138–144
Reinhardt HC, Aslanian AS, Lees JA, Yaffe MB (2007) p53-deficient cells rely on ATM- and ATR-mediated checkpoint signaling through the p38MAPK/MK2 pathway for survival after DNA damage. Cancer Cell. 11(2):175–189
Rocha CRR, Silva MM, Quinet A, Cabral-Neto JB, Menck CFM (2018) DNA repair pathways and cisplatin resistance: an intimate relationship. Clinics (Sao Paulo). 73(suppl 1):e478s
Roos WP, Kaina B (2006) DNA damage-induced cell death by apoptosis. Trends Mol Med. 12(9):440–450

Sampath J, Sun D, Kidd VJ, Grenet J, Gandhi A, Shapiro LH et al (2001) Mutant p53 cooperates with ETS and selectively up-regulates human MDR1 not MRP1. J Biol Chem. 276(42):39359–39367
Sancho M, Leiva D, Lucendo E, Orzaez M (2022) Understanding MCL1: from cellular function and regulation to pharmacological inhibition. FEBS J. 289(20):6209–6234
Sau A, Lau R, Cabrita MA, Nolan E, Crooks PA, Visvader JE et al (2016) Persistent activation of NF-kappaB in BRCA1-deficient mammary progenitors drives aberrant proliferation and accumulation of DNA damage. Cell Stem Cell. 19(1):52–65
Schwickart M, Huang X, Lill JR, Liu J, Ferrando R, French DM et al (2010) Deubiquitinase USP9X stabilizes MCL1 and promotes tumour cell survival. Nature. 463(7277):103–107
Sears R, Nuckolls F, Haura E, Taya Y, Tamai K, Nevins JR (2000) Multiple Ras-dependent phosphorylation pathways regulate Myc protein stability. Genes Dev. 14(19):2501–2514
Shahar N, Larisch S (2020) Inhibiting the inhibitors: Targeting anti-apoptotic proteins in cancer and therapy resistance. Drug Resist Updat. 52:100712
Shaulian E, Karin M (2001) AP-1 in cell proliferation and survival. Oncogene. 20(19):2390–2400
Shen S, Kepp O, Kroemer G (2012) The end of autophagic cell death? Autophagy. 8(1):1–3
Sigal A, Rotter V (2000) Oncogenic mutations of the p53 tumor suppressor: the demons of the guardian of the genome. Cancer Res. 60(24):6788–6793
Singh R, Letai A, Sarosiek K (2019) Regulation of apoptosis in health and disease: the balancing act of BCL-2 family proteins. Nat Rev Mol Cell Biol. 20(3):175–193
Smith LT, Mayerson J, Nowak NJ, Suster D, Mohammed N, Long S et al (2006) 20q11.1 amplification in giant-cell tumor of bone: Array CGH, FISH, and association with outcome. Genes Chromosomes Cancer. 45(10):957–966
Smith SM, Wachter K, Burris HA 3rd, Schilsky RL, George DJ, Peterson DE et al (2021) Clinical cancer advances 2021: ASCO's report on progress against cancer. J Clin Oncol. 39(10):1165–1184
Snezhkina AV, Kudryavtseva AV, Kardymon OL, Savvateeva MV, Melnikova NV, Krasnov GS et al (2019) ROS generation and antioxidant defense systems in normal and malignant cells. Oxid Med Cell Longev. 2019:6175804
Stephanou A, Brar BK, Knight RA, Latchman DS (2000) Opposing actions of STAT-1 and STAT-3 on the Bcl-2 and Bcl-x promoters. Cell Death Differ. 7(3):329–330
Stephanou A, Scarabelli TM, Brar BK, Nakanishi Y, Matsumura M, Knight RA et al (2001) Induction of apoptosis and Fas receptor/Fas ligand expression by ischemia/reperfusion in cardiac myocytes requires serine 727 of the STAT-1 transcription factor but not tyrosine 701. J Biol Chem. 276(30):28340–28347
Stine ZE, Walton ZE, Altman BJ, Hsieh AL, Dang CV (2015) MYC, metabolism, and cancer. Cancer Discov. 5(10):1024–1039
Su H, Yang JR, Xu T, Huang J, Xu L, Yuan Y et al (2009) MicroRNA-101, down-regulated in hepatocellular carcinoma, promotes apoptosis and suppresses tumorigenicity. Cancer Res. 69(3):1135–1142
Sun SC (2017) The non-canonical NF-kappaB pathway in immunity and inflammation. Nat Rev Immunol. 17(9):545–558
Suzanne M, Steller H (2013) Shaping organisms with apoptosis. Cell Death Differ. 20(5):669–675
Szremska AP, Kenner L, Weisz E, Ott RG, Passegue E, Artwohl M et al (2003) JunB inhibits proliferation and transformation in B-lymphoid cells. Blood. 102(12):4159–4165
Takaoka A, Tamura T, Taniguchi T (2008) Interferon regulatory factor family of transcription factors and regulation of oncogenesis. Cancer Sci. 99(3):467–478
Tang D, Kang R, Berghe TV, Vandenabeele P, Kroemer G (2019) The molecular machinery of regulated cell death. Cell Res. 29(5):347–364
Tenev T, Bianchi K, Darding M, Broemer M, Langlais C, Wallberg F et al (2011) The Ripoptosome, a signaling platform that assembles in response to genotoxic stress and loss of IAPs. Mol Cell. 43(3):432–448
Thompson CB (1995) Apoptosis in the pathogenesis and treatment of disease. Science. 267(5203):1456–1462

Tolomeo M, Cascio A (2021) The multifaced role of STAT3 in cancer and its implication for anti-cancer therapy. Int J Mol Sci. 22(2):603

Tong J, Zheng X, Tan X, Fletcher R, Nikolovska-Coleska Z, Yu J et al (2018) Mcl-1 Phosphorylation without degradation mediates sensitivity to HDAC inhibitors by liberating BH3-only proteins. Cancer Res. 78(16):4704–4715

Tonon G, Wong KK, Maulik G, Brennan C, Feng B, Zhang Y et al (2005) High-resolution genomic profiles of human lung cancer. Proc Natl Acad Sci U S A. 102(27):9625–9630

Trop-Steinberg S, Azar Y (2017) AP-1 expression and its clinical relevance in immune disorders and cancer. Am J Med Sci. 353(5):474–483

Turkmen S, Riehn M, Klopocki E, Molkentin M, Reinhardt R, Burmeister T (2011) A BACH2-BCL2L1 fusion gene resulting from a t(6;20)(q15;q11.2) chromosomal translocation in the lymphoma cell line BLUE-1. Genes Chromosomes Cancer. 50(6):389–396

Varley JM, McGown G, Thorncroft M, James LA, Margison GP, Forster G et al (1999) Are there low-penetrance TP53 Alleles? evidence from childhood adrenocortical tumors. Am J Hum Genet. 65(4):995–1006

Vasan N, Baselga J, Hyman DM (2019) A view on drug resistance in cancer. Nature. 575(7782):299–309

Vishnoi K, Viswakarma N, Rana A, Rana B (2020) Transcription factors in cancer development and therapy. Cancers (Basel). 12(8):2296

Vitale I, Pietrocola F, Guilbaud E, Aaronson SA, Abrams JM, Adam D et al (2023) Apoptotic cell death in disease-Current understanding of the NCCD 2023. Cell Death Differ. 30(5):1097–1154

Walther W, Kobelt D, Bauer L, Aumann J, Stein U (2015) Chemosensitization by diverging modulation by short-term and long-term TNF-alpha action on ABCB1 expression and NF-kappaB signaling in colon cancer. Int J Oncol. 47(6):2276–2285

Wang JM, Chao JR, Chen W, Kuo ML, Yen JJ, Yang-Yen HF (1999) The antiapoptotic gene mcl-1 is up-regulated by the phosphatidylinositol 3-kinase/Akt signaling pathway through a transcription factor complex containing CREB. Mol Cell Biol. 19(9):6195–6206

Wang DJ, Ratnam NM, Byrd JC, Guttridge DC (2014) NF-kappaB functions in tumor initiation by suppressing the surveillance of both innate and adaptive immune cells. Cell Rep. 9(1):90–103

Wang K, Yin XM, Chao DT, Milliman CL, Korsmeyer SJ (1996) BID: a novel BH3 domain-only death agonist. Genes Dev. 10(22):2859–2869

Weaver AN, Yang ES (2013) Beyond DNA Repair: Additional Functions of PARP-1 in Cancer. Front Oncol. 3:290

Wei CL, Wu Q, Vega VB, Chiu KP, Ng P, Zhang T et al (2006) A global map of p53 transcription-factor binding sites in the human genome. Cell. 124(1):207–219

Wilkie-Grantham RP, Matsuzawa S, Reed JC (2013) Novel phosphorylation and ubiquitination sites regulate reactive oxygen species-dependent degradation of anti-apoptotic c-FLIP protein. J Biol Chem. 288(18):12777–12790

Wolpaw AJ, Dang CV (2018) MYC-induced metabolic stress and tumorigenesis. Biochim Biophys Acta Rev Cancer. 1870(1):43–50

Wong RS (2011) Apoptosis in cancer: from pathogenesis to treatment. J Exp Clin Cancer Res. 30(1):87

Wu Q, Luo W, Li W, Wang T, Huang L, Xu F (2021) First-generation EGFR-TKI plus chemotherapy versus EGFR-TKI alone as first-line treatment in advanced NSCLC with EGFR activating mutation: a systematic review and meta-analysis of randomized controlled trials. Front Oncol. 11:598265

Xia Y, Shen S, Verma IM (2014) NF-kappaB, an active player in human cancers. Cancer Immunol Res. 2(9):823–830

Xia L, Tan S, Zhou Y, Lin J, Wang H, Oyang L et al (2018) Role of the NFkappaB-signaling pathway in cancer. Onco Targets Ther. 11:2063–2073

Xiao Z, Chen Z, Gunasekera AH, Sowin TJ, Rosenberg SH, Fesik S et al (2003) Chk1 mediates S and G2 arrests through Cdc25A degradation in response to DNA-damaging agents. J Biol Chem. 278(24):21767–21773

Xie Y, Tobin LA, Camps J, Wangsa D, Yang J, Rao M et al (2013) MicroRNA-24 regulates XIAP to reduce the apoptosis threshold in cancer cells. Oncogene. 32(19):2442–2451

Xu X, Lai Y, Hua ZC (2019) Apoptosis and apoptotic body: disease message and therapeutic target potentials. Biosci Rep. 39(1)

Xu Y, Yang X, Xiong Q, Han J, Zhu Q (2023) The dual role of p63 in cancer. Front Oncol. 13:1116061

Yang A, Kaghad M, Wang Y, Gillett E, Fleming MD, Dotsch V et al (1998) p63, a p53 homolog at 3q27-29, encodes multiple products with transactivating, death-inducing, and dominant-negative activities. Mol Cell. 2(3):305–316

Yang J, Liu X, Bhalla K, Kim CN, Ibrado AM, Cai J et al (1997) Prevention of apoptosis by Bcl-2: release of Cytochrome c from mitochondria blocked. Science. 275(5303):1129–1132

Yang Y, Yang Y, Yang J, Zhao X, Wei X (2020) Tumor microenvironment in ovarian cancer: function and therapeutic strategy. Front Cell Dev Biol. 8:758

Yang Y, Yuan H, Zhao L, Guo S, Hu S, Tian M et al (2022) Targeting the miR-34a/LRPPRC/MDR1 axis collapse the chemoresistance in P53 inactive colorectal cancer. Cell Death Differ. 29(11):2177–2189

You H, Pellegrini M, Tsuchihara K, Yamamoto K, Hacker G, Erlacher M et al (2021) Correction: FOXO3a-dependent regulation of Puma in response to cytokine/growth factor withdrawal. J Exp Med. 218(2). https://doi.org/10.1084/jem.2006035301132021c

Youle RJ, Strasser A (2008) The BCL-2 protein family: opposing activities that mediate cell death. Nat Rev Mol Cell Biol. 9(1):47–59

Yu J, Fang T, Yun C, Liu X, Cai X (2022) Antibody-drug conjugates targeting the human epidermal growth factor receptor family in cancers. Front Mol Biosci. 9:847835

Yuan J, Kroemer G (2010) Alternative cell death mechanisms in development and beyond. Genes Dev. 24(23):2592–2602

Yun CW, Lee SH (2018) The roles of autophagy in cancer. Int J Mol Sci. 19(11)

Zack TI, Schumacher SE, Carter SL, Cherniack AD, Saksena G, Tabak B et al (2013) Pan-cancer patterns of somatic copy number alteration. Nat Genet. 45(10):1134–1140

Zha H, Aime-Sempe C, Sato T, Reed JC (1996) Proapoptotic protein Bax heterodimerizes with Bcl-2 and homodimerizes with Bax via a novel domain (BH3) distinct from BH1 and BH2. J Biol Chem. 271(13):7440–7444

Zhang W, Cheng GZ, Gong J, Hermanto U, Zong CS, Chan J et al (2008) RACK1 and CIS mediate the degradation of BimEL in cancer cells. J Biol Chem. 283(24):16416–16426

Zhang H, Kreis J, Schelhorn SE, Dahmen H, Grombacher T, Zuhlsdorf M et al (2023) Mapping combinatorial drug effects to DNA damage response kinase inhibitors. Nat Commun. 14(1):8310

Zuo Y, Xiang B, Yang J, Sun X, Wang Y, Cang H et al (2009) Oxidative modification of caspase-9 facilitates its activation via disulfide-mediated interaction with Apaf-1. Cell Res. 19(4):449–457

# Mechanisms of PANoptosis and Its Implications in Cancer Progression and Treatment

10

Shraman Jana, Satyajit Halder, Ankita Bhattacharya, Mihir K. Bhattacharya, and Kuladip Jana

**Abstract**

Activation of the innate immune system triggers inflammatory cell death in response to microbial infections or disruptions in cellular homeostasis. PANoptosis represents a novel and distinct inflammatory programmed cell death pathway governed by complex PANoptosome assemblies and emphasizes the substantial interplay and cross-talk among pyroptosis (P), apoptosis (A), or necroptosis (N). This study briefly overviews the basic processes underlying apoptosis, pyroptosis, and necroptosis, along with their extensive cross-talk events within PANoptosis. The PANoptosis concept and the PANoptosome complex's assembly process, which triggers cell death, will be explained. It emphasizes the notable contribution of PANoptosis to a range of cancer types, such as digestive cancers, breast cancers, and gliomas, illustrating its impact on tumor development and patient prognosis. Furthermore, the study discusses recent advancements in cancer treatment, highlighting the promise of targeting PANoptosis to enhance anti-tumor immune responses and diverse approaches for modulating PANoptosis pathways to achieve therapeutic benefits. In the future, continued research on PANoptosis will deepen our understanding of the core

S. Jana
IISER Kolkata, Campus Rd, Mohanpur, Haringhata Farm, West Bengal, India

S. Halder · K. Jana (✉)
Division of Molecular Medicine, Bose Institute, Unified Academic Campus, Kolkata, West Bengal, India
e-mail: kuladip@jcbose.ac.in

A. Bhattacharya
Department of Biochemistry, K.P.C Medical College & Hospital, Kolkata, India

M. K. Bhattacharya
ICMR-National Institute of Cholera & Enteric Diseases (NICED), Kolkata, West Bengal, India



K. Jana (ed.), *Apoptosis and Human Health: Understanding Mechanistic and Therapeutic Potential*, https://doi.org/10.1007/978-981-97-7905-5_10

mechanisms governing cell death and contribute crucial scientific insights to cancer research.

**Keywords**

PANoptosis · PANoptosome · Pyroptosis · Apoptosis · Necroptosis · Caspase · ZBP1 · AIM2 · MLKL · RIPK1 · RIPK3 · Cancer · Anti-tumor immunity

## Abbreviations

| | |
|---|---|
| iNOS | Inducible nitric oxide synthase |
| IRF1 | Interferon regulatory factor 1 |
| NLRP3/12 | NLR family pyrin domain-containing 3/12 |
| NEI | Nuclear export inhibitor |
| NO | Nitric oxide |
| P | Phosphorylation |
| TAK1i, TAK1 | Inhibition |
| TNF | Tumor necrosis factor |
| ZBP1 | Z-DNA-binding protein 1 |
| RIPK | Receptor interacting serine/threonine kinase |
| IRAK-M | Interleukin-1 receptor-associated kinase-M |
| STING | Stimulator of interferon genes |
| TBK1 | TANK-binding kinase 1 |
| diABZI | Diamidobenzimidazole |
| ADAR1 | Adenosine deaminase acting on RNA 1 |
| TCGA | The cancer genome atlas |
| GEO | Gene expression omnibus |
| MAP3K7 | Mitogen-activated protein kinase kinase kinase 7 |
| NFS1 | Cysteine desulfurase |
| AICAR | 5-Aminoimidazole-4-carboxamide ribonucleotide |

## 10.1 Introduction

Programmed cell death (PCD) is necessary to maintain homeostasis in multicellular organisms. It regulates excessive cell proliferation and host-immune defense against invading pathogens. Apoptosis, pyroptosis, and necroptosis are the most extensively studied cell death processes and have immense physiological implications. These processes are characterized by distinct molecular pathways that trigger cell death through various mechanisms (Bedoui et al. 2020). Apoptosis is a non-inflammatory, caspase-dependent mode of cell death characterized by apoptosome formation (Kiraz et al. 2016). Pyroptosis is induced by inflammasomes in which caspase-1 (CASP1) and gasdermin-D (GSDMD)-dependent pores form in the plasma membrane. This leads to cell lysis and the release of inflammatory substances (Bergsbaken et al. 2009). Necroptosis is associated with plasma membrane rupture and release of cellular contents. It is mainly regulated by two proteins, i.e., receptor-interacting protein kinase 3 (RIPK3) and mixed lineage kinase domain-like pseudokinase (MLKL). Various cytokines like TNF-α, Toll-like receptor (TLR) ligands, and chemotherapeutic drugs induce necroptosis (Gong et al. 2019). Initially, it was widely believed that the three programmed cell death processes work independently. But recently, it was reported that pyroptosis, apoptosis, and necroptosis exhibit significant overlap and cross-talk. This gave rise to the concept of a total cell death termed PANoptosis (P for pyroptosis, A for apoptosis, and N for necroptosis). This concept cannot be wholly explained by the above individual pathways alone (Malireddi et al. 2019). PANoptosis involves several interconnected upstream receptors and molecular signals, which converge into a deadly complex called PANoptosome; PANoptosome is a central platform that recruits a cascade of downstream effectors to initiate all three PCD pathways (Christgen et al. 2020). Recent studies have reported that PANoptosis involves various diseases, including infectious diseases, autoinflammatory diseases, and especially cancer (Sun et al. 2024). Studies on the alterations of genes associated with PANoptosis in cancer have identified several clusters. Each of them is associated with distinct immune microenvironment, rates of survival, and sensitivities to chemotherapeutic drugs (Wang et al. 2022; Ren et al. 2023). Therefore, understanding the relationship between PANoptosis and cancer could offer better insights into cancer occurrence and progression. These will ultimately facilitate the harnessing of new therapeutic avenues in the future.

PANoptosis plays a significant role in modulating innate immune cells, which detect and eliminate many pathogens. Moreover, multiple studies have shown that cancer patients express varying patterns of important pro-inflammatory cytokines. This suggests that PANoptosis might play a role in modulating anti-tumor immune responses (Malireddi et al. 2021). Still, the specific details of the PANoptosis pathway remain somewhat unclear. The therapeutic potential of targeting PANoptosis in cancer is still being explored. However, it holds promise for developing new treatments and improving existing therapies. It has been found that phosphorylated NFS1 inhibits PANoptosis in colorectal cancer and reduces its sensitivity to oxaliplatin (Lin et al. 2022). Therefore, NFS1 inhibition could stimulate apoptosis in colorectal cancer and improve the outcome of chemotherapy. More research on PANoptosis is necessary to figure out novel diagnostic and prognostic markers for cancer. Additionally, it could offer deeper insights into how immune responses are modulated in cancer.

*In this chapter*, we will describe the basic mechanisms of each PCD process and their significant cross-talk events in PANoptosis. We will then focus on four currently known PANoptosomes that induce PANoptosis. At last, we will discuss the implications of PANoptosis in tumor progression and the potential of effective PANoptosis-based cancer therapies.

## 10.2 Mechanisms of PCD Pathways and Applications in Cancer

i. *Apoptosis*: Apoptosis is a tightly controlled cell death process that does not elicit inflammatory responses. It can be induced by various stimuli, including pathogen infections, DNA damage, hypoxia, and specific tumor-associated proteins (Levine 1997). Two major pathways of apoptosis are the intrinsic pathway and the extrinsic pathway. The intrinsic pathway occurs in the mitochondria and is triggered by any intracellular oxidative stress or DNA damage. This results in mitochondrial outer membrane permeabilization (MOMP) and cytochrome c release in the cytosol. It is regulated by BCL-2 family proteins such as Bax, Bak, Bad, etc. (Tait and Green 2013). Afterward, cytochrome c forms a complex with apoptotic protease activating factor (APAF-1) and pro-caspase-9, known as the apoptosome. This complex activates pro-caspase-9 to form mature caspase-9. The mature caspase-9 activates pro-caspase-3 and pro-caspase-7, which perform proteolytic cleavage and subsequent cell death (Zou et al. 1997). The extrinsic pathway involves transmembrane death receptors such as tumor necrosis factor receptor (TNFR), TNF-related apoptosis-inducing ligand (TRAIL) receptor, etc., which bind to their respective ligands. This recruits FAS-associated protein-containing the death domain (FADD) and activates pro-caspase-8, forming a death-inducing signaling complex (DISC). Finally, mature caspase-8 activates the same executioner caspases and triggers downstream processes (Kiraz et al. 2016). During apoptosis, cells undergo specific structural alterations, such as shrinkage, nucleus fragmentation, and membrane blebs formation. Subsequently, the intracellular contents are packaged within tiny vesicles called apoptotic bodies. Phagocytic cells engulf and degrade these bodies without triggering any inflammatory reaction (D'arcy 2019). For many years, the utility of apoptosis modulators in clinical settings has been the subject of active research. Specific inhibitors of BCL-2 and BCL-XL, antagonists of inhibitors of apoptosis (IAP) proteins, and SMAC mimetics are some ways to target the intrinsic pathway of apoptosis. Methods for triggering cancer cell death through the extrinsic apoptotic pathway employ death receptor agonists like TNFα and TNF-related apoptosis-inducing ligands (TRAIL) (Carneiro and El-Deiry 2020).

ii. *Pyroptosis*: Pyroptosis is a lytic and inflammatory form of cell death. It gets triggered by the assembly and activation of inflammasomes. Inflammasomes are large protein complexes that generally include sensor proteins, adaptor proteins, and caspase enzymes (Malik and Kanneganti 2017). Generally, pattern recognition receptors (PRRs), like nucleotide-binding domain-like receptors (NLRs), pyrin, and absent in melanoma 2 (AIM2) act as the sensor proteins (Sharma and Kanneganti 2016). These inflammatory ligands trigger the activation and oligomerization of sensors and recruit adaptor proteins like apoptosis-associated speck-like proteins containing CARD (ASC) (Broz and Dixit 2016). Pyroptosis is activated via two signaling pathways. The canonical pathway is mediated by caspase-1 and inflammasome, whereas the non-canonical pathway is mediated by caspase-4, caspase-5, and caspase-11. In the canonical pathway, assembly of the inflammasome occurs, activating caspase-1 in the presence of various pathogen-associated molecular patterns (PAMPs) and damage-associated molecular patterns (DAMPs). Activated caspase-1 cuts and activates Gasdermin-D (GSDMD), which then executes cell death. It also activates IL-1β and IL-18, triggering inflammatory reactions (Liu et al. 2016). On the other hand, in the non-canonical pathways, human caspase-4, caspase-5, and murine caspase-11 directly recognize bacterial lipopolysaccharide (LPS) and activate GSDMD to drive pyroptosis (Shi et al. 2017). Both pathways release DAMPs, but only the canonical pathway releases active IL-1β and IL-18. The correlation between GSDM genes with tumor characteristics and their involvement in carcinogenesis makes them a promising therapeutic target across various types of cancer. Potential targets for regulating pyroptotic cell death in cancer involve agents like ferrostatin-1 and cisplatin, which promote pyroptosis, as well as agents like necrostatin-1, which inhibits pyroptosis (Tan et al. 2023) (Fig. 10.1).

iii. *Necroptosis*: The TNF/TNFR signaling pathway is mainly involved in necroptosis (Fuchs and Steller 2015). RIPK1 is the key upstream regulator of the TNF-mediated pathway. Without activated caspase-8, the TNF superfamily receptors get stimulated and activate RIPK1. RIPK1 then interacts with downstream molecules, i.e., RIPK3 and its substrate MLKL, to form a necrosome. RIPK3 phosphorylates MLKL, which now translocates to the plasma membrane and forms a multimeric pore. This disrupts membrane integrity, creates pores, and results in necroptosis (Wang et al. 2014). Thus, necroptosis is a rapid and lytic form of cell death that releases various DAMPs to trigger an innate immune response. Necroptosis can both promote and suppress tumorigenic activities within cancer cells. It can initiate an inflammatory response, attracting immune cells to eliminate the tumor. Conversely, it can also promote tumor metastasis by facilitating the detachment of cancer cells from the primary tumor and their invasion into other tissues (Qin et al. 2019). Specific applications of necroptosis in cancer treatment involve necroptosis-inducing medications and necroptosis inhibitors, including RIPK1 inhibitors, RIPK3 inhibitors, and MLKL inhibitors (Fig. 10.1).

## 10.3 Cross-Talk Events in PCD Pathways and the Emergence of PANoptosis

Growing evidence has suggested that the three modes of programmed cell death pathways do not operate completely in parallel. Rather, they cross-regulate each other and form a triangle of cell death. Caspases exist in two categories, i.e., inflammatory caspases (caspase-1, 4, 5, 11) and apoptotic caspases (caspase-3, 6, 7, 8, 9, 10) (Kesavardhana et al. 2020). Caspase-1 is one of the major molecules known to bridge apoptosis and pyroptosis. Usually, it activates GSDMD and takes part in pyroptosis. But, it can also initiate apoptosis via the caspase-3/7 axis without GSDMD (Tsuchiya et al. 2019). During apoptosis, caspase-3 and caspase-7 inhibit apoptosis by processing GSDMD, which reveals a close link between apoptosis and

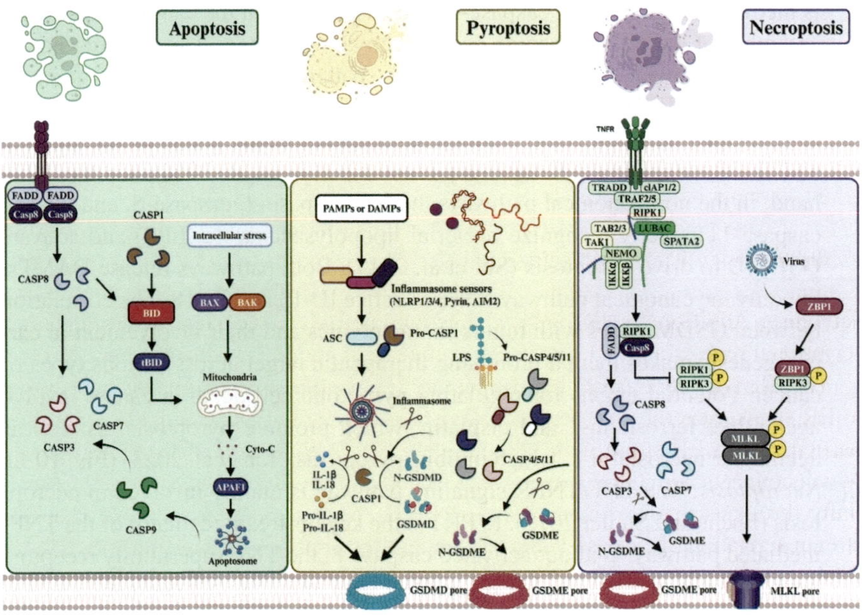


**Fig. 10.1** *Mechanistic overview of three conventional PCD pathways*. In extrinsic apoptosis, activation of death receptors recruits FADD and caspase-8, which are involved in forming the death-inducing signaling complex (DISC). Activated caspase-8 then further activates caspase-3 and -7, leading to GSDME-mediated pyroptosis. In intrinsic apoptosis, caspase-8 and caspase-1 cleave Bid into tBID, leading to mitochondrial outer membrane permeabilization (MOMP). As a result, cytochrome c is released, apoptosome forms, and caspase-9 gets activated, subsequently activating caspase-3 and -7. In pyrotosis, the canonical pathway involves the inflammasome to activate caspase-1 in the presence of PAMPs and DAMPs. Caspase-1 activates Gasdermin-D (GSDMD), IL-1β, and IL-18, triggering inflammation and cell death. In the non-canonical pathway, human caspase-4, caspase-5, and murine caspase-11 detect bacterial LPS, activate GSDMD, and induce pyroptosis. MLKL is the key executor of necroptosis and connects it to pyroptosis. Upon activation of RIPK1-RIPK3 or ZBP1, MLKL gets phosphorylated, forming membrane pores and triggering necroptosis. This disrupts the plasma membrane, causing potassium efflux and cellular stress. ZBP1 can itself activate the NLRP3 inflammasome in response to viral infections. Additionally, TNF binding to its receptor leads to caspase-8 mediated activation of caspase-3 and -7, resulting in GSDME-mediated pyroptosis. Figure created using BioRender (https://biorender.com/)

pyroptosis. Caspase-3 cleaves and activates gasdermin-E (Gasdermin-E) to induce secondary necrosis or pyroptosis. Gasdermin-Eswitchescaspase-3 mediated apoptosis to pyroptosis (Wang et al. 2017). Another key protein of the extrinsic apoptosis pathway is caspase-8. It is also considered a molecular switch for apoptosis, necroptosis, and pyroptosis (Fritsch et al. 2019). The intersection of pyroptosis and necroptosis primarily occurs in regulating NLRP3 inflammasome in pyroptosis and RIPK3/MLKL in necroptosis. Studies suggest that NLRP3 inflammasome and MLKL have a mutual regulatory loop of activation (Guo et al. 2019). It was previously thought that necroptosis is an alternative route of cell death in the absence of caspase-8-dependent apoptosis. Another novel finding is that caspase-8 and RIPK3 can assemble to form a necrosome when the level of RIP1 exceeds around 1000 molecules per cell. This results in the simultaneous initiation of apoptosis and necroptosis. On the contrary, only necroptosis occurs when the RIP1 level is too high (Li et al. 2021). Similarly, RIP3 also determines whether a cell will take apoptotic (when RIP3 is low) or necroptotic (when RIP3 is high) fate (Zhang et al. 2009). All these events conclude that there is a great deal of cross-talk and mechanistic interconnectedness between the above death mechanisms. So, in 2019, this connection between apoptosis, pyroptosis, and necroptosis was coined as "PANoptosis" (Malireddi et al. 2019). Several multi-protein complexes exist to regulate programmed cell death via mutual interactions among three types of protein domains, i.e., sensing, assembly, and catalytic domains (Samir et al. 2020). Several domains, including CARD, DEATH, PYD, and receptor-interacting protein homotypic interaction motif (RHIM), fall under the category of assembly domains. They play a critical role in forming programmed cell death-executing complexes. Pyroptosis is mediated by assembling several molecules that form inflammasomes via interactions between these death domains (Malik and Kanneganti 2017). Similarly, CARD domain facilitates interactions between APAF-1, caspase 9, and cytochrome c to form apoptosome (Li et al. 1997). Also, necroptosis is mediated by interactions between RHIM domains of RIPK1 and RIPK3, which ultimately form necrosomes without CASP8 (Vanden Berghe et al. 2016). Therefore, all these multimeric complexes harbor key proteins like FADD, RIPK1, and NLRP3 (Fritsch et al. 2019). This ultimately leads to the concept of a PANoptosome, which contains vital molecules that activate all three pathways and execute the process of PANoptosis (Christgen et al. 2020). Apart from experimental findings, the protein–protein interactions between the components of these three death pathways have also been visually assessed with the help of the STRING database. This offers robust evidence for cross-talk and redundancies between the three cell death pathways, supporting the study of PANoptosis. In summary, PANoptosome acts as a molecular scaffold to activate the required machinery for executing apoptosis, apoptosis, and necroptosis.

## 10.4 Mechanisms of PANoptosis and Assembly of PANoptosome

In 2020, Christgen et al. introduced the concept of the PANoptosome, in which molecules associated with these three conventional death pathways can assemble to form a singular complex (Christgen et al. 2020). Each PANoptosome consists of three essential components, i.e., sensors for identifying PAMPs or DAMPs like ZBP1, AIM2, and NLRP3; adaptors acting as connecting links like ASC and FADD; and catalytic executioners like CASP1, CASP8, RIPK1 and RIPK3 (Samir et al. 2020). Upon activation through several upstream receptors, it facilitates homotopic or heterotopic interactions of vital proteins of apoptosis, pyroptosis, and necroptosis, resulting in cell death (Samir et al. 2020). Still, mechanisms of recognizing microbial infections and their interplay remain elusive. Although four main upstream components of the PANoptosome have been identified to date, there are likely other upstream molecules that can induce PANoptosome formation and drive PANoptosis (Cai et al. 2023).

a) *ZBP1 PANoptosome*: The first PANoptosome to be discovered was the ZBP1 PANoptosome, consisting of ZBP1which harbors two different domains (Zα domain and RHIM domain). Zα domain contains two parts, i.e., Zα1 and Zα2, which recognize and bind Z-DNA and Z-RNA, respectively (Takaoka et al. 2007). Other components undergo homotypic interaction with ZBP1 due to the presence of the same RHIM domain. To protect against viral infections like Influenza A Virus (IAV) and Herplex Simplex Virus 1(HSV1), ZBP1 detects viral RNA and triggers PANoptosis. Studies have revealed that ZBP1 PANoptosome consists of receptors (ZBP1), inflammasome (NLRP3), adaptors (ASC and FADD), and catalytic effectors (RIPK3, RIPK1, caspase-8, and caspase-1) (Samir et al. 2020). ZBP1 recruits RIPK3 via homotypic RHIM domain interaction. This leads to MLKL phosphorylation and recruitment of RIPK1 through homotypic RHIM domain interaction. After that, RIPK1 binds FADD and recruits caspase-8, activating downstream executioner molecules, i.e., caspase-3 and caspase-7. ASC is another important component of PANoptosome, which associates with caspase-8 and activates GSDMD in pyroptosis (Zheng and Kanneganti 2020). So, MLKL, caspase-3/7, and GSDMD, the three final executioner proteins, are together activated during PANoptosome formation to execute inflammatory cell death.
b) *AIM2 PANoptosome*: Recently, the AIM2 PANoptosome has been identified, comprising AIM2, ZBP1, pyrin, ASC, caspase-1, caspase-8, RIPK3, and FADD (Lee et al. 2021). AIM2 inflammasome detects double-stranded DNA and protects against various infectious diseases and tumors. Pathogens like HSV1 and Francisella activate AIM2, which leads to AIM2 PANoptosome assembly and the onset of PANoptosis (Rathinam et al. 2010). In the AIM2 PANoptosome, AIM2 acts as an upstream receptor to activate pyrin and ZBP1 and enhance their expression levels via interferon (IFN) type I signaling. Activated ZBP1 promotes necroptosis by recruiting MLKL and RIPK3, while pyrin activates caspase-1 and caspase-8 to mediate pyroptosis and apoptosis, respectively. Therefore, these events collectively complete PANoptosome assembly and induce cell death (Christgen et al. 2020) (Fig. 10.2).
c) *RIPK1 PANoptosome*: The RIPK1 PANoptosome was initially discovered during Yersinia infection. Yersinia can trigger RIPK1-mediated PANoptosis by simultaneously activating proteins of all three death pathways. RIPK1 PANoptosome is almost identical to that of the ZBP1 PANoptosome in terms of all components except the sensor, which is RIPK1 itself. The total assembly procedure somehow closely matches that of ZBP1 (Lee et al. 2021). In contrast to ZBP1-mediated PANoptosis, removing RIPK1 does not completely halt cell death in this case. This indicates that some molecules in the RIPK1 PANoptosome harbor functional redundancies (Fig. 10.2).
d) *NLRP12 PANoptosome*: The NLRP12 PANoptosome, a recently identified PANoptosome, has shed new light on the role of NLRP12, which belongs to the NLR protein family. NLRP12 is an innate immune cytoplasmic receptor that has

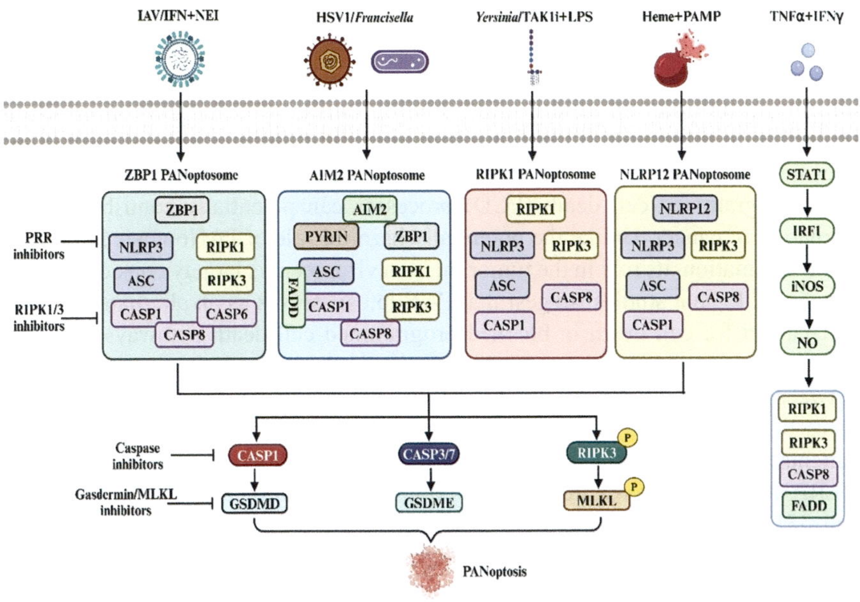


**Fig. 10.2** *Assembly of diverse PANoptosome complexes to execute PANoptosis.* PANoptosomes are heteromeric complexes that assemble in response to diverse pathogenic and sterile cellular insults by recruitment of various innate immune sensors, adaptors, caspases, and RIPKs to drive PANoptosis. During IAV infection or IFN+ NEI stimulation, the upstream sensor ZBP1 forms the ZBP1-PANoptosome (ZBP1, NLRP3, ASC, caspases-1, −8, −6-, RIPK1, and RIPK3). The canonical AIM2 sensor can detect pathogens like HSV1 or Francisella, which trigger the assembly of AIM2-PANoptosomes (AIM2-ZBP1-pyrin-ASC-caspase 1/8-FADD-RIPK1/3). RIPK1-PANoptosome formation (RIPK1, NLRP3, ASC, caspase-1, RIPK3, and caspase-8) following Yersinia infection or TAK1 inhibition plus LPS priming. Additionally, heme with PAMPs activate NLRP12-PANoptosome (NLRP12, NLRP3, ASC, Capase-1, Caspase-8, and RIPK3) formation. The TNF plus IFN-γ-induced PANoptosis sensors have yet to be identified, but this pathway is regulated by JAK/STAT-IRF1-iNOS-NO signaling. This leads to the activation of many caspases and RIPKs, which then feed forward to activate downstream executioners like GSDMD, GSDME, and pMLKL, driving lytic cell death and inflammatory cytokine release (PANoptosis). Figure created using BioRender (https://biorender.com/)

been reported to activate inflammasome and PANoptosome under the influence of heme and PAMPs (Sundaram et al. 2023). The NLRP12 PANoptosome comprises NLRP12, ASC, caspase 8, and RIPK3. Upstream signals from TLR2 and TLR4 induce the expression of NLRP12. Still, further studies are necessary to elucidate the assembly mechanism of the NLRP12 PANoptosome. Additionally, NLRP12 has been implicated in the pathogenesis of multiple diseases. This highlights the therapeutic potential of targeting NLRP12 for treating various diseases (Fig. 10.2).

## 10.5 PANoptosis in Cancer

While there has been extensive investigation into PANoptosis in the context of infections, its role in cancer still requires extensive scientific investigation. Nevertheless, substantial evidence has indicated that PANoptosis significantly regulates tumorigenesis (Jiang et al. 2021). The inability to regulate individual or combined programmed cell death (PCD) processes can potentially contribute to the formation of cancer. Unlike the potentially harmful role of PANoptosis in promoting inflammation, its role in the tumor microenvironment is largely considered to be beneficial. Recent studies suggest that PANoptosis serves as an alternative backup mechanism for cell death in blocked programmed cell death pathways in cancer cells (Hanahan and Weinberg 2000). Moreover, there is a significant correlation between PANoptosis genes and PANoptosis scores with patient survival in several types of cancer, which aids in predicting response to immunotherapy in cancer patients. Genes associated with PANoptosis have been demonstrated to mutate with NLRP3, having the most frequent mutations. Additionally, ZBP1, AIM2, NLRP3, and GSDMD exhibit the highest levels of copy number variation. In tumor samples, downregulation is observed in NLRP3, CASP7, RIP1, and RIP3 (Wang et al. 2022). Therefore, PANoptosis might favor specific cancer types, providing a perspective on identifying relevant therapeutic targets (Table 10.1).

- *PANoptosis in Digestive System Cancers*: Currently, extensive research is underway to investigate the role of PANoptosis in digestive system cancers. The PANoptosis-related risk score (PANS), corresponding to genes linked to PANoptosis, has been extensively explored to delineate the relationship between PANS and gastric cancer prognosis, tumor microenvironment, response to immunotherapy, and chemotherapy drugs. This provides a solid framework for anticancer therapy (Liu et al. 2024). In another study, more than 1000 individuals with gastric cancer were found to possess three distinct PANoptosis patterns, with each one having distinct clinical, pathological, and morphological traits. Analyzing these patterns will allow for the prediction of response to immunotherapy and survival, offering a framework for categorizing patients across different types of cancer (Pan et al. 2022). A recent study by Song et al. analyzed 69 PANoptosis-associated differentially expressed genes in hepatocellular carcinoma (HCC) associated with drug resistance and tumor growth. Based on molecular subtyping, the team constructed an HPAN-index for predicting response to immunotherapy and drug sensitivity (Song et al. 2023). Interestingly, studies have revealed a significant association between PANoptosis and the survival and immune response in liver hepatocellular carcinoma (LIHC). The molecular patterns identified in PANoptosis may serve as predictive indicators for the immune and prognostic statuses of LIHC patients (Zhang et al. 2023). In the case of colorectal cancer, it has been found that IRF1 could significantly prevent AOM/DSS-induced colorectal cancer in mice by inducing PANoptosis (Karki et al. 2020). IRF1, an upstream regulator of PANoptosis, might be a potent target for treating colorectal cancer. The regulatory interaction between ADAR1 and ZBP1

**Table 10.1** Correlation of PANoptosis with prognostic and therapeutic outcomes in cancer

| Category | Experimental analysis done | Analysis outcomes | References |
|---|---|---|---|
| Glioma | Analysis of three databases covering 1203 lower-grade glioma patient samples | Each PANcluster possesses unique molecular, clinical, and immunological status, and the pattern of PANoptosis can predict clinical survival outcomes. | Abulaiti et al. (2023) |
| Gastric and pan-cancer | Collection and analysis of RNA-seq data, along with associated pathological information of samples from various sources | The PANoptosis pattern score correlates with immunotherapy response rates and prognosis. | Pan et al. (2022) |
| Colon cancer | Bioinformatic approach to analyze PANoptosis-associated genes and data of cancer genomic studies and database repositories. | Clusters of genes linked with PANoptosis are associated with the survival of patients, functionality of the immune system, and tumorigenic signaling pathways. | Wang et al. (2022) |
| Breast cancer | Analysis of a particular scRNA-seq dataset GSE176078 retrieved from several clinical databases | Correlation of PANoptosis with improved prognosis and increased expression of genes linked to immunological checkpoints | He et al. (2023) |
| Cutaneous melanoma | Collecting and analyzing TCGA datasets | Genes linked to PANoptosis (ZBP1, MAP3K7, and RBCK1) hold promise for analyzing risks and predicting overall survival, reflecting the patients' immune status, and providing a new framework for personalized approaches in cancer therapeutics. | Wang et al. (2023) |
| Colorectal cancer | Knockout of ADAR1 | ZBP1-mediated PANoptosis is triggered, which helps in combating tumorigenesis | Karki et al. (2021) |
| Colorectal cancer | Induction of IRF1 | IRF1 triggers PANoptosis in colon cells, aiding in their defense against AOM/DSS-induced tumor formation | Karki et al. (2020) |
| Colorectal cancer | Knockout of NFS1 | Increases intracellular oxidative stress-induced PANoptosis and improve the effectiveness of platinum-based chemotherapy | Lin et al. (2022) |

has also been instrumental in harnessing therapeutic strategies against colorectal cancer by using a combination of interferon (IFN) and a nuclear export inhibitor (Karki et al. 2021). Recently, a correlation has been found between patient survival, activity of the immune system, and tumorigenic pathways in colon cancer with the PANoptosis-associated gene clusters. Additionally, a computed risk score can predict patients at risk of a worse prognosis. Several factors, including cancer stem cell index, the number of immune cells, checkpoint expression, and effectiveness of immunotherapy and chemotherapy drugs, are all associated with this score (Wang et al. 2022).

- *PANoptosis in Breast Cancer*: Three main types of interplay occur between PANoptosis and breast cancer: promoting apoptosis to induce cell death in breast cancer cells, the combined synergistic impact of pyroptosis and necroptosis together in triggering cell death, and the suppression of apoptosis-triggered cell death in breast cancer by necroptosis (Liu et al. 2024). Through a bioinformatic study, researchers have constructed prognostic models and risk stratification of patients to identify PANoptosis-associated genes with significant prognostic value, thereby offering promising new avenues for breast cancer immunotherapy (He et al. 2023). A highly accurate predictive model has been developed recently based on PANoptosis-related score (PRS) to understand the relevance of PANoptosis-related gene signature to modulating tumor microenvironment and drug sensitivity in breast cancer (Yu et al. 2024).
- *PANoptosis in Gliomas*: Several studies have established the link between PANoptosis and gliomas. Extensive evaluation has revealed genetic and molecular factors linked with PANoptosis-High and PANoptosis-Low prognostic clusters in gliomas, possibly improving the prognosis for patient populations via risk stratification and selection of treatment strategies. A study of 1203 glioma samples obtained from three public databases has focused on the implications of PANoptosis in shaping tumor microenvironment and modulating patient prognosis. Additionally, they identified unique PAN clusters, each characterized by unique molecular, immunological, and clinical profiles (Abulaiti et al. 2023). An artificial neural network (ANN)model has been devised by Chen et al. for differentiating PANoptosis-related tumor subgroups, providing some level of prognostic indication for gliomas (Chen et al. 2022). It is noteworthy that overexpression of some key proteins in PANoptosis, like caspase-3/7, GSDMD, and RIPK1/3, contribute to poor prognosis in gliomas, which contradicts the viewpoint that PANoptosis has a beneficial role against tumorigenesis. Hence, PANoptosis may not consistently exert an anti-cancer effect across all tumor types, and further studies are required to discover the underlying mechanisms.

## 10.6 Involvement of Innate Immune Cells in PANoptosis and Its Therapeutic Interventions in Cancer

The body's innate immunity is intrinsically designed to inhibit microbial pathogenesis or the spread of tumors. Upon invasion, it detects tumor cells or conserved microbial molecules. It quickly initiates a response by producing inflammatory mediators and activating programmed cell death pathways, including pyroptosis, apoptosis, necroptosis, and PANoptosis (Place et al. 2021). AIM2 is reported to regulate the sensors (Pyrin and ZBP1) for identifying potential PAMPs or DAMPs, thereby executing PANoptosis and associated inflammatory responses to protect the host. CASP6 promotes the interaction of RIPK3 and ZBP1, activating inflammasomes, regulating cell death, and protecting the host (Zheng et al. 2020). Therefore, research has focused on the involvement of CASP6 in PANoptosis and its effect on regulating tumorigenesis, pointing to it as a possible therapeutic strategy for cancer treatment (Qi et al. 2023). Additionally, the score to estimate levels of PANoptosis is closely linked with the status of tumor-immune microenvironment, expression of genes involved in immune checkpoints, and the amount of infiltration of various cells of innate immunity, such as macrophages, dendritic cells (DCs), natural killer (NK) cells, neutrophils, and various subsets of T cells (Zhuang et al. 2023). Below, we will discuss these cells and their significance in driving PANoptosis.

i. *Macrophages*: Macrophages are implicated in innate immune responses and are characterized by their large phagocytic capacity to recognize and clear pathogens, foreign particles, or cellular debris. In general, they are well known for being involved in PANoptosis. For instance, research demonstrates that the risk-assessment metric score, known as PANscore, is utilized to understand the status of PANoptosis. This score shows a significant association with immunosuppressive activities of M2 macrophages, TGF-β over-expression, and increased immune checkpoint activity (Pan et al. 2022). Another research revealed that CASP6 facilitates innate immune responses and activates PANoptosis by increasing the differentiation capacity of M2 macrophages (Zheng et al. 2020). This emphasizes the significance of macrophages in PANoptosis activation.
ii. *Dendritic Cells (DCs)*: Dendritic cells (DCs) are antigen-presenting cells that play a vital role in innate and adaptive immunity. DCs potentially contribute to PANoptosis by secretion of cytokines, expression of death receptors, and formation of PANoptsome. For instance, DCs have been observed to upregulate the expression of cytosolic sensors of the innate immune system, like AIM2, ZBP1, and RIPK1. These molecules play a role in PANoptosome assembly, promoting PANoptosis activation (Ocansey et al. 2024). DCs also release interferon-γ (IFN-γ), which has been shown to facilitate PANoptosis in murine models. Low IFN-γ levels inhibit the activation of several specific executioner molecules of PANoptosis, i.e., GSDMD, CASP3, and MLKL, and also reduce levels of IL-1β expression (Xiao et al. 2023). Similarly, in another study, it was found that the synergistic effect of TNF-α promotes activation of these similar death executioner molecules, undergoing PANoptosisin several cancers, both in vitro and xenograft mice models (Malireddi et al. 2021), and therefore represents a promising therapeutic approach (Fig. 10.3).
iii. *Natural killer (NK) cells*: Natural killer (NK) cells are a heterogeneous population that is involved in cytotoxicity, killing of tumor cells, and production of interferon-gamma (IFN-γ). They can produce a suitable microenvironment for potent T-cell immune responses (Guillerey et al. 2016). Additionally, they can stimulate other immune cells, which produce a synergistic immune response for tumor eradication. Experimental findings suggest that NLRP3 inhibits NK cell-mediated regulation of carcinogenesis and metastasis within the lung tumor microenvironment. This is demonstrated by over-activation of NK cells during tumor progression and better anti-metastatic response (Gao et al. 2024). Also,

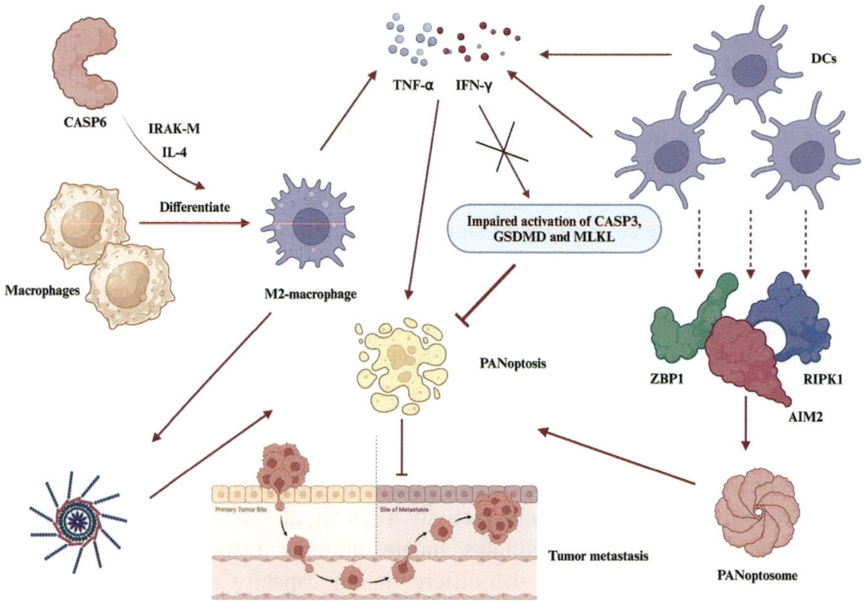


**Fig. 10.3** *Participation of macrophages and DCs in PANoptosis*. CASP6 is vital for activating inflammasomes and driving PANoptosis. It contributes to these processes by facilitating the transformation of macrophages into M2 macrophages. CASP6 also aids in macrophage activation by inhibiting IRAK-M, and increased IL-4 favors M2 differentiation, leading to increased CASP6 expression. Dendritic cells (DCs) assist in assembling the PANoptosome to initiate PANoptosis by overexpressing key cytosolic PANoptosis sensors. IFN-γ deficiency impedes the activation of executioner molecules of PANoptosis, i.e., GSDMD, CASP3, and MLKL, and decreases IL-1β expression. Moreover, the synergistic application of TNF-α and IFN-γ also drives PANoptosis, inhibiting tumor progression and metastasis. Figure created using BioRender (https://biorender.com/)

the levels of tumor-infiltrating circulating NK cells positively correlate with the survival advantages in hepatocellular carcinoma (HCC) (Sajid et al. 2022).

iv. *Neutrophils*: Neutrophils are pivotal in the primary defense against invading pathogens. STING is a master regulator of type I interferon signaling, which induces PANoptosis and mediates immune responses to eradicate tumors. It has been demonstrated that diamidobenzimidazole, an inducer of STING, promotes the activation of neutrophils to execute PANoptosis. Subsequently, STING dimerizes and phosphorylatesTBK1/IRF3, thereby initiating type I IFN response, activation of NF-kB, and formation of pro-inflammatory cytokines such as TNFα and IL-6 (Messaoud-Nacer et al. 2022). Activation of the IFNAR1/IFNAR2 signaling pathway activates ZBP1 and effectors of PANoptosis, i.e., RIPK3, ASC, and CASP8, leading to apoptotic cell death. This downstream cascade of events amplifies the neutrophil response and inflammatory cycle, thereby triggering PANoptosis (Fig. 10.4).

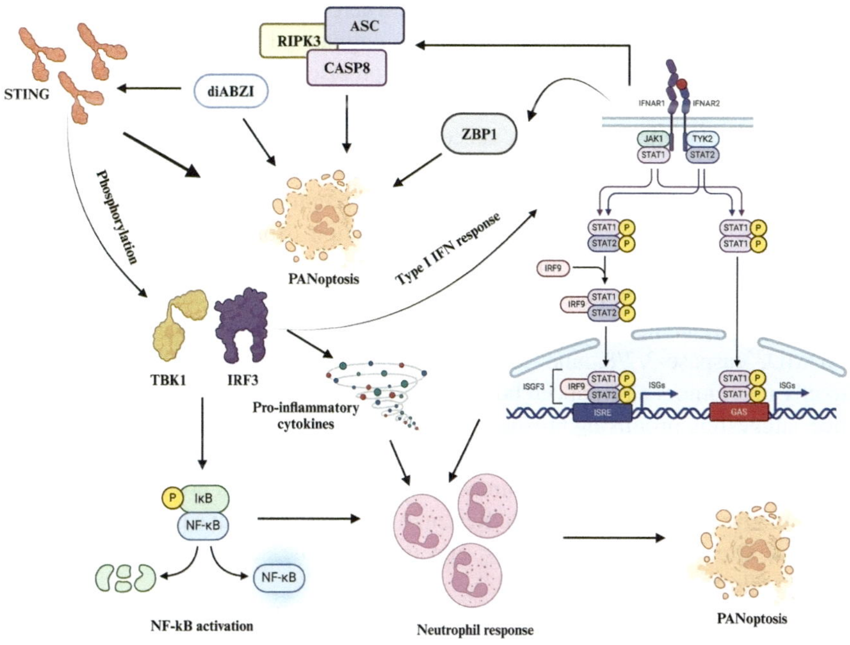


**Fig. 10.4** *PANoptosis and participation of neutrophils*. STING inducers initiate a sequence of downstream events that trigger PANoptosis. Dimerization of STING phosphorylates TBK1/IRF3, inducing type I interferon response and activation of NF-kB, which favors the formation of pro-inflammatory cytokines. Activation of ZBP1 and catalytic executioners of PANoptosis, i.e., RIPK3, ASC, and CASP8, are mediated by activation of IFNAR1/IFNAR2 signaling pathway. All these events subsequently induce apoptosis. Figure created using BioRender (https://biorender.com/)

It is crucial to explore various immune cell types and their roles in promoting or inhibiting tumor growth, especially their interaction mechanisms with PANoptosis. Additionally, understanding the dynamic processes contributing to heterogeneity and metabolic adaptability within the tumor microenvironment (TME) could enhance patients' response to immune-targeted therapies.

## 10.7 Implications of PANoptosis in Cancer Therapy

Apoptosis, necroptosis, and pyroptosis, the three classical modes of programmed cell death (PCD) pathways, have been extensively investigated in cancer therapeutics to inhibit cancer progression. But still, drugs that target individual PCD pathways have shown limited efficacy due to evasion and resistance by cancer cells. Hence, promoting tumor cell death via PANoptosis as a therapeutic approach for controlling cancer growth has garnered significant attention in research for quite some time.

Malignant melanoma is a typical example where PANoptosis has been studied as a cancer therapy. A group of researchers has engineered biocompatible folic acid-cholesterol-sodium alginate nanoparticles to precisely target melanoma cells and subsequently deliver improved efficiency by delivering methine (MET) and doxorubicin (DOX). The drug treatment successfully blocked melanoma progression by inducing PANoptosis. This establishes a groundwork for effectively systemically triggering PANoptosisin, although the upstream receptors and the underlying mechanisms behind the process need further elucidation (Song et al. 2021). There is considerable evidence that tumor necrosis factor-alpha (TNF-α) and interferon-gamma (IFN-γ) induce cell death in many cancer cell types and are involved in PANoptosis signaling. Their combined action can activate crucial components like GSDMD, caspase-3/7/8, and MLKL. Co-administration of TNF-α and IFN-γ directly into tumors has reduced both the weight and volume of xenograft tumors in mice, suggesting promising prospects for future clinical use (Malireddi et al. 2021). Sulconazole, an FDA-approved drug, demonstrates a variety of anticancer effects. It triggers multiple types of programmed cell death and inhibits proliferation, migration, and glycolysis in cancer cells. Research demonstrates that sulconazole triggers oxidative stress, inhibits glycolysis, and increases the sensitivity of cancer cells toward radiotherapy (Liu et al. 2023).

Another therapeutic aspect of PANoptosis is its ability to mediate strong anti-tumor immunity. Therapies involving immune checkpoint inhibitors (ICIs) can potentially improve the anti-tumor efficacy of several PCD pathways, even in ICI-resistant cancers (Tang et al. 2020).Yet, only a handful of studies have delved into the combination of PANoptosis and ICIs, which could offer promising new avenues for patients with refractory tumors. As CTLs are greatly involved in anti-tumor immunity, a pertinent example comes from a study investigating the function of GSDMD for regulating PANoptosis in CTLs. GSDMD is necessary for an optimal CTL response against tumor cells (Xi et al. 2019). On the other hand, over-expression of GSDME also promotes phagocytosis by tumor-associated macrophages (TAMs) and infiltration of natural killer (NK) cells. This suggests a strong association between the anti-tumor effects of GSDME and the activity of NK and CD8+ T cells (Zhang et al. 2020). Additionally, creating a PANoptosis signature is important for tumor immunity, as it increases the effectivity of immune response against tumors in the case of prostate cancer. PANoptosis enhances anti-tumor immunity and correlates positively with T cells' and NK cells' infiltration levels within the tumor microenvironment. Furthermore, the PANoptosis signature is positively associated with levels of immune checkpoints such as CD274, LAG3, and CXCR4 (Yi et al. 2023). The link between components of the immune system and PANoptosis provides promising therapeutic and prognostic possibilities in cancer.

Focus can be centered on the upstream signaling pathways, sensors, and components within the PANoptosome, such as caspases, ZBP1, AIM2, NLRP3, and RIPK1/3. Drugs that selectively increase the expression of cell surface PANoptosis-associated receptors, such as ZBP1, may exhibit potent anti-tumor effects (Baik et al. 2021). Zhang et al. identified a small molecule, Curaxin CBL0137, which activated ZBP1-mediated necroptosis in cancer-associated fibroblasts by triggering Z-DNA formation. Additionally, it reversed unresponsiveness against immune checkpoint blockade in murine melanoma models (Zhang et al. 2022). The caspase-3/GSDME signal pathway can be utilized as a balance point to reduce anti-apoptotic properties and strengthen anti-tumor immunity of tumors. Chemotherapeutic drugs like cisplatin have been reported to improve pathologic remission rates in TNBC patients by inducing pyroptosis through the NLRP3/caspase-1/GSDMD signaling pathway (Yan et al. 2021). Numerous clinical studies have investigated small-molecule RIPK1 inhibitors as potent emerging therapeutic agents against multifaceted diseases, including cancer, due to their favoring role in tumorigenesis (Shi et al. 2022). Moreover, focusing on the assembly process of the PANoptosome can also be an option to consider. PANoptosis can also be modulated by regulating the assembly of these proteins. Although, currently, not much research on drugs has been done to address this issue, it is surely worthy of exploration.

Lastly, there is a focus on combining PANoptosis with other advanced therapies, offering a novel approach to cancer treatment (Table 10.2). Immunogenic PANoptosis-initiated cancer sono-immune-reediting nano therapy has gained attention as a promising emerging technology for boosting the immune response against cancer. A recent study has demonstrated that a tumor's innate or acquired mechanisms of resistance towards immunotherapies could be effectively addressed through an immune-modulating anti-tumor strategy leveraging engineered extracellular vesicles (EVs) to induce PANoptotic cell death pathways (Zhou et al. 2024). Moreover, in contrast to conventional chemotherapy, therapies that target PANoptosis can minimize damage to healthy cells, decrease side effects after treatment, and improve treatment precision. Combining PANoptosis-targeted therapy with other treatment options, such as immunotherapy and radiotherapy, creates a more holistic treatment approach that widens the scope to induce cancer cell death and addresses drug resistance, providing personalized treatment tailored to each patient. Targeting specific proteins within the pathway with precise interventions holds promise for making breakthroughs in cancer treatment.

## 10.8 Conclusion and Future Directions

PANoptosis stands as one of the most intricate forms of programmed cell death known to date, which is intertwined with various human diseases, and its emerging significance in cancer is increasingly recognized. As tumors undergo simultaneous occurrences of multiple forms of cell death, delving into the interactions among these processes during tumor development will be crucial for understanding cancer pathogenesis. This exploration will offer essential insights and aid the development of effective anti-tumor medications. Despite impressive progress in understanding PANoptosis, some intriguing questions remain to be addressed. ZBP1 receptors are involved in triggering PANoptosis in case of viral or bacterial infections and cancer, but what connections and distinctions exist between these processes? Apart from known PANoptosomes, are other unknown proteins or molecules involved in

**Table 10.2** PANoptosis-associated therapeutic targets in recent studies

| Type of cancer | Model of study | Potential therapeutic approach | Major results | References |
|---|---|---|---|---|
| Colorectal cancer | Clinical samples, in vitro cell lines, BALB/c nude mice for in vivo studies | Application of oxaliplatin | Deficiency of NFS1 and treatment with oxaliplatin treatment induced PANoptosis. | Lin et al. (2022) |
| Low-grade glioma, renal cell carcinoma, and melanoma | Clinical details of samples from databases and in vitro cell lines | Administering KPT-335 in the presence or absence of IFN-γ, regulating ZBP1 in melanoma cells. | CASP3, CASP4, and AIM2 antagonize renal-cell carcinoma. ZBP1, GSDMD, CASP8, and ADAR inhibit low-grade glioma. GSDMD, CASP8, and ZBP1promote skin cutaneous melanoma. | Mall et al. (2022) |
| Adrenocortical carcinoma | TCGA database, in vitro SW-13 cells and BALB/c nude mice for in vivo studies | Administration of cucurbitacin E (CurE), either alone or in combination with mitotane | CurE as a CDK1 inhibitor, shows promise as a safe and effective drug for ACC treatment by modulating EMT, the G2/M checkpoint, and PANoptosis | Ren et al. (2022) |
| Esophageal cancer | In vitro cancer cell lines | Administration of sulconazole | Sulconazole triggers oxidative stress, inhibits glycolysis, and increases the sensitivity of cancer cells towards radiotherapy. Additionally, it restrains the proliferative and migratory abilities of tumor cells. | Liu et al. (2023) |
| Lung | Data acquisition from TCGA and GEO databases, in vitro studies on H1299 and H441 cells | Chemotherapeutic agents (such as AICAR and gemcitabine) or by inhibiting FADD. | Elevated FAAD levels are associated with improved responses to chemotherapies but diminished responses to immunotherapy. Blocking FADD hinders the proliferation of tumor cells, while its knockdown promotes apoptosis and pyroptosis. | Wei et al. (2023) |

(continued)

**Table 10.2** (continued)

| Type of cancer | Model of study | Potential therapeutic approach | Major results | References |
|---|---|---|---|---|
| KRAS-mutant PDAC | In vitro cells | Treatment with SEP, a multi-specific platinum complex | Performs DNA damage and induces mitochondrial superoxide anion radicals, affecting redox homeostasis and tumor metabolism, thereby triggering PANoptosis | Yuan et al. (2022) |

mediating PANoptosome assembly? The duality of PANoptosis in shaping tumor growth and progression creates obstacles in developing targeted therapies. While numerous questions about PANoptosis remain unanswered, the emergence of this concept and the advancements in research have undoubtedly opened a novel field of exploration for researchers. Rigorous investigation and clinical trials are necessary to validate both the safety and effectiveness of PANoptosis-targeted cancer treatments in the future.

## 10.9 Some Potential Future Questions for Further Research

1. What is the molecular makeup of the PANoptosomes that are currently unknown, like the complex that triggers PANoptosis induced by the synergistic effect of TNF-α and IFN-γ?
2. What are the assembly procedures of PANoptosome complex structures at the atomic level?
3. In cases where genetic ablation of key molecules (caspase 3, caspase 8, etc.) associated with PANoptosis proves to be lethal, how can suitable experimental models be made?
4. As a result of functional redundancies between cell death pathways, several therapies fail to exhibit desired efficacies. So, to address this issue, how can we design advanced therapeutics that will specifically target PANoptosis?

## References

Abulaiti A, Maimaiti A, Yiming N, Fu Q, Li S, Li Y, Wang Y, Zhou Q (2023) Molecular subtypes based on PANoptosis-related genes and tumor microenvironment infiltration characteristics in lower-grade glioma. Funct Integr Genomics 23(2):84

Baik JY, Liu Z, Jiao D, Kwon H-J, Yan J, Kadigamuwa C, Choe M, Lake R, Kruhlak M, Tandon M (2021) ZBP1 not RIPK1 mediates tumor necroptosis in breast cancer. Nat Commun 12(1):2666

Bedoui S, Herold MJ, Strasser A (2020) Emerging connectivity of programmed cell death pathways and its physiological implications. Nat Rev Mol Cell Biol 21(11):678–695

Bergsbaken T, Fink SL, Cookson BT (2009) Pyroptosis: host cell death and inflammation. Nat Rev Microbiol 7(2):99–109
Broz P, Dixit VM (2016) Inflammasomes: mechanism of assembly, regulation and signalling. Nat Rev Immunol 16(7):407–420
Cai H, Lv M, Wang T (2023) PANoptosis in cancer, the triangle of cell death. Cancer Med 12(24):22206–22223
Carneiro BA, El-Deiry WS (2020) Targeting apoptosis in cancer therapy. Nat Rev Clin Oncol 17(7):395–417
Chen G, He Z, Jiang W, Li L, Luo B, Wang X, Zheng X (2022) Construction of a machine learning-based artificial neural network for discriminating PANoptosis related subgroups to predict prognosis in low-grade gliomas. Sci Rep 12(1):22119
Christgen S, Zheng M, Kesavardhana S, Karki R, Malireddi RKS, Banoth B, Place DE, Briard B, Sharma BR, Tuladhar S (2020) Identification of the PANoptosome: a molecular platform triggering pyroptosis, apoptosis, and necroptosis (PANoptosis). Front Cell Infect Microbiol 10:237
D'arcy MS (2019) Cell death: a review of the major forms of apoptosis, necrosis and autophagy. Cell Biol Int 43(6):582–592
Fritsch M, Günther SD, Schwarzer R, Albert M-C, Schorn F, Werthenbach JP, Schiffmann LM, Stair N, Stocks H, Seeger JM (2019) Caspase-8 is the molecular switch for apoptosis, necroptosis and pyroptosis. Nature 575(7784):683–687
Fuchs Y, Steller H (2015) Live to die another way: modes of programmed cell death and the signals emanating from dying cells. Nat Rev Mol Cell Biol 16(6):329–344
Gao J, Xiong A, Liu J, Li X, Wang J, Zhang L, Liu Y, Xiong Y, Li G, He X (2024) PANoptosis: bridging apoptosis, pyroptosis, and necroptosis in cancer progression and treatment. Cancer Gene Ther:1–14
Gong Y, Fan Z, Luo G, Yang C, Huang Q, Fan K, Cheng H, Jin K, Ni Q, Yu X (2019) The role of necroptosis in cancer biology and therapy. Mol Cancer 18:1–17
Guillerey C, Huntington ND, Smyth MJ (2016) Targeting natural killer cells in cancer immunotherapy. Nat Immunol 17(9):1025–1036
Guo C, Fu R, Zhou M, Wang S, Huang Y, Hu H, Zhao J, Gaskin F, Yang N, Fu SM (2019) Pathogenesis of lupus nephritis: RIP3 dependent necroptosis and NLRP3 inflammasome activation. J Autoimmun 103:102286
Hanahan D, Weinberg RA (2000) The hallmarks of cancer. Cell 100(1):57–70
He P, Ma Y, Wu Y, Zhou Q, Du H (2023) Exploring PANoptosis in breast cancer based on scRNA-seq and bulk-seq. Front Endocrinol 14:1164930
Jiang M, Qi L, Li L, Wu Y, Song D, Li Y (2021) Caspase-8: a key protein of cross-talk signal way in "PANoptosis" in cancer. Int J Cancer 149(7):1408–1420
Karki R, Sharma BR, Banoth B, Malireddi RKS, Samir P, Tuladhar S, Mummareddy H, Burton AR, Vogel P, Kanneganti T-D (2020) Interferon regulatory factor 1 regulates PANoptosis to prevent colorectal cancer. JCI Insight 5(12):e136720
Karki R, Sundaram B, Sharma BR, Lee S, Malireddi RKS, Nguyen LN, Christgen S, Zheng M, Wang Y, Samir P (2021) ADAR1 restricts ZBP1-mediated immune response and PANoptosis to promote tumorigenesis. Cell Rep 37(3):109858
Kesavardhana S, Malireddi RKS, Kanneganti T-D (2020) Caspases in cell death, inflammation, and pyroptosis. Annu Rev Immunol 38:567–595
Kiraz Y, Adan A, Kartal Yandim M, Baran Y (2016) Major apoptotic mechanisms and genes involved in apoptosis. Tumour Biol 37:8471–8486
Lee S, Karki R, Wang Y, Nguyen LN, Kalathur RC, Kanneganti T-D (2021) AIM2 forms a complex with pyrin and ZBP1 to drive PANoptosis and host defence. Nature 597(7876):415–419
Levine AJ (1997) p53, the cellular gatekeeper for growth and division. Cell 88(3):323–331
Li P, Nijhawan D, Budihardjo I, Srinivasula SM, Ahmad M, Alnemri ES, Wang X (1997) Cytochrome c and dATP-dependent formation of Apaf-1/caspase-9 complex initiates an apoptotic protease cascade. Cell 91(4):479–489

Li X, Zhong C-Q, Wu R, Xu X, Yang Z-H, Cai S, Wu X, Chen X, Yin Z, He Q (2021) RIP1-dependent linear and nonlinear recruitments of caspase-8 and RIP3 respectively to necrosome specify distinct cell death outcomes. Protein Cell 12(11):858–876
Lin J-F, Hu P-S, Wang Y-Y, Tan Y-T, Yu K, Liao K, Wu Q-N, Li T, Meng Q, Lin J-Z (2022) Phosphorylated NFS1 weakens oxaliplatin-based chemosensitivity of colorectal cancer by preventing PANoptosis. Signal Transduct Target Ther 7(1):54
Liu L-X, Heng J-H, Deng D-X, Zhao H, Zheng Z-Y, Liao L-D, Lin W, Xu X-E, Li E-M, Xu L-Y (2023) Sulconazole induces panoptosis by triggering oxidative stress and inhibiting glycolysis to increase radiosensitivity in esophageal cancer. Mol Cell Proteomics 22(6):100551
Liu X, Miao M, Sun J, Wu J, Qin X (2024) PANoptosis: a potential new target for programmed cell death in breast cancer treatment and prognosis. Apoptosis 29(3):277–288
Liu Z, Sun L, Peng X, Zhu J, Wu C, Zhu W, Huang C, Zhu Z (2024) PANoptosis subtypes predict prognosis and immune efficacy in gastric cancer. Apoptosis:1–17
Liu X, Zhang Z, Ruan J, Pan Y, Magupalli VG, Wu H, Lieberman J (2016) Inflammasome-activated gasdermin D causes pyroptosis by forming membrane pores. Nature 535(7610):153–158
Malik A, Kanneganti T-D (2017) Inflammasome activation and assembly at a glance. J Cell Sci 130(23):3955–3963
Malireddi R, Karki R, Sundaram B, Kancharana B, Lee S, Samir P, Kanneganti T-D (2021) Inflammatory cell death, PANoptosis, mediated by cytokines in diverse cancer lineages inhibits tumor growth. Immunohorizons 5(7):568–580
Malireddi RKS, Kesavardhana S, Kanneganti T-D (2019) ZBP1 and TAK1: master regulators of NLRP3 inflammasome/pyroptosis, apoptosis, and necroptosis (PAN-optosis). Front Cell Infect Microbiol 9:406
Mall R, Bynigeri RR, Karki R, Malireddi RKS, Sharma BR, Kanneganti T-D (2022) Pancancer transcriptomic profiling identifies key PANoptosis markers as therapeutic targets for oncology. NAR Cancer 4(4):zcac033
Messaoud-Nacer Y, Culerier E, Rose S, Maillet I, Rouxel N, Briault S, Ryffel B, Quesniaux VFJ, Togbe D (2022) STING agonist diABZI induces PANoptosis and DNA mediated acute respiratory distress syndrome (ARDS). Cell Death Dis 13(3):269
Ocansey DKW, Qian F, Cai P, Ocansey S, Amoah S, Qian Y, Mao F (2024) Current evidence and therapeutic implication of PANoptosis in cancer. Theranostics 14(2):640
Pan H, Pan J, Li P, Gao J (2022) Characterization of PANoptosis patterns predicts survival and immunotherapy response in gastric cancer. Clin Immunol 238:109019
Place DE, Lee S, Kanneganti T-D (2021) PANoptosis in microbial infection. Curr Opin Microbiol 59:42–49
Qi L, Wang L, Jin M, Jiang M, Li L, Li Y (2023) Caspase-6 is a key regulator of cross-talk signal way in PANoptosis in cancer. Immunology 169(3):245–259
Qin X, Ma D, Tan Y, Wang H, Cai Z (2019) The role of necroptosis in cancer: a double-edged sword? Biochim Biophys Acta (BBA) Rev Cancer 1871(2):259–266
Rathinam VAK, Jiang Z, Waggoner SN, Sharma S, Cole LE, Waggoner L, Vanaja SK, Monks BG, Ganesan S, Latz E (2010) The AIM2 inflammasome is essential for host defense against cytosolic bacteria and DNA viruses. Nat Immunol 11(5):395–402
Ren H, Kang N, Yin S, Xu C, Qu T, Dai D (2023) Characteristic of molecular subtypes based on PANoptosis-related genes and experimental verification of hepatocellular carcinoma. Aging (Albany NY) 15(10):4159
Ren L, Yang Y, Li W, Zheng X, Liu J, Li S, Yang H, Zhang Y, Ge B, Zhang S (2022) CDK1 serves as a therapeutic target of adrenocortical carcinoma via regulating epithelial–mesenchymal transition, G2/M phase transition, and PANoptosis. J Transl Med 20(1):444
Sajid M, Liu L, Sun C (2022) The dynamic role of NK cells in liver cancers: role in HCC and HBV associated HCC and its therapeutic implications. Front Immunol 13:887186
Samir P, Malireddi RKS, Kanneganti T-D (2020) The PANoptosome: a deadly protein complex driving pyroptosis, apoptosis, and necroptosis (PANoptosis). Front Cell Infect Microbiol 10:238
Sharma D, Kanneganti T-D (2016) The cell biology of inflammasomes: mechanisms of inflammasome activation and regulation. J Cell Biol 213(6):617–629

Shi J, Gao W, Shao F (2017) Pyroptosis: gasdermin-mediated programmed necrotic cell death. Trends Biochem Sci 42(4):245–254

Shi K, Zhang J, Zhou E, Wang J, Wang Y (2022) Small-molecule receptor-interacting protein 1 (RIP1) inhibitors as therapeutic agents for multifaceted diseases: current medicinal chemistry insights and emerging opportunities. J Med Chem 65(22):14971–14999

Song F, Wang C-G, Mao J-Z, Wang T-L, Liang X-L, Hu C-W, Zhang Y, Han L, Chen Z (2023) PANoptosis-based molecular subtyping and HPAN-index predicts therapeutic response and survival in hepatocellular carcinoma. Front Immunol 14:1197152

Song M, Xia W, Tao Z, Zhu B, Zhang W, Liu C, Chen S (2021) Self-assembled polymeric nanocarrier-mediated co-delivery of metformin and doxorubicin for melanoma therapy. Drug Deliv 28(1):594–606

Sun X, Yang Y, Meng X, Li J, Liu X, Liu H (2024) PANoptosis: mechanisms, biology, and role in disease. Immunol Rev 321(1):246–262

Sundaram B, Pandian N, Mall R, Wang Y, Sarkar R, Kim HJ, Malireddi RKS, Karki R, Janke LJ, Vogel P (2023) NLRP12-PANoptosome activates PANoptosis and pathology in response to heme and PAMPs. Cell 186(13):2783–2801

Tait SWG, Green DR (2013) Mitochondrial regulation of cell death. Cold Spring Harb Perspect Biol 5(9):a008706

Takaoka A, Wang Z, Choi MK, Yanai H, Negishi H, Ban T, Lu Y, Miyagishi M, Kodama T, Honda K (2007) DAI (DLM-1/ZBP1) is a cytosolic DNA sensor and an activator of innate immune response. Nature 448(7152):501–505

Tan J, Zhuo Z, Si Y (2023) Application of pyroptosis in tumor research. Oncol Lett 26(3):1–8

Tang R, Xu J, Zhang B, Liu J, Liang C, Hua J, Meng Q, Yu X, Shi S (2020) Ferroptosis, necroptosis, and pyroptosis in anticancer immunity. J Hematol Oncol 13:1–18

Tsuchiya K, Nakajima S, Hosojima S, Thi Nguyen D, Hattori T, Manh Le T, Hori O, Mahib MR, Yamaguchi Y, Miura M (2019) Caspase-1 initiates apoptosis in the absence of gasdermin D. Nat Commun 10(1):2091

Vanden Berghe T, Hassannia B, Vandenabeele P (2016) An outline of necrosome triggers. Cell Mol Life Sci 73:2137–2152

Wang Y, Gao W, Shi X, Ding J, Liu W, He H, Wang K, Shao F (2017) Chemotherapy drugs induce pyroptosis through caspase-3 cleavage of a gasdermin. Nature 547(7661):99–103

Wang X, Sun R, Chan S, Meng L, Xu Y, Zuo X, Wang Z, Hu X, Han Q, Dai L (2022) PANoptosis-based molecular clustering and prognostic signature predicts patient survival and immune landscape in colon cancer. Front Genet 13:955355

Wang H, Sun L, Su L, Rizo J, Liu L, Wang L-F, Wang F-S, Wang X (2014) Mixed lineage kinase domain-like protein MLKL causes necrotic membrane disruption upon phosphorylation by RIP3. Mol Cell 54(1):133–146

Wang W, Zhou Q, Lan L, Xu X (2023) PANoptosis-related prognostic signature predicts overall survival of cutaneous melanoma and provides insights into immune infiltration landscape. Sci Rep 13(1):8449

Wei S, Chen Z, Ling X, Zhang W, Jiang L (2023) Comprehensive analysis illustrating the role of PANoptosis-related genes in lung cancer based on bioinformatic algorithms and experiments. Front Pharmacol 14:1115221

Xi G, Gao J, Wan B, Zhan P, Xu W, Lv T, Song Y (2019) GSDMD is required for effector CD8+ T cell responses to lung cancer cells. Int Immunopharmacol 74:105713

Xiao H, Zhao Q, Yuan J, Liang W, Wu R, Wen Y, Du S, Wang Y, Zhao S, Lang Y (2023) IFN-γ promotes PANoptosis in Pasteurella multocida toxin-induced pneumonia in mice. Vet Microbiol 285:109848

Yan H, Luo B, Wu X, Guan F, Yu X, Zhao L, Ke X, Wu J, Yuan J (2021) Cisplatin induces pyroptosis via activation of MEG3/NLRP3/caspase-1/GSDMD pathway in triple-negative breast cancer. Int J Biol Sci 17(10):2606

Yi X, Li J, Zheng X, Xu H, Liao D, Zhang T, Wei Q, Li H, Peng J, Ai J (2023) Construction of PANoptosis signature: novel target discovery for prostate cancer immunotherapy. Mol Ther Nucleic Acids 33:376–390

Yu T, Cheng W, Zhang J, Wang T, Liu Y, Duan Y, Hu A, Feng J, Li M, Li Y (2024) Identification of a PANoptosis-related gene signature for predicting the prognosis, tumor microenvironment and therapy response in breast cancer. J Cancer 15(2):428

Yuan X, Zhang S, Zhong X, Yuan H, Song D, Wang X, Yu H, Guo Z (2022) The induction of PANoptosis in KRAS-mutant pancreatic ductal adenocarcinoma cells by a multispecific platinum complex. Sci China Chem 65(10):1978–1984

Zhang D-W, Shao J, Lin J, Zhang N, Lu B-J, Lin S-C, Dong M-Q, Han J (2009) RIP3, an energy metabolism regulator that switches TNF-induced cell death from apoptosis to necrosis. Science 325(5938):332–336

Zhang T, Yin C, Fedorov A, Qiao L, Bao H, Beknazarov N, Wang S, Gautam A, Williams RM, Crawford JC (2022) ADAR1 masks the cancer immunotherapeutic promise of ZBP1-driven necroptosis. Nature 606(7914):594–602

Zhang Z, Zhang F, Pang P, Li Y, Chen X, Sun S, Bian Y (2023) Identification of PANoptosis-relevant subgroups to evaluate the prognosis and immune landscape of patients with liver hepatocellular carcinoma. Front Cell Devel Biol 11:1210456

Zhang Z, Zhang Y, Xia S, Kong Q, Li S, Liu X, Junqueira C, Meza-Sosa KF, Mok TMY, Ansara J (2020) Gasdermin E suppresses tumour growth by activating anti-tumour immunity. Nature 579(7799):415–420

Zheng M, Kanneganti T (2020) The regulation of the ZBP1-NLRP3 inflammasome and its implications in pyroptosis, apoptosis, and necroptosis (PANoptosis). Immunol Rev 297(1):26–38

Zheng M, Karki R, Vogel P, Kanneganti T-D (2020) Caspase-6 is a key regulator of innate immunity, inflammasome activation, and host defense. Cell 181(3):674–687

Zhou L, Lyu J, Liu F, Su Y, Feng L, Zhang X (2024) Immunogenic PANoptosis-initiated cancer sono-immune reediting nanotherapy by iteratively boosting cancer immunity cycle. Adv Mater 36(2):2305361

Zhuang L, Sun Q, Huang S, Hu L, Chen Q (2023) A comprehensive analysis of PANoptosome to prognosis and immunotherapy response in pan-cancer. Sci Rep 13(1):3877

Zou H, Henzel WJ, Liu X, Lutschg A, Wang X (1997) Apaf-1, a human protein homologous to C. Elegans CED-4, participates in cytochrome c–dependent activation of caspase-3. Cell 90(3):405–413

# Role of Apoptosis in Neurodegeneration: Therapeutic Targets and Strategies

11

Lokesh Murumulla and Suresh Challa


**Abstract**

Neurodegeneration, characterized by the progressive loss of structure and function of neurons, poses a significant challenge to healthcare. A fundamental aspect of neurodegeneration is the dysregulation of apoptosis, the programmed cell death mechanism crucial for maintaining tissue homeostasis. Dysfunctional apoptotic pathways can lead to the untimely death of neurons, contributing to the pathogenesis of disorders such as Alzheimer's disease, Parkinson's disease, Huntington's disease, and amyotrophic lateral sclerosis (ALS). Apoptosis is initiated through various signaling pathways, including the intrinsic (mitochondrial) and extrinsic (death receptor) pathways. In neurodegenerative diseases, these pathways can become activated due to various factors, such as protein misfolding, oxidative stress, DNA damage, inflammation, neuroinflammatory feedback loop, mitochondrial dysfunction, loss of neurotrophic factors, and genetic mutations. All these mechanisms stand as potential therapeutic targets in apoptosis-induced neurodegeneration and altering these targets like enhancing the activity of antiapoptotic factors like Bcl-2 and Bcl-XL, modulating specific genes involved, promoting neurotrophic factors, maintaining healthy mitochondria, and reducing neuroinflammation and target protein clearance mechanisms by using different strategies like gene therapy, small molecule inhibitors, stem cell therapies, and immunomodulation either individually or by combining these approaches stand as a valuable strategy against it. This chapter will also briefly explore the potential of natural bioactive compounds and other active medicinal compounds as interventional agents targeting therapeutic pathways associated with apoptosis and neurodegeneration.



L. Murumulla · S. Challa (✉)
Cell Biology Division, National Institute of Nutrition, Hyderabad, India
e-mail: suresh.challa@icmr.gov.in







K. Jana (ed.), *Apoptosis and Human Health: Understanding Mechanistic and Therapeutic Potential*, https://doi.org/10.1007/978-981-97-7905-5_11

**Keywords**

Apoptosis · Neurodegeneration · Alzheimer's disease · Parkinson's disease · Huntington's disease · Amyotrophic lateral sclerosis · Natural bioactive compounds

## 11.1 Introduction

Neurodegenerative disorders, a formidable class of conditions that includes Alzheimer's, Parkinson's, and Huntington's diseases, share a distressing hallmark—gradual neuronal loss. The hallmark of neurodegenerative disorders lies in the substantial loss of neurons. This progressive deterioration, marked by the diminishing structure and function of neurons, presents a formidable challenge to the field of healthcare (Saleem 2021). This cadre of disorders represents a significant and escalating global health challenge, as it systematically erodes the intricate architecture of the central nervous system. From the poignant cognitive deficits characterizing Alzheimer's to the distinctive motor symptoms of tremors and rigidity in Parkinson's, these maladies cast a profound and often debilitating influence on the lives of millions worldwide. Apoptosis, positioned at the epicentre of this intricate narrative, assumes a dual role as both a protagonist and antagonist in the unfolding chronicle of neurodegeneration when dysregulated. The initiation of neuronal cell damage hinges on abnormal neuroinflammation, autophagy dysregulation, and the disruption of apoptosis, a fundamental programmed cell death mechanism vital for sustaining tissue homeostasis in the context of neurodegeneration (Wang et al. 2020). The elaborate molecular choreography of apoptosis, governing the orchestrated demise of cells, holds profound implications for unraveling the complex initiation and progression of neurodegenerative disorders. Understanding this intricate dance of cellular events provides valuable insights that could pave the way for targeted therapeutic interventions, offering hope in mitigating the devastating effects of neurodegeneration on individuals and societies at large. Apoptosis is triggered by the engagement of cell-surface death receptors, such as Fas, with their ligands (the extrinsic pathway) or by the permeabilization of the mitochondrial outer membrane induced by BCL-2-family proapoptotic proteins (the intrinsic pathway). Both intrinsic and extrinsic pathways culminate in the activation of caspases, a family of proteases responsible for executing the final stages of cell demise in the apoptosis process (Lossi 2022). The apoptotic process involves multiple signalling pathways, with apoptotic elements interacting with trophic factors and various signalling molecules, including Ras-ERK, JNK, GSK-3β, BDNF/TrkB/CREB, and PI3K/AKT/mTOR (Abdulhussein et al. 2021; Kumari et al. 2023). In neurodegenerative diseases, these pathways can become activated due to a variety of factors, such as protein misfolding, oxidative stress, DNA damage, inflammation, neuroinflammatory feedback loop, mitochondrial dysfunction, loss of neurotrophic factors, and genetic mutations (Lossi 2022; Vitale et al. 2023). We delve into the intricacies of apoptotic pathways, navigating the delicate equilibrium of pro- and antiapoptotic factors. Beyond the context of cellular demise, our narrative extends to illuminate the auspicious horizons of therapeutic intervention—a beacon of hope for those contending with the inexorable progression of neurodegenerative disorders.

## 11.2 Molecular Mechanisms in Neurodegeneration

Apoptosis, at its fundamental level, represents a highly orchestrated sequence of events meticulously controlled by a diverse array of molecular components. Starting from the signalling pathways that announce the initiation of the process to the subsequent activation of caspases, the key agents responsible for cellular fate, each stage of this process occurs with remarkable precision. When examining neurodegenerative diseases, this intricate molecular process is situated within the context of various cellular stressors (Hetz et al. 2020; Merelli et al. 2021; Olajide et al. 2021; Murumulla et al. 2023). Factors such as misfolded proteins, mitochondrial dysfunction, and genetic abnormalities assume crucial roles in instigating the apoptotic cascade. In the intricate landscape of molecular events governing cellular fate, the Bcl-2 family stands out prominently as a central regulator. Comprising both pro- and antiapoptotic members, this family significantly influences the delicate equilibrium that dictates the fine line between cellular survival and demise. Simultaneously, the caspases, armed with their proteolytic capabilities, play a pivotal role in executing the conclusive and irreversible steps of cellular dismantling. The complexity of these molecular interactions extends into the mitochondrial realm, where cytochrome c takes on the role of a critical messenger signalling cellular fate. Additionally, the activation of p53, a tumor suppressor protein responding to cellular stress, and the unfolded protein response (UPR) contribute further nuances to the intricate molecular choreography. These molecular players collectively shape the destiny of neurons, responding dynamically to the challenges presented by stress and insults. The orchestration of these molecular elements goes beyond a mere regulatory role; it intricately governs the intricate balance between cellular survival and programmed cell death.

Delving into the intricacies of apoptotic signalling cascades, this section elucidates how dysregulation of apoptosis contributes to neuronal loss in neurodegenerative diseases. Discussions will include the involvement of key players such as caspases, Bcl-2 family proteins, and mitochondria in apoptotic pathways specific to the central nervous system. Apoptosis can be initiated by various factors, including cellular stress, DNA damage, and the activation of death receptors. In neurodegenerative diseases, factors such as misfolded proteins, oxidative stress, and mitochondrial dysfunction often serve as triggers. In some cases, extrinsic apoptotic pathways are activated through death receptors, such as Fas and TNF (Tumor necrosis factor) receptors, leading to the activation of caspase cascades.

### 11.2.1 Role of Bcl-2 Family Proteins

Proapoptotic proteins, such as Bax and Bak, play a crucial role in the regulation of cellular fate by promoting the release of cytochrome c from the mitochondria. This action initiates the intrinsic apoptotic pathway, a fundamental process that culminates in programmed cell death. On the other hand, antiapoptotic proteins, exemplified by Bcl-2 and Bcl-xL, function as guardians of cellular integrity. These proteins exert their protective effect by inhibiting the release of cytochrome c, thereby acting as crucial regulators that impede the progression of apoptosis (Tsujimoto 1998). The delicate balance between proapoptotic and antiapoptotic proteins intricately determines the cell's decision to undergo apoptosis or survive, highlighting these molecular players' pivotal role in cellular homeostasis and the maintenance of tissue integrity.

### 11.2.2 Mitochondrial Dysfunction

In the intricate cascade of apoptosis, the release of cytochrome c serves as a pivotal event. When proapoptotic signals surpass the counteracting efforts of antiapoptotic defenses, cytochrome c is liberated from the mitochondria and translocates into the cytoplasm (Wang and Youle 2009). Subsequently, this released cytochrome c collaborates with apoptotic protease-activating factor 1 (Apaf-1) and procaspase-9 to assemble the apoptosome—a macromolecular complex that initiates the caspase cascade. Dysregulation associated with neurodegeneration extends to mitochondrial dysfunction, producing reactive oxygen species (ROS). While ROS normally functions as signaling molecules, excessive production harms cellular components such as DNA, proteins, and lipids. This abundance of ROS leads to decreased cellular mechanism efficacy, initiation of inflammatory pathways, excitotoxicity, protein aggregation, and apoptosis (Wu et al. 2019; Dionísio et al. 2021; Kleele et al. 2021). Within the intricate network of mitochondrial proteostasis, the enzyme pitrilysin metallopeptidase 1 (PITRM1) plays a pivotal role. PITRM1 is responsible for degrading mitochondrial targeting sequences and digesting a mitochondrial fraction of amyloid beta (Aβ), highlighting its significance in mitigating mitochondrial dysfunction and oxidative stress, crucial factors in the mechanisms of neurodegeneration (Brunetti et al. 2021).

### 11.2.3 Caspase Activation

The initiation phase of apoptosis unfolds with the activation of caspase-9 within the apoptosome—a macromolecular complex formed by the collaboration of cytochrome c, apoptotic protease-activating factor 1 (Apaf-1), and procaspase-9. Once activated, caspase-9 serves as a pivotal trigger, cleaving and activating downstream effector caspases, with caspase-3 taking the central stage in the execution phase. Activated caspase-3, in turn, plays a decisive role in the orchestrated breakdown of cellular components, inducing characteristic morphological changes associated with apoptosis (Brentnall et al. 2013). Its enzymatic activity involves cleavage key cellular proteins, leading to DNA fragmentation and cell shrinkage.

### 11.2.4 DNA Fragmentation

DNA fragmentation, a characteristic feature of apoptosis, is induced by apoptotic stimuli across diverse cell types. The signal pathways leading to DNA fragmentation predominantly involve the family of cysteinyl aspartate-specific protease caspases. These caspases are crucial in mediating apoptotic signals, ultimately resulting in DNA fragmentation (Kitazumi and Tsukahara 2011). This intricate process is characterized by the cleavage of DNA into distinct fragments, forming a pattern reminiscent of a ladder when analyzed through gel electrophoresis. The characteristic DNA laddering observed in this stage is a hallmark of apoptosis, reflecting cellular demise's regulated and programmed nature (Matassov et al. 2004).

### 11.2.5 Involvement of p53

The activation of p53 serves as a crucial molecular checkpoint in cellular health. Upon activation, p53 engages in intricate transcriptional regulation, directly influencing the intrinsic apoptotic pathway. p53 serves as a key regulator of cell death, activating both the classical extrinsic and intrinsic apoptosis pathways and synapse-specific cell death mechanisms. An association has been identified between Aβ and elevated levels of the p53 upregulated modulator of apoptosis protein (PUMA), a BH3-only protein that hinders antiapoptotic BCL-2 proteins. Moreover, phosphorylated p53 modifications have been detected in the substantia nigra tissue of Parkinson's disease-affected patients (Wolfrum et al. 2022; Li et al. 2023).

### 11.2.6 Antiapoptotic Pathways

Within the intricate network of cellular signaling, the PI3K/Akt pathway emerges as a sentinel promoting cell survival by intricately inhibiting proapoptotic factors. Among the various signaling pathways, the PI3K/AKT/mTOR pathway is crucial in numerous human diseases, encompassing ischemic brain injury, neurodegenerative diseases, and tumors. This pathway, predominantly regulated by phosphatidylinositol 3-kinase (PI3K) and Akt, intricately manages cellular decision-making and is pivotal in promoting cell survival. (Xu et al. 2020). Meanwhile, the Bcl-2-Associated Athanogene (BAG) proteins contribute to this regulatory dance by interacting with antiapoptotic members of the Bcl-2 family. Specifically, BAG proteins serve as molecular accomplices, preventing the inhibition of antiapoptotic Bcl-2 family members by their proapoptotic counterparts. This interaction adds another layer of complexity to the cellular signaling milieu, further emphasizing the nuanced interplay between pro-survival and proapoptotic pathways in determining the cell's fate.

### 11.2.7 Inflammation and Apoptosis Crosstalk

Neuroinflammation is a common contributor to neuronal damage and the progression of neurodegeneration. Characterized by the activation of microglia, the neuroimmune cells of the central nervous system (CNS), these cells are identified as active contributors to neuronal damage (Ramírez Hernández et al. 2022). Microglial NLRP3 has been identified as a sustained source of neuroinflammation, potentially driving progressive dopaminergic neuropathology (Gordon et al. 2018). A wealth of evidence from postmortem examinations, brain imaging, epidemiological analyses, and animal studies collectively underscores the involvement of both innate and adaptive immunity in the neurodegenerative process. Importantly, misfolded protein aggregates emerge as a critical trigger for inducing neuroinflammation in this complex landscape (Gelders et al. 2018). The activation of nuclear factor-kappa B (NF-κB) is a pivotal player, particularly in the context of inflammation and apoptosis (Muhammad et al. 2019). As a transcription factor, NF-κB orchestrates inflammatory pathways, influencing the delicate balance of apoptotic signaling within the cell. Notably, NF-κB activation showcases a remarkable crosstalk with apoptotic pathways, revealing a complex interplay between inflammation and cell survival. In certain scenarios, NF-κB activation takes on a pro-survival role by inhibiting apoptosis, underscoring the dynamic and multifaceted nature of its involvement in neurodegenerative processes. This intricate relationship highlights the importance of understanding the contextual nuances of NF-κB activation to unravel its impact on the fate of cells in the face of inflammation and apoptotic stimuli.

### 11.2.8 Endoplasmic Reticulum (ER) Stress

Cellular stress, triggered by the abnormal accumulation of unfolded or misfolded proteins at the endoplasmic reticulum (ER), is a hallmark of neurodegenerative disorders. ER proteostasis surveillance, overseen by the unfolded protein response (UPR), comprises signal transduction pathways that gauge the fidelity of protein folding within the ER lumen and are associated with the accumulation of misfolded proteins (Hetz, Zhang). In the intricate web of molecular events, the UPR plays a dual role, acting as a cellular safeguard against stress while also being implicated in apoptotic pathways under prolonged or severe ER stress conditions. This involves the activation of the UPR pathways, including PKR-like ER kinase (PERK), inositol-requiring transmembrane kinase/endoribonuclease 1α (IRE1α), and activating transcription factor 6 (ATF6) (Ghemrawi and Khair 2020). Additionally, the UPR influences synaptic plasticity and memory, not solely through ER stress-dependent mechanisms but also through ER stress-independent pathways (Gerakis and Hetz 2018). Understanding these molecular mechanisms becomes paramount in the context of neurodegenerative diseases, where the burden of misfolded proteins is heightened. The UPR's involvement in apoptotic pathways highlights its potential contribution to neuronal death, emphasizing its significance as a target for therapeutic interventions.

### 11.2.9 Autophagy

Autophagy, a cellular process that directs cytoplasmic material to lysosomes for degradation, encompasses macroautophagy, chaperone-mediated autophagy, and microautophagy (Fleming et al. 2022). It is a crucial mechanism for maintaining neuronal proteostasis, ensuring the removal of misfolded proteins and dysfunctional organelles. Autophagy, specifically mitophagy, selectively targets and removes damaged mitochondria. This process is crucial for maintaining a healthy mitochondrial pool, preventing the release of harmful reactive oxygen species (ROS), and preserving cellular energy production. In the context of neurodegeneration, the intricate relationship between autophagy and the progression of various neurological disorders has garnered significant attention. Dysregulated autophagy is linked to increased reactive oxygen species (ROS) and cytosolic calcium, contributing to dopaminergic neurodegeneration (Jung et al. 2018). Autophagy deficiency is associated with elevated apoptosis, highlighting its role in maintaining cellular homeostasis. The selenoprotein GPX4 (glutathione peroxidase 4), initially known as PHGPX (phospholipid hydroperoxide glutathione peroxidase), plays a pivotal role in utilizing glutathione and interacting with the autophagic degradation pathway, influencing cell fate during oxidative stress. Impaired GPX4 function is implicated in neurodegeneration (Xie et al. 2023). Furthermore, staufen1 (STAU1), a stress granule (SG) protein, interacts with mTOR, driving hyperactivation and inhibiting autophagic flux in various neurodegenerative models (Paul et al. 2023). However, chronic or overwhelming stress conditions can lead to dysregulation of autophagy, contributing to disease progression. Balancing autophagic responses in the face of stress is a crucial aspect of neuroprotection. Neuroinflammation is a common feature in neurodegenerative diseases, and autophagy intersects with the immune response. Autophagy helps regulate the clearance of inflammatory mediators and dysfunctional immune cells, influencing the overall inflammatory state in the central nervous system. Dysfunctional autophagy can exacerbate neuroinflammation, further propagating neurodegenerative processes (Yang et al. 2020).

## 11.3 Therapeutic Targets

This section delves into potential therapeutic interventions by identifying crucial molecular targets within apoptotic pathways. The analysis spans from caspase inhibitors to modulators of Bcl-2 family proteins, providing a comprehensive examination of pharmacological targets with implications for targeted drug development.

### 11.3.1 Caspases

Caspases emerge as central players in the intricate execution of apoptosis. Functioning as proteases, they play a crucial role in cleaving essential cellular proteins, ultimately leading to cell death. In neurodegenerative diseases, the heightened activation of caspases contributes significantly to neuronal death. Researchers target caspases, specifically focusing on caspase-3 and caspase-9, aiming to impede the apoptotic cascade and thereby preserve neuronal integrity (Unnisa et al. 2023). Researchers employ caspase inhibitors to actively regulate apoptotic pathways, designed as both broad-spectrum and selective. These inhibitors, administered pharmacologically, are tailored to block the activity of caspases, presenting a strategic approach to modulating the apoptotic process (Kim et al. 2007).

### 11.3.2 Bcl-2 Family Proteins

The Bcl-2 family intricately regulates the mitochondrial pathway of apoptosis, exerting control over the release of cytochrome c and influencing the activation of caspases. In neurodegenerative conditions, an observable imbalance in the expression of antiapoptotic (e.g., Bcl-2) and proapoptotic (e.g., Bax, Bak) members of the Bcl-2 family underscores their significance. Phosphorylation at the MAPK/ERK site in rat neuronal cells induces alterations in the expression of apoptotic proteins Bcl-2 and Bax, ultimately alleviating apoptosis (Guo et al. 2020). Targeting apoptosis activation, particularly by modulating the Bax/Bcl-2 ratio and pathways such as c-Jun N-terminal kinases (JNK) and p38, presents promising therapeutic targets (Bekker et al. 2021). Therapeutic strategies to modulate this balance seek to prevent mitochondrial dysfunction and subsequent apoptosis. Ongoing investigations explore small molecules or peptides targeting Bcl-2 family members. Notably, BH3 mimetics mimic the action of proapoptotic proteins, fostering apoptosis in affected cells as a potential therapeutic avenue.

### 11.3.3 Mitochondrial Dysfunction

The mPTP stands as a pivotal regulator of both mitochondrial function and apoptosis. Its opening triggers the release of apoptotic factors, ultimately leading to cell death (Ayyalasomayajula et al. 2022; Bandaru et al. 2022, 2023). Implicated in neurodegenerative diseases, dysregulation of the mPTP compromises mitochondrial integrity. Therapeutic endeavors focused on modulating the mPTP aim to sustain mitochondrial function and prevent the demise of neurons. Compounds that target the opening of mPTP, such as cyclosporine A, are being explored as potential neuroprotective agents (Gao et al. 2013). The nuclear factor erythroid 2-related factor 2 (Nrf2) plays a pivotal role in regulating the metabolism of glutathione, iron, and lipids and maintaining mitochondrial function. This multifaceted regulation is crucial for protecting against neurodegeneration (Abdalkader et al. 2018). These compounds are designed to stabilize mitochondrial membranes and inhibit apoptotic signaling, offering a prospective avenue for therapeutic intervention.

### 11.3.4 NMDA Receptors

The overactivation of NMDA receptors is implicated in excitotoxicity, a process directly linked to neuronal death in various neurodegenerative disorders. Strategically modulating the activity of NMDA receptors emerges as a viable approach to preventing excitotoxic damage and the subsequent initiation of apoptotic pathways. Investigations into neuroprotective strategies focus on NMDA receptor antagonists, such as memantine. By selectively blocking NMDA receptors, these drugs aim to alleviate excitotoxicity and mitigate the onset of apoptosis in affected neurons.

### 11.3.5 Targeting ER Stress and UPR

Endoplasmic reticulum (ER) stress and the unfolded protein response (UPR) have emerged as promising therapeutic targets against apoptosis in neurodegenerative diseases. Dysregulation in protein folding and the accumulation of misfolded proteins trigger ER stress, activating the UPR to restore cellular homeostasis. However, prolonged or severe ER stress can lead to apoptotic pathways, contributing to neuronal death. By targeting key components of the UPR, such as PERK, IRE1α, and ATF6, researchers aim to modulate the delicate balance between adaptive responses and apoptotic signaling. By inhibiting mTOR, researchers aim to mitigate ER stress and enhance autophagic processes, offering a potential therapeutic strategy to alleviate cell death and promote neuronal survival (Wan et al. 2020). Strategies that selectively intervene in the UPR pathways hold the potential to mitigate neuronal apoptosis, providing a novel avenue for therapeutic intervention in neurodegenerative disorders (Ramos Rego et al. 2021).

### 11.3.6 Autophagy and Neuroinflammation

Targeting autophagy and neuroinflammation emerges as a promising therapeutic strategy against apoptosis in neurodegenerative disorders. Dysregulation of autophagy, a cellular recycling process, and heightened neuroinflammation contribute to the progression of neuronal damage and cell death. Interventions that enhance autophagic flux or modulate inflammatory responses are under investigation to mitigate apoptotic pathways. By promoting efficient clearance of cellular debris through autophagy and dampening excessive neuroinflammatory reactions, researchers aim to disrupt the cascade of events leading to apoptosis. FGF21 exhibits neuroprotective effects by mitigating neuroinflammation and oxidative stress. This is achieved by regulating the NF-κB pathway and activating the AMPKα/AKT pathway, enhancing mitochondrial protection in neurons (Kang et al. 2020). Meanwhile,

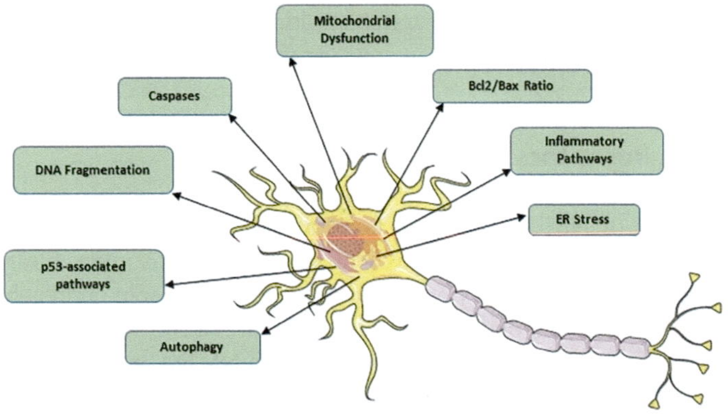


**Fig. 11.1** Figure delineating therapeutic targets linked to apoptosis-driven neurodegeneration

ACT001 intervenes in the neuroinflammatory process by restraining NFκBnuclear translocation in microglia cells, achieved through inhibiting AKT phosphorylation. This dual mechanism results in the decreased activation of the NLRP3 inflammasome and, consequently, a downregulation of the microglial neuroinflammatory response (Cai et al. 2022). These findings underscore the potential of targeting these pathways as a therapeutic strategy against neurodegeneration, emphasizing the interconnected roles of neuroinflammation, oxidative stress, and mitochondrial function in the progression of neuronal damage. These dual targets offer a multifaceted approach, addressing both the intracellular recycling mechanisms and the inflammatory milieu, with the potential to pave the way for innovative therapeutic interventions in neurodegenerative diseases (Fig. 11.1).

## 11.4 Emerging Strategies

Beyond traditional pharmacological approaches, this section delves into innovative strategies for addressing neurodegeneration. The focus includes the evolving field of gene therapy utilizing CRISPR/Cas9 or RNA interference (RNAi) to manipulate apoptotic regulators and exploring the roles of neuroprotective compounds, lifestyle modifications, and precision medicine in mitigating apoptosis-associated neurodegeneration.

### 11.4.1 Gene Therapy

Gene therapy involves introducing, removing, or altering genetic material within cells to treat or prevent diseases. In the context of neurodegeneration, it targets numerous genes involved in apoptotic and neurodegenerative pathways (Pardo and van Duijn 2005; Van Giau et al. 2019). Gene therapy enables the precise manipulation of specific genes implicated in apoptosis. This may involve enhancing the expression of antiapoptotic genes or silencing proapoptotic genes, offering a targeted and personalized approach. Techniques like CRISPR/Cas9 allow precise genome editing, potentially correcting genetic abnormalities associated with neurodegenerative disorders.

### 11.4.2 CRISPR/Cas9 Technology

CRISPR/Cas9 is a revolutionary gene-editing technology that allows precise modification of DNA sequences within the genome. CRISPR/Cas9 is a powerful tool for editing genes associated with apoptotic pathways in neurodegeneration. Researchers selectively target and modify specific genes involved in apoptosis regulation, enhancing the expression of antiapoptotic genes or inhibiting proapoptotic genes. This technology provides unprecedented accuracy and efficiency in genome editing. It allows scientists to investigate the direct impact of specific genetic alterations on apoptotic processes, paving the way for the development of targeted therapies tailored to individual patient's genetic makeup.

### 11.4.3 RNA Interference (RNAi)

RNA interference is a natural cellular process regulating gene expression by silencing or "knocking down" specific mRNA molecules, preventing them from being translated into proteins. In neurodegeneration, RNAi can selectively inhibit gene expression in apoptotic pathways. Small RNA molecules, such as small interfering RNAs (siRNAs) or short hairpin RNAs (shRNAs), are designed to target and degrade the mRNA of specific apoptotic regulators. RNAi provides a targeted and temporary way to modulate gene expression. By silencing key apoptotic genes, researchers explore the therapeutic potential of temporarily dampening apoptotic pathways without permanent genetic modifications, allowing precise control over the intervention.

### 11.4.4 Neuroprotective Compounds

Neuroprotective compounds are substances that can potentially preserve the structure and function of neurons, particularly under conditions of stress or injury. These compounds act on various cellular processes, including apoptotic pathways, to promote neuronal survival. For example, antioxidants like resveratrol may counteract oxidative stress, a common trigger for apoptosis. By reducing oxidative damage, neuroprotective compounds contribute to overall neuronal resilience and may slow down the progression of neurodegeneration. Various neuroprotective compounds are under investigation to elucidate their mode of action and enhance their bioavailability. Researchers are exploring naturally occurring plants and their associated active compounds, as well as small molecules, mimetics, structural analogs, and more (Ayyalasomayajula and Suresh 2018; Ayyalasomayajula et al. 2019; Wang et al. 2019; Javed et al. 2020; Park et al. 2020; Lee et al. 2021; Chen et al. 2022; Iorio et al. 2022; Eddin et al. 2023). The goal is to harness the therapeutic potential of these compounds against apoptosis and neurodegeneration. This comprehensive approach involves studying diverse sources and types of neuroprotective agents to uncover effective strategies for mitigating neuronal damage and advancing potential treatments for neurodegenerative disorders.

### 11.4.5 Lifestyle Modifications

Lifestyle modifications encompass changes in daily habits, such as diet, exercise, and sleep patterns. Healthy lifestyle choices have been linked to a reduced risk and progression of neurodegenerative diseases. Regular physical activity, a balanced diet rich in antioxidants and omega-3 fatty acids, and sufficient sleep can modulate various factors, including inflammation and oxidative stress, influencing apoptotic pathways. These modifications contribute to a neuroprotective environment, potentially delaying the onset or progression of neurodegeneration.

### 11.4.6 Precision Medicine

Precision medicine involves tailoring medical treatment and interventions to the individual characteristics of each patient, considering genetic, environmental, and lifestyle factors. In the context of neurodegeneration, precision medicine allows for personalized treatment strategies based on each patient's unique genetic and molecular profile. By understanding the specific apoptotic pathways and genetic factors contributing to an individual's condition, clinicians can design targeted therapies addressing the root causes of neurodegeneration, potentially enhancing treatment efficacy.

These emerging strategies collectively offer a multifaceted approach to tackling apoptosis in neurodegeneration. By leveraging the precision of gene therapy, the protective effects of neuroprotective compounds, the influence of lifestyle choices, and the personalized nature of precision medicine, researchers and clinicians aim to develop innovative and tailored interventions, revolutionizing the landscape of neurodegenerative disease treatment (Fig. 11.2).

## 11.5 Challenges

Recognizing the intricacies of translating bench research to clinical applications, this section confronts challenges in developing effective therapeutics targeting apoptosis in neurodegenerative disorders. Moreover, it outlines future directions,

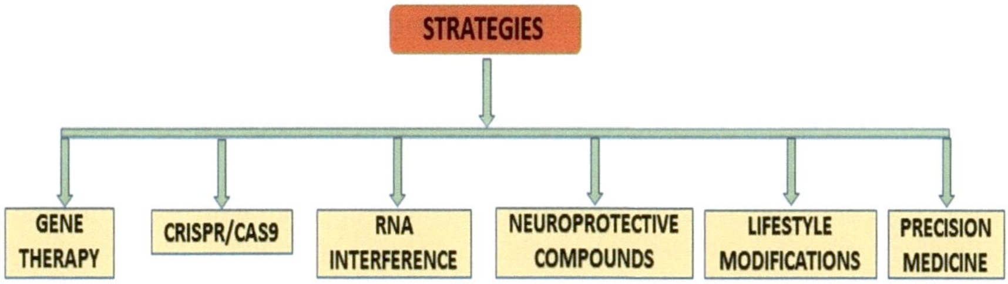


**Fig. 11.2** Figure depicting strategies to mitigate apoptosis-induced neurodegeneration

emphasizing the need for interdisciplinary collaborations and personalized approaches to treatment.

### 11.5.1 Heterogeneity of Neurodegenerative Diseases

Neurodegenerative diseases are inherently heterogeneous, displaying diverse clinical manifestations and underlying molecular pathologies. This heterogeneity poses a significant challenge in identifying common apoptotic mechanisms applicable across different disorders. Tailoring therapeutic strategies to the specific characteristics of each disease subtype becomes imperative, requiring a nuanced understanding of the molecular underpinnings unique to conditions like Alzheimer's, Parkinson's, and Huntington's.

### 11.5.2 Complex Interplay of Pathogenic Factors

The intricate interplay of genetic, environmental, and lifestyle factors in neurodegeneration complicates elucidating direct cause-effect relationships in apoptotic pathways. Untangling this complex web of influences is essential for developing targeted interventions. Researchers must grapple with deciphering the contribution of each factor and understanding their synergistic or antagonistic effects on apoptotic processes.

### 11.5.3 Limited Biomarkers for Early Detection

The absence of robust and specific biomarkers for early detection of neurodegenerative diseases hinders identifying individuals at risk before substantial neuronal damage occurs. Early intervention is pivotal in neurodegenerative diseases. The lack of reliable biomarkers complicates the timely implementation of therapeutic strategies, making it challenging to halt or slow the progression of apoptosis in its initial stages.

### 11.5.4 Blood-Brain Barrier (BBB) Challenges

The selective permeability of the blood-brain barrier presents a hurdle in delivering therapeutic agents to the central nervous system. Developing effective therapies requires overcoming the BBB barrier to ensure that potential interventions can reach target cells in the brain. Strategies for enhancing drug delivery to the brain without compromising the integrity of the barrier are essential.

## 11.6 Future Directions

### 11.6.1 Advancements in Biomarker Discovery

Invest in extensive research to identify specific and sensitive biomarkers associated with apoptotic processes in neurodegenerative diseases. Robust biomarkers can facilitate early diagnosis, track disease progression, and serve as indicators of treatment efficacy, enabling personalized and timely interventions.

### 11.6.2 Integration of Omics Technologies

Embrace the integration of genomics, proteomics, and metabolomics to gain a holistic understanding of the molecular landscape underlying apoptotic dysregulation in neurodegeneration. Comprehensive omics approaches provide a systems-level view, unraveling complex molecular networks and identifying potential therapeutic targets for intervention.

### 11.6.3 Multimodal Imaging Techniques

Advances in neuroimaging technologies, such as positron emission tomography (PET) and functional magnetic resonance imaging (fMRI), can be leveraged to visualize and quantify apoptotic processes in vivo. Real-time imaging of apoptotic events can offer insights into the spatiotemporal dynamics of neuronal death, aiding in the development of targeted therapies and monitoring treatment responses.

### 11.6.4 Precision Medicine and Patient Stratification

Emphasize precision medicine approaches, incorporating genetic profiling, to stratify patients based on their molecular subtype and tailor therapies accordingly. Recognizing the diversity within neurodegenerative diseases allows for the development of targeted interventions that account for each patient's unique genetic and molecular characteristics.

### 11.6.5 Collaborative and Interdisciplinary Research

Foster collaboration between diverse scientific disciplines, bringing together neurobiology, genetics, pharmacology, and data science experts. The complexity of neurodegenerative diseases demands interdisciplinary approaches to address the challenges comprehensively. Cross-disciplinary collaboration can accelerate the translation of research findings into clinically relevant applications.

### 11.6.6 Ethical Considerations and Societal Engagement

Promote open dialogue on the ethical implications of gene-based therapies and engage the public in discussions surrounding cutting-edge technologies like CRISPR/Cas9. Ethical considerations are integral to developing and applying novel therapeutic strategies. The inclusion of diverse perspectives ensures that societal concerns and values are considered in the advancement of apoptotic-targeted therapies.

## 11.7 Conclusion

In summarizing the key discoveries, this section emphasizes the pivotal role of apoptosis in neurodegenerative processes and the potential effectiveness of precisely targeted therapeutic approaches. The chapter concludes by advocating for sustained research efforts aimed at unraveling the intricacies of apoptotic pathways with the ultimate goal of developing effective treatments for neurodegenerative disorders. This book chapter presents a comprehensive exploration of the complex interplay between apoptosis and neurodegeneration, providing a foundation for researchers, clinicians, and policymakers to advance our understanding and formulate transformative therapies for these debilitating conditions. As we wrap up our analysis of apoptosis within the context of neurodegeneration, a profound sense of both challenge and promise permeates the scientific landscape. The molecular dynamics of apoptosis, guiding the decline of neurons amid the complexity of neurodegenerative diseases, unfolds as a nuanced narrative. Challenges, such as the heterogeneity of disorders and the elusive quest for biomarkers, cast shadows on our understanding, yet concurrently act as stimuli for innovation and exploration. In addressing these challenges, our envisioned future directions illuminate the pathway forward with a beacon of hope. The potential for identifying robust biomarkers, integrating omics technologies, and embracing multimodal imaging techniques heralds a new era of precision and clarity. The convergence of disciplines, where genetics, neurobiology, and pharmacology intersect, magnifies the collective wisdom derived from collaborative efforts. The promise of gene-based therapies, driven by technologies such as CRISPR/Cas9, stands as evidence of the indomitable spirit of scientific innovation. However, the ethical considerations accompanying these advancements underscore the imperative need for a considerate and all-encompassing dialogue, ensuring that progress aligns harmoniously with societal values. As we embark on this extraordinary journey, it becomes apparent that exploring apoptosis in neurodegeneration transcends the confines of laboratory benches and clinical settings. It is a journey that calls upon the collective brilliance of scientists, ethicists, clinicians, and society. The challenges we confront are formidable, yet they constitute an integral part of the evolution of our understanding and the formulation of transformative therapies. In the culmination of our discussion, the impactful dynamics of apoptosis resonate not only with the melancholy tones of cellular demise, but also with the resilient chords of scientific determination. The symphony of future possibilities resonates with the potential to redefine the narrative of neurodegenerative diseases. In the pursuit of unravelling the mysteries of apoptosis, we stand on the cusp of ground-breaking discoveries that promise to alleviate the burdens borne by countless individuals affected by these relentless disorders.

## References


Abdalkader M et al (2018) Targeting Nrf2 to suppress ferroptosis and mitochondrial dysfunction in neurodegeneration. Front Neurosci 12:466. https://doi.org/10.3389/fnins.2018.00466

Abdulhussein D et al (2021) Apoptosis in health and diseases of the eye and brain. Adv Protein Chem Struct Biol 126:279–306. https://doi.org/10.1016/bs.apcsb.2021.01.008

Ayyalasomayajula N, Suresh C (2018) Mechanistic comparison of current pharmacological treatments and novel phytochemicals to target amyloid peptides in Alzheimer's and neurodegenerative diseases. Nutr Neurosci 21(10):682–694. https://doi.org/10.1080/1028415X.2017.1345425

Ayyalasomayajula N et al (2019) Mitigative effects of epigallocatechin gallate in terms of diminishing apoptosis and oxidative stress generated by the combination of lead and amyloid peptides in human neuronal cells. J Biochem Mol Toxicol 33(11):1–9. https://doi.org/10.1002/jbt.22393

Ayyalasomayajula N et al (2022) Mitochondria-mediated moderation of apoptosis by EGCG in cytotoxic neuronal cells induced by lead (Pb) and amyloid peptides. Biol Trace Elem Res 200(8):3582–3593. https://doi.org/10.1007/s12011-021-02959-w

Bandaru LJM et al (2022) Mechanisms associated with the dysregulation of mitochondrial function due to lead exposure and possible implications on the development of Alzheimer's disease. BioMetals 35(1). https://doi.org/10.1007/s10534-021-00360-7

Bandaru LJM et al (2023) Exposure of combination of environmental pollutant, lead (Pb) and β-amyloid peptides causes mitochondrial dysfunction and oxidative stress in human neuronal cells. J Bioenerg Biomembr 55(1):79–89. https://doi.org/10.1007/s10863-023-09956-9

Bekker M et al (2021) Can the interplay between autophagy and apoptosis be targeted as a novel therapy for Parkinson's disease? Neurobiol Aging 100:91–105. https://doi.org/10.1016/j.neurobiolaging.2020.12.013

Brentnall M et al (2013) Caspase-9, caspase-3 and caspase-7 have distinct roles during intrinsic apoptosis. BMC Cell Biol 14:32. https://doi.org/10.1186/1471-2121-14-32

Brunetti D et al (2021) Role of PITRM1 in mitochondrial dysfunction and neurodegeneration. Biomedicines 9(7). https://doi.org/10.3390/biomedicines9070833

Cai L et al (2022) ACT001 attenuates microglia-mediated neuroinflammation after traumatic brain injury via inhibiting AKT/NFκB/NLRP3 pathway. Cell Commun Signal: CCS 20(1):56. https://doi.org/10.1186/s12964-022-00862-y

Chen B et al (2022) Neuroprotective effects of natural compounds on neurotoxin-induced oxidative stress and cell apoptosis. Nutr Neurosci 25(5):1078–1099. https://doi.org/10.1080/1028415X.2020.1840035

Dionísio PA, Amaral JD, Rodrigues CMP (2021) Oxidative stress and regulated cell death in Parkinson's disease. Ageing Res Rev 67:101263. https://doi.org/10.1016/j.arr.2021.101263

Eddin LB et al (2023) Limonene, a monoterpene, mitigates rotenone-induced dopaminergic neurodegeneration by modulating neuroinflammation, Hippo signaling and apoptosis in rats. Int J Mol Sci 24(6). https://doi.org/10.3390/ijms24065222

Fleming A et al (2022) The different autophagy degradation pathways and neurodegeneration. Neuron 110(6):935–966. https://doi.org/10.1016/j.neuron.2022.01.017

Gao J et al (2013) Mitochondrial permeability transition pore in inflammatory apoptosis of human conjunctival epithelial cells and T cells: effect of cyclosporin A. Invest Ophthalmol Vis Sci 54(7):4717–4733. https://doi.org/10.1167/iovs.13-11681

Gelders G, Baekelandt V, Van der Perren A (2018) Linking neuroinflammation and neurodegeneration in Parkinson's disease. J Immunol Res 2018:4784268. https://doi.org/10.1155/2018/4784268

Gerakis Y, Hetz C (2018) Emerging roles of ER stress in the etiology and pathogenesis of Alzheimer's disease. FEBS J 285(6):995–1011. https://doi.org/10.1111/febs.14332

Ghemrawi R, Khair M (2020) Endoplasmic reticulum stress and unfolded protein response in neurodegenerative diseases. Int J Mol Sci 21(17). https://doi.org/10.3390/ijms21176127

Gordon R et al (2018) Inflammasome inhibition prevents α-synuclein pathology and dopaminergic neurodegeneration in mice. Sci Transl Med 10(465). https://doi.org/10.1126/scitranslmed.aah4066

Guo Y et al (2020) A relationship between MAPK/ERK pathway expression and neuronal apoptosis in rats with white matter lesions. Eur Rev Med Pharmacol Sci 24(8):4412–4419. https://doi.org/10.26355/eurrev_202004_21023

Hetz C, Zhang K, Kaufman RJ (2020) Mechanisms, regulation and functions of the unfolded protein response. Nat Rev Mol Cell Biol 21(8):421–438. https://doi.org/10.1038/s41580-020-0250-z

Iorio R, Celenza G, Petricca S (2022) Multi-target effects of ß-caryophyllene and carnosic acid at the crossroads of mitochondrial dysfunction and neurodegeneration: from oxidative stress to microglia-mediated neuroinflammation. Antioxidants (Basel, Switzerland) 11(6). https://doi.org/10.3390/antiox11061199

Javed H et al (2020) α-bisabolol, a dietary bioactive phytochemical attenuates dopaminergic neurodegeneration through modulation of oxidative stress, neuroinflammation and apoptosis in rotenone-induced rat model of Parkinson's disease. Biomol Ther 10(10). https://doi.org/10.3390/biom10101421

Jung S, Chung Y, Oh YJ (2018) Breaking down autophagy and the ubiquitin proteasome system. Parkinsonism Relat Disord 46 Suppl 1:S97–S100. https://doi.org/10.1016/j.parkreldis.2017.07.026

Kang K et al (2020) FGF21 attenuates neurodegeneration through modulating neuroinflammation and oxidant-stress. Biomed Pharmacother 129:110439. https://doi.org/10.1016/j.biopha.2020.110439

Kim K-W et al (2007) The apoptotic effects of oxidative stress and antiapoptotic effects of caspase inhibitors on rat notochordal cells. Spine 32(22):2443–2448. https://doi.org/10.1097/BRS.0b013e318157395a

Kitazumi I, Tsukahara M (2011) Regulation of DNA fragmentation: the role of caspases and phosphorylation. FEBS J 278(3):427–441. https://doi.org/10.1111/j.1742-4658.2010.07975.x

Kleele T et al (2021) Distinct fission signatures predict mitochondrial degradation or biogenesis. Nature 593(7859):435–439. https://doi.org/10.1038/s41586-021-03510-6

Kumari S, Dhapola R, Reddy DH (2023) Apoptosis in Alzheimer's disease: insight into the signaling pathways and therapeutic avenues. Apoptosis 28(7–8):943–957. https://doi.org/10.1007/s10495-023-01848-y

Lee C-T, Hsieh C-F, Wang J-Y (2021) Bisphenol a induces autophagy defects and AIF-dependent apoptosis via HO-1 and AMPK to degenerate N2a neurons. Int J Mol Sci 22(20). https://doi.org/10.3390/ijms222010948

Li H et al (2023) New insights into the roles of p53 in central nervous system diseases. Int J Neuropsychopharmacol 26(7):465–473. https://doi.org/10.1093/ijnp/pyad030

Lossi L (2022) The concept of intrinsic versus extrinsic apoptosis. Biochem J 479(3):357–384. https://doi.org/10.1042/BCJ20210854

Matassov D et al (2004) Measurement of apoptosis by DNA fragmentation. Methods Mol Biol (Clifton, N.J.) 282:1–17. https://doi.org/10.1385/1-59259-812-9:001

Merelli A et al (2021) Hypoxia, oxidative stress, and inflammation: three faces of neurodegenerative diseases. J Alzheimers Dis: JAD 82(s1):S109–S126. https://doi.org/10.3233/JAD-201074

Muhammad T et al (2019) Hesperetin, a citrus flavonoid, attenuates LPS-induced neuroinflammation, apoptosis and memory impairments by modulating TLR4/NF-κB signaling. Nutrients 11(3). https://doi.org/10.3390/nu11030648

Murumulla L, Bandaru LJM, Challa S (2023) Heavy metal mediated progressive degeneration and its noxious effects on brain microenvironment. Biol Trace Elem Res 202(4):1411–1427. https://doi.org/10.1007/s12011-023-03778-x

Olajide OJ, Suvanto ME, Chapman CA (2021) Molecular mechanisms of neurodegeneration in the entorhinal cortex that underlie its selective vulnerability during the pathogenesis of Alzheimer's disease. Biology Open 10(1). https://doi.org/10.1242/bio.056796

Pardo LM, van Duijn CM (2005) In search of genes involved in neurodegenerative disorders. Mutat Res 592(1–2):89–101. https://doi.org/10.1016/j.mrfmmm.2005.06.006

Park H-A et al (2020) Anti-apoptotic effects of carotenoids in neurodegeneration. Molecules (Basel, Switzerland) 25(15). https://doi.org/10.3390/molecules25153453

Paul S et al (2023) Staufen impairs autophagy in neurodegeneration. Ann Neurol 93(2):398–416. https://doi.org/10.1002/ana.26515

Ramírez Hernández E et al (2022) Neuroinflammation and galectins: a key relationship in neurodegenerative diseases. Glycoconj J 39(5):685–699. https://doi.org/10.1007/s10719-022-10064-w

Ramos Rego, I. et al. (2021) 'TRAP1 in oxidative stress and neurodegeneration.', Antioxidants (Basel, Switzerland), 10(11). doi:https://doi.org/10.3390/antiox10111829

Saleem S (2021) Apoptosis, autophagy, necrosis and their multi galore crosstalk in neurodegeneration. Neuroscience 469:162–174. https://doi.org/10.1016/j.neuroscience.2021.06.023

Tsujimoto Y (1998) Role of Bcl-2 family proteins in apoptosis: apoptosomes or mitochondria? Genes Cells 3(11):697–707. https://doi.org/10.1046/j.1365-2443.1998.00223.x

Unnisa A, Greig NH, Kamal MA (2023) Inhibition of caspase 3 and caspase 9 mediated apoptosis: a multimodal therapeutic target in traumatic brain injury. Curr Neuropharmacol 21(4):1001–1012. https://doi.org/10.2174/1570159X20666220327222921

Van Giau V et al (2019) Analysis of 50 neurodegenerative genes in clinically diagnosed early-onset Alzheimer's disease. Int J Mol Sci 20(6). https://doi.org/10.3390/ijms20061514

Vitale I et al (2023) Apoptotic cell death in disease-current understanding of the NCCD 2023. Cell Death Differ 30(5):1097–1154. https://doi.org/10.1038/s41418-023-01153-w

Wan H et al (2020) WDR45 contributes to neurodegeneration through regulation of ER homeostasis and neuronal death. Autophagy 16(3):531–547. https://doi.org/10.1080/15548627.2019.1630224

Wang C, Youle RJ (2009) The role of mitochondria in apoptosis*. Annu Rev Genet 43:95–118. https://doi.org/10.1146/annurev-genet-102108-134850

Wang L et al (2019) Epigallocatechin-3-gallate stimulates autophagy and reduces apoptosis levels in retinal Müller cells under high-glucose conditions. Exp Cell Res 380(2):149–158. https://doi.org/10.1016/j.yexcr.2019.04.014

Wang X-X et al (2020) Inflammation, apoptosis and autophagy as critical players in vascular dementia. Eur Rev Med Pharmacol Sci 24(18):9601–9614. https://doi.org/10.26355/eurrev_202009_23048

Wolfrum P et al (2022) The function of p53 and its role in Alzheimer's and Parkinson's disease compared to age-related macular degeneration. Front Neurosci 16:1029473. https://doi.org/10.3389/fnins.2022.1029473

Wu Y, Chen M, Jiang J (2019) Mitochondrial dysfunction in neurodegenerative diseases and drug targets via apoptotic signaling. Mitochondrion 49:35–45. https://doi.org/10.1016/j.mito.2019.07.003

Xie Y et al (2023) GPX4 in cell death, autophagy, and disease. Autophagy 19(10):2621–2638. https://doi.org/10.1080/15548627.2023.2218764

Xu F et al (2020) Roles of the PI3K/AKT/mTOR signalling pathways in neurodegenerative diseases and tumours. Cell Biosci 10(1):54. https://doi.org/10.1186/s13578-020-00416-0

Yang Y et al (2020) Autophagy promotes mammalian survival by suppressing oxidative stress and p53. Genes Dev 34(9–10):688–700. https://doi.org/10.1101/gad.335570.119

# Role of Apoptosis in Diabetes: Therapeutic Targets and Strategies

# 12

Sourav Bar, Soumik Dhara, Satyajit Halder, Kuladip Jana, and Sudipta Kumar Ghorai


**Abstract**

Diabetes mellitus is a significant global health concern, requiring a comprehensive exploration of its underlying mechanisms for effective management. Apoptosis, a genetically regulated cell death, emerges as a pivotal player in diabetes pathophysiology, influencing pancreatic beta cells, insulin-responsive tissues, and the progression of diabetic complications. A comprehensive overview of apoptosis in diabetes has been discussed, covering its relevance in pancreatic beta cells, insulin-responsive tissues, and diabetic complications. The intricate mechanisms, including oxidative stress, endoplasmic reticulum-induced stress, and inflammatory pathways, drive programmed cell death in pancreatic beta cells. The consequences of beta cell apoptosis on insulin secretion and glucose homeostasis are crucial and highlight the pivotal role of these cells in diabetes pathogenesis. Apoptotic pathways play a major role in insulin sensitivity and the interaction between apoptosis, inflammation, and insulin resistance in insulin-responsive tissues such as muscle, adipose tissue, and the liver. Additionally, the relationship between apoptosis and diabetic complications includes several pathophysiological aspects like nephropathy, retinopathy, and neuropathy. Omics technologies have revolutionized our understanding of apoptosis in diabetes, revealing intricate molecular mechanisms underlying the disease. Genomics, transcriptomics, proteomics, and metabolomics have identified genetic variants, gene expression changes, protein modifications, metabolic dysregulation, and epigenetic alterations associated with apoptotic pathways in diabetic tissues.



S. Bar · S. Dhara · S. K. Ghorai
Coastal Environmental Studies Research Centre of Egra SSB College, Vidyasagar University, Midnapore, West Bengal, India

S. Halder · K. Jana (✉)
Division of Molecular Medicine, Bose Institute, Unified Academic Campus, Kolkata, West Bengal, India
e-mail: kuladip@jcbose.ac.in







K. Jana (ed.), *Apoptosis and Human Health: Understanding Mechanistic and Therapeutic Potential*, https://doi.org/10.1007/978-981-97-7905-5_12

**Keywords**

Diabetes · Insulin · Beta cells · Apoptosis · Glucose · Omics · Pathophysiology

## 12.1 Introduction

An increased blood glucose level due to abnormal insulin action, secretion, or both characterizes diabetes mellitus, a chronic metabolic illness (American Diabetes Association 2021). The International Diabetes Federation (2019) has evaluated that 700 million people are expected to have diabetes by 2045, creating serious problems for healthcare systems.

Tissue homeostasis, apoptosis, or programmed cell death removes unhealthy or superfluous cells (Elmore 2007). It was found that type 1 and type 2 diabetes are pathophysiologically associated with apoptosis deregulation (Huang and Sheu 2019). In type I diabetes, pancreatic beta cells are destroyed by autoimmune activation through apoptotic pathways, resulting in insulin deficiency (Eizirik and Cnop 2020). This process is marked by several key events like the activation of proapoptotic signals such as tumor necrosis factor-alpha (TNF-α) and Fas ligand (FasL) and the release of cytokines by the penetration of immune cells (Eizirik and Mandrup-Poulsen 2012). In type 2 diabetes, apoptosis can be seen in pancreatic beta cells and insulin-responsive tissues such as the liver, skeletal muscle, and adipose tissue (Maedler and Donath 2009; Hotamisligil 2017). Impaired insulin signaling and escalated hyperglycemia are caused by abnormal apoptosis in insulin-responsive tissues due to chronic low-grade inflammation and insulin resistance (Gregor and Hotamisligil 2011).

Understanding the molecular mechanisms behind diabetes-related apoptosis is essential. This understanding is crucial for developing targeted therapeutics to prevent diabetic complications (El Assaad et al. 2003). As we know, apoptosis maintains tissue homeostasis, and the malfunction of apoptotic pathways leads to the progression of diabetes (Butler et al. 2003). The molecular crosstalk between programmed cell death and inflammation in the pancreatic islets of individuals having type II diabetes has been investigated in recent studies. Apoptotic pathways are triggered by chronic low-grade inflammation, causing malfunction and loss of beta cells (Marchetti et al. 2021). Recently, several microRNA-based therapies have been developed to regulate apoptosis against insulin resistance. It has been found that specific microRNAs have critical roles in the regulation of apoptosis in insulin-responsive tissues (Gonzalez-Rodriguez et al. 2022).

## 12.2 Apoptosis in Pancreatic Beta Cells

### 12.2.1 Overview of Pancreatic Beta Cells and their Role in Insulin Production

Pancreatic beta cells, which play a major role in blood glucose homeostasis, lie within the islets of Langerhans. These specialized pancreas cells occupy about

60–80% of the total cells in the islet. These beta cells synthesize and secrete a hormone called insulin, which is crucial for regulating blood glucose homeostasis (Aguayo-Mazzucato and Bonner-Weir 2018). The heterogeneity within the beta cells of the pancreas has been identified through single-cell RNA sequencing. The distinct subpopulations of pancreatic beta cells have unique transcriptional profiles that distinguish their functional attributes. This heterogeneous subpopulation of pancreatic cells has diverse functions in glucose metabolism (Zakharov et al. 2020). Several complex interconnected molecular events control insulin secretion. ATP-sensitive potassium channels are vital in coupling insulin release with glucose metabolism. These channels act as vital intermediates to respond according to precise insulin concentration for regulating glucose homeostasis in the blood. This indicates the intricate signaling pathways within beta cells (Rorsman and Ashcroft 2018).

## 12.2.2 Mechanisms Leading to Apoptosis in Pancreatic Beta Cells

The beta cells of the pancreas, which are vulnerable to apoptosis, perform a crucial role in insulin production and glucose homeostasis in blood. The dysfunction of mitochondria in pancreatic beta cells is induced by oxidative stress. Cellular antioxidant defenses cannot cope with excessive reactive oxygen species (ROS) (Dinić et al. 2022). On the other hand, stress related to the endoplasmic reticulum (ER) also significantly contributes to the apoptosis of beta cells in the pancreas. The unfolded protein response (UPR) and misfolded protein accumulation trigger this ER stress-induced apoptosis (Cunha et al. 2009). Moreover, apoptosis in the pancreatic beta cells is driven by pro-inflammatory cytokines like interleukin-1β (IL-1β) (Fig. 12.3) and tumor necrosis factor-alpha (TNF-α) through the activation of apoptotic signaling pathways (Salpeter et al. 2010).

### 12.2.2.1 Oxidative Stress and Mitochondrial Dysfunction

Oxidative stress triggers mitochondrial malfunction which is responsible for initiating a series of molecular cascades leading to apoptosis of beta cells. The elevated glucose level in the blood causes diabetes, which results in the excess generation of reactive oxygen species (ROS) within beta cells of the pancreas (Eguchi et al. 2021). Several major changes in the structure and function of mitochondria, such as fragmentation and compromised respiratory capacity, can be the key consequences of this oxidative stress (Eguchi et al. 2021). Additionally, the overproduction of reactive oxygen species (ROS) induces the activation of mitochondrial permeability transition pores (mPTP). An intricate cascade of molecular mechanisms within the mitochondria is initiated by the overproduction of reactive oxygen species (ROS), leading to the activation of mitochondrial permeability transition pores (mPTP) (Fig. 12.1). The oxidative stress due to increased ROS levels within the mitochondria initiates the opening of mPTP (Eguchi et al. 2021). This is responsible for oxidizing vital thiol groups in mitochondrial proteins, forming the mPTP structure. This oxidation induces conformational changes to facilitate mPTP opening.

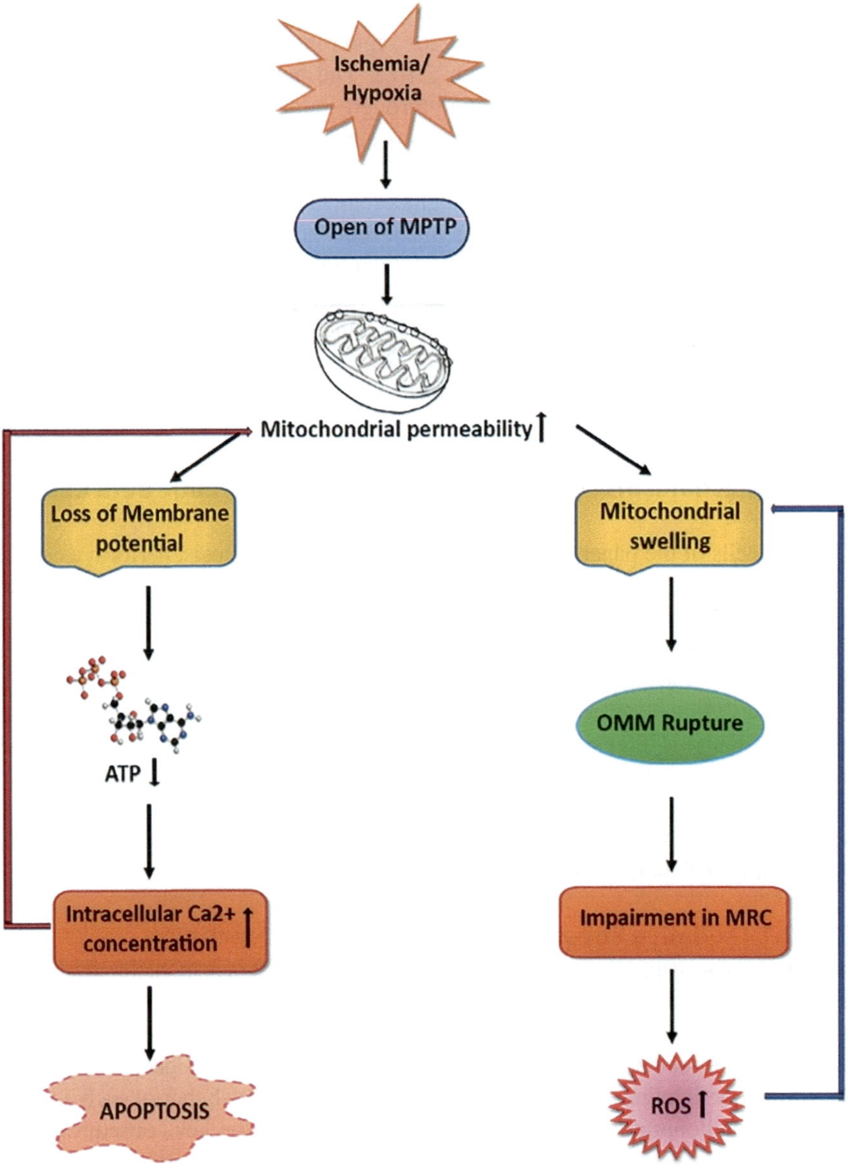


**Fig. 12.1** The molecular communication between mitochondrial dysfunction, oxidative stress, and apoptosis in beta cells of the pancreas is the major key factor responsible for diabetes

Therefore, the membrane potential of mitochondria is lost. The opening of the mitochondrial permeability transition pore (mPTP) helps to allow the unrestricted movement of molecules and ions through the inner membrane of mitochondria (Eguchi et al. 2021). This disrupts the electrochemical gradient, crucial for generating ATP and properly functioning mitochondria (Fig. 12.1). The opening of mPTP liberates the proapoptotic agents, like cytochrome c, from the intermembrane space of mitochondria to the cytoplasm. These molecular events ultimately lead to the cellular demise of beta cells by initiating a series of caspase-mediated apoptotic signaling pathways (Eguchi et al. 2021). Mitochondrial dysfunction due to oxidative stress prevents the autophagy of damaged mitochondria. This intensifies cellular stress and enhances apoptotic signaling pathways. The functional impairment of beta cells seen in diabetes is caused by the misbalance of autophagy-mitophagy mechanisms (Eguchi et al. 2021).

#### 12.2.2.2 Endoplasmic Reticulum Stress

Endoplasmic reticulum (ER) stress is pivotal in determining the fate of pancreatic beta cells. The misregulation in protein folding causes the formation of misfolded proteins, which initiates a series of molecular cascades, ultimately resulting in cellular apoptosis. The extensive requirement of protein synthesis and folding beyond the capabilities of ER is needed for disruptions in the equilibrium of beta cell functioning, such as excessive nutrient intake or genetic susceptibilities (Cunha et al. 2009). The aggregation of misfolded proteins in the inner lumen of the endoplasmic reticulum (ER) initiates a well-conserved cascade of signaling pathways called the unfolded protein response (UPR) (Fig. 12.2). The UPR reduces the overall synthesis of proteins by enhancing the intracellular concentration of chaperone proteins to restore ER balance and enhance the breakdown of misfolded proteins via ER-associated degradation (ERAD) pathways (Cunha et al. 2009). The unfolded protein response (UPR) promotes apoptosis instead of promoting cell survival when the stress associated with the endoplasmic reticulum (ER) remains unsolved. Long-term exposure to ER stress can initiate apoptotic signaling pathways. The high expression of C/EBP homologous protein (CHOP) and the activation of caspases ultimately lead to the apoptosis of beta cells in the pancreas (Hu et al. 2019).

However, the endoplasmic reticulum (ER) has several distinct unfolded protein responses (UPR) pathways with three major stress sensors: PERK, IRE1α, and ATF6. The inositol-requiring enzyme 1 (IRE1) and protein kinase RNA-like endoplasmic reticulum kinase (PERK) pathways help to make the choices of beta cells against ER stress (Fig. 12.2). IRE1α has a pivotal role across plant and animal cells. IRE1α (inositol-requiring enzyme 1 alpha) acts as a type I transmembrane protein with kinase/nuclease activities and activates upon ER stress. It enhances inter-organelle interaction during stress through membrane contacts (Zhou et al. 2021). The absence of PERK in myocytes has shown a significantly high protective effect against programmed cell death during high glucose concentration in the blood compared to the absence of ATF6 or IRE in these same myocytes. Fractioning the subcellular components of primary cardiomyocytes has revealed that PERK is a constituent within MAMs (mitochondria-associated endoplasmic reticulum membranes) instead of ATF6 and IRE1. It serves as a bridge for cellular communication between the endoplasmic reticulum (ER) and mitochondria (Liu et al. 2013). Any disturbance in activating these specific UPR pathways can lead to the dysfunction and the decline of beta cells seen in diabetes.

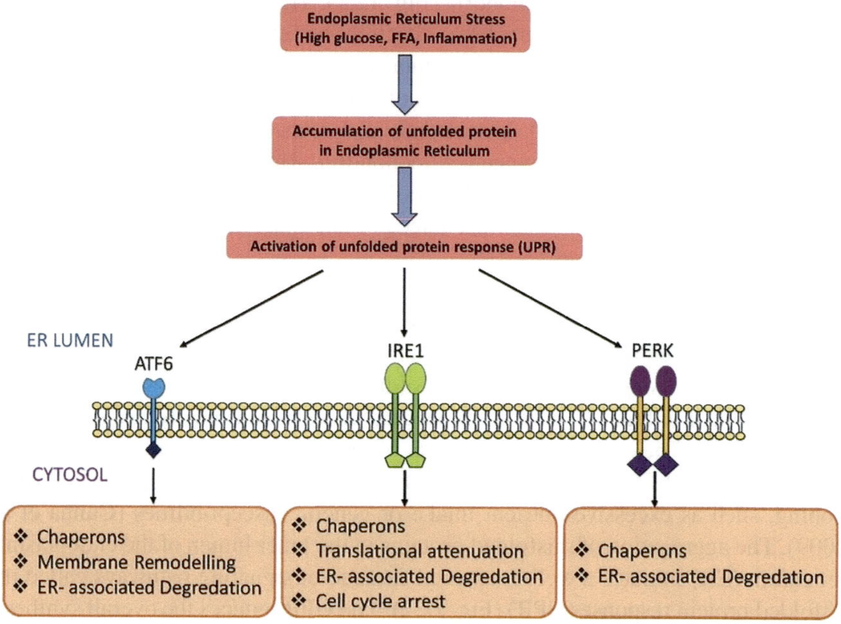


**Fig. 12.2** Endoplasmic reticulum stress-induced apoptosis in pancreatic beta cell includes several UPR pathways

#### 12.2.2.3 Inflammatory Pathways and Cytokines

In diabetic conditions, inflammatory pathways can also initiate apoptosis in pancreatic beta cells. The major pro-inflammatory cytokines are tumor necrosis factor-alpha (TNF-α), interleukin-1β (IL-1β), and interleukin-6 (IL-6), which coordinate a series of molecular cascades responsible for the death of beta cells. The different cells of the immune system secrete cytokines such as TNF-α, IL-1β, and IFN-γ, which have the potential to activate several pathways like STAT-1 and NF-κB, leading to beta-cell apoptosis in type 1 diabetes. Therefore, nitric oxide is generated, ER calcium level is depleted, and mitogen-activated protein kinases are activated. IL-1β acts as a potent inducer (Fig. 12.3) for apoptosis in beta cells by activating the nuclear factor-kappa B (NF-κB) signalling pathway (Cnop et al. 2005). This can lead to the amplification of genes like inducible nitric oxide synthase (iNOS) and Fas ligand (FasL) which are associated with apoptosis, ultimately destroying the beta cells. Then again, type 2 diabetes involves chronic exposure to free fatty acids and high glucose that result in dysfunction and apoptosis of beta cells. However, elevated blood glucose levels trigger apoptosis through various mechanisms like ER stress and free fatty acids-induced apoptosis, independent of NF-κB and nitric oxide (Cnop et al. 2005). TNF-α affects beta cells by initiating apoptosis through the induction of mitogen-activated protein kinase (MAPK) pathways. This is responsible for the generation of reactive oxygen species (ROS) as well as the initiation of

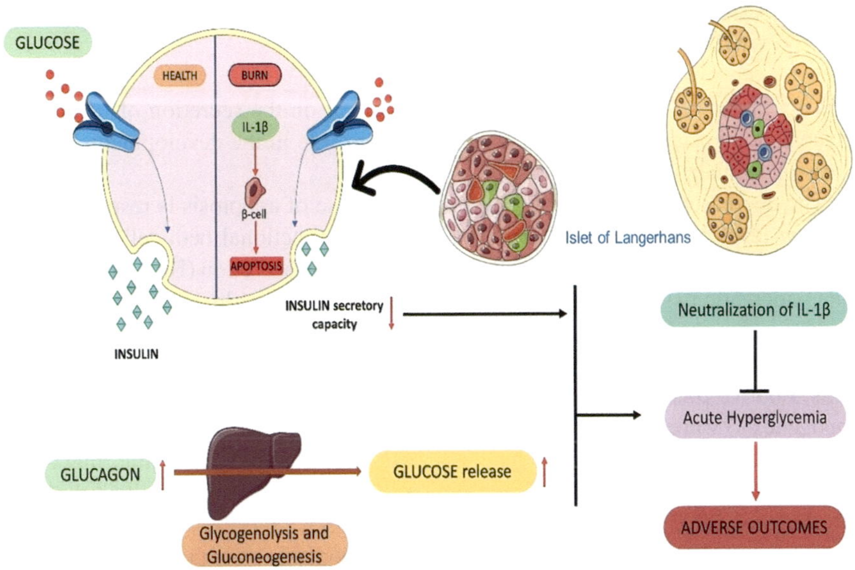


**Fig. 12.3** Cytokine-induced apoptosis in pancreatic beta cells

endoplasmic reticulum (ER) stress. These can ultimately activate the apoptotic pathways in the beta cells of the pancreas (Ou et al. 2020). Postburn hyperglycemia may be caused by several mechanisms: acute pancreatic beta cell apoptosis induced by interleukin-1β (IL-1β). The findings of the study on humans as well as mice models indicate that IL-1β contributes to the destruction of pancreatic beta cells (Fig. 12.3), impairing insulin secretion. It also affects gluconeogenesis and glycogenolysis, processes vital to glucose control. Gluconeogenesis and glycogenolysis are involved in glucose production following burn injuries, contributing to hyperglycemia (Fig. 12.3). These findings highlight the link between inflammation, altered pancreatic function, and hyperglycemia as the result of burn injuries (Li et al. 2019).

IL-6 induces apoptosis in beta cells by enhancing low-level inflammation within the pancreatic islets. It induces the Janus kinase (JAK)/signals transducer and activator of transcription (STAT) signaling pathways to amplify the production of pro-inflammatory agents, enhancing the dysfunction of beta cells and subsequent cell death (Akbari and Hassan-Zadeh 2018). Interleukin 6 (IL-6) has a multifaceted function in the pathophysiology of type 2 diabetes (T2D). However, the abnormal increase of IL-6 is responsible for the progression of T2D. It has anti-inflammatory properties and also can enhance glucose metabolism. IL-6 exerts different biological effects through two distinct signaling pathways, namely classic signaling and trans-signaling. Specifically, inhibition of trans-signaling of IL-6 has a promising result for targeted therapy in T2D management (Akbari and Hassan-Zadeh 2018).

### 12.2.3 Consequences of Beta Cell Apoptosis on Insulin Secretion and Glucose Homeostasis

The apoptosis of beta cells has a significant effect on the secretion of insulin and glucose homeostasis. Thus, it plays an important role in the development and progression of diabetes.

The decreased number of beta cell mass because of apoptosis is responsible for the low level of insulin secretion. The decline of functional beta cells affects the homeostasis of blood glucose levels, impairing insulin secretion (Butler et al. 2003). The pancreatic tissue obtained from autopsies of 124 individuals helped to analyze the differences between lean and obese individuals with and without type 2 diabetes or impaired fasting glucose (IFG). The rates of replication and apoptosis in beta cells, relative β-cell volume, and the formation of new islets from exocrine ducts (neogenesis) were investigated. Therefore, it was concluded that a reduction in the volume of beta cells in cases of type 2 diabetes was marked by the abnormal elevation of apoptosis. At the same time, replication and neogenesis remained the same (Butler et al. 2003).

Insulin production can be decreased by the apoptosis of beta cells that can disturb the regulation of glucose levels, causing hyperglycemia. Beta cell apoptosis plays a critical role in the decline of insulin secretion stimulated by glucose stimulation. This can exert a negative effect on individuals with diabetes by worsening hyperglycemia (Marchetti et al. 2007). The absence of factors from beta cells enhances peripheral insulin resistance, magnifying the metabolic imbalances linked with diabetes (Shantikumar et al. 2012). After the loss of beta cells, compensatory mechanisms may initially maintain normal blood glucose levels. However, if beta cell apoptosis persists, these compensatory mechanisms are surpassed. This can gradually impair blood glucose regulation, and diabetes will be evident (Butler et al. 2003).

## 12.3 Apoptosis in Insulin-Responsive Tissues

### 12.3.1 Apoptotic Pathways in Insulin-Responsive Tissues (E.G., Liver, Muscle, Adipose Tissue)

The muscle, liver, and adipose tissues are the major tissues that can respond to insulin. These tissues play a critical role in maintaining glucose homeostasis. Any disturbance in apoptotic pathways within these insulin-responsive tissues can significantly affect metabolic equilibrium. Several complex mechanisms control apoptosis in insulin-responsive tissues, which is critical in developing and progressing metabolic disorders.

The abnormal level of apoptosis in hepatocytes can initiate resistance toward insulin, a distinct characteristic feature of type II diabetes. The apoptotic signaling pathways are initiated by activating death receptors like tumor necrosis factor receptor 1 (TNFR1) and Fas within the liver. Apoptosis induced by the stress associated with the endoplasmic reticulum (ER) in hepatocytes facilitates insulin resistance by interfering with insulin signaling pathways (Feldstein et al. 2003). Liver biopsy samples from individuals who were diagnosed with nonalcoholic steatohepatitis (NASH) showed elevated levels of TUNEL- (terminal deoxynucleotidyl transferase dUTP nick end labeling) positive cells as compared to those with simple steatosis and control individuals. The individuals with NASH had also been observed to have increased TUNEL-positive cell counts as compared to those with alcoholic hepatitis. The presence of active caspases 3 and 7 in the liver samples of NASH individuals was confirmed by immunohistochemical analysis that marked the molecular cascade of apoptosis in this condition. However, a positive relationship was observed between hepatocyte apoptosis and hepatic fibrosis and inflammatory activity. The Fas receptor expression was significantly higher in hepatocytes within NASH liver individuals than in control individuals (Feldstein et al. 2003).

Apoptosis occurring in skeletal muscle has an intense connection with insulin resistance and prevents glucose absorption. Factors like inflammation, oxidative stress, and mitochondrial dysfunction act as promoters of apoptosis in muscle cells. Activating proapoptotic proteins like caspases is responsible for the decline of muscle cells and reduces sensitivity toward insulin (Santos-Silva et al. 2015). Malfunctioning adipose tissue is responsible for increased sensitivity towards type 2 diabetes (T2D). The expansion capacity of subcutaneous adipose tissue is constrained, which can lead to fat accumulation in tissues and organs, which initiates the development of T2D. However, intramuscular lipids alone may not directly induce resistance against insulin. However, several lipid intermediates, such as fatty acyl-CoAs, diacylglycerols (DAG), and ceramides, and their localization and turnover have crucial roles. Comprehending the regulation of adipogenesis, adipose tissue expansibility, and metabolic disruption is difficult. The restricted adipose tissue expansibility and consequent ectopic lipid accumulation can contribute to resistance against insulin in the liver and muscle (Brøns and Grunnet 2017). The apoptosis of adipocytes disturbs the balance within adipose tissue and leads to metabolic dysfunction. Expanding adipose tissue induces both hypoxic conditions and nutrient scarcity. These ultimately result in the apoptosis of adipocytes. Therefore, several factors originating from adipocytes, such as free fatty acids and adipokines, are released after these events. Finally, insulin resistance and inflammation throughout the body can increase and help sustain metabolic disruptions (Kanda et al. 2006).

### 12.3.2 Impact of Apoptosis on Insulin Sensitivity

The apoptosis of adipocytes has a connection with the overexpression of inflammatory cytokines like TNF-α, MCP-1, and IL-6. These cytokines can activate various signaling pathways, which include two major players, namely, c-Jun N-terminal kinase (JNK) and inhibitor of kappa B kinase (IKK). IKK and JNK phosphorylate on serine residues of insulin receptor substrate-1 (IRS-1), thus blocking insulin signaling. Furthermore, adipokines like adiponectin and leptin, which control the sensitivity towards insulin in peripheral tissues, are released during apoptosis in adipocytes (Cinti et al. 2005).

Increased oxidative stress and impaired function of mitochondria can induce the activation of stress-responsive enzymes like p38 mitogen-activated protein kinase (MAPK) and JNK, establishing the connection with muscle cell apoptosis. This activation leads to the phosphorylation of the serine residue of IRS-1, and insulin signaling is interrupted. Therefore, the muscle cells, which enter into apoptosis, release cytokines and chemokines like inflammatory molecules, which can intensify insulin resistance by enhancing the entry of macrophages and inflammation within muscle tissues (Santos-Silva et al. 2015).

Apoptosis of hepatocytes in the liver can initiate endoplasmic reticulum (ER) stress, which induces the unfolded protein response (UPR). As a result, the level of C/EBP homologous protein (CHOP) is increased, and activation of JNK and IKK is also increased. These molecular players interrupt insulin signaling by inducing the phosphorylation of serine residue in IRS-1 (Fig. 12.4). This stimulates lipogenesis and gluconeogenesis in hepatocytes. The apoptosis in hepatocytes can result in the release of fetuin-A, a hepatokine, initiating insulin resistance in peripheral tissues (Feldstein et al. 2003).

Stressors like endoplasmic reticulum (ER) stress, oxidative stress, and inflammatory cytokines trigger both the intrinsic mitochondrial pathway and the extrinsic death receptor pathway of apoptosis in beta cells. Therefore, the activation of caspases and the cleavage of crucial apoptotic proteins occurs. The beta cells release danger-associated molecular patterns (DAMPs) and inflammatory cytokines due to apoptotic signaling pathways, contributing to insulin resistance and systemic inflammation in peripheral tissues (Butler et al. 2003).

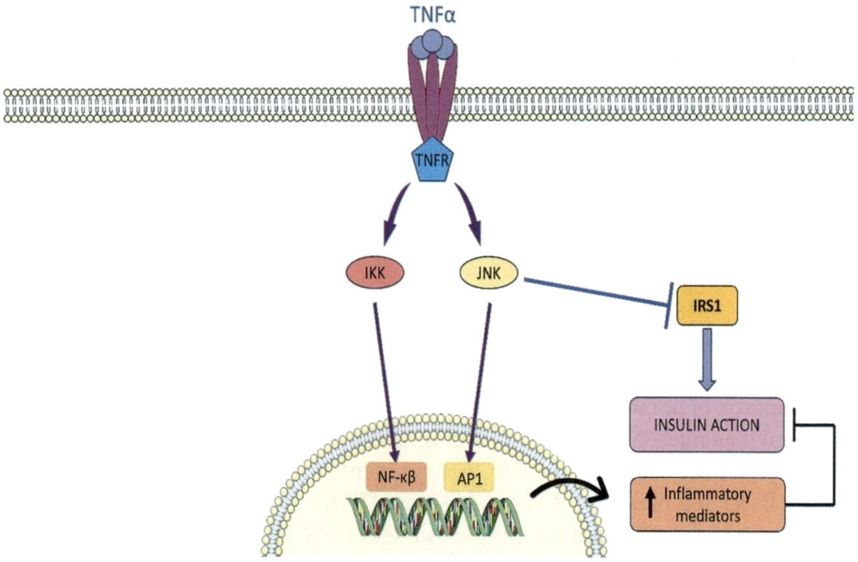


**Fig. 12.4** Regulation of insulin action by cytokine-induced signaling pathways

### 12.3.3 Link Between Inflammation, Apoptosis, and Insulin Resistance

The complex molecular connections among apoptosis, inflammation, and insulin resistance help to understand the initiation and progression of metabolic disorders. The interaction of different molecular components controls this intricate system. One of the reasons for insulin resistance is the failure of nutrient metabolism regulation. A significant increase in inflammation and macrophage-related genes in white adipose tissue (WAT) precedes a significant increase in insulin secretion in mouse obesity models. Rosiglitazone, an insulin-sensitizing medication, decreases the expression of these genes. Furthermore, obese mice exhibit macrophage infiltration in WAT, suggesting obesity contributes to insulin resistance. Insulin resistance associated with obesity may be caused by persistent inflammation within adipose tissue (Xu et al. 2003).

A cascade of signaling is started due to the secretion of inflammatory cytokines, such as IL-6 and TNF-α, that eventually target IRS-1, one of the main regulators of insulin signaling. Stress-responsive kinases, such as JNK and IKK, are activated by these cytokines and phosphorylate IRS-1 on serine residues (Fig. 12.4). After phosphorylation, IRS-1 cannot perform its regular function normally, which hinders the transmission of insulin signals downstream and reduces the amount of glucose absorbed and used by insulin-sensitive tissues (Hotamisligil 2017). Moreover, the cellular processes involved in the beginning of proapoptotic pathways in response to inflammatory signals are intricately linked. TNFR1, for instance, is the receptor of TNF-α, and the ligand-receptor binding recruits adaptor proteins like FADD and initiates the extrinsic apoptotic cascade. When caspase-8 is stimulated as a result of this activation, the caspase cascade is initiated. At the same time, TNF-α can trigger the intrinsic apoptotic pathway by breaking down mitochondria and letting cytochrome c leak into the cytoplasm. Following caspase-9 activation by cytochrome c, the apoptotic signal is amplified, and cell death occurs (Hotamisligil 2017). Furthermore, NF-κB, considered an essential regulator of inflammation, has a significant effect on apoptotic pathways. Antiapoptotic genes like Bcl-2 and XIAP, which suppress apoptotic signals and improve cell survival, are stimulated by NF-κB to initiate transcription. However, under chronic inflammatory conditions, proapoptotic components may also be upregulated by NF-κB-mediated transcriptional changes, favoring apoptosis and accelerating tissue damage (Hotamisligil 2017).

Furthermore, noncoding RNAs—such as microRNAs and long noncoding RNAs (lncRNAs) play a crucial role in modulating apoptotic and inflammatory processes in insulin-sensitive tissues. These noncoding RNAs fine-tune cellular responses by influencing posttranscriptional gene expression related to inflammation and apoptosis (Xu et al. 2003).

## 12.4 Apoptosis and Diabetic Complications

### 12.4.1 Relationship Between Apoptosis and Diabetic Complications

#### 12.4.1.1 Nephropathy

A comprehensive comprehension of the dynamic correlation between apoptosis and diabetic consequences, including retinopathy, neuropathy, and nephropathy, is essential for comprehending the causes of these fatal diseases (Fig. 12.5). There are several ways that apoptosis influences both the onset and progression of diabetic complications.

The inflammation and oxidative stress induced by Hyperglycemia in the renal tubules and glomeruli trigger apoptotic pathways that result in podocyte loss, tubular cell death, and extracellular matrix growth. The result of these events leads to progressive renal failure, tubulointerstitial fibrosis, and glomerulosclerosis, which are characteristic of diabetic nephropathy (Abrass 1995). Reactive oxygen species (ROS) and advanced glycation end products (AGEs) produced during hyperglycemic conditions induce apoptotic pathways in diabetic nephropathy including the extrinsic death receptor pathway which requires interaction between death receptor and death ligands where members of the Bcl-2 family proteins control the permeability of the mitochondrial membrane. Through the stimulation of proapoptotic genes and inflammatory cytokines, the pro-inflammatory cytokine TNF-α and its downstream effector, NF-κB, play critical roles in increasing the death of renal cells (Pourghasem et al. 2015). Activating the receptor for AGEs (RAGE) by AGEs in diabetic nephropathy starts intracellular signaling pathways that result in

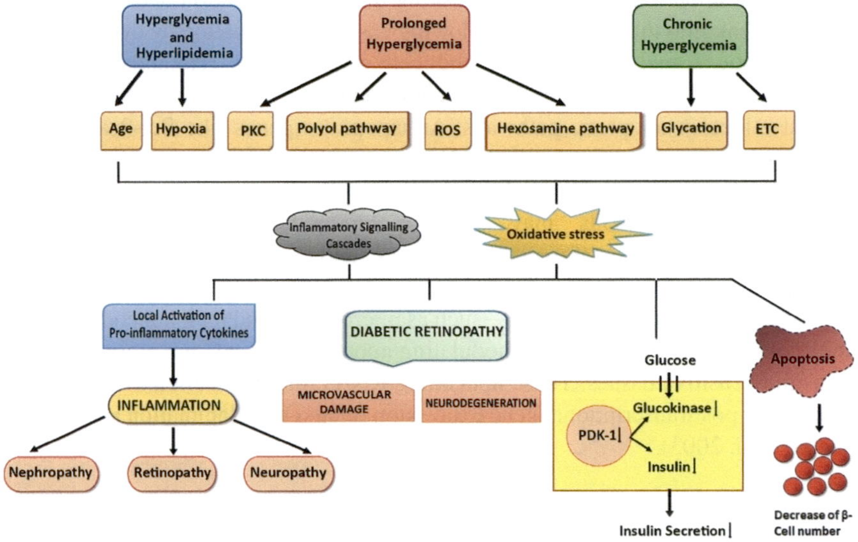


**Fig. 12.5** Diabetic complications caused by several apoptosis-inducing factors

inflammation and oxidative stress, ultimately responsible for the programmed cell death of renal cells. Apoptotic pathways are related to the renin-angiotensin system (RAS), where angiotensin II initiates apoptosis by stimulating proapoptotic genes and suppressing antiapoptotic proteins. Furthermore, apoptosis can be accelerated by the dysfunction of microRNAs, such as miR-21 and miR-192, which alter the gene expression that promotes apoptosis in renal cells (Wolf and Ziyadeh 1999).

#### 12.4.1.2 Retinopathy

Chronic hyperglycemia and other related metabolic disorders cause oxidative stress, vascular dysfunction, and retina inflammation. These factors ultimately result in the death of retinal cells and the degeneration of the neurons. During apoptosis in retinal endothelial cells, pericytes, and neurons, several processes like microvascular injury, capillary dropout, and retinal ischemia are intensified, leading to vision loss and the development of diabetic retinopathy (Tang and Kern 2011). The polyol pathway accelerates retinal cell death by depositing sorbitol and generating ROS when aldose reductase is active in hyperglycemic conditions. The production of AGEs and the activation of protein kinase C (PKC) isoforms contribute to inflammatory responses and oxidative stress, intensifying retinal cell death (Rübsam et al. 2018). Altered signaling pathways combine in the diabetic retina to trigger retinal cell death and neurodegeneration. The activated Wnt/β-catenin signaling pathway increases the activation of proapoptotic proteins and decreases antiapoptotic proteins, eventually leading to apoptosis. Moreover, the deregulation of the PI3K/Akt/mTOR pathway causes a rise in retinal cell death. In diabetic retinopathy, epigenetic modifications such as DNA methylation and histone acetylation also influence the gene expressions involved in apoptosis (Rübsam et al. 2018).

Infections or autoimmune immune responses can cause type 1 diabetes. Chronic inflammation caused by defective glucose metabolism and escalating insulin resistance contributes to type 2 diabetes. Diabetic retinopathy (DR) usually develops during an extended time of uncontrolled diabetes. The first stage is marked by damage to the retinal microvasculature caused by a low-grade activation of white blood cells that persists over an extended period. This causes blockages in tiny blood vessels, reducing blood flow to the retina. Macrophage-restricted protein tyrosine phosphatase 1B (PTP1B) may have a role in microvascular problems in the retina by controlling inflammation associated with insulin resistance. Blockages in retinal microvessels have also been connected to CCR5 + CD11b + monocyte macrophages (Forrester et al. 2020).

#### 12.4.1.3 Neuropathy

Persistent hyperglycemia causes neuroinflammation, endoplasmic reticulum stress, and mitochondrial dysfunction in peripheral nerves, accelerating apoptosis and axonal degradation. The typical signs of diabetic neuropathy are peripheral neuropathy, sensory impairments, and neuropathic pain. These are caused by apoptotic cell death in sensory neurons and Schwann cells, which affects nerve function (Smith and Singleton 2012). Peripheral neurons undergo apoptosis and inflammatory reactions when Toll-like receptors (TLRs) are activated by endogenous ligands, such as high-mobility group box 1 (HMGB1). Furthermore, in diabetic neuropathy, disruption of autophagy, a cellular mechanism essential in maintaining cellular homeostasis, contributes to neuronal death. In peripheral nerves, epigenetic changes that affect apoptotic pathways include histone deacetylation and noncoding RNA-mediated regulation (Smith and Singleton 2012).

The most common diabetic consequence, diabetic neuropathy, is caused by persistent hyperglycemia, which starts a series of metabolic changes. These changes include enhanced polyol pathway activity, elevated glycation end-product generation, and excess cytokine production, activated protein kinase C, and elevated oxidative stress. Even though neuropathy and other diabetes consequences involve metabolic pathways, they show unique histological and biochemical characteristics. Degeneration of distant nerve fibres is caused by reduced vascular supply, autoregulation, and the vulnerability of long axons. The bone marrow-derived cellular components influence the development of neuropathy (Yagihashi et al. 2011).

### 12.4.2 Mechanisms Contributing to Apoptosis Implicated in Diabetic Complications

#### 12.4.2.1 Apoptosis in Diabetic Nephropathy

Oxidative stress induced by hyperglycemia activates a number of molecular events which lead to renal cell death. The excessive poly(ADP-ribosyl)ation of nuclear proteins and concomitant decrease of cellular NAD+ and ATP are the result of the hyperactivation of the nuclear enzyme poly(ADP-ribose) polymerase-1 (PARP-1). This loss of energy increases cellular stress and accelerates apoptosis (Sifuentes-Franco et al. 2018). Furthermore, in diabetic nephropathy, the renin-angiotensin system (RAS) plays a critical role in the process of renal cell death. One of the main components of RAS is angiotensin II which activates NF-κB and leads to the generation of reactive oxygen species (ROS). Renal cell apoptotic signalling pathways are boosted and apoptosis-related genes are upregulated as a result of this activation (Sifuentes-Franco et al. 2018).

#### 12.4.2.2 Apoptosis in Diabetic Retinopathy

The apoptotic markers Bcl-2 and Bax are present in the conjunctiva of diabetic patients with and without retinopathy. Various cell types in normal conjunctiva have shown positive expression of Bax, although no Bcl-2 staining has been observed. Bax expression is very robust across numerous cell types among diabetic individuals without retinopathy, while Bcl-2 expression is either insufficient or very weak. In diabetic retinopathy patients, Bax expression is frequently elevated. For instance, in diabetic conjunctiva, no discernible relationship exists between Bax and Bcl-2 expression. The conjunctiva may be the location of programmed cell death in diabetic individuals, indicated by Bax in areas with high apoptosis rates (Khalfaoui et al. 2010). Apoptosis related to diabetic retinopathy is initiated to some extent by disruptions of the vascular endothelial growth factor (VEGF) signaling pathway.

Hypoxia-induced (Fig. 12.5) elevations of VEGF cause ischemia and increased vascular permeability in the retina, which drives the activation of apoptotic pathways in retinal cells. Furthermore, autophagy is a vital biological process that maintains retinal homeostasis. Any abnormalities in this process may contribute to retinal cell death in diabetic retinopathy (Tang and Kern 2011).

Death of retinal cells in diabetic retinopathy has been associated with activating the mammalian target of the rapamycin (mTOR) signaling pathway that governs cell metabolism and development. The onset and progression of diabetic retinopathy are promoted by dysregulated mTOR activity, which interrupts cellular homeostasis and favors retinal cell death via apoptosis. The complicated system of vascularization that is essential for providing energy to the retina is disrupted by diabetic retinopathy, which lowers oxygen levels and subsequently leads to degeneration of the neurons. Chronic hyperglycemia affects several processes that contribute to retinal neurodegeneration in diabetic retinopathy, including oxidative stress, autophagy, cell death, and the production of pro-inflammatory cytokines. All these events are fundamentally regulated by the mTOR pathway (Casciano et al. 2022).

#### 12.4.2.3 Apoptosis in Diabetic Neuropathy

The activation of the receptor for advanced glycation end products (RAGE) in diabetic neuropathy is a result of oxidative stress and inflammation driven by hyperglycemia. This stimulation triggers neuronal apoptosis and neuro-inflammatory reactions. RAGE signaling promotes the production of pro-inflammatory cytokine and oxidative stress by activating the mitogen-activated protein kinase (MAPK) and nuclear factor-kappa B (NF-κB) pathways. In diabetic neuropathy, these mechanisms accelerate neuronal death and axonal degradation (Smith and Singleton 2012). Neuronal cell death in diabetic neuropathy is believed to be associated with an imbalance in calcium regulation and activation of the calcium/calmodulin-dependent protein kinase II (CaMKII) pathway. High calcium concentrations in cells cause CaMKII to become activated, subsequently altering several proteins involved in pathways that lead to apoptosis and, eventually, the death of neurons (Vincent et al. 2011).

Acutely disconnected and undamaged dorsal root ganglion (DRG) neurons from diabetic and normal rats have been employed to conduct several investigations. From these investigations, results have indicated that diabetic rats having DRG neurons have increased rates of death between three and six weeks of age. The investigations have also shown an increased baseline mitochondrial membrane potential (delta psi) and a delayed recovery with glutamate-induced depolarization. Nerve conduction velocities, delta psi levels, and apoptosis rates have returned to normal after two weeks of insulin implant therapy to restore blood sugar levels. Moreover, cytochrome C has been seen to move from mitochondria to the cytoplasm in diabetic rats and decrease Bcl-2, a protein that promotes cell survival. These outcomes highlight that mitochondrial dysfunction-associated apoptosis may play a critical role in the progression of diabetic sensory neuropathy (Srinivasan et al. 2000).

### 12.4.3 Potential Therapeutic Targets to Prevent or Mitigate Diabetic Complications through Apoptosis Modulation

Antioxidants such as alpha-lipoic acid (ALA) and N-acetylcysteine (NAC) have shown potential in preclinical studies by efficiently scavenging free radicals and lowering oxidative stress-induced apoptosis in various diabetic-affected tissues. These efforts give a chance to minimize apoptosis and complications related to diabetes (Rochette et al. 2014).

Since immunological dysfunction and inflammation are markers of type 2 diabetes (T2D), anti-cytokine medications may be useful as treatments. The anti-cytokine medicines and disease-modifying antirheumatic medications (DMARDs) are very effective in combatting type 2 diabetes. Modulating pro-inflammatory cytokines and signaling pathways is another method for minimizing apoptosis and diabetes consequences. The efficaciousness of drugs such as infliximab and etanercept has been proven to inhibit TNF-α to reduce inflammation and programmed cell death in disorders including diabetic retinopathy and nephropathy (Velikova et al. 2021).

Targeted therapy against apoptosis in diabetes consequences can be accomplished by directly regulating apoptotic signaling pathways. It has been investigated if inhibitors of caspases, important apoptotic executioners, could be used as therapeutic drugs to stop apoptotic cell death in diabetic retinopathy, neuropathy, and nephropathy (Shi et al. 2018). The rat liver cells are affected by high glucose levels, and Naringenin, a flavanone, has protective advantages. High exposure to glucose induces a decline in cell viability, abnormalities in mitochondrial functioning, and initiation of apoptotic pathways associated with mitochondrial proteins. Naringenin efficiently neutralized these effects by lowering the generation of reactive oxygen species, preserving antioxidant defenses, and preventing the migration of mitochondrial proteins implicated in apoptosis. Naringenin can scavenge reactive oxygen species and control the mitochondrial apoptotic pathway, making it a promising preventative therapy against apoptosis driven by high glucose levels (Kapoor et al. 2013).

An elevated level of glucose causes several signaling pathways, including the AGE/RAGE, polyol, PKC, and hexosamine pathways, to become activated (Fig. 12.5). In an activated state, these pathways trigger inflammation, oxidative stress, and cell death, all enhancing vascular dysfunction in diabetic patients. Polyphenols are rich in plant-based foods, including vegetables, fruits, whole grains, and nuts. The anti-inflammatory, antioxidant, and anti-cell death properties of polyphenols are considered to be responsible for their beneficial effects (Mohd Nor et al. 2022).

Apoptosis regulation in diabetes complications is significantly influenced by epigenetic alterations, such as histone acetylation and DNA methylation. By focusing on epigenetic regulators such as histone deacetylases (HDACs) and DNA methyltransferases (DNMTs), diabetes consequences can be minimized and apoptotic pathways can be modulated. Genetic, epigenetic, and environmental variables, including age, nutrition, and habits, all affect the development of type 2 diabetes mellitus (T2DM). The metabolic consequences of insulin shortage can be reversed by restoring the activity of β-cells. Islet cells require pancreatic duodenal homeobox 1 (PDX-1) to mature and function properly. DNA methylation levels of PDX-1 become elevated (Fig. 12.5) in T2DM, which may cause β-cell dysregulation. Clinical methods employing PDX-1 for diabetes control explore the epigenetic pathways underlying PDX-1 in T2DM, focusing on DNA methylation (Liu et al. 2021).

## 12.5 Emerging Research and Technologies

### 12.5.1 Cutting-Edge Research on Apoptosis in Diabetes

To minimize the risk and slow down the progression of type 2 diabetes, it may be essential to look into the genetic factors behind pancreatic beta-cell death or to further research into preventing beta-cell apoptosis. Conventional drugs such as biguanides can reduce the beta-cell apoptosis caused by lipotoxicity by lowering proapoptotic PERK/CHOP signaling and ER stress response. By stimulating beta-cell proliferation and reducing apoptosis, newer pharmacological alternatives, such as SGLT2 inhibitors and GLP-1R agonists, such as exendin-4, have shown significant therapeutic improvements to combat type 2 diabetes. Nevertheless, it is still unclear whether these therapies would protect beta-cells in clinical environments or the specific processes that promote their healing. To find novel therapeutic targets, it is essential to investigate the in-depth processes leading to beta-cell death in type 2 diabetes (You et al. 2022).

Several investigations have found the promising effect of puerarin (PR) on beta cell death (apoptosis) in a mouse model of type 2 diabetes (T2DM). The mouse model has been stimulated using streptozotocin injection and a high-fat diet. PR treatment lowers blood sugar and fat levels while raising insulin levels. It also inhibits cell death markers, including caspase proteins and AIF, within the pancreas. A specific staining technique (TUNEL) has confirmed reduced beta cell death with PR. Additionally, pancreatic tissue has appeared healthier following PR administration. These findings suggest the potential of PR to safeguard beta cells from apoptosis in T2DM mice by regulating cell death-related proteins (Liang et al. 2019).

High blood sugar (hyperglycemia) and the death of insulin-producing cells (beta cell apoptosis) are hallmarks of diabetes. New diabetes therapies might focus on preventing cell death (apoptosis) or supporting factors protecting the pancreas. Recent research suggests the Hippo signaling pathway, particularly the kinases MST1 and LATS2, could be a target for treatment. By inhibiting apoptosis and promoting tissue repair, this pathway has potential benefits in diabetes. A clearer understanding of the role of the Hippo pathway in diabetes-induced cell death is essential for developing effective therapies to stop or slow beta cell loss. Manipulating components of this pathway offers a promising avenue for future diabetes treatments aimed at preventing cell death (Kilanowska and Ziółkowska 2022).

Several investigations have revealed how diabetes affects cell death (apoptosis) and cell division (mitosis) in a specific area of the rat brain (hippocampal CA1 region). Diabetes has been induced in rats using a streptozotocin injection. As compared to healthy controls, diabetic rats have displayed a significant rise in both apoptotic cells and the number of star-shaped brain cells (astrocytes). Interestingly, these diabetic rats have shown increased cell division activity. Findings have concluded that diabetes might trigger cell death and cell proliferation in the hippocampal CA1 region, potentially leading to imbalances impairing normal function (Ünver Saraydin et al. 2021).

TRAIL, a protein related to tumor necrosis factor (TNF), has complex effects depending on cell type and surrounding signals. TRAIL might protect against type 1 and type 2 diabetes (T1DM and T2DM). Studies show that blocking or lacking TRAIL increases diabetic complications in animals, while giving TRAIL or its genes prevents or improves T1DM and T2DM. Diabetic patients have lower circulating TRAIL levels. TRAIL can directly affect various tissues involved in diabetes development. At the same time, some studies disagree on the impact of TRAIL on fat tissue health. Overall, evidence supports further investigation of TRAIL-based interventions in diabetic patients with heart and metabolism problems (Koliaki and Katsilambros 2022).

Investigations have revealed the potential of enalapril and paricalcitol, alone and combined, to protect the kidneys in rats with streptozotocin (STZ)-induced diabetes. Male rats receive STZ to induce diabetes, followed by treatment with enalapril, paricalcitol, or both for four weeks. All treatments have been significantly found to improve kidney function, reduce harmful fat breakdown (lipid peroxidation), boost antioxidant activity, and control blood sugar levels. Additionally, treatment prevents damage to kidney and pancreas tissues observed under a microscope (histopathological changes). Analysis of protein markers within the kidney has revealed decreased inflammatory molecules and cell death signals (apoptosis markers) alongside increased proteins that promote cell survival (antiapoptotic proteins). Overall, both enalapril and paricalcitol, administered individually or together, prevent kidney damage in STZ-diabetic rats by improving blood sugar control, enhancing the body's antioxidant defenses, and suppressing oxidative stress, inflammation, and cell death pathways. Interestingly, combining the treatments offers no additional advantages compared to using them alone (Ahmed et al. 2019).

### 12.5.2 Apoptosis in Diabetes Through the Lens of Omics

#### 12.5.2.1 Genomics

Chronic, low-grade inflammation, a well-known marker of type 2 diabetes (T2D), affects genes within pancreatic islets, the major structure for insulin production. The mouse islets have been exposed to cytokines that can mimic this inflammatory state in T2D. Genes whose expression significantly changes (more than two-fold increase) have been analyzed for connections to insulin response in human T2D patients using data from the Genetics Underlying Diabetes in Hispanics (GUARDIAN)

Consortium. Seven genes (Arap3, F13a1, Klhl6, Nid1, Pamr1, Ripk2, Steap4) (Fig. 12.6) have shown significant connections to cytokine exposure and human insulin response. Further experiments have explored how these genes impact various stresses, including inflammation, mitochondrial dysfunction, glucotoxicity, and lipotoxicity. RT-PCR has shown that most genes have responded significantly to at least one form of stress, except for Arap3. Notably, Thapsigargin, a chemical miming endoplasmic reticulum (ER) stress, can increase the expression of Pamr1 and Klhl6. Three genes, namely F13a1, Ripk2, and Steap4, have confirmed the initial findings using microarray by demonstrating a strong response to cytokine exposure. These genes might affect how chronic inflammation disrupts islet function during T2D development (Sharma et al. 2015).

### 12.5.2.2 Transcriptomics

The traditional methods have found genetic variations associated with T2D. However, these methods haven't provided a complete picture of the inherited risk. Breakthroughs in gene sequencing techniques, like RNA sequencing (RNA-seq) (Fig. 12.6) and single-cell analysis, have transformed the field of transcriptomics. Studying the transcriptome, which includes all the RNA molecules in a cell, offers exciting possibilities for uncovering novel biomarkers. These biomarkers could help identify people at high risk of developing T2D and its complications. This could significantly improve strategies for early detection, treatment, and prevention. The techniques related to transcriptomics include looking at large sets of genes (expression arrays), reading individual gene sequences (RNA-seq), and even analyzing

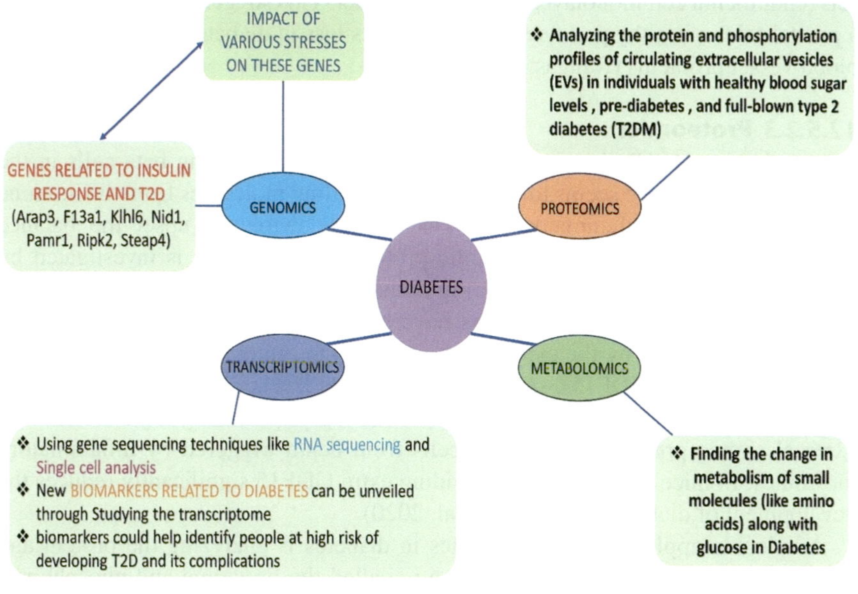


**Fig. 12.6** Apoptosis in diabetes through omics

gene activity in single cells. Using these techniques, several findings reveal how the body's inflammatory response contributes to the development of T2D. Findings have also identified the activation of signaling pathways typically linked to cancer, along with alterations in cell division processes and the malfunction of a cellular "clean up" system in T2D patients. However, there are still challenges in how these studies are designed, what samples are used, which methods are chosen, and how the results are interpreted. While these technologies offer a window into the molecular workings of T2D and potentially new ways to diagnose and treat it, larger studies are needed before these findings can be used in clinical practice (Tonyan et al. 2022).

The study of gene activity (transcriptomics) in white blood cells, specifically neutrophils, has revealed novel genes involved in long-term inflammation and fat breakdown. These genes behave differently in T2D cells as compared to healthy cells. Interestingly, the cells also adjust their gene activity in response to increasing amounts of a specific molecule (RvE1) (Kleinstein et al. 2021).

The specialized cells called pericytes contribute to diabetic retinopathy (DR) in mice. The activity of genes (transcriptome) in pericytes isolated from the retina of diabetic mice has been analyzed. Using a technique called RNA sequencing, gene activity is compared in both diabetic pericytes and healthy controls. Findings related to this have revealed the changes in hundreds of genes, with 217 genes becoming more active and 495 genes becoming less active in diabetes. More specifically, genes like Ext2, B3gat3, and Notch3 have shown significant changes. Further studies have confirmed that Notch3 activity has decreased in human retinal pericytes exposed to high sugar levels, mimicking diabetic conditions. Another interesting finding is that the inactivation of Notch3 in pericytes causes the lining of blood vessels (endothelial cell monolayer) to become leakier. This suggests that Notch3 plays a pivotal role in healthy pericyte function, and its loss in diabetes may contribute to blood vessel problems (Rangasamy et al. 2020).

#### 12.5.2.3 Proteomics

Type 1 diabetes (T1D) involves slowly losing insulin-producing beta cells in the pancreas. This damage seems to be caused by signaling molecules from the immune system called pro-inflammatory cytokines. These cytokines upset the balance between antiapoptotic and proapoptotic proteins. This process is investigated by analyzing all the proteins (proteome) in human pancreatic islets treated with specific cytokines. Over 11,000 proteins have been identified, and nearly 400 were significantly affected by the cytokine treatment. One protein, called growth/differentiation factor 15 (GDF15), stands out because its levels decrease during inflammation and are lowered in T1D islets. Interestingly, providing GDF15 from an external source protects human islet cells from death triggered by inflammatory molecules. In mice prone to T1D, providing extra GDF15 significantly reduces the development of diabetes (Nakayasu et al. 2020).

A valuable application of proteomics in diabetes is analyzing the protein and phosphorylation profiles (combined picture called the proteome and phosphoproteome) of circulating extracellular vesicles (EVs) (Fig. 12.6) in individuals with healthy blood sugar levels (normal glucose tolerance, NGT), prediabetes (prediabetes, PDM), and full-blown type 2 diabetes (T2DM) (Fig. 12.6). The distinct sets of proteins and phosphorylation patterns within EVs from each group are identified. Interestingly, EVs from diabetic individuals carry more specific signaling molecules (phosphorylated kinases). This raises the possibility that these molecules might contribute to spreading diabetes-related effects throughout the body. A shift in the types of integrin proteins present in EVs from those with prediabetes and diabetes has been observed. Unexpectedly, EVs from prediabetic and diabetic individuals have higher levels of proteins involved in energy production (oxidative phosphorylation, OXPHOS). Normally, this process is reduced in diabetic tissues. This finding suggests a potential link between abnormal EV-mediated secretion of OXPHOS components and the development of diabetes. Finally, a decrease in liver-specific markers within EVs from prediabetic and diabetic individuals has been noted. This hints at an alteration in liver EV function that might occur early in the disease process. Overall, the research using proteomics provides novel insights into the role of EVs in diabetes progression. It also suggests potential avenues for future investigations into how diabetes develops and how it might be treated (Nunez Lopez et al. 2022).

High blood sugar levels can arise from a combination of different factors, including lifestyle choices, genes, changes in gene expression (epigenetics), and abnormal protein profiles (proteome). Modifications made to proteins after they are built (posttranslational modifications) and how proteins interact with each other (protein-protein interactions) also significantly influence the development of diabetes. Recent technological breakthroughs have given scientists powerful tools to study these protein changes systematically, providing new insights into diabetes. One such tool is affinity purification mass spectrometry (AP-MS), which, combined with advanced computer analysis, allows the mapping of large networks of interacting proteins. This is particularly helpful for studying key players in maintaining healthy blood sugar levels (glucose homeostasis) (Kalita et al. 2020).

#### 12.5.2.4 Metabolomics

Diabetes, a widespread noncontagious disease, results from problems with insulin production or use, leading to high blood sugar (hyperglycemia) and various health complications. Traditionally, research has focused on glucose levels in diabetes. However, metabolomics, the study of a broader range of small molecules within cells (Fig. 12.6), offers a powerful new tool. Metabolomics plays an important role in uncovering fresh insights into diabetes. It focuses on certain key areas. How the body processes certain amino acids might be linked to diabetes development. Metabolomics can highlight how the body produces insulin. Studying metabolite profiles may help us understand the initiation and development of complications like neuropathy or kidney disease. Metabolomics can be used to assess the effectiveness of new treatments for diabetes (Dunn 2013).

Diabetic retinopathy (DR) is a leading cause of blindness worldwide, especially among middle-aged and older adults with diabetes. While current tests can offer some clues about a person's risk, there is a critical need for better markers in the bloodstream (circulating biomarkers) to predict DR development. Recent breakthroughs in a field called metabolomics have opened exciting possibilities. Metabolomics allows the study of small molecules within cells, potentially revealing new DR biomarkers. It highlights the identification of efficient biomarkers and prediction of molecular targets by combining genome-wide association studies (GWAS) with metabolomics data (Jian et al. 2022). Interestingly, some molecules, like alanine, lactate, and glutamine, have emerged as potential biomarkers across different sample types like blood and retinal tissue. Alanine might be involved in regulating blood sugar control and insulin production. Lactate could contribute to oxidative stress and oxygen deprivation in the retina. Abnormal levels of glutamine might indicate problems with how cells recycle glutamate, impacting the activation and growth of specific retinal cells called Müller cells (Du et al. 2022).

Researchers examined how varying arginine levels, an amino acid, affect insulin-producing beta cells (MIN6 cells) in the pancreas. The cells have been grown in cultures containing low (LA), standard (SA), or high (HA) concentrations of arginine for 24 hours. The findings reveal that high arginine (HA) is responsible for lowering the amount of insulin constantly released by the cells, while both low (LA) and high arginine (HA) decrease insulin secretion triggered by glucose (the glucose-stimulated insulin secretions or GSIS) as compared to the standard amount. Analyzing gene activity (transcriptomics) shows that low arginine (LA) prevents energy production through cellular respiration (oxidative phosphorylation) and ATP generation but enhances DNA repair and processing of messenger RNA (mRNA). On the other hand, high arginine (HA) impacts the handling of ammonium ions and the export of mRNA. Interestingly, low and high arginine levels influence genes involved in DNA replication, cell cycle progression, and the response to oxidative stress. Further analysis of cellular metabolites (metabolomics) reveals that low arginine (LA) primarily affects the metabolism of glycerophospholipids (fats containing glycerol) and amino acids. In contrast, high arginine (HA) mainly impacts amino acid metabolism. By combining these findings (transcriptomics and metabolomics), the researchers have highlighted disruptions in carbohydrate, fat, and amino acid metabolism, along with increased cell death (apoptosis) and oxidative stress, as potential contributors to the reduced insulin secretion observed with both low and high arginine treatments (Xu et al. 2022).

Findings reveal how human kidney cells (HK-2 cells) responded to conditions mimicking diabetic nephropathy (kidney damage in diabetes). The cells are exposed to high sugar levels (hyperglycemia) and low oxygen levels (hypoxia), replicating a diabetic-like environment compared with normal controls. Using sophisticated techniques to analyze cellular chemicals (metabolomics), over 700 metabolites are identified within the cells (intracellular) and nearly 300 outside the cells (extracellular). Their analysis has revealed significant changes in several metabolic pathways, including those involving carbohydrates, sugars with five carbon atoms (pentose phosphate pathway), the amino acids arginine and proline, purines (components of RNA and DNA), and fats (lipids). Additionally, several studies have identified the activation of pathways related to protecting cells from osmotic stress (caused by high sugar levels) and the breakdown of fatty acids for energy (β-oxidation). These findings align with previous research conducted in living organisms (in vivo studies), further emphasizing the importance of high blood sugar and low oxygen in the development of diabetic nephropathy (Valdés et al. 2021).

## 12.6 Conclusion

The complex relationship between diabetes and apoptosis highlights the multifaceted nature of this disease. Apoptosis emerges as a key player in the pathophysiology of diabetes, from the initial dysregulation of pancreatic beta cell apoptosis leading to impaired secretion of insulin to the later involvement of apoptosis in the progression of diabetic complications, such as nephropathy, neuropathy, and retinopathy. Various molecular processes, such as inflammatory pathways, oxidative stress, mitochondrial dysfunction, endoplasmic reticulum stress, and epigenetic regulation that play critical roles in the dysregulation of apoptosis in diabetes, are discussed in this chapter. Various novel approaches must be discovered to mitigate diabetic complications and improve clinical outcomes for people with diabetes by analyzing these molecular intricacies and identifying possible therapeutic targets. By translating new research findings into real-world treatments, future studies hold the potential to revolutionize personalized diabetes care. This, in turn, could significantly improve the lives of millions battling this global health crisis.

## References

Abrass CK (1995) Diabetic nephropathy. Mechanisms of mesangial matrix expansion. West J Med 162(4):318

Aguayo-Mazzucato C, Bonner-Weir S (2018) Pancreatic β cell regeneration as a possible therapy for diabetes. Cell Metab 27(1):57–67

Ahmed OM, Ali TM, Abdel Gaid MA, Elberry AA (2019) Effects of enalapril and paricalcitol treatment on diabetic nephropathy and renal expressions of TNF-α, p53, caspase-3 and Bcl-2 in STZ-induced diabetic rats. PLoS One 14(9):e0214349

Akbari M, Hassan-Zadeh V (2018) IL-6 signalling pathways and the development of type 2 diabetes. Inflammopharmacology 26:685–698

American Diabetes Association (2021) Diagnosis and classification of diabetes mellitus. Diabetes Care 44(Supplement_1):S15–S33

Brøns C, Grunnet LG (2017) Mechanisms in endocrinology: skeletal muscle lipotoxicity in insulin resistance and type 2 diabetes: a causal mechanism or an innocent bystander? Eur J Endocrinol 176(2):R67–R78

Butler AE, Janson J, Bonner-Weir S, Ritzel R, Rizza RA, Butler PC (2003) β-Cell deficit and increased β-cell apoptosis in humans with type 2 diabetes. Diabetes 52(1):102–110

Casciano F, Zauli E, Rimondi E, Mura M, Previati M, Busin M, Zauli G (2022) The role of the mTOR pathway in diabetic retinopathy. Front Med 9:973856

Cinti S, Mitchell G, Barbatelli G, Murano I, Ceresi E, Faloia E et al (2005) Adipocyte death defines macrophage localization and function in adipose tissue of obese mice and humans. J Lipid Res 46(11):2347–2355

Cnop M, Welsh N, Jonas JC, Jorns A, Lenzen S, Eizirik DL (2005) Mechanisms of pancreatic β-cell death in type 1 and type 2 diabetes: many differences, few similarities. Diabetes 54(suppl_2):S97–S107
Cunha DA, Ladriere L, Ortis F, Igoillo-Esteve M, Gurzov EN, Lupi R et al (2009) Glucagon-like Peptide-1 agonists protect pancreatic β-cells from Lipotoxic endoplasmic reticulum stress through upregulation of BiP and JunB. Diabetes 58(12):2851–2862
Dinić S, Arambašić Jovanović J, Uskoković A, Mihailović M, Grdović N, Tolić A et al (2022) Oxidative stress-mediated beta cell death and dysfunction as a target for diabetes management. Front Endocrinol 13:1006376
Du X, Yang L, Kong L, Sun Y, Shen K, Cai Y et al (2022) Metabolomics of various samples advancing biomarker discovery and pathogenesis elucidation for diabetic retinopathy. Front Endocrinol 13:1037164
Dunn WB (2013) Diabetes-the role of metabolomics in the discovery of new mechanisms and novel biomarkers. Curr Cardiovasc Risk Rep 7:25–32
Eguchi N, Vaziri ND, Dafoe DC, Ichii H (2021) The role of oxidative stress in pancreatic β cell dysfunction in diabetes. Int J Mol Sci 22(4):1509
Eizirik DL, Cnop M (2020) ER stress in pancreatic β cells: the thin red line between adaptation and failure. Sci Signal 13(622):eaaz0104
Eizirik DL, Mandrup-Poulsen T (2012) A choice of death—the signal-transduction of immune-mediated β-cell apoptosis. Diabetologia 55(1):1753–1777
El Assaad W, Buteau J, Peyot ML, Nolan C (2003) Saturated fatty acids synergize with elevated glucose to cause pancreatic β-cell death. Endocrinology 144(9):4154–4163
Elmore S (2007) Apoptosis: a review of programmed cell death. Toxicol Pathol 35(4):495–516
Feldstein AE, Canbay A, Angulo P, Taniai M, Burgart LJ, Lindor KD, Gores GJ (2003) Hepatocyte apoptosis and fas expression are prominent features of human nonalcoholic steatohepatitis. Gastroenterology 125(2):437–443
Forrester JV, Kuffova L, Delibegovic M (2020) The role of inflammation in diabetic retinopathy. Front Immunol 11:583687
Gonzalez-Rodriguez A, Unamuno X, M. T. M. G. R, Valverde AM, Riezu-Boj JI (2022) Role of microRNAs in the crosstalk between obesity, diabetes, and cardiovascular disease. Front Endocrinol 13:123456
Gregor MF, Hotamisligil GS (2011) Inflammatory mechanisms in obesity. Annu Rev Immunol 29(1):415–445
Hotamisligil GS (2017) Inflammation, Metaflammation and Immunometabolic disorders. Nature 542(7640):177–185
Hu H, Tian M, Ding C, Yu S (2019) The C/EBP homologous protein (CHOP) transcription factor functions in endoplasmic reticulum stress-induced apoptosis and microbial infection. Front Immunol 9:423124
Huang T, Sheu W (2019) Life or death decision in diabetic islet cells. World J Diabetes 10(5):277–282
International Diabetes Federation (2019) IDF Diabetes Atlas, 9th edn. International Diabetes Federation, Brussels. Retrieved from https://www.diabetesatlas.org/en/
Jian Q, Wu Y, Zhang F (2022) Metabolomics in diabetic retinopathy: from potential biomarkers to molecular basis of oxidative stress. Cells 11(19):3005
Kalita B, Bano S, Vavachan VM, Taunk K, Seshadri V, Rapole S (2020) Application of mass spectrometry based proteomics to understand diabetes: a special focus on interactomics. Biochim Biophys Acta (BBA)-Proteins Proteom 1868(10):140469
Kanda H, Tateya S, Tamori Y, Kotani K, Hiasa KI, Kitazawa R et al (2006) MCP-1 contributes to macrophage infiltration into adipose tissue, insulin resistance, and hepatic steatosis in obesity. J Clin Invest 116(6):1494–1505
Kapoor R, Rizvi F, Kakkar P (2013) Naringenin prevents high glucose-induced mitochondria-mediated apoptosis involving AIF, Endo-G and caspases. Apoptosis 18:9–27
Khalfaoui T, Basora N, Ouertani-Meddeb A (2010) Apoptotic factors (Bcl-2 and Bax) and diabetic retinopathy in type 2 diabetes. J Mol Histol 41:143–152

Kilanowska A, Ziółkowska A (2022) Apoptosis in type 2 diabetes: can it be prevented? Hippo pathway prospects. Int J Mol Sci 23(2):636
Kleinstein SE, McCorrison J, Ahmed A, Hasturk H, Van Dyke TE, Freire M (2021) Transcriptomics of type 2 diabetic and healthy human neutrophils. BMC Immunol 22(1):37
Koliaki C, Katsilambros N (2022) Repositioning the role of tumor necrosis factor-related apoptosis-inducing ligand (TRAIL) on the TRAIL to the development of diabetes mellitus: an update of experimental and clinical evidence. Int J Mol Sci 23(6):3225
Li J, Xu J, Qin X, Yang H, Han J, Jia Y et al (2019) Acute pancreatic beta cell apoptosis by IL-1β is responsible for postburn hyperglycemia: evidence from humans and mice. Biochim Biophys Acta (BBA)-Mol Basis Dis 1865(2):275–284
Liang T, Xu X, Ye D, Chen W, Gao B, Huang Y (2019) Caspase/AIF/apoptosis pathway: a new target of puerarin for diabetes mellitus therapy. Mol Biol Rep 46:4787–4797
Liu ZW, Zhu HT, Chen KL, Dong X, Wei J, Qiu C, Xue JH (2013) Protein kinase RNA-like endoplasmic reticulum kinase (PERK) signaling pathway plays a major role in reactive oxygen species (ROS)-mediated endoplasmic reticulum stress-induced apoptosis in diabetic cardiomyopathy. Cardiovasc Diabetol 12:1–16
Liu J, Lang G, Shi J (2021) Epigenetic regulation of PDX-1 in type 2 diabetes mellitus. Diabet Metab Syndrom Obes:431–442
Maedler K, Donath MY (2009) Beta-cells in type 2 diabetes: a loss of function and mass. Horm Res 69(1):211–221
Marchetti P, Bugliani M, Lupi R, Marselli L, Masini M, Boggi U et al (2007) The endoplasmic reticulum in pancreatic beta cells of type 2 diabetes patients. Diabetologia 50:2486–2494
Marchetti P, Bugliani M, Lupi R, Marselli L, Masini M, Boggi U, Filipponi F, Weir GC, Eizirik DL (2021) The endoplasmic reticulum in pancreatic Beta cells of type 2 diabetes patients. Diabetologia 64(12):2610–2622
Mohd Nor NA, Budin SB, Zainalabidin S, Jalil J, Sapian S, Jubaidi FF, Mohamad Anuar NN (2022) The role of polyphenol in modulating associated genes in diabetes-induced vascular disorders. Int J Mol Sci 23(12):6396
Nakayasu ES, Syed F, Tersey SA, Gritsenko MA, Mitchell HD, Chan CY et al (2020) Comprehensive proteomics analysis of stressed human islets identifies GDF15 as a target for type 1 diabetes intervention. Cell Metab 31(2):363–374
Nunez Lopez YO, Iliuk A, Petrilli AM, Glass C, Casu A, Pratley RE (2022) Proteomics and phosphoproteomics of circulating extracellular vesicles provide new insights into diabetes pathobiology. Int J Mol Sci 23(10):5779
Ou Y, Zheng Z, Niu B, Su J, Su H (2020) Different MAPK signal transduction pathways play different roles in the impairment of glucose-stimulated insulin secretion in response to IL-1β. Mol Med Rep 22(4):2973–2980
Pourghasem M, Shafi H, Babazadeh Z (2015) Histological changes of kidney in diabetic nephropathy. Caspian J Intern Med 6(3):120
Rangasamy S, Monickaraj F, Legendre C, Cabrera AP, Llaci L, Bilagody C et al (2020) Transcriptomics analysis of pericytes from retinas of diabetic animals reveals novel genes and molecular pathways relevant to blood-retinal barrier alterations in diabetic retinopathy. Exp Eye Res 195:108043
Rochette L, Zeller M, Cottin Y, Vergely C (2014) Diabetes, oxidative stress and therapeutic strategies. Biochim Biophys Acta (BBA)-Gen Sub 1840(9):2709–2729
Rorsman P, Ashcroft FM (2018) Pancreatic β-cell electrical activity and insulin secretion: of mice and men. Physiol Rev 98(1):117–214
Rübsam A, Parikh S, Fort PE (2018) Role of inflammation in diabetic retinopathy. Int J Mol Sci 19(4):942
Salpeter SJ, Klein AM, Huangfu D, Grimsby J, Dor Y (2010) Glucose and aging control the quiescence period that follows pancreatic beta cell replication. Development 137(19):3205–3213
Santos-Silva JC, Ribeiro RA, Vettorazzi JF, Irles E, Rickli S, Borck PC et al (2015) Taurine supplementation ameliorates glucose homeostasis, prevents insulin and glucagon hypersecretion, and controls β, α, and δ-cell masses in genetic obese mice. Amino Acids 47:1533–1548

Shantikumar S, Caporali A, Emanueli C (2012) Role of microRNAs in diabetes and its cardiovascular complications. Cardiovasc Res 93(4):583–593

Sharma PR, Mackey AJ, Dejene EA, Ramadan JW, Langefeld CD, Palmer ND et al (2015) An islet-targeted genome-wide association scan identifies novel genes implicated in cytokine-mediated islet stress in type 2 diabetes. Endocrinology 156(9):3147–3156

Shi GJ, Shi GR, Zhou JY, Zhang WJ, Gao CY, Jiang YP et al (2018) Involvement of growth factors in diabetes mellitus and its complications: a general review. Biomed Pharmacother 101:510–527

Sifuentes-Franco S, Padilla-Tejeda DE, Carrillo-Ibarra S, Miranda-Díaz AG (2018) Oxidative stress, apoptosis, and mitochondrial function in diabetic nephropathy. Int J Endocrinol 2018

Smith AG, Singleton JR (2012) Diabetic neuropathy. Continuum Lifelong Learn Neurol 18(1):60–84

Srinivasan S, Stevens M, Wiley JW (2000) Diabetic peripheral neuropathy: evidence for apoptosis and associated mitochondrial dysfunction. Diabetes 49(11):1932–1938

Tang J, Kern TS (2011) Inflammation in diabetic retinopathy. Prog Retin Eye Res 30(5):343–358

Tonyan ZN, Nasykhova YA, Danilova MM, Barbitoff YA, Changalidi AI, Mikhailova AA, Glotov AS (2022) Overview of transcriptomic research on type 2 diabetes: challenges and perspectives. Gene 13(7):1176

Ünver Saraydin S, Özdenoglu Kutlu B, Saraydın D (2021) Effects of diabetes on apoptosis and mitosis in rat hippocampus. Biotech Histochem 96(6):460–467

Valdés A, Lucio-Cazaña FJ, Castro-Puyana M, García-Pastor C, Fiehn O, Marina ML (2021) Comprehensive metabolomic study of the response of HK-2 cells to hyperglycemic hypoxic diabetic-like milieu. Sci Rep 11(1):5058

Velikova TV, Kabakchieva PP, Assyov YS, Georgiev TA (2021) Targeting inflammatory cytokines to improve type 2 diabetes control. Biomed Res Int 2021

Vincent AM, Callaghan BC, Smith AL, Feldman EL (2011) Diabetic neuropathy: cellular mechanisms as therapeutic targets. Nat Rev Neurol 7(10):573–583

Wolf G, Ziyadeh FN (1999) Molecular mechanisms of diabetic renal hypertrophy. Kidney Int 56(2):393–405

Xu H, Barnes GT, Yang Q, Tan G, Yang D, Chou CJ et al (2003) Chronic inflammation in fat plays a crucial role in the development of obesity-related insulin resistance. J Clin Invest 112(12):1821–1830

Xu L, Lin X, Li X, Hu Z, Hou Q, Wang Y, Wang Z (2022) Integration of transcriptomics and metabolomics provides metabolic and functional insights into reduced insulin secretion in MIN6 β-cells exposed to deficient and excessive arginine. FASEB J 36(3):e22206

Yagihashi S, Mizukami H, Sugimoto K (2011) Mechanism of diabetic neuropathy: where are we now and where to go? J Diabet Investig 2(1):18–32

You S, Zheng J, Chen Y, Huang H (2022) Research progress on the mechanism of beta-cell apoptosis in type 2 diabetes mellitus. Front Endocrinol 13:976465

Zakharov PN, Hu H, Wan X, Unanue ER (2020) Single-cell RNA sequencing of murine islets shows high cellular complexity at all stages of autoimmune diabetes. J Exp Med 217(6)

Zhou Z, Wang Q, Michalak M (2021) Inositol requiring enzyme (IRE), a multiplayer in sensing endoplasmic reticulum stress. Anim Cells Syst 25(6):347–357

# Role of Apoptosis in Cardiovascular Diseases: Therapeutic Targets and Strategies

13

Abhik Kar, Arkadeep Mitra, and Sagartirtha Sarkar

**Abstract**

Among noncommunicable diseases, cardiovascular diseases (CVDs) are one of the major contributors to the global death burden. In 2021, it accounted for almost 20.5 million deaths worldwide. Cell death of the cardiomyocytes has been recognized as one of the fundamental incidents that precede heart failure. Ascribed to the resolute job of interminable pumping owing to its constituent cardiomyocyte cells, the heart becomes vulnerable to any injury as these cardiomyocytes are terminally differentiated and are very rarely compensable. Cardiomyocytes primarily die through apoptosis, necrosis, necroptosis, autophagy, pyroptosis, or ferroptosis, the reasons for which can be attributed to the amassed toxic levels of ROS and metabolites that onset a vast array of pathological signaling pathways inside the cells. Adverse ventricular remodeling following cardiac injury during pathological cardiac hypertrophy and myocardial infarction is associated with concomitant cell loss which persuaded researchers to focus on the intricacies of varied forms of cell death in the heart. As CVDs pose a greater threat in the post-COVID era, a precise molecular detailing of the intricacies of the cell death mechanisms would enlighten the cellular milieu of the injured heart that in turn would pave the path for future therapeutic interventions to ameliorate cardiac pathophysiology preceding heart failure.

A. Kar · S. Sarkar (✉)
Department of Zoology, University of Calcutta, Kolkata, India

A. Mitra
Department of Zoology, University of Calcutta, Kolkata, India

Department of Zoology, City College (Affiliated to University of Calcutta), Kolkata, India





K. Jana (ed.), *Apoptosis and Human Health: Understanding Mechanistic and Therapeutic Potential*, https://doi.org/10.1007/978-981-97-7905-5_13

**Keywords**

Cardiovascular diseases · Apoptosis · Autophagy · Pyroptosis · Ferroptosis · Cell death inhibition · Cardiomyocyte renewal

## 13.1 Introduction

The heart is the first functional organ to develop (Donovan and Cascella 2022), underlining the indispensable need to transport gasses and other materials to and from all of the organs during the development of a prenatal body. The adult mammalian heart tissue is primarily composed of cardiomyocytes, fibroblasts, endothelial cells, and pericytes. However, cardiomyocytes alone claim the majority of the tissue volume, about 70–85% (Zhou and Pu 2016). Amidst all the constituent cell types, only these cardiomyocytes are endowed with contraction and relaxation cycles that empower the heart with its unique pumping functions (Kartha 2021). Interestingly, cardiomyocytes present in old age are the same cells that are present during early adulthood. This is because robust proliferation of the embryonic and neonatal cardiomyocytes is needed for the formation of the four heart chambers to account for the increasing systemic circulation. However, this proliferative capacity diminishes with adulthood where the same cells shift towards hypertrophic expansion to accommodate any increase in workload (Zhao et al. 2020). This added workload on the adult heart can either be a sudden incident or chronic accumulation of stressors. Over the years, our laboratory has dissected out a variety of such stressors aggravating these two etiologically different forms of cardiac pathophysiology (Mitra et al. 2013). Both of the pathological scenarios become detrimental for any healthy heart leading to cell death which ultimately results in heart failure. By virtue of a cardiac injury, the heart creates a fibrotic scar which serves as a patch in place for the lost cardiomyocytes. However, this fibrotic scar tissue does not retain the contractility of the former cardiomyocytes. Eventually, the heart tries to compensate for the regressed functionality by other means that in turn leads to heart failure (Richardson et al. 2015). Due to the lack of proliferative capacity or a resident stem cell population, death of cardiomyocytes becomes unarguably the crux of all cardiac pathophysiologies. Heart failure has already been announced as an emerging epidemic two and a half decades ago, contributing significantly to the global mortality rate every year. The World Health Organization estimated about 17.9 million deaths from cardiovascular diseases (CVDs) in 2019 representing 32% of the global mortality (World Health Organization 2023). Moreover, the World Heart Federation had found that more than half a billion people around the world continued to be affected by CVDs, which accounted for 20.5 million deaths in 2021. Therefore, the need of the hour is a clear understanding of the intricacies underlining various cell death pathways in a diseased heart for the identification of any novel therapeutic intervention. The present chapter will discuss a diverse array of cell death pathways associated with the diseased heart.

## 13.2 Why Is Cell Death Important?

Cell death plays an integral part in sculpting all things living. In other words, cell death restrains the surfeit of all corporeal processes, viz., ontogeny, metamorphosis, and tissue homeostasis (Hinchliffe 1981; Hinsull and Bellamy 1981; Suzanne and Steller 2013). In the failing heart, the apoptotic cell population is lower among the dying cardiomyocytes. However, the apoptotic cardiomyocyte is 10 to 100-fold higher than observed in the controlled heart. Therefore, a low but continuous loss of cardiomyocytes over time condemns an affected heart to failure (Wencker et al. 2003). Discussion over cell loss in the heart cannot ignore the event of cellular necrosis, as it is reported to be far more pronounced in the failing heart (Bennett 2002). Although necrosis has traditionally been thought to be unregulated, recent evidence suggests that a considerable proportion of necrotic loss is vigorously implemented by a cell in a controlled manner. This form of necrosis is sometimes referred to as regulated necrosis or necroptosis (Guo et al. 2022). Contrary to the killing nature of the aforementioned cell death forms, the heart employs autophagy to maintain cellular homeostasis as long as possible under any circumstances. However, under overwhelming stress, these autophagic measures advocate for cell death (Rothermel and Hill 2007). All the varied forms of cell death discussed so far, during their affair, release one or more pro-inflammatory factors which in turn magnify the cell death load in terms of pyroptosis that further promotes clearance of the dead cells. This clearance of the dead cells becomes of utmost importance for the time being as the heart generates a fibrotic patch in the cleared space as a measure to compensate for the lost cells (Naryzhnaya et al. 2022). However, this mechanism of patch-compensation instead of apt cellular regeneration ensures the functional decompensation of the heart, ultimately leading to heart failure. A relatively newer form of cell death has been found by the scientific community, namely ferroptosis, which is an improper iron metabolism and accumulation of lipid peroxides leading the cell towards a programmed cell death with a distinctly different phenotype than conventional apoptosis (Wang et al. 2022b; Naryzhnaya et al. 2022). All these cell death pathways are mosaic with each other and join forces together to aggravate the process of CVDs. Moreover, they are linked by multiple biochemical and functional connections, i.e., crosstalk across the pivotal nodal proteins which exert a critical function in aggravating the affair of CVDs in its totality. A notion can be extrapolated from such connectedness among the varied forms of cell death that there might be a single unified death mechanism and all of these forms are just but a part of it.

## 13.3 How Do Cells Die During CVDs?

Our human civilization from the post-industrialization period had shifted from a physically arduous to a sedentary job with long working hours accompanied by minimal to no leisure time for physical or recreational activities along with improper diet, which may explain the concurrence of CVDs. To be precise, physical dormancy, consumption of alcohol, unbalanced diet, intake of too much-saturated fats,

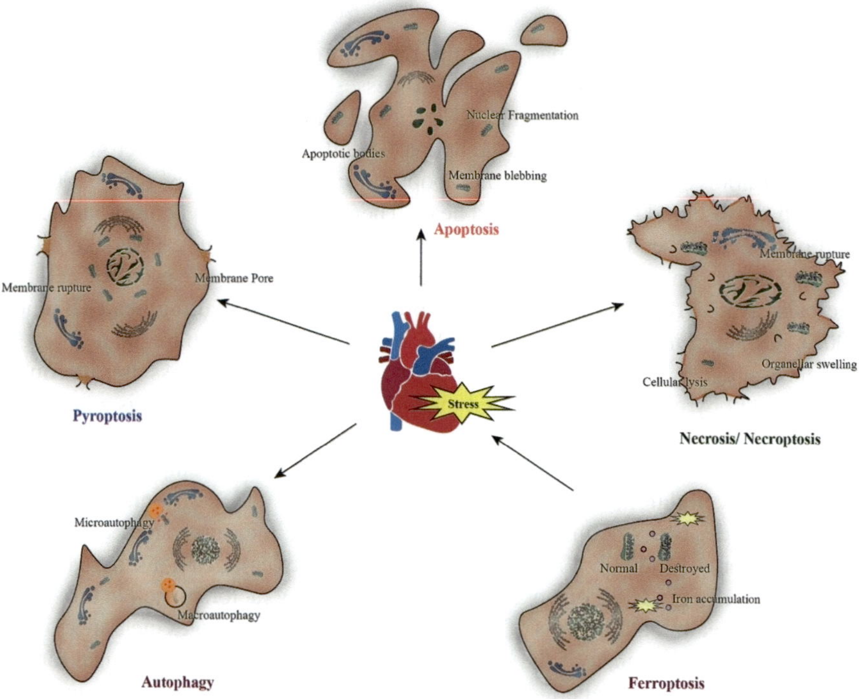


**Fig. 13.1** Varied forms of cell death during CVDs

added sugars, and smoking have been found to be correlated with the onset of atherosclerosis and several distinct metabolic disorders such as dysmetabolic syndrome, diabetes mellitus, and hypertension that are particularly insidious in patients suffering from CVDs (Benjamin et al. 2018; US Preventive Services Task Force et al. 2018). The INTERHEART study has found several formidable yet manageable risk factors to be associated with MI like dyslipidemia, hypertension, diabetes, obesity, psychosocial factors, alcohol consumption, and physical inactivity (Yusuf et al. 2004). Depending upon the severity of the incoming stress on the myocardium, one or both modes of cell death mechanism become prevalent. Cell death in the heart is an intricate molecular interplay encompassing a substantial number of players. An array of cell death mechanisms employed by the diseased heart is discussed below (Fig. 13.1 and Table 13.1).

### 13.3.1 Apoptosis

Apoptosis or programmed cell death has been vastly described among all forms of cell death during cardiac pathophysiology. Furthermore, this is unequivocally the paramount prerequisite during organ development and cellular regulation. It plays a

**Table 13.1** Different forms of cell death contributing to the amassed cell loss in the heart during CVDs

| Cell death isoforms | Cellular characteristics | Molecular phenotype | Key effectors | Classification |
|---|---|---|---|---|
| Apoptosis | Blebbing of the plasma membrane, chromatin condensation, nuclear fragmentation, formation of apoptotic bodies | Phosphatidylserine exposure, DNA fragmentation, caspase activation | BCL2 and caspase family, p53, Cyt-c | Programmed cell death |
| Necrosis | Swelling, ruptured plasma membrane, cellular content release | ATP depletion, mPTP activation | Not specified | Accidental cell death |
| Necroptosis | Swelling, ruptured plasma membrane, chromatin condensation, cellular content release | Reduced ATP; induced levels of RIP1, RIP3, MLKL phosphorylation, and ROS; DAMPs release | RIP1, RIP3, MLKL | Programmed cell death |
| Autophagy | Accumulation of double-membraned autolysosomes | LC3-I to LC3-II conversion; p62 degradation; induced autophagic flux and lysosomal activity | Beclin1, ATG5/7, AMPK, mTOR | Programmed cell death |
| Pyroptosis | Less swelling, pyrosome formation prior plasma membrane rupture, cell content release, unaffected mitochondrial integrity | Caspase-1 and GSDMDN activation; GSDMDN induced pore formation and IL-1β release | Caspase1/11; GASDMD | Programmed cell death |
| Ferroptosis | Smaller mitochondria, shrinkage or disappearance of cristae, mitochondrial outer membrane rupture, normal nucleus, no chromatin condensation. | Accumulation of iron and ROS; iron-dependent lipid peroxidation | GPX4, NCOA4, TFRC, ACSL4, NRF2 | Programmed cell death |

crucial role in a diverse range of physiological and pathological ordeals. Kerr and colleagues put forward the concept of "apoptosis" to the scientific community in 1972 (Kerr et al. 1972). Since then, the understanding of the vast knowledge about this particular type of cell demise has been expanded subsequently. The initial denotation of apoptosis relied on morphological features like chromatin condensation, nuclear fragmentation, cell shrinkage, and shedding of apoptotic bodies that are ultimately evacuated through phagocytosis. Nevertheless, the considerable advancement in deciphering the molecular mechanisms underlying cell death in the past decade enabled the Nomenclature Committee on Cell Death (NCCD) to recommend apoptosis as a caspase-dependent, genetically governed form of cell death (Galluzzi et al. 2015). A modest increase in apoptotic cardiomyocytes has been recorded during heart failure indicating its important role in the pathogenesis of heart failure.

In general, the apoptotic process can broadly be classified into two modes, i.e., either extracellular (extrinsic) or intracellular (intrinsic) (Fig. 13.2). Extrinsic apoptotic signals move along with the death receptor-mediated pathway which is

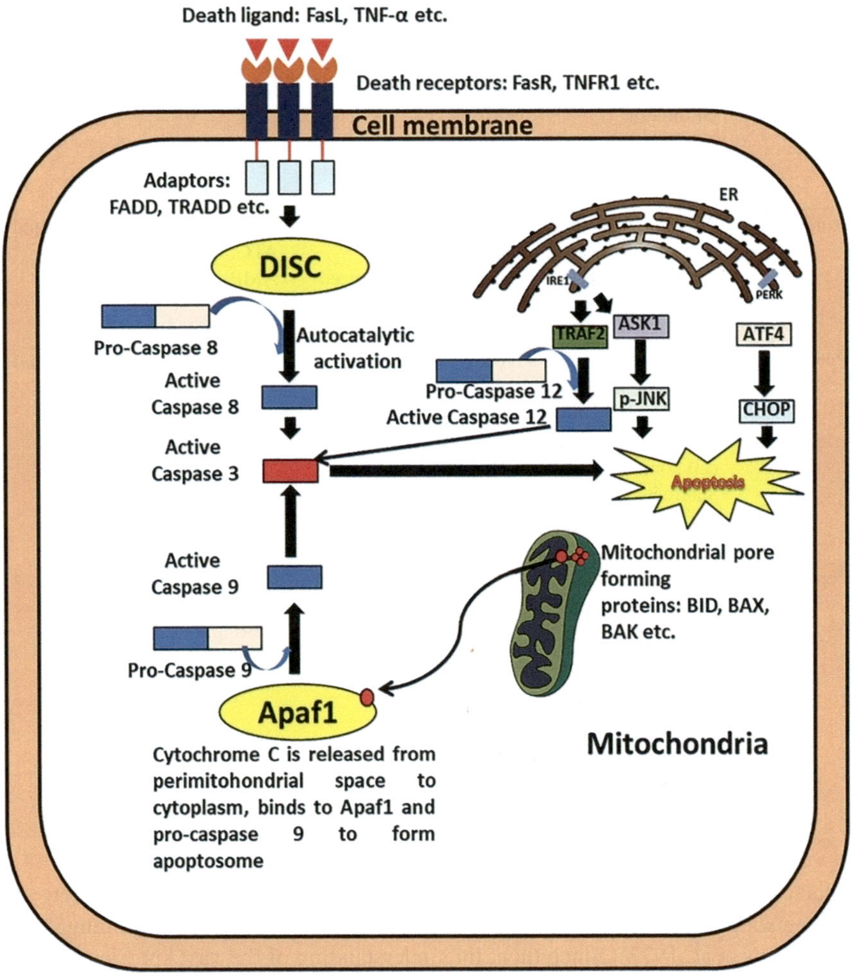


**Fig. 13.2** Schematic diagram of apoptosis in the heart

actuated by the association of a death ligand such as Fas ligand (FasL) or tumor necrosis factor-alpha (TNF-α) to a membrane-bound death receptor like FasR or TNFR. This association leads to the recruitment of the Fas-associated death domain (FADD) and results in the formation of a death-inducing signaling complex (DISC) which concomitantly activates caspase-8 followed by activation of the downstream effector caspases 3 (Galluzzi et al. 2015). On the other hand, the intrinsic mode of apoptosis depends on mitochondrial outer membrane permeabilization (MOMP) or permeability transition pore (PT-pore), which results in cytosolic release of cytochrome c (CytC), ATP, and apoptotic peptidase-activating factor 1 (APAF1) which in turn activates caspase-9 and finally caspase 3 (Li et al. 1997; Ow et al. 2008). Formation of PT-pore demands either of two Bcl-2 family members, viz., B-cell CLL/lymphoma 2 (BCL2)-associated X protein (BAX) and BCL2-antagonist/killer 1 (BAK1), whose pore-forming activity is impeded by other components of the family, including BCL2 itself, BCL2-like 1 (BCL2L1, also known as BCL-XL), and myeloid cell leukemia 1 (MCL1). The mitochondrial release of apoptotic factors is governed by the balance among the proteins in the Bcl-2 family (Adams and Cory 1998; Kim et al. 2009). A key proapoptotic player, Bcl-2-interacting protein, i.e., BH3-interacting domain death agonist or Bid, arbitrates the molecular connection between extrinsic and intrinsic modes of apoptosis. Bid under normal circumstances is located in the cytosol, but under stress conditions activated caspase-8 cleaves Bid to generate "truncated Bid" (tBid) that translocates to the mitochondria and influences cytochrome c release (Luo et al. 1998). Apart from the classical external and internal modes of cell death, activated CRYAB protein has been found to bind to the mitochondrial VDAC1 channel, following disruption of cytochrome c release into the cytosol during myocardial infarction (MI) (Mitra et al. 2013). Adult cardiomyocytes being postmitotic cells show lower potential for proliferation and only respond with proliferative prowess under stressful conditions. However, inhibition of this stress concomitantly reduces the proliferative prowess of the cardiomyocytes (Chatterjee et al. 2011). Literature suggests that chronic stress on the heart leading to pathological cardiac hypertrophy makes the cardiomyocytes more prone to apoptosis and consequently escalates the odds of heart failure. Although necrosis was thought to be the sole cause of cell death during MI, recent evidence has shown that the acute stress response during MI leads to the onset of Endoplasmic reticulum (ER) stress-induced apoptosis (Mitra et al. 2014). Furthermore, apoptosis also plays an important role in the cell loss following an acute myocardial infarction and subsequently participates in ventricular remodeling that again leads to the development of symptomatic heart failure (Teringova and Tousek 2017). Nonetheless, as cardiomyocytes have no or at most a very limited proliferative capacity, inhibition of apoptosis becomes a lucrative target as a therapeutic intervention for CVDs for obvious reasons (Haunstetter and Izumo 2000).

### 13.3.2 Necrosis or Necroptosis

Although necrosis has been thought to be a chaotic and unregulated process of cell death earlier, accumulated evidence over the past decades has suggested it to be a highly regulated process and is now called regulated necrosis or necroptosis (Crompton et al. 1988; Kung et al. 2011). From a morphological perspective in diseased myocardium, necrotic cells are typically marked by the swelling of several organelles, including the endoplasmic reticulum and mitochondria. This is accompanied by the rupture of the plasma membrane and subsequent cellular lysis (Sarhan et al. 2018). Biochemical and molecular detailing of the process has shown that necrosis is highly ordered and accomplished through proper mechanisms as opposed to the unorganized dissolution of cells (Syntichaki and Tavernarakis 2003). In general, programmed necrosis has two major forms, (1) mitochondrial-dependent and (2) death receptor-dependent. A plethora of factors like oxidative stress, elevated levels of phosphate, and adenine nucleotide (ATP and ADP) can and do influence $Ca^{2+}$ sensitivity to pore opening (Crompton et al. 1988) (Fig. 13.3).

Hypoxic cardiomyocytes already have a decreased level of available oxygen, leading to decreased ATP production which consequently leads to the functional collapse of the energy-driven sodium pump in the cell membrane that triggers an excess intrusion of calcium and water, resulting in the cellular swelling and disengagement of ribosomes from the endoplasmic reticulum. Moreover, this surplus of cytosolic calcium and ROS leads to mitochondrial desecration (Konstantinidis et al. 2012). In contrast to the outer mitochondrial membrane (OMM) permeabilization during apoptosis, necrosis shows a distinct difference in the opening of mPTP situated in the mitochondrial inner membrane (IMM) (Goldenthal 2016). Cells tend to undergo necrosis when put under extensive stress like depleted oxygen levels as observed during acute stress of MI (Tavernarakis 2007). A shift towards anaerobic glycolysis for ATP production during MI leads to an $H^+$ surplus that causes acidosis. The ion pumps on the plasma membrane initiate the removal of excess $H^+$ by exchanging $Na^+$, which in turn results in increased levels of intracellular $Na^+$. However, the $Na^+/Ca^{2+}$ exchanger working in the reverse mode is less competent in emigrating intracellular $Ca^{2+}$, resulting in elevated levels of cytoplasmic $Ca^{2+}$ (Whelan et al. 2010). Rising levels of $Ca^{2+}$ within mitochondria result in the induction of a $Ca^{2+}$-dependent dehydrogenase, leading to the reduction of NADH and electron spurt down the electron transport chain and finally elevated levels of ROS, accompanied by a reduction in ATP production. Furthermore, $Ca^{2+}$ surplus in the mitochondria leads to the cyclophilin-D-regulated opening of the mPTPs which causes swelling of the mitochondria and ultimately calls forth cellular necrosis or necroptosis in the infarct heart (Baines et al. 2005). Although regulated, necroptosis on the other hand is primarily a caspase-independent process governed by death receptors and implemented via the induction of the necrosome complex, consisting of RIPK1, RIPK3, and MLKL. The process of necroptosis initiates through the attachment of cell death ligands, such as TNFα, lymphotoxin-α (LTα), Fas ligand (FasL), CD40 ligand (CD40L), and TNF-related apoptosis-inducing ligand (TRAIL), to the complimentary death receptors (Linkermann and Green 2014).

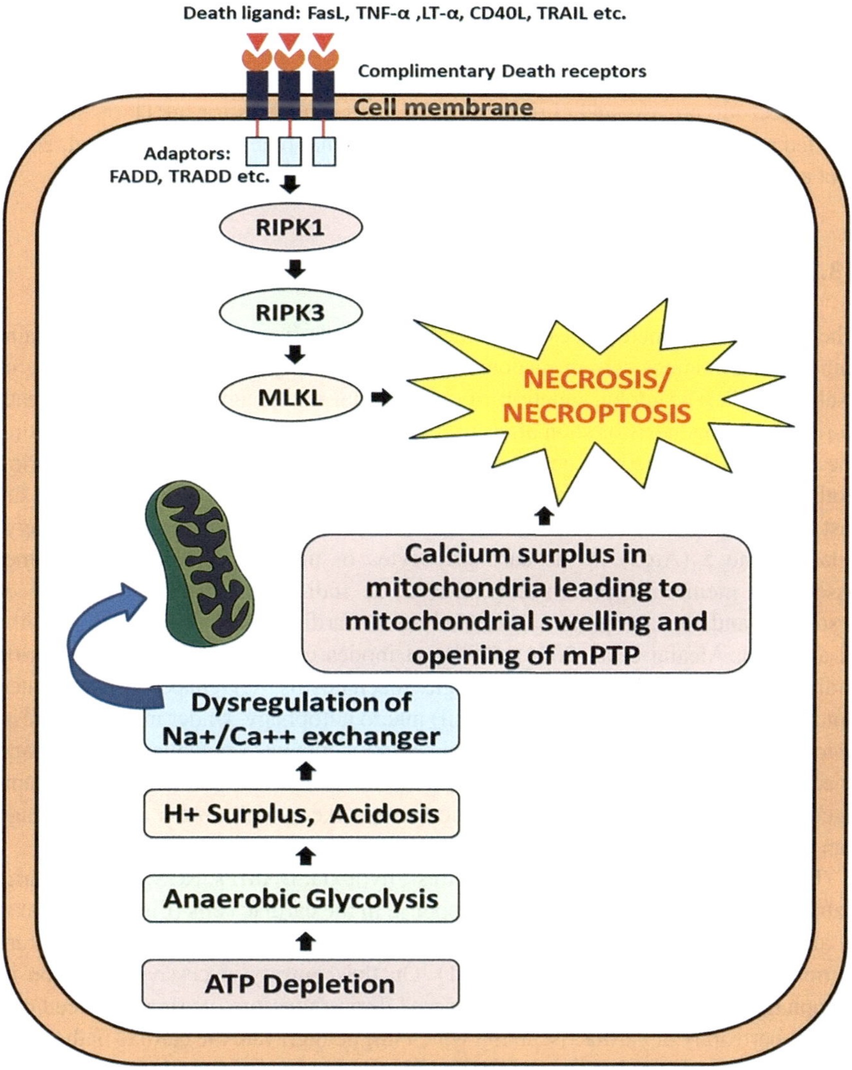


**Fig. 13.3** Schematic diagram of necrosis/necroptosis in the heart

This activates the homologous receptor-interacting protein kinase 1 (RIPK1) and concomitantly RIPK3 (He et al. 2009), which in turn activates a pseudokinase MLKL that, upon oligomerization, permeabilizes the cell membrane causing cellular swelling and necroptotic cell death (He et al. 2009; Wang et al. 2014). Furthermore, the necroptotic release of intracellular content including the damage-associated molecular patterns (DAMPs) plays a pivotal role in initiating the inflammatory response and immunosurveillance in postischemic heart failure (Lichý et al. 2019). Apart from acute MI, the reperfusion procedure also induces necrotic cell death (Chiong et al. 2011). Furthermore, the chronic stress on the heart leading to $Ca^{2+}$ overload inside the cardiomyocytes also contributes to mPTP opening and necrosis (Guerra et al. 1999; Nakagawa et al. 2005). The opening of mPTP is central to the necrotic process and clinical application of inhibitors for mPTP opening has yielded reduced necrosis with promising translational potential (Baines et al. 2005; Piot et al. 2008).

### 13.3.3 Autophagy

The only programmed process of intracellular catabolism that is initiated to maintain cellular homeostasis in response to a wide variety of stress is autophagy or "self-eating". The global deletion of an essential autophagy gene leads to death during embryogenesis or soon after birth (Mizushima and Levine 2010), indicating the indispensable need for autophagy. Existence of an autophagic response during both the normal state and cardiac pathophysiology has been recognized since the last decade (Nemchenko et al. 2011; Chai et al. 2022). Eliminating the autophagy-related gene 5 (Atg5) in the cardiomyocytes or modification in the lysosome-associated membrane protein-2 (LAMP-2) is indispensable for the merger of lysosome and the autophagosome, leading to cardiac pathophysiology (Taneike et al. 2010; Alcalai et al. 2021). Different modes of lysosomal cargo separation individualize three types of autophagic mechanisms: (i) chaperone-moderated autophagy, (ii) micro autophagy, and (iii) macro autophagy. Under moderate stress, autophagic measures recycle cytoplasmic components like organelles, proteins, and specifically remove damaged mitochondria (Kim et al. 2007). On the contrary, damaged mitochondria release proapoptotic factors like cytochrome c and autophagy can prevent such activation of apoptosis (Hamacher-Brady et al. 2007).

The vast array of stimuli like shear stress, hypoxia, DAMPs, ROS, mitochondrial damage, etc. prompts an autophagic response in the cardiac cells (Fig. 13.4) (Nakai et al. 2007; Kroemer et al. 2010) and reduction of autophagy is a frequent feature during heart failure (Hahn et al. 2021). On the contrary, excessive induction of autophagic response has also been observed during pressure overload-induced cardiac hypertrophy or cardiac ischemia which might aggravate the cardiac pathophysiology (Zhu et al. 2007; Nah et al. 2022). Thus, upon receiving pathological stimuli the heart initiates several measures to maintain cellular homeostasis. However, depending upon the cellular microenvironment and time of activation, these measures either become beneficial or detrimental. During pathological cardiac hypertrophy, chronic stress on the myocardium turns on several autophagy inducers such as Atg5, Atg16, LC3-II, and beclin-1 through the PKC and ERK1/2 pathways. Furthermore, pressure overload cardiac hypertrophy also induces Toll-like receptor 9 (TLR9) which, apart from inflammatory response, contributes towards autophagic induction in the cardiomyocytes (Oka et al. 2012). Moreover, programmed cell death 5 (PDCD5) has been found to play a crucial role during pathological cardiac hypertrophy and also promotes autophagy in the cell (An et al. 2012). A controlled level of autophagy is essential for maintaining cellular homeostasis, whereas

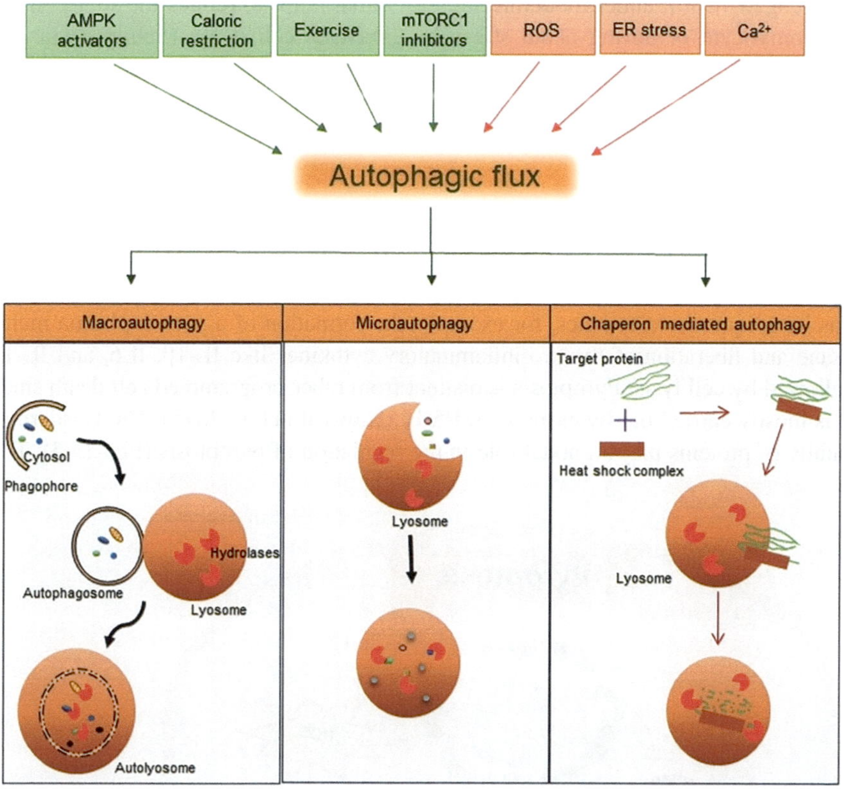


**Fig. 13.4** Schematic diagram of autophagy in the diseased myocardium

frustrated autophagic activity leads to pathogenesis towards heart failure (Rifki and Hill 2012). On the other hand, the acute stress on the heart in terms of an ischemic insult also increases cellular autophagic responses. Studies on autophagy inhibition during cardiac ischemia resulted in an increased infarct patch on the heart (Kanamori et al. 2011). Such cardioprotection in the ischemic milieu was bestowed through the induction of the AMPK-mTOR signaling pathway in cardiomyocytes (Li et al. 2016). Furthermore, in the hypoxic cardiomyocytes, the phosphatase and tensin homolog 10 (PTEN)-PI3K-Akt signaling pathway has been identified as a key contributing factor in the initiation of cardiac autophagy (Zhang et al. 2017). On the contrary, depending upon the severity of the ischemic insult, an overwhelming autophagic response was also found to foster cell demise and may exacerbate heart function (Li et al. 2018; Xiao et al. 2018). Apart from these CVDs, the autophagic response has been evident in other cardiac disease forms such as diabetic cardiomyopathy, atherosclerosis, cardiac arrhythmia, cardiac reperfusion injury, etc. Studying autophagy in various cardiac disease forms points towards a single notion that autophagic mechanisms exist to maintain cellular homeostasis in a normal cell and depending upon the severity of the pathological stimuli, autophagy becomes a dual-edged sword; under moderate stress, it either helps to reduce the stress on the cardiomyocyte or during acute stress it aggravates cell death (Schiattarella and Hill 2016).

### 13.3.4 Pyroptosis

The etymology of the word "pyroptosis" comes from the Greek vocable, "pyro", denoting fire or fever, and "ptosis", meaning falling or dropping (Cookson and Brennan 2001). This is a unique type of pro-inflammatory-programmed cell death that has its unique characteristics, for example, the formation of a porous plasma membrane and liberation of the pro-inflammatory cytokines like IL-1β, IL6, and IL-18 followed by cell lysis. Pyroptosis is distinct from other programmed cell death since it is mostly carried out by caspase- 1/4/5/11 (Kayagaki et al. 2015). The Gasdermin family of proteins plays a nodal role in the regulation of pyroptosis (Fig. 13.5). The

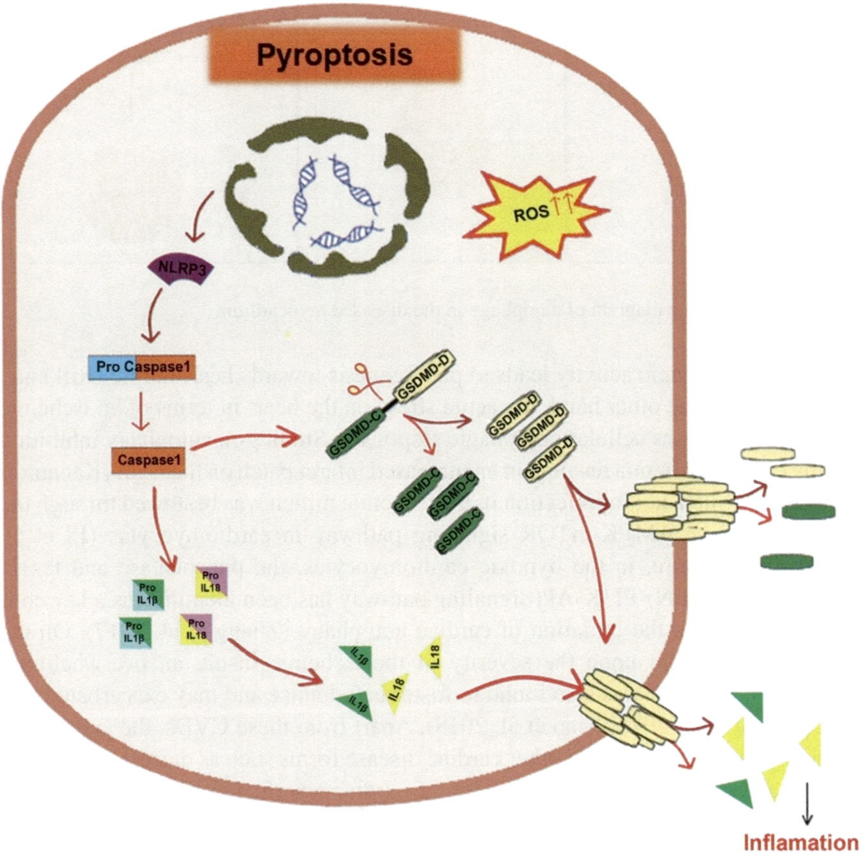


**Fig. 13.5** Schematic diagram of pyroptosis during CVDs

process of pyroptosis follows a bi-pronged approach. It is initiated by increased ROS, leading to the activation of NLRP3. Such activation of NLRP3 results in the structural alteration of the NLRP3 inflammasome which leads to Caspase-1 activation. In one instance, this active Caspase-1 cleaves Gasdermin D to form a polypeptide containing the nitrogen-terminated structurally active form which in turn leads to the perforation of the myocardial membrane. As a result, cellular constituents are distributed, triggering an inflammatory response. Conversely, this activated form of caspase-1 is also responsible for the cleavage of the precursors of IL-1β and IL-18 to form the activated IL-1β and IL-18, which upon extracellular release recruit inflammatory cells and ultimately amplify the inflammatory response (Ji et al. 2021). Pyroptosis has been associated with many forms of cardiovascular disorders as it leads to cardiomyocyte death and promotes cardiac remodeling towards heart failure. Chronic pressure overload-induced pathological cardiac hypertrophy often presents hyperactivated NLRP3-inflammasome and other factors promoting pyroptosis. Blockade to Caspase-1 and IL-1β expression through Caspase-1 inhibitors can diminish the pathological cardiac hypertrophy induced by angiotensin II (Bai et al. 2018). Intriguingly, beyond the conventional mode of pyroptotic cascade activation, protein S-nitrosylation can also induce pyroptosis and thus aggravate the cardiac pathophysiology. This is evidenced by the S-nitrosylation of muscle LIM protein that escalates the interaction between toll-like receptor 3 (TLR3) and receptor-interacting protein kinase 3 (RIP3), which in turn leads to NLRP3 inflammasome formation, ultimately culminating towards development of pathological hypertrophy (Bai et al. 2018; Tang et al. 2020). Exogenous induction of pyroptosis can also occur through silica nanoparticles which augment intracellular ROS leading to the activation of NLRP3/Caspase-1/GSDMD signaling cascade in cardiomyocytes (Wang et al. 2022a). However, the acute insult on the myocardium leading to necrosis releases a large amount of ATP and damage-associated molecular patterns (DAMPs) which are activators of NLRP3 inflammasome that ignites pyroptosis inside the cardiomyocytes (Liu et al. 2021). Reoxygenation also induces GSDMD-mediated pyroptosis and cardiac inflammatory response (Ji et al. 2021). Moreover, myocardial injury has always been associated with the fibrotic response. The proportion of cardiac fibrosis influences the unfavorable result in most of the cardiac diseases. The magnitude of fibrosis is further influenced by pyroptosis in the cardiac fibroblast which has been associated with NLRP3 inflammasome activation, IL-1β and IL-18 release that mainly occurs during myocardial remodeling and repair after myocardial infarction (Kawaguchi et al. 2011). Pyroptosis has involved itself with several stages of pathological cardiac hypertrophy, cardiomyocyte death, cardiac fibrosis, and excessive inflammation leading to myocardial dysfunction and these factors often overlap to promote adverse cardiac remodeling towards heart failure. Targeted inhibitors have been developed for reducing the pyroptosis to ameliorate pathological cardiac remodeling (Chai et al. 2022). Investigations focusing on betterment of cardiac remodeling in heart failure from pyroptosis have a promising likelihood.

### 13.3.5 Ferroptosis

Ferroptosis is relatively a new character that has been introduced to the cataclysm of cell death (Dixon et al. 2012). This is a form of regulated cell death owing to the improper iron metabolism along with excessive accumulation of lipid peroxides and ROS (Fig. 13.6). Growing numbers of evidence are continuously accumulating in support of ferroptosis in development of many forms of cardiovascular disease. For instance, elevated levels of ferroptosis, regulated by distinct signaling and metabolic pathways, have the potential to play a key role in the onset of ischemic heart disease, cardiac damage, cardiomyopathy, and heart failure (Fang et al. 2023). Heart is susceptible to damage by iron overload (Kremastinos and Farmakis 2011). Overall, such iron overload can be categorized into two general types. Preeminent cause of iron overload is genetic disorders that lead to abnormal levels of dietary iron absorption. Conversely, the subservient cause of iron overload manifests as a consequence of frequent blood transfusions, toxicity induced as an effect of pharmacological intervention or superfluous ingestion of iron. Primarily, intracellular iron overload is due to hereditary hemochromatosis, which is the lion's share among

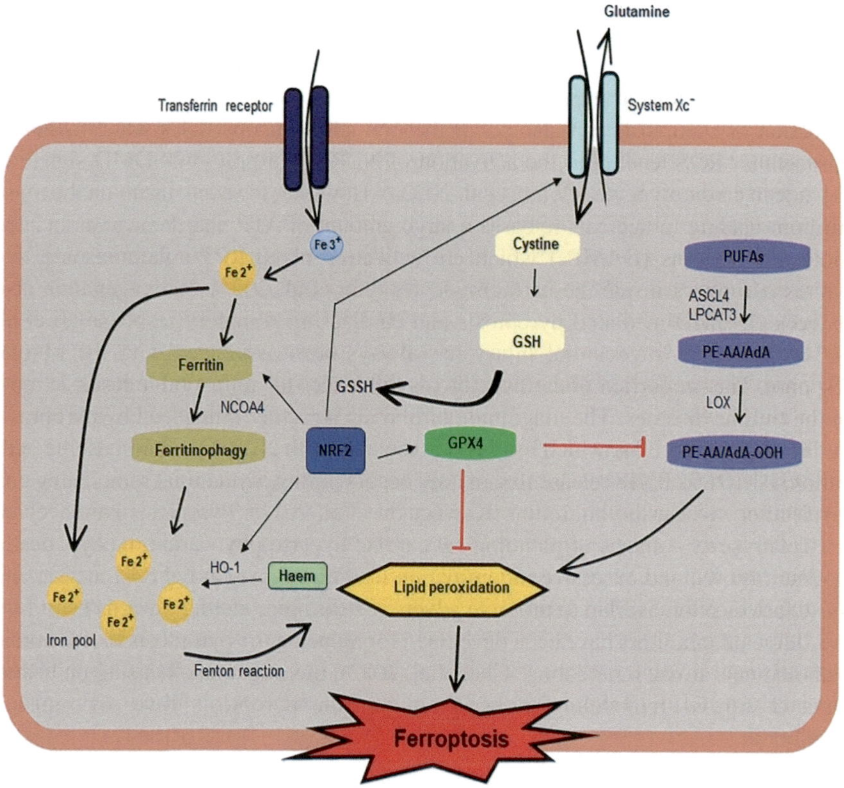


**Fig. 13.6** Schematic diagram of ferroptosis in the diseased heart

inherited disorders due to the varied extent of mutations influencing either the fabrication or activity of the hepcidin, the iron-regulator peptide. Hepcidin has the potential to induce iron accumulation of varying magnitude in multiple organ systems, including the heart (Gulati et al. 2014). Patients carrying the C282Y mutation in HFE (encoding hereditary hemochromatosis protein) tend to show increased risk of cardiac disease-related mortality compared to the noncarriers (Tuomainen et al. 1999).

Another genetic anomaly is Friedreich ataxia, which is characterized as an autosomal neurodegenerative disorder, capable of affecting both the neuronal and non-neuronal tissues, including cardiac tissue. This is caused by homozygous increase of the GAA triplet repeat located in the first intron of the mitochondrial frataxin gene and the deficiency of which leads to an elevated level of ROS generation, iron accumulation in the mitochondria, and finally increase in lipid peroxidation (Tsou et al. 2011). Apart from the genetic predisposition, the heart faces iron accumulation from blood transfusion, which is a common complication among patients with thalassemia, sickle cell anemia, myelodysplastic syndromes, pure red cell aplasia, and leukemia (Sheth 2019). Moreover, population-based studies have correlated high intake of dietary iron and an elevated risk of cardiac diseases and cardiovascular mortality in the general populace (Han et al. 2020). Apart from improper iron metabolism, glutathione exhaustion, cysteine insufficiency, and inhibition of a key enzyme, GPX4 (phospholipid hydroperoxide glutathione peroxidase 4) plays an important part in aggravating ferroptosis in the heart. A key cellular antioxidant is glutathione, composed of the three key amino acids glycine, cysteine, and glutamic acid. Among them, cysteine is the rate-limiting precursor in glutathione production (Stockwell et al. 2017). This glutathione acts as a cofactor for GPX4, and together, they scavenge the harmful by-products of iron-dependent lipid peroxidation (Yang et al. 2014). The key molecular phenotype of ferroptosis is amassing lipid peroxides. These degrade into reactive toxic aldehydes or hydroxy fatty acids, such as malondialdehyde (MDA). This in turn incites cellular toxicity and triggers a cascade leading to cell death ultimately. The most susceptible lipids to peroxidation are the PUFAs (polyunsaturated fatty acids). Increased ROS reacts with the PUFAs in lipid membranes and induces lipid peroxidation (Yang et al. 2016). Recent evidence has shown that most of the etiologically different forms of CVDs show induction of ferroptosis to some extent. During the early and middle stages of MI, the key anti-ferroptotic enzyme GPX4 has been found to contribute to the ferroptosis of cardiomyocytes, leading to cell death under metabolic stress, such as cysteine deprivation (Park et al. 2019). In addition, ferroptosis often initiates inflammation, thus compounding the cardiac dysfunction and poor myocardial remodeling following MI (Huang and Frangogiannis 2018). The percutaneous coronary intervention is the general clinical choice to restore blood flow for MI patients. Unfortunately, reperfusion further adds damage to the patient's hearts. Such reperfusion further contributes to the ROS, lipid peroxides, and accumulated iron pool, indicating that reperfusion injury may cause ferroptosis (Son et al. 2020; Shan et al. 2021). On the other hand, chronic stress on the heart during pathological cardiac hypertrophy manifests by accumulation of oxidative load regulated through the JNK/p53 signaling cascade leading to ferroptosis. However, the pyroptosis-inducing factor MLK3 also contributes to ferroptosis which further aggravates myocardial fibrosis, ultimately contributing to the progression towards heart failure (Wang et al. 2020). Nonetheless, contrary to previously identified forms of cell demise, viz., apoptosis, necrosis, pyroptosis, and autophagy, ferroptosis is a form of regulated cell death that is dependent on iron. This process is characterized by two distinct biochemical features: the amassment of iron within the cell and lipid peroxidation. Despite such findings, the precise molecular mechanism through which reactive oxygen species (ROS) induces ferroptosis remains elusive, albeit parameters such as ROS levels, intracellular iron concentration, cell viability, and presence of certain marker proteins have been utilized to assess ferroptosis experimentally. However, there is a lack of precise methods for detecting the progression of ferroptosis in vivo. The development of a specific probe or a kit related to ferroptosis could potentially prevent and even help in designing treatment of CVDs (Wang et al. 2022b).

## 13.4 Does Cellular Cross-Talk Influence Cell Death in CVDs?

The cardiac muscle is a uniquely specialized tissue comprising a variety of cell types and a well-balanced intercellular communication among all the cell types is indispensable for proper maintenance of cardiac homeostasis and function. On the other hand, under pathological conditions, these interactions can also cause adverse cardiac remodeling, speed up the disease process, and eventually lead to heart failure (Fig. 13.7).

Cellular crosstalk between cardiomyocytes and cardiac fibroblasts, the most studied intercellular communication, imparts a pivotal role in deciding cardiomyocyte serendipity and function, especially under chronic or acute stress milieu. The chief mode of interaction between cardiomyocytes and fibroblasts (and vice versa) involves paracrine signaling (Cartledge et al. 2015; Datta et al. 2017), electrotonic coupling via gap junctions (Quinn et al. 2016), mechanical stimuli (Engler et al. 2008), and exosome-mediated communication (Engler et al. 2008; Nguyen et al. 2021). In contrast to the proliferative phenotype, embryonic cardiomyocytes when cocultured with adult cardiac fibroblasts showed a hypertrophic growth with enlarged cell size and sarcomeric organization. Therefore, the paracrine effectors secreted from cardiac fibroblasts have a distinct influence over the cardiomyocyte's fate during development in a manner that differs in the adult (Kakkar and Lee 2010). Chronic stress on the heart inculcates hypertrophied cardiomyocytes which adversely communicate with neighboring fibroblasts to culminate in excess collagen synthesis and ultimately lead to severely compromised cardiac function (Cartledge et al. 2015; Datta et al. 2017). On the other hand, stress-mediated production of certain cytokines like TGFβ, IL-33, IL-6 TNFα, IGF1, and ET-1 by the fibroblasts directly affects cardiomyocyte mortality (Nicin et al. 2022; Piek et al. 2016).

Although the crosstalk between cardiomyocytes and endothelial cells is indispensable for proper cardiac development, it plays a key role throughout the onset,

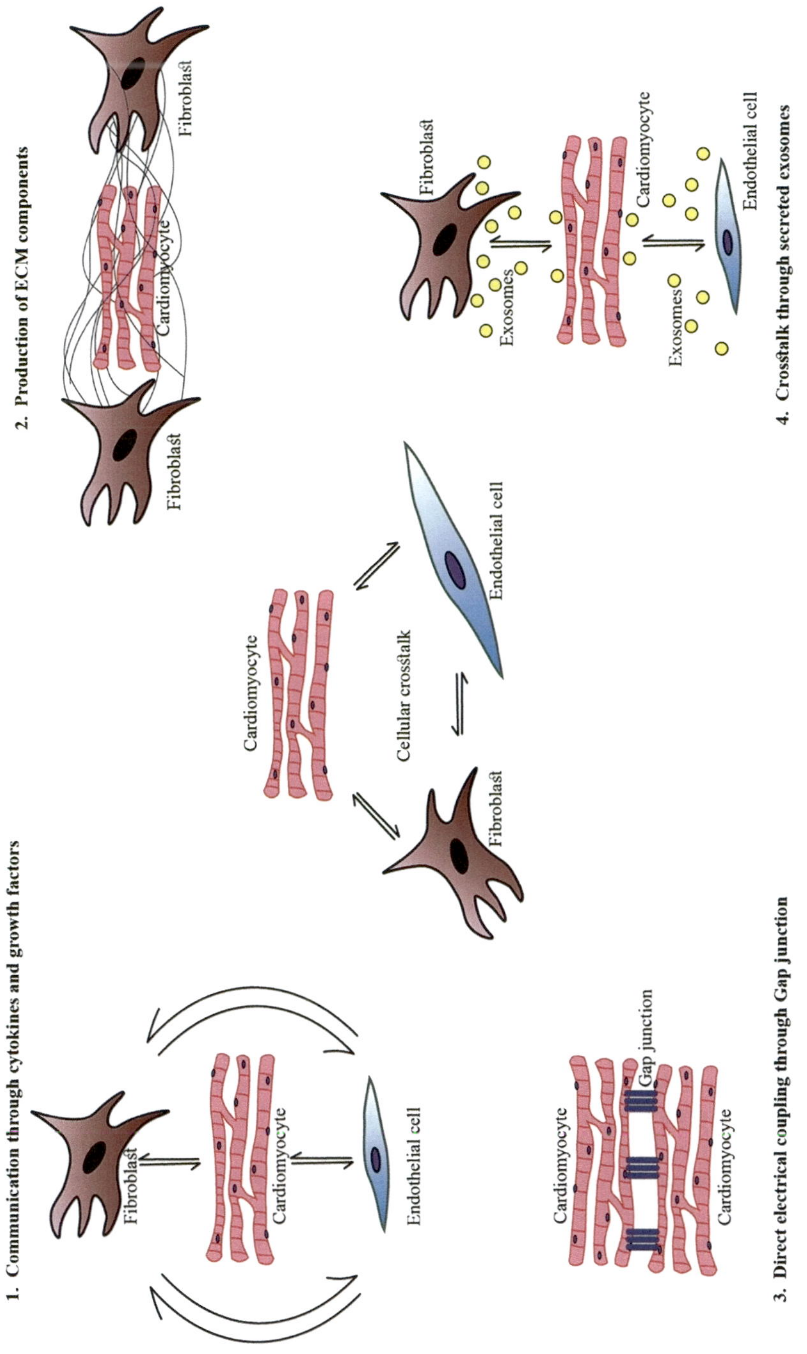


**Fig. 13.7** Schematic diagram of intercellular communication during CVDs

progression, and even in aggravation of cardiac pathophysiology, via paracrine signaling and direct cell-to-cell contact (Colliva et al. 2020). The majority of the signals from cardiomyocytes to endothelial cells regulate the "gatekeeper" activity of the endothelium in modulating the absorption of fatty acids from the circulation. Fatty acids circulate inside lipoproteins and must be hydrolyzed to free fatty acids by lipoprotein lipase (LPL) (Olivecrona and Olivecrona 2010), whose prime source is cardiomyocytes. However, heparanase secreted from the endothelial cells helps internalize these LPLs with the help of high-density lipoprotein binding protein 1 (GPIHBP1) (Camps et al. 1990). Alterations in fatty acid metabolism are a major driving force leading to heart failure. Preference for substrate utilization shifts from fatty acid to glucose during pressure overloaded pathological cardiac hypertrophy and leads to elevated lipotoxicity and ultimately cardiomyocyte loss (Edelberg et al. 1998; Banerjee et al. 2020). The intricated microenvironment of the heart encompasses communications between cardiomyocytes and noncardiomyocyte cells, viz., cardiac fibroblasts and endothelial cells. They play the role of "sentinel", amplify inflammation, and influence the fate of the cardiomyocytes (Bowers et al. 2022; Farbehi et al. 2019). However, whether such intracellular communications have any direct impact on cardiomyocyte cell loss has not been adequately studied and warrants further investigation.

## 13.5 The Pertinent Question: Does Thwarting Cell Demise Revive the Ailing Heart?

Attaining adulthood has left the cardiomyocytes with very little proliferative ability and a chronic or acute insult on the myocardium uses this situation as leverage to massacre chiefly the cardiomyocytes. Other neighboring cells further aggravate this onslaught. Such a scenario has left the scientist with only two options, either devise a blockade to this very complicated process of cell death or explore ways for cellular regeneration.

### 13.5.1 Blockade to Cell Death to Improve Prospects for CVDs

The induction of apoptosis might not always lead to immediate cell demise in the myocardium. Earlier observations in the end stage of cardiomyopathic hearts had identified the presence of normal nuclei with increased Cytochrome C release and activated caspase-3 (Narula et al. 1999; Reed and Paternostro 1999). Contrastingly, these cells were undergoing DNA repair (Kanoh et al. 1999), indicating that under chronic stress, cells of the myocardium persist in a proapoptotic phase for a good amount of time prior to actual demise and such a phenomenon is often termed as "interrupted apoptosis". Interruption to apoptosis has emerged as an enigma for cardiac pathophysiology. Notably, the inhibition of proteolytic functions of activated caspase 3 holds prospects as otherwise activated caspase 3 promotes contractile dysfunction in the cardiac cells (Kanoh et al. 1999; Yang et al. 2013).

Furthermore, mixed results have surfaced regarding the effects of caspase-3 inhibitors and only a few have progressed to clinical trials owing to the prevailing challenges faced amidst the varied types of caspase inhibitors, such as inadequate efficacy, poor target specificity, or adverse side effects (Dhani et al. 2021). Although the inhibition of caspase activation is a promising and possible therapeutic approach against CVDs, the risk of carcinogenesis as a result of cell death inhibition needs serious consideration. Contrary to a direct downstream inhibition of caspases, some upstream proteins of the cell death pathway might become lucrative such as hindering TNF-α expression has improved ventricular functions during a chronic insult on the heart (Bergman and Holycross 1996). On the other hand, using inhibitors of the nodal regulator for necroptosis, viz., RIP1 or an open-mPTP during percutaneous coronary intervention, has shown promise by reducing necrotic cell death (Lim et al. 2007; Piot et al. 2008). Cardiac pathophysiology is an umbrella term encompassing a diverse array of stress responses, and based on the etiology of the insult, the disease manifestation varies greatly, yet ultimately leading to the same fate of cell death and heart failure. Over the decades, apoptosis has been well characterized but most of the other death mechanisms remained to be deciphered thoroughly for the sake of finding a proper and effective therapeutic target to reduce cellular loss in human patients.

### 13.5.2 Balance Between Cell Death and Cell Regeneration Dictates Cardiomyocyte Cell Fate in CVDs

So far, cell death is winning the competition. Although arduous scientific work globally, including our laboratory, has observed the existence of regenerative potential in the endogenous cardiomyocytes of the affected myocardium (Chen et al. 2013; Aguirre et al. 2014; Nakada et al. 2017), cell death prowess overpowers during CVDs (Datta Chaudhuri et al. 2022; Naqvi et al. 2021; Chatterjee et al. 2011). Precisely, competition between two antagonistic transcription factors, viz., p53 and p65, for binding with the common coactivator p300 during pathological cardiac hypertrophy decides the cardiomyocyte fate. However, with the amelioration of the pathological stimuli, the proliferative potential of the cardiomyocytes went down along with the cellular apoptosis (Chatterjee et al. 2011) indicating a cooperative balance between the two. Other recorded instances where cardiomyocytes had reentered the cell cycle include hypoxia, regulated substrate utilization in PDK4 knockout animals, etc. (Cardoso et al. 2020). Once thought terminally differentiated, cardiomyocytes show some regenerative potential under pathological scenarios that might contribute towards maintaining tissue homeostasis by countering myocyte loss.

### 13.5.3 Cardiomyocyte Regeneration: A Strategy to Counter-Balance Cell Death During CVDs

The last resort in treating a compromised heart is via a heart transplantation procedure. However, donor constraints and complexity of the surgery pose serious restrictions globally. Although the past few decades have witnessed a noteworthy development in temporary artificial cardiac assist devices, viz., left ventricular assist devices (LVADs) and total artificial heart systems (TAHs), their use has certain limitations like infections, thromboembolism, and durability. Owing to the challenges of the present technologies, research in cardiomyocyte regeneration has gained momentum. The earlier notion of terminally differentiated cardiomyocytes in the heart has been challenged due to the groundbreaking discoveries related to cellular regeneration in the mammalian heart (Chen et al. 2013; Aguirre et al. 2014; Nakada et al. 2017). The annual rate of adult mammalian cardiomyocyte turnover ranges from 0.5% to <1% (Bergmann et al. 2009, 2015; Senyo et al. 2013). Although the cardiomyocyte turnover rate becomes higher post-injury (Hsieh et al. 2007), this renewal rate is not enough to counter the cell loss in the damaged heart. Our laboratory has ventured on the thin line between cellular regeneration and cell loss during pathological cardiac hypertrophy and found that cardiomyocyte fate is sealed towards cell death during such chronic insults on the myocardium. However, there is undeniable cellular regenerative prowess in the backdrop of the overpowering molecular onslaught inside the cardiomyocytes (Chatterjee et al. 2011). Altogether, these discoveries have directed the scientific community to think about strategizing ways to tinker with this thin line of balance toward cellular regeneration to counter cell loss during CVDs. Cardiomyocyte-targeted gene therapy could become one such praxis owing to the present encouraging discoveries in this field. Application of cardiomyocyte-targeted overexpression of Nemo S-nitrosyl mutant in a pathologically hypertrophied heart was able to shift this balance towards cellular regeneration in a preclinical model that has debated strongly in support of potential cardiomyocyte regeneration (Datta Chaudhuri et al. 2022). Non-cardiomyocytes have been reprogrammed to function as a replacement for the lost cardiomyocytes by introducing cardiogenic transcription factors (Hsieh et al. 2007; Ieda 2011), miRNAs (Jayawardena et al. 2012), and epigenetic modifications (Dal-Pra et al. 2017). Notably, the last few decades have witnessed a considerable shift in the conviction of adult stem cells and their ability to transdifferentiate into new cardiomyocytes (Fig. 13.8). The benefits of bone-marrow-derived stem cells and cardiosphere-derived cell (a heterogeneous consortium encompassing subpopulations of c-kit+, CD31/CD34+, and CD90+ cells) therapies may be ascribed to paracrine effects on the heart tissue, rather than spawning of newer cardiomyocytes (Michler 2018; López et al. 2020). Moreover, the development of pluripotent stem-cell-derived cardiac progenitor cells (PSC-CMs) and induced pluripotent stem cells (iPSCs) has eased the clinical trials and showed promising results through better systolic function following treatment with a cell patch (Menasché et al. 2018). Partial maturation of PSC-CMs can be realized by prolonging the culture period (Lundy et al. 2013), electrostimulation, or via tinkering the metabolic pathways

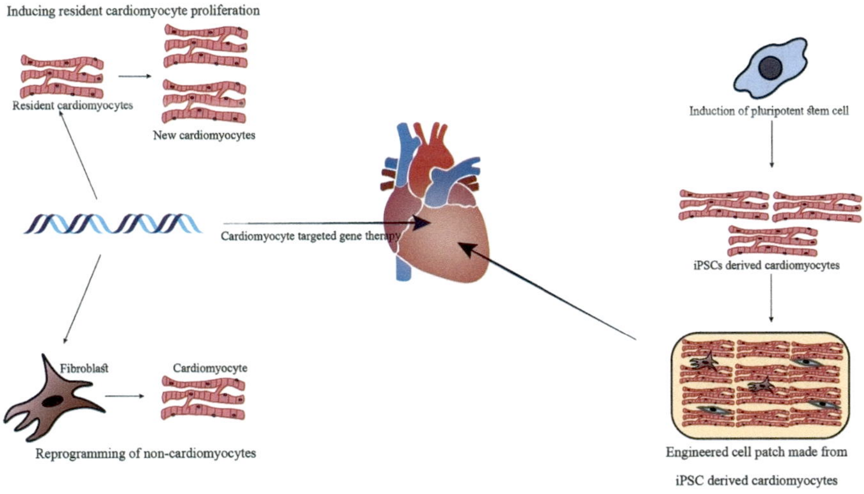


**Fig. 13.8** Schematic diagram of cardiomyocyte renewal strategies in the affected myocardium

(Nunes et al. 2013; Feyen et al. 2020). Though the stem cell therapy to counter cell loss in CVDs was initially encouraging, cardiac stem cell therapy has significant limitations due to its homing efficacy and formation of cardiac syncytium (Sano et al. 2020; Tao et al. 2018). Moreover, the presence of cardiogenic stem cells has been highly debated off late (He et al. 2020). Cardiac regeneration is an umbrella term encompassing the notions of resident cardiomyocyte regeneration prowess, cell-based therapy, cellular reprogramming, and cardiac tissue-targeted engineering approaches to revamp the injured heart. The past two decades of scientific achievements have provided a strong and rigorous foundation in our understanding of regenerative cardio-biology (Garbern and Lee 2022), which might come forward as a revolutionary advancement in the field of cardiac intervention to fight against cardiomyocyte cell loss in different cardiovascular diseases worldwide.

## 13.6 Concluding Remarks

From the various aspects of cardiac cell death discussed in this chapter, it is quite obvious that the development of any counteracting mechanisms against cell death will be beneficial for treating heart failure. However, there might be a possibility of increased tumorigenic potential in the extra-cardiac and neuronal tissues surrounding the heart with this approach. If this strategy is applied in a nontargeted way, this will lead to devastating consequences in patients. Selective inhibition of cardiac cell death through the development of drugs targeted to cardiomyocytes alone may solve this issue.

Furthermore, the onset of cardiac cell death by different processes is quite obscure. Hence, the timing of administering any anti-cell death drug is very critical.

For this, the molecular milieu at different stages of heart failure needs to be examined using conventional as well as high-throughput approaches. Nonetheless, drugs or surgical interventions such as artificial cardiac assist devices and heart transplantation will remain the linchpin of heart failure therapy, unless there is a revolutionary advancement in the field of regenerative cardio-biology; only then there might come a time when regenerative-cardiomyocyte biology would impart perpetual clinical panacea for countless patients.

## References


Adams JM, Cory S (1998) The Bcl-2 protein family: arbiters of cell survival. Science 281(5381):1322–1326

Aguirre A et al (2014) In vivo activation of a conserved microRNA program induces mammalian heart regeneration. Cell Stem Cell 15(5):589–604

Alcalai R et al (2021) Cardiomyopathy: consequences of impaired autophagy in the heart. J Am Heart Assoc 10(17):e018829

An L et al (2012) Involvement of autophagy in cardiac remodeling in transgenic mice with cardiac specific over-expression of human programmed cell death 5. PLoS One 7(1):e30097

Bai Y et al (2018) Caspase-1 regulate AngII-induced cardiomyocyte hypertrophy via upregulation of IL-1β. Biosci Rep 38(2). https://doi.org/10.1042/BSR20171438

Baines CP et al (2005) Loss of cyclophilin D reveals a critical role for mitochondrial permeability transition in cell death. Nature 434(7033):658–662

Banerjee D et al (2020) Metabolic impairment in response to early induction of C/EBPβ leads to compromised cardiac function during pathological hypertrophy. J Mol Cell Cardiol 139:148–163

Benjamin EJ et al (2018) Heart disease and stroke statistics-2018 update: a report from the American Heart Association. Circulation 137(12):e67–e492

Bennett MR (2002) Apoptosis in the cardiovascular system. Heart 87(5):480–487

Bergman MR, Holycross BJ (1996) Pharmacological modulation of myocardial tumor necrosis factor alpha production by phosphodiesterase inhibitors. J Pharmacol Exp Ther 279(1):247–254

Bergmann O et al (2009) Evidence for cardiomyocyte renewal in humans. Science 324(5923):98–102

Bergmann O et al (2015) Dynamics of cell generation and turnover in the human heart. Cell 161(7):1566–1575

Bowers SLK, Meng Q, Molkentin JD (2022) Fibroblasts orchestrate cellular crosstalk in the heart through the ECM. Nat Cardiovas Res 1(4):312–321

Camps L et al (1990) Lipoprotein lipase: cellular origin and functional distribution. Am J Phys 258(4 Pt 1):C673–C681

Cardoso AC et al (2020) Mitochondrial substrate utilization regulates cardiomyocyte cell cycle progression. Nat Metab 2(2):167–178

Cartledge JE et al (2015) Functional crosstalk between cardiac fibroblasts and adult cardiomyocytes by soluble mediators. Cardiovasc Res 105(3):260–270

Chai R et al (2022) Cardiac remodeling in heart failure: role of pyroptosis and its therapeutic implications. Front Cardiovasc Med 9:870924

Chatterjee A et al (2011) Analysis of p53 and NF-κB signaling in modulating the cardiomyocyte fate during hypertrophy. J Cell Physiol 226(10):2543–2554

Chen J et al (2013) mir-17-92 cluster is required for and sufficient to induce cardiomyocyte proliferation in postnatal and adult hearts. Circ Res 112(12):1557–1566

Chiong M et al (2011) Cardiomyocyte death: mechanisms and translational implications. Cell Death Dis 2(12):e244

Colliva A et al (2020) Endothelial cell-cardiomyocyte crosstalk in heart development and disease. J Physiol 598(14):2923–2939

Cookson BT, Brennan MA (2001) Pro-inflammatory programmed cell death. Trends Microbiol 9:113–114

Crompton M, Ellinger H, Costi A (1988) Inhibition by cyclosporin A of a Ca2+-dependent pore in heart mitochondria activated by inorganic phosphate and oxidative stress. Biochem J 255(1):357–360

Dal-Pra S et al (2017) Demethylation of H3K27 Is essential for the induction of direct cardiac reprogramming by miR combo. Circ Res 120(9):1403–1413

Datta Chaudhuri R et al (2022) Cardiomyocyte-specific regression of nitrosative stress-mediated S-Nitrosylation of IKKγ alleviates pathological cardiac hypertrophy. Cell Signal 98:110403

Datta R et al (2017) Myocyte-derived Hsp90 modulates collagen upregulation via biphasic activation of STAT-3 in fibroblasts during cardiac hypertrophy. Mol Cell Biol 37(6). https://doi.org/10.1128/MCB.00611-16

Dhani S, Zhao Y, Zhivotovsky B (2021) A long way to go: caspase inhibitors in clinical use. Cell Death Dis 12(10):949

Dixon SJ et al (2012) Ferroptosis: an iron-dependent form of nonapoptotic cell death. Cell 149(5):1060–1072

Donovan MF, Cascella M (2022) Embryology, weeks 6–8. In: StatPearls. StatPearls Publishing, Treasure Island

Edelberg JM et al (1998) PDGF mediates cardiac microvascular communication. J Clin Invest 102(4):837–843

Engler AJ et al (2008) Embryonic cardiomyocytes beat best on a matrix with heart-like elasticity: scar-like rigidity inhibits beating. J Cell Sci 121(Pt 22):3794–3802

Fang X et al (2023) The molecular and metabolic landscape of iron and ferroptosis in cardiovascular disease. Nat Rev Cardiol 20(1):7–23

Farbehi N et al (2019) Single-cell expression profiling reveals dynamic flux of cardiac stromal, vascular and immune cells in health and injury. eLife 8. https://doi.org/10.7554/eLife.43882

Feyen DAM et al (2020) Metabolic maturation media improve physiological function of human iPSC-derived cardiomyocytes. Cell Rep 32(3):107925

Galluzzi L et al (2015) Essential versus accessory aspects of cell death: recommendations of the NCCD 2015. Cell Death Differ 22(1):58–73

Garbern JC, Lee RT (2022) Heart regeneration: 20 years of progress and renewed optimism. Dev Cell 57(4):424–439

Goldenthal MJ (2016) Mitochondrial involvement in myocyte death and heart failure. Heart Fail Rev 21(2):137–155

Guerra S et al (1999) Myocyte death in the failing human heart is gender dependent. Circ Res 85(9):856–866

Gulati V et al (2014) Cardiac involvement in hemochromatosis. Cardiol Rev 22(2):56–68

Guo X, Chen Y, Liu Q (2022) Necroptosis in heart disease: molecular mechanisms and therapeutic implications. J Mol Cell Cardiol 169:74–83

Hahn VS et al (2021) Myocardial gene expression signatures in human heart failure with preserved ejection fraction. Circulation 143(2):120–134

Hamacher-Brady A et al (2007) Response to myocardial ischemia/reperfusion injury involves Bnip3 and autophagy. Cell Death Differ 14(1):146–157

Han M et al (2020) Dietary iron intake and risk of death due to cardiovascular diseases: a systematic review and dose-response meta-analysis of prospective cohort studies. Asia Pac J Clin Nutr 29(2):309–321

Haunstetter A, Izumo S (2000) Toward antiapoptosis as a new treatment modality. Circ Res 86(4):371–376

He S et al (2009) Receptor interacting protein kinase-3 determines cellular necrotic response to TNF-alpha. Cell 137(6):1100–1111

He L et al (2020) Heart regeneration by endogenous stem cells and cardiomyocyte proliferation: controversy, fallacy, and progress. Circulation 142(3):275–291

Hinchliffe JR (1981) Cell death in embryogenesis. In: Cell death in biology and pathology. Springer Netherlands, Dordrecht, pp 35–78
Hinsull SM, Bellamy D (1981) Tissue homeostasis and cell death. In: Cell death in biology and pathology. Springer Netherlands, Dordrecht, pp 123–144
Hsieh PCH et al (2007) Evidence from a genetic fate-mapping study that stem cells refresh adult mammalian cardiomyocytes after injury. Nat Med 13(8):970–974
Huang S, Frangogiannis NG (2018) Anti-inflammatory therapies in myocardial infarction: failures, hopes and challenges. Br J Pharmacol 175(9):1377–1400
Ieda M (2011) Direct reprogramming of fibroblasts into functional cardiomyocytes by defined factors. Nihon rinsho Jpn J Clin Med 69(Suppl 7):524–527
Jayawardena TM et al (2012) MicroRNA-mediated in vitro and in vivo direct reprogramming of cardiac fibroblasts to cardiomyocytes. Circ Res 110(11):1465–1473
Ji N et al (2021) Pyroptosis: a new regulating mechanism in cardiovascular disease. J Inflamm Res 14:2647–2666
Kakkar R, Lee RT (2010) Intramyocardial fibroblast myocyte communication. Circ Res 106(1):47–57
Kanamori H et al (2011) Autophagy limits acute myocardial infarction induced by permanent coronary artery occlusion. Am J Physiol Heart Circ Physiol 300(6):H2261–H2271
Kanoh M et al (1999) Significance of myocytes with positive DNA in situ nick end-labeling (TUNEL) in hearts with dilated cardiomyopathy: not apoptosis but DNA repair. Circulation 99(21):2757–2764
Kartha CC (2021) Structure and function of cardiomyocyte. In: Cardiomyocytes in health and disease. Springer International Publishing, Cham, pp 3–12
Kawaguchi M et al (2011) Inflammasome activation of cardiac fibroblasts is essential for myocardial ischemia/reperfusion injury. Circulation 123(6):594–604
Kayagaki N et al (2015) Caspase-11 cleaves gasdermin D for non-canonical inflammasome signalling. Nature 526(7575):666–671
Kerr JF, Wyllie AH, Currie AR (1972) Apoptosis: a basic biological phenomenon with wide-ranging implications in tissue kinetics. Br J Cancer 26(4):239–257
Kim I, Rodriguez-Enriquez S, Lemasters JJ (2007) Selective degradation of mitochondria by mitophagy. Arch Biochem Biophys 462(2):245–253
Kim H et al (2009) Stepwise activation of BAX and BAK by tBID, BIM, and PUMA initiates mitochondrial apoptosis. Mol Cell 36(3):487–499
Konstantinidis K, Whelan RS, Kitsis RN (2012) Mechanisms of cell death in heart disease. Arterioscler Thromb Vasc Biol 32(7):1552–1562
Kremastinos DT, Farmakis D (2011) Iron overload cardiomyopathy in clinical practice. Circulation 124(20):2253–2263
Kroemer G, Mariño G, Levine B (2010) Autophagy and the integrated stress response. Mol Cell 40(2):280–293
Kung G, Konstantinidis K, Kitsis RN (2011) Programmed necrosis, not apoptosis, in the heart. Circ Res 108(8):1017–1036
Li P et al (1997) Cytochrome c and dATP-dependent formation of Apaf-1/caspase-9 complex initiates an apoptotic protease cascade. Cell 91(4):479–489
Li Q et al (2016) AMPK-mediated cardioprotection of atorvastatin relates to the reduction of apoptosis and activation of autophagy in infarcted rat hearts. Am J Transl Res 8(10):4160–4171
Li J et al (2018) Role of autophagy in proteostasis: friend and foe in cardiac diseases. Cells 7(12). https://doi.org/10.3390/cells7120279
Lichý M et al (2019) Different signalling in infarcted and non-infarcted areas of rat failing hearts: A role of necroptosis and inflammation. J Cell Mol Med 23(9):6429–6441
Lim SY et al (2007) The cardioprotective effect of necrostatin requires the cyclophilin-D component of the mitochondrial permeability transition pore. Cardiovasc Drugs Ther 21(6):467–469
Linkermann A, Green DR (2014) Necroptosis. New Engl J Med 370(5):455–465
Liu W et al (2021) Pyroptosis inhibition improves the symptom of acute myocardial infarction. Cell Death Dis 12(10):852

López E et al (2020) The immunomodulatory signature of extracellular vesicles from cardiosphere-derived cells: a proteomic and miRNA profiling. Front Cell Dev Biol 8:321
Lundy SD et al (2013) Structural and functional maturation of cardiomyocytes derived from human pluripotent stem cells. Stem Cells Dev 22(14):1991–2002
Luo X et al (1998) Bid, a Bcl2 interacting protein, mediates cytochrome c release from mitochondria in response to activation of cell surface death receptors. Cell 94(4):481–490
Menasché P et al (2018) Transplantation of human embryonic stem cell-derived cardiovascular progenitors for severe ischemic left ventricular dysfunction. J Am Coll Cardiol 71(4):429–438
Michler RE (2018) The current status of stem cell therapy in ischemic heart disease. J Card Surg 33(9):520–531
Mitra A et al (2013) Role of α-crystallin B as a regulatory switch in modulating cardiomyocyte apoptosis by mitochondria or endoplasmic reticulum during cardiac hypertrophy and myocardial infarction. Cell Death Dis 4(4):e582
Mitra A et al (2014) Cardioprotective role of P38 MAPK during myocardial infarction via parallel activation of α-crystallin B and Nrf2. J Cell Physiol 229(9):1272–1282
Mizushima N, Levine B (2010) Autophagy in mammalian development and differentiation. Nat Cell Biol 12(9):823–830
Nah J et al (2022) Tfeb-mediated transcriptional regulation of autophagy induces autosis during ischemia/reperfusion in the heart. Cells 11(2). https://doi.org/10.3390/cells11020258
Nakada Y et al (2017) Hypoxia induces heart regeneration in adult mice. Nature 541(7636):222–227
Nakagawa T et al (2005) Cyclophilin D-dependent mitochondrial permeability transition regulates some necrotic but not apoptotic cell death. Nature 434(7033):652–658
Nakai A et al (2007) The role of autophagy in cardiomyocytes in the basal state and in response to hemodynamic stress. Nat Med 13(5):619–624
Naqvi N et al (2021) Mechanism-based cardiac regeneration strategies in mammals. Front Cell Dev Biol 9:747842
Narula J et al (1999) Apoptosis in heart failure: release of cytochrome c from mitochondria and activation of caspase-3 in human cardiomyopathy. Proc Natl Acad Sci USA 96(14):8144–8149
Naryzhnaya NV et al (2022) Pyroptosis is a drug target for prevention of adverse cardiac remodeling: the crosstalk between pyroptosis, apoptosis, and autophagy. J Biomed Res 36(6):375–389
Nemchenko A et al (2011) Autophagy as a therapeutic target in cardiovascular disease. J Mol Cell Cardiol 51(4):584–593
Nguyen BY, Azam T, Wang X (2021) Cellular signaling cross-talk between different cardiac cell populations: an insight into the role of exosomes in the heart diseases and therapy. Am J Physiol Heart Circ Physiol 320(4):H1213–H1234
Nicin L et al (2022) Fibroblast-mediated intercellular crosstalk in the healthy and diseased heart. FEBS Lett 596(5):638–654
Nunes SS et al (2013) Biowire: a platform for maturation of human pluripotent stem cell-derived cardiomyocytes. Nat Methods 10(8):781–787
Oka T et al (2012) Mitochondrial DNA that escapes from autophagy causes inflammation and heart failure. Nature 485(7397):251–255
Olivecrona G, Olivecrona T (2010) Triglyceride lipases and atherosclerosis. Curr Opin Lipidol 21(5):409–415
Ow Y-LP et al (2008) Cytochrome c: functions beyond respiration. Nat Rev Mol Cell Biol 9(7):532–542
Park T-J et al (2019) Quantitative proteomic analyses reveal that GPX4 downregulation during myocardial infarction contributes to ferroptosis in cardiomyocytes. Cell Death Dis 10(11):835
Piek A, de Boer RA, Silljé HHW (2016) The fibrosis-cell death axis in heart failure. Heart Fail Rev 21(2):199–211
Piot C et al (2008) Effect of cyclosporine on reperfusion injury in acute myocardial infarction. N Engl J Med 359(5):473–481
Quinn TA et al (2016) Electrotonic coupling of excitable and nonexcitable cells in the heart revealed by optogenetics. Proc Natl Acad Sci USA 113(51):14852–14857

Reed JC, Paternostro G (1999) Postmitochondrial regulation of apoptosis during heart failure. Proc Natl Acad Sci USA 96(14):7614–7616
Richardson WJ et al (2015) Physiological Implications of Myocardial Scar Structure. Compr Physiol 5(4):1877–1909
Rifki OF, Hill JA (2012) Cardiac autophagy: good with the bad. J Cardiovasc Pharmacol 60(3):248–252
Rothermel BA, Hill JA (2007) Myocyte autophagy in heart disease: friend or foe? Autophagy 3(6):632–634
Sano T et al (2020) Stem cell therapy in heart disease: limitations and future possibilities. Acta Med Okayama 74(3):185–190
Sarhan M et al (2018) Origin and consequences of necroinflammation. Physiol Rev 98(2):727–780
Schiattarella GG, Hill JA (2016) Therapeutic targeting of autophagy in cardiovascular disease. J Mol Cell Cardiol 95:86–93
Senyo SE et al (2013) Mammalian heart renewal by pre-existing cardiomyocytes. Nature 493(7432):433–436
Shan X et al (2021) The protective effect of cyanidin-3-glucoside on myocardial ischemia-reperfusion injury through ferroptosis. Oxidative Med Cell Longev 2021:8880141
Sheth S (2019) Strategies for managing transfusional iron overload: conventional treatments and novel strategies. Curr Opin Hematol 26(3):139–144
Son E et al (2020) The optimal model of reperfusion injury using H9c2 transformed cardiac myoblasts. Korean J Physiol Pharmacol 24(2):173–183
Stockwell BR et al (2017) Ferroptosis: a regulated cell death nexus linking metabolism, redox biology, and disease. Cell 171(2):273–285
Suzanne M, Steller H (2013) Shaping organisms with apoptosis. Cell Death Differ 20(5):669–675
Syntichaki P, Tavernarakis N (2003) The biochemistry of neuronal necrosis: rogue biology? Nat Rev Neurosci 4(8):672–684
Taneike M et al (2010) Inhibition of autophagy in the heart induces age-related cardiomyopathy. Autophagy 6(5):600–606
Tang X et al (2020) SNO-MLP (S-nitrosylation of muscle LIM protein) facilitates myocardial hypertrophy through TLR3 (toll-like receptor 3)-mediated RIP3 (receptor-interacting protein kinase 3) and NLRP3 (NOD-like receptor pyrin domain containing 3) inflammasome activation. Circulation 141(12):984–1000
Tao Z et al (2018) Stem cell homing: a potential therapeutic strategy unproven for treatment of myocardial injury. J Cardiovasc Transl Res 11(5):403–411
Tavernarakis N (2007) Cardiomyocyte necrosis: alternative mechanisms, effective interventions. Biochim Biophys Acta 1773(4):480–482
Teringova E, Tousek P (2017) Apoptosis in ischemic heart disease. J Transl Med 15(1):87
Tsou AY et al (2011) Mortality in Friedreich ataxia. J Neurol Sci 307(1–2):46–49
Tuomainen TP et al (1999) Increased risk of acute myocardial infarction in carriers of the hemochromatosis gene Cys282Tyr mutation: a prospective cohort study in men in eastern Finland. Circulation 100(12):1274–1279
US Preventive Services Task Force et al (2018) Risk assessment for cardiovascular disease with nontraditional risk factors: US Preventive Services Task Force Recommendation Statement. JAMA 320(3):272–280
Wang H et al (2014) Mixed lineage kinase domain-like protein MLKL causes necrotic membrane disruption upon phosphorylation by RIP3. Mol Cell 54(1):133–146
Wang J et al (2020) Pyroptosis and ferroptosis induced by mixed lineage kinase 3 (MLK3) signaling in cardiomyocytes are essential for myocardial fibrosis in response to pressure overload. Cell Death Dis 11(7):574
Wang F et al (2022a) Silica nanoparticles induce pyroptosis and cardiac hypertrophy via ROS/NLRP3/Caspase-1 pathway. Free Radic Biol Med 182:171–181
Wang K et al (2022b) Emerging roles of ferroptosis in cardiovascular diseases. Cell Death Discovery 8(1):394

Wencker D et al (2003) A mechanistic role for cardiac myocyte apoptosis in heart failure. J Clin Invest 111(10):1497–1504
Whelan RS, Kaplinskiy V, Kitsis RN (2010) Cell death in the pathogenesis of heart disease: mechanisms and significance. Annu Rev Physiol 72:19–44
World Health Organization (2023) World health statistics 2023: monitoring health for the SDGs, Sustainable Development Goals. World Health Organization
Xiao C et al (2018) Transplanted mesenchymal stem cells reduce autophagic flux in infarcted hearts via the exosomal transfer of miR-125b. Circ Res 123(5):564–578
Yang B, Ye D, Wang Y (2013) Caspase-3 as a therapeutic target for heart failure. Expert Opin Ther Targets 17(3):255–263
Yang WS et al (2014) Regulation of ferroptotic cancer cell death by GPX4. Cell 156(1-2):317–331
Yang WS et al (2016) Peroxidation of polyunsaturated fatty acids by lipoxygenases drives ferroptosis. Proc Natl Acad Sci USA 113(34):E4966–E4975
Yusuf S et al (2004) Effect of potentially modifiable risk factors associated with myocardial infarction in 52 countries (the INTERHEART study): case-control study. Lancet 364(9438):937–952
Zhang Z et al (2017) Knockdown of MicroRNA-122 protects H9c2 cardiomyocytes from hypoxia-induced apoptosis and promotes autophagy. Med Sci Monit 23:4284–4290
Zhao M-T et al (2020) Cardiomyocyte proliferation and maturation: two sides of the same coin for heart regeneration. Front Cell Dev Biol 8:594226
Zhou P, Pu WT (2016) Recounting cardiac cellular composition. Circ Res 118:368–370
Zhu H et al (2007) Cardiac autophagy is a maladaptive response to hemodynamic stress. J Clin Invest 117(7):1782–1793

# Role of Apoptosis in Parasitic Infections: Therapeutic Targets and Strategies

# 14

Tapasi Das, Souravi Roy, Pijush K. Das, and Anindita Ukil

**Abstract**

Apoptosis represents an intricately orchestrated biological phenomenon and a crucial component of the host's immune response against invading pathogens. Upon intrusion by infectious agents, cells undergo apoptosis as a deliberate mechanism to hinder disease progression within the host. Parasitic infections balance the host's immune responses and the parasite's survival strategies. Hence, regulation of apoptosis emerges as a great strategy for parasites, as supported by countless studies. Parasites interfere with the host's distinct apoptotic pathways to ensure their continued existence. Some parasites delay host cell apoptosis to secure their shelter and source of nourishment, facilitating intracellular parasite survival. Others promote apoptosis to release themselves near neighboring host cells, which uptake the parasites and ensure their propagation and dissemination within the host. This interplay between cellular apoptosis and parasitic survival strategies has prompted researchers to seek ways to control parasitic diseases. Potential drug targets can be identified by determining pivotal molecular components within signaling pathways governing apoptosis. This enables the development of pharmaceutical agents to inhibit the parasite's advantageous manipulation of host cell apoptosis. This chapter extensively explores various parasites' strategies for survival by exploiting host cell apoptotic mechanisms, as well as insights into drug target identification and formulation of effective therapeutic interventions.

T. Das · S. Roy · A. Ukil (✉)
Department of Biochemistry, University of Calcutta, Kolkata, India

P. K. Das
Infectious Diseases and Immunology Division, CSIR- Indian Institute of Chemical Biology, Kolkata, India



K. Jana (ed.), *Apoptosis and Human Health: Understanding Mechanistic and Therapeutic Potential*, https://doi.org/10.1007/978-981-97-7905-5_14

**Keywords**

Apoptosis · Parasites · Infection · Host-parasite interactions · Drug target identification · Therapeutic interventions

## 14.1 What Is Apoptosis?

For a long period, researchers have been fascinated by cellular defense mechanisms in the context of disease comprehension—defensive strategies within cells help in restraining the whole biological system from succumbing to such diseases. Apoptosis, also well known as "programmed cell death" or "cellular suicide," is a meticulously curated and highly conserved process among multicellular organisms that not only helps the biological system in the removal of diseased, damaged, and unwanted cells but also in biological functions such as development, aging, tissue homeostasis, etc.

The term "apoptosis" was first coined in 1972 to signify a morphologically distinct form of cell death (Kerr et al. 1972). The commencement of apoptosis is characterized by physiological markers such as cell and nucleus shrinkage, nuclear chromatin condensation, and breakage of the nucleus. Detachment of dying cells from the surrounding tissue follows the process. Eventually, the apoptotic cell shrinks and is engulfed without discharging any intracellular materials that would otherwise cause inflammation. Additionally, some biochemical markers of apoptosis, such as internucleosomal DNA fragmentation and activation of caspase proteins (Saraste and Pulkki 2000), will be often discussed in the later part of this chapter. To get triggered, this intricately designed process depends on some immediate environmental factors. Though apoptosis often occurs normally during aging or to maintain cell populations in tissues, it also occurs alongside immune reactions and in cells damaged by disease or noxious agents (Norbury and Hickson 2001).

Upon detecting damage or pathological factors via several intracellular sensors, a cell can inaugurate its death, a mechanism known as the intrinsic apoptosis pathway. This pathway relies upon mitochondria-released factors and is initiated either by negative signals, such as the absence of cytokines, hormones, and growth factors in the immediate environment of the cell, or by positive signals, such as exposure to hypoxia, toxins, radiation, reactive oxygen species (ROS), viruses, and a variety of toxic agents (Elmore 2007). The initiation of apoptosis largely depends on the activation of some cysteine-aspartic proteases known as caspases, which are of two types such as initiator caspases (caspase 8 and 9) and executioner caspases (caspase 3, 6, and 7). Caspase 9 controls this intrinsic apoptosis pathway by binding to adapter protein apoptotic protease activating factor 1 (APAF1) following exposure to its caspase recruitment domain (CARD domain). Upon induction of apoptosis, the mitochondrial membrane is triggered, and the permeability transition pore opens, resulting in the leakage of cytochrome c. Cytochrome c interacts with and causes structural changes in APAF, forming an apoptosome complex. This complex recruits and activates several procaspase 9 proteins, activating the executioner procaspase 3, thus ultimately inducing apoptosis. Besides the intrinsic pathway, apoptosis can also occur due to interaction between a cell of the immune system and a damaged cell, which is generally called the extrinsic pathway of apoptosis. The extrinsic pathway is triggered by the interaction between NK cells or macrophage-released death ligands and death receptors (DRs) on the target cell membrane. This initiates the pathway by activating procaspase 8, converting it into caspase 8. DRs recruit monomeric procaspase 8 to a death-inducing signal complex (DISC) via FAS-associated death domain (FADD) or TNFR-associated death domain (TRADD). Active caspase 8 (D'arcy 2019) (Fig. 14.1).

Apart from apoptosis, there are some other well-known programmed cell death mechanisms. One such regulated form of cell death is pyroptosis, which relies on the activation of caspase-1 instead of apoptotic executor proteins caspase-3 and 7 and is primarily seen in inflammatory cells such as macrophages triggered by pathogenic infection (Bergsbaken et al. 2009). Another form of programmed cell death is ferroptosis, which is largely dependent on iron accumulation and is characterized by the accumulation of lethal lipid species derived from the peroxidation of lipids (Li et al. 2020). Necroptosis is another programmed form of cell death that is regulated by DRs and relies on two proteins such as receptor-interacting protein kinases (RIPKs) and its substrate, mixed lineage kinase domain-like pseudokinase (MLKL). It mimics both apoptosis and necrosis by showing morphological characteristics such as cell swelling followed by rupturing of the plasma membrane (Dhuriya and Sharma 2018) (Fig. 14.2). Though all these types of programmed cell deaths are now known to us, apoptosis is by far the most studied cell death phenomenon

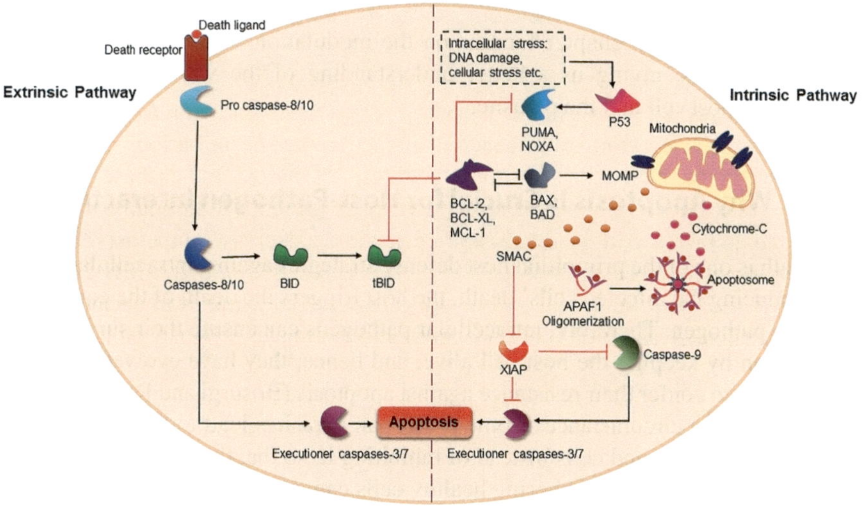


**Fig. 14.1** Schematic representation of the intrinsic and extrinsic pathways of apoptosis. The extrinsic pathway (left) is triggered by extracellular signals binding to death receptors, leading to caspase-8 and caspase-3 activation, tBID formation, and apoptosis. The intrinsic pathway (right) involves mitochondrial events, including the release of cytochrome c, activation of caspase-9 and caspase-3, apoptosome formation, and imbalance between proapoptotic (BAX) and antiapoptotic (BCL-2) proteins

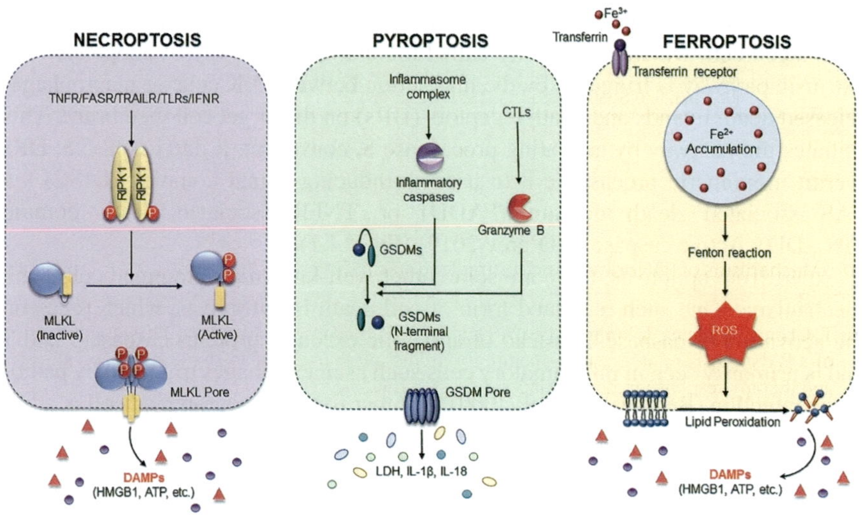


**Fig. 14.2** Nonapoptotic pathways of programmed cell death. Necroptosis (left) is triggered by the activation of RIPKs and MLKL formation, followed by loss of plasma membrane integrity, release of damage-associated molecular patterns (DAMPs), and inflammatory responses. Pyroptosis (middle) is initiated by activating inflammasome complex and GSDM pore formation resulting in pro-inflammatory cytokine release. Ferroptosis (right) is characterized by dysregulation of iron homeostasis, induction of lipid peroxidation, and lethal accumulation of lipid reactive oxygen species (ROS)

worldwide. This book chapter focuses on the modulation of apoptosis in several parasitic diseases, giving us a better understanding of the war of survivability between the host cell and the parasite.

## 14.2 Why Apoptosis Is Crucial for Host-Pathogen Interaction?

Cell death is one of the primordial host defense strategies against intracellular infections. Inducing the infected cells' death, the host triggers the death of the equivalent infecting pathogen. Therefore, intracellular pathogens can ensure their survival and propagation by keeping the host cell alive, and hence, they have evolved different mechanisms to confer their resistance against apoptosis (Bosurgi and Rothlin 2021).

Under certain circumstances, some protozoan parasites lead to the induction of apoptosis of the infected cells instead of inhibiting it, so that the consequent uptake of the infected cells by neighboring healthy cells can ensure the propagation of the parasite within the host. This induced host cell death contributes to disease acceleration. During their evolution, parasites have developed various mechanisms to induce or avoid host cell apoptosis to survive and complete their life cycle.

Signaling pathways related to apoptosis are highly regulated, demonstrating that this event is finely tuned according to the biological environment of the cell. Early apoptosis of host cells could facilitate their fight against infection by intracellular pathogens, and it could also favor the penetration of the parasite. Late apoptosis of cells could benefit the host as it can clear excess cells, thereby avoiding the detrimental effects of excessive inflammatory response in the tissue. This could also benefit the pathogen by restricting the potential for the host to develop a protective immune response (Bienvenu et al. 2010).

A major issue that has not yet been resolved is the helpful/harmful ratio obtained with antiapoptotic treatment during infection. Since apoptosis is common to almost all human cells, it could be speculated that the therapy using an antiapoptotic drug would benefit damaged cells. However, it could also induce deleterious effects on tissues, where apoptosis is essential to maintaining cell homeostasis. Conversely, the treatment that can cause the induction of infected-cell apoptosis could also be responsible for harmful effects on noninfected cells. Therefore, detailed knowledge regarding the role of cell death is needed in the context of parasitic infection.

## 14.3 Apoptosis Inhibition and Induction in Parasitic Diseases

Since infected host cells try to commit to apoptosis as a "spiteful" action to die along with the intracellular parasites, the parasites obey several brilliant strategies to "pacify" the host cells by thwarting apoptosis to safeguard their niche (Fig. 14.3). Additionally, the parasites would wish to maintain the virulence in the host to a threshold level so as not to push the host to the brink of death, for which parasites are also found to induce apoptosis (Fig. 14.4). In this chapter, we will discuss some of the parasites and their master plans for both inhibiting and inducing apoptosis.

### 14.3.1 *Leishmania sp.*

*Leishmania*, responsible for the parasitic disease leishmaniasis, is an intracellular parasite that particularly depends metabolically on the principal host cell macrophages and other cells such as neutrophils, dendritic cells, etc. *Leishmania* must downregulate the host cell defense mechanism to ensure long-term survivability and growth. For the parasites, the inhibition of host cell apoptosis is an important step. One study emphasized how *Leishmania donovani*-infected macrophages upregulate suppressors of cytokine signaling (SOCS) proteins to prevent oxidative burst-mediated host cell apoptosis (Srivastav et al. 2014). The parasite is also reported to induce the expression of antioxidant enzyme heme oxygenase-1 which neutralizes ROS and inhibits apoptosis (Saha et al. 2019; Kim et al. 2006a). Macrophages infected with *L. donovani* have exhibited diminished apoptosis of host cells and heightened survival of the parasite. This phenomenon is facilitated through the modulation of AKT signaling, leading to the suppression of its downstream pro-apoptotic effector FOXO-1, and simultaneously activation of β-catenin, resulting in decreased levels of proinflammatory cytokines and host cell apoptosis (Gupta et al. 2016). The parasitic infection has also shown a significant downregulation of

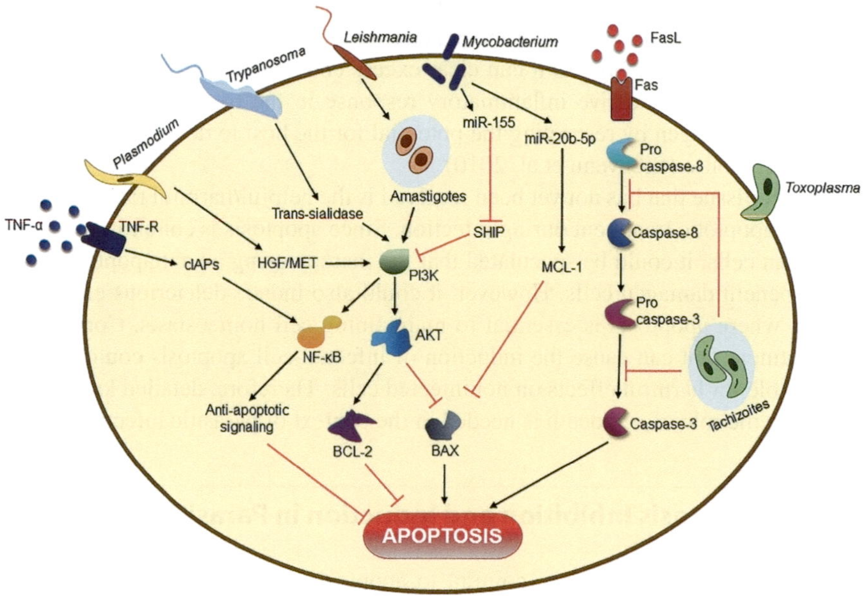


**Fig. 14.3** Mechanisms of apoptosis inhibition by parasites. The illustration depicts how intracellular parasites *Leishmania, Toxoplasma, Trypanosoma, Plasmodium, and Mycobacterium* interfere with host cell apoptosis pathways by modulating multiple signaling pathways, including Fas/FasL, TNF/TNF-R, PI3K/AKT pathway

programmed cell death receptor-1 (PD-1), which otherwise would have activated apoptosis by removing the inhibitory effect from the proapoptotic protein BAD (Roy et al. 2017). Studies have shown that this parasite induces the expression of an extremely labile protein known as myeloid cell leukemia-1 (MCL-1) to maintain mitochondrial membrane integrity and prevent the release of apoptogenic proteins (Giri et al. 2016). Another study follows up, revealing that L. donovani increases the expression of translationally controlled tumor protein (TCTP), which helps in the stabilization of MCL-1, thus consistently preventing host cell apoptosis (Giri et al. 2022). The parasite has also been reported to inhibit bone marrow-derived macrophage (BMDM) apoptosis induced by macrophage-colony-stimulating factor (M-CSF) deprivation. It is also demonstrated that upon *L. donovani* infection, macrophage apoptosis is inhibited. Still, the production of TNF-α, TGF-β, IL-6, and GM-CSF increases, and simultaneously, secretion of M-CSF and IL-1β decreases (Moore and Matlashewski 1994). *Leishmania* promastigotes transmitted to the host by the bite of sandflies have been reported to activate several signaling pathways that inhibit apoptosis, such as NF-kB, PI3K, and p38 MAPK pathways. (Ruhland et al. 2007). *L. donovani* also leads to inhibition of PKC-mediated induction of c-fos, which triggers the expression of proapoptotic molecules, thus preventing apoptosis (Moore et al. 1993). Further studies with L. *major* infected macrophages cultured without M-CSF showed a delay in apoptosis through the inhibition of mitochondrial cytochrome-c release and caspase 3 activation in an NF-κB independent

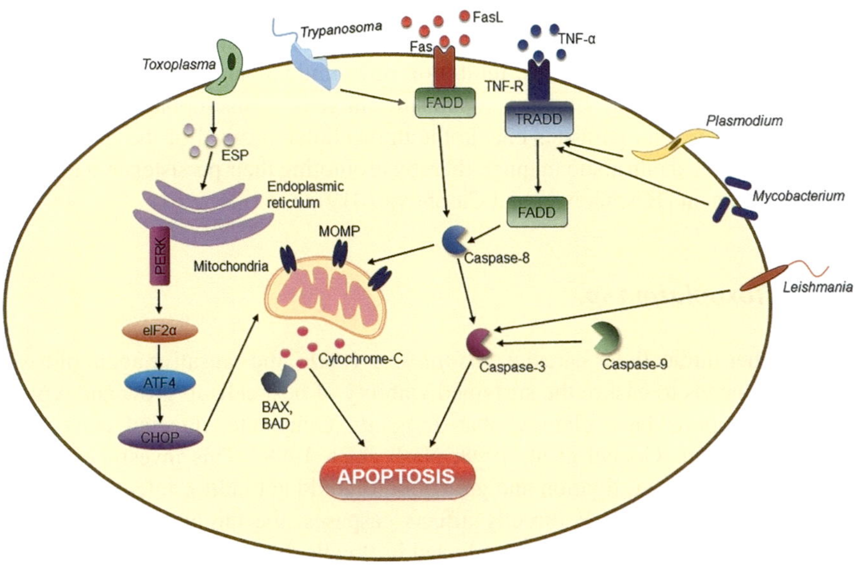


**Fig. 14.4** Mechanisms of apoptosis induction by parasites. The illustration depicts how intracellular parasites *Leishmania, Toxoplasma, Trypanosoma, Plasmodium, and Mycobacterium* induce host cell apoptosis for their advantage, such as through inducing ER stress and activating Fas/FasL, TNF/TNF-R pathway and causing mitochondrial dysfunction

manner (Akarid et al. 2004). *L. major* infection in neutrophils has also shown a decrease in caspase 3 activity and, thus, inhibition of spontaneous apoptosis of these cells (Aga et al. 2002). *L. Mexicana*-infected dendritic cells showed inhibition of camptothecin-induced apoptosis in monocyte-derived dendritic cells (Valdés-Reyes et al. 2009). Another study with *L. mexicana* demonstrated its ability to inhibit apoptosis of monocyte-derived dendritic cells by blocking the proteolytic cleavage of caspase 3, thus favoring parasite persistence inside host cells (Gutiérrez-Kobeh et al. 2013). *L. Mexicana* has also been reported to inhibit camptothecin-induced apoptosis through downregulation of p38 and JNK phosphorylation, upregulation of glutathione which aids in neutralizing ROS, and induction of antiapoptotic protein Bcl-2 expression in monocyte-derived dendritic cells (Rodríguez-González et al. 2022).

In the context of *L. major* infection, some studies have demonstrated an elevated apoptosis rate in infected phagocytes, coupled with direct cell-to-cell transfer to newly recruited phagocytes. This transfer process involves the uptake of cellular material from the original infected host cell, observed both in vivo and in vitro. Furthermore, investigations have revealed that *L. major* proliferation actively modulates host cell metabolism and induces host cell death, facilitating the efficient dissemination of the pathogen to new phagocytes (Baars et al. 2023). Intriguingly, some studies have delved into the occurrence of regulated cell death (RCD) in *Leishmania* parasites, a mechanism employed to regulate parasite density and prevent hyperparasitism-induced host death (Bruchhaus et al. 2007; Lüder et al. 2010).

Several studies have indicated that the phagocytosis of apoptotic *Leishmania* by host macrophages leads to the upregulated expression of antiinflammatory cytokines such as IL-10 and TGF-β, inhibition of proinflammatory cytokines such as TNF-α, and a reduced presentation of parasitic antigens by macrophages. In essence, apoptotic *Leishmania* parasites enable the intracellular survival of the most adaptable parasites in an altruistic manner, thereby promoting their persistence within the host environment (Basmaciyan and Casanova 2019).

### 14.3.2 *Toxoplasma sp.*

As any other intracellular parasite, *Toxoplasma gondii*, the causative agent of toxoplasmosis, needs to ensure the sustained viability of host cells to grow and replenish. Some reports have claimed that *T. gondii* tachyzoites-infected cells resist apoptotic stimuli (Goebel et al. 1998; Nash et al. 1998). This investigation also reported that Fas/FasL ligation and granzyme B could not induce infected host cell apoptosis. As granzyme B directly affects caspases, the inhibition of granzyme B-mediated apoptosis in *T. gondii*-infected host cells demonstrates that inhibitions occur at the level of downstream caspases (Nash et al. 1998). Further reports have shown that upon induction of apoptosis, *T. gondii*-infected cells fail to activate caspases 8, 9, and 3 (Goebel et al. 2001). Inhibitions of both intrinsic and extrinsic apoptosis during *T. gondii* infection are demonstrated to depend upon NF-κB activation, which further leads to transcriptional upregulation of antiapoptotic genes (Kim et al. 2006b). Since NF-κB regulates the expression of both antiapoptotic proteins and proinflammatory cytokines, hence NF-κB mediated signaling is important in host immune response against *T. gondii* infection. *T. gondii* has been previously reported to upregulate the expression of anti-apoptotic Bcl-2 family (Molestina et al. 2003). Studies have suggested that NF-κB is behind this upregulation of Bcl-2 family proteins and other inhibitors of apoptosis (IAP), resulting in the blockade of apoptosis (Hwang et al. 2010). *T. gondii* also inhibits cytochrome c release, allowing parasite-dependent degradation of proapoptotic protein Bad and Bax and increased transcription of antiapoptotic Bfl-1 and Bcl-2 (Carmen et al. 2006). Some investigations have found HeLa cells infected with *T. gondii* exhibiting downregulated phospho-JNK levels upon exposure to UV, thus blockade of JNK-mediated transmission of apoptotic stimuli required to induce cytochrome c in these cells (Carmen et al. 2006).

On the other hand, a study conducted on *T. gondii* infection revealed the upregulation of proapoptotic genes, including DNA damage-inducible transcript 3 (DDIT3), growth arrest, and DNA damage-inducible α (GADD45A), caspase-3, and high-temperature requirement protease A2 (HtrA2), along with the downregulation of antiapoptosis genes such as poly (adenosine diphosphate (ADP)-ribose polymerase family member 3 (PARP3), Bcl-2, and inhibitor of apoptosis protein (IAP) repeat containing 5 (BIRC5). Additionally, TNF, TRAF1, TRAF2, TNF receptor superfamily member 10b (TNFRSF10b), disabled homolog 2 (DAB2)-interacting protein (DAB2IP), and inositol 1,4,5-trisphosphate receptor type 3

(ITPR3) were found to be enriched upstream of TNF, TNF-related apoptosis-inducing ligand (TRAIL), and endoplasmic reticulum (ER) stress pathways. Interestingly, TRAIL-receptor 2 (TRAIL-R2), previously overlooked, emerged as a significant membrane receptor influenced by *T. gondii* (Du et al. 2022). Another research demonstrated that treatment with ESP (excretory-secretory products) from *T. gondii* regulates endoplasmic stress (ER)-related protein expression positively correlated with mitochondrial dysfunction in FHs 74 Int (small intestinal epithelial) cells. Preconditioning with the ER stress inhibitor 4-PBA attenuated levels of ER stress, Bcl-2, and apoptosis-related proteins, including p-IRE1α, p-PERK, p-eIF2α, cleaved ATF6, CHOP, Bax, Bak, Bim, cleaved PARP, and cleaved caspase-3, as well as mitochondrial membrane potential depolarization in *T. gondii*-infected or ESP-treated FHs 74 Int cells. Remarkably, the study demonstrates that *T. gondii* infection and ESP treatment modulate mitochondrial apoptosis in FHs 74 Int cells through the PERK/eIF2α/ATF4/CHOP, ATF6/CHOP, and IRE1α/caspase-12 pathways. Intracellular signaling pathways regulated by ER stress responses are pivotal in understanding small intestinal immunopathology after oral *T. gondii* infection (Wang et al. 2021).

### 14.3.3 *Trypanosoma sp.*

*Trypanosoma,* a group of unicellular parasitic protozoans responsible for sleeping sickness, accomplishes the inhibition of host cell apoptosis through a complex network of strategies. Studies have found two *T. cruzi*-secreted proteins that contribute to the inhibition of host cell apoptosis. One of these factors, transsialidase, mimics the host nerve growth factor (NGF) and binds to the host NGF receptor TrkA, which induces PI3K-dependent apoptosis in infected neurons (Chuenkova and Pereira 2003; Chuenkova and PereiraPerrin 2004; Chuenkova and PereiraPerrin 2005). Another factor, cruzipain (CZ), protects neonatal cardiomyocytes from serum deprivation-induced apoptosis without its catalytic activity. CZ-treated cells show increased Bcl-2 levels and arginase activation (Aoki et al. 2004), preventing apoptosis from depleting arginine levels (Esch et al. 1998). CZ treatment also activates several signaling pathways, such as c-Jun N-terminal kinase (JNK), p38 mitogen-activated protein kinase (MAPK), extracellular-signal-regulated kinases 1/2 (Erk1/2), and PI3K/Akt, though only the activation PI3K/Akt and Erk1/2 aids in *T. cruzi* dependent inhibition of apoptosis. On the other hand, MAPK-dependent activation Erk1/2 elevates Bcl-1 levels in the cell. Along with that, PI3K inactivates the proapoptotic protein Bad (del Pilar Aoki et al. 2006). Another study showed that *T. cruzi* utilizes NF-κB instead of PI3K, to inhibit TNF-α induces cardiomyocyte apoptosis (Petersen et al. 2006). In addition to that *T. cruzi* also inhibits the extrinsic pathway of apoptosis through interfering with Fas/FasL interaction and subsequent inhibition of caspase-8. They also found that *T. cruzi* inhibits the activation of caspase 8 following the Fas/FasL interaction by blocking the degradation of c-FLIP, an inhibitor of caspase 8 activation by the DISC activated by FasL binding (Hashimoto et al. 2005).

*Trypanosoma* would induce host cell apoptosis as a means to limit their defensive strategies to keep surviving in the host along with spreading infection. During the acute phase of *T. cruzi* infection, the parasites activate activation-induced cell death (AICD) in CD4+ T cells, eliminating them. Studies in experimentally infected mice have revealed heightened expression and activity of Fas/FasL in CD+ T cells, potentially contributing to the deleterious impact of AICD on infection by promoting T cell elimination and aiding parasite escape (Decote-Ricardo et al. 2017). B-cell apoptosis is also prevalent in mice with acute *T. cruzi* infection, marked by increased Fas expression in these cells (Zuniga et al. 2000). One study has interestingly shown thattrans-sialidase can be present in the bloodstream and induce apoptosis in immune system cells during *T. cruzi* infection, even at a distance from the site of infection (Mucci et al. 2006).

### 14.3.4 *Plasmodium sp.*

*Plasmodium*, the causative agent of malaria, has shown interesting dynamics associated with inhibiting or inducing apoptosis. Following invasion within the host hepatocytes, *Plasmodium* sporozoites triggers the release of hepatocyte growth factor (HGF), which results in substantiation of the parasite's residence and subsequent differentiation and multiplication. Signaling initiated by the interaction between HGF and its receptor Met has been proved to inhibit host cell apoptosis by inhibiting the PI3K/Akt pathway during the early stage of *Plasmodium berghei* infection (Xiao et al. 2001). Though PI3K is shown to be involved in inhibiting host hepatocyte apoptosis, another study claims that the PI3K activation is also not responsible for saving the hepatocytes from their apoptotic fate (Van De Sand et al. 2005). One study has observed increased numbers of detached hepatocytes filled with *P. berghei* merozoites before erythrocyte infection. These detached cells showed multiple characteristics of apoptotic cells such as cytoplasmic cytochrome c release, loss of mitochondrial membrane potential, nuclear condensation, and loss of phosphatidylserine in the outer leaflet of the plasma membrane. One study has concluded that even when the earlier stages of hepatocytic infection depend on apoptosis inhibition, merozoites do induce nonapoptotic and nonnecrotic cell death to promote the formation of vesicles, which would transport merozoites to the bloodstream, where they would be released causing infection of erythrocytes (Sturm et al. 2006). One recent study has reported the importance of glucose transporters (GLUTs) in the survival of *Plasmodium*, as treatment with the GLUT inhibitor WZB117 inhibits glycolysis, induces generation of ROS, and induces apoptosis (Wei et al. 2018).

Plasmodium induces apoptosis to prevent excessive proliferation within the host and minimize stress to maintain a sustainable host environment for prolonged survival. This self-induced apoptosis, eryptosis when occurring in infected red blood cells (RBCs), is crucial for controlling parasite numbers. Infected apoptotic RBCs are cleared by the spleen, thereby regulating the parasite load and preventing anemia. However, this process must be balanced, as excessive RBC apoptosis threatens parasite development. *Plasmodium*-infected RBCs often evade splenic clearance by sequestering in organs like the brain, leading to complications such as cerebral malaria, blood-brain barrier dysfunction, and organ failure. Host cell apoptosis is strongly associated with severe malaria symptoms. Malaria also results in reduced lymphocyte counts (Wei et al. 2018). $CD4^+$ T cells, which offer protection against the parasite, have been a focal point of research. Studies indicate that *P. vivax* infection induces apoptosis in $CD4^+$ T cells through two pathways such as activation of the TNFR1 DR (extrinsic pathway), amplified by Bid, and decreased expression of the antiapoptotic protein Bcl-2 (intrinsic pathway). T lymphocyte apoptosis appears to be a parasite-triggered immune evasion strategy, enabling parasite persistence while limiting immunopathology (Hojo-Souza et al. 2015). Also, acute and chronic *P. falciparum* infections are associated with elevated mononuclear cell apoptosis levels. Remarkably, high spontaneous apoptosis rates are observed in short-term lymphocyte cultures from individuals with clinical malaria and asymptomatic carriers (Kakani et al. 2016).

### 14.3.5 *Mycobacterium sp.*

An earlier study has reported that *Mycobacterium*, the pathogen causing tuberculosis, can induce robust production of antiinflammatory cytokine IL-10 which leads to an upregulated expression of Bcl-2 family proteins and downregulation of caspase-1, P53, nitric oxide (NO), and TNF-α production (Balcewicz-Sablinska et al. 1999). Another study proved this fact by using transgenic mice expressing human IL-10 infected with *M. avium*, displaying reduced apoptosis, along with decreased NO and TNF-α production (Feng et al. 2002). Il-10 has also been shown to inhibit *Mycobacterium*-induced alveolar macrophage apoptosis by inhibiting TNF-α production, which is because IL-10 induces Bcl-3 protein level, which in turn reduces NF-κB signaling and downregulates TNF-α production (Patel et al. 2009). One study has shown the importance of IL-17 cytokine in *Mycobacterium* pathogenesis, where IL-17 blocks BMDM apoptosis upon infection via suppressing the intrinsic pathway of apoptosis through blocking nuclear translocation of p53, thus upregulating Bcl-2 antiapoptotic proteins and downregulating Bax proapoptotic proteins, along with the inhibition of cytochrome c release and caspase 3 activations(Cruz et al. 2015). Since microRNAs (miR) are shown to play important roles in the post-transcriptional regulation of genes, several studies have been done to check their role in *Mycobacterium* pathogenesis. A study has shown the upregulated expression of miR-582-5p in monocytes of tuberculosis patients compared to control individuals. Transfection with this microRNA to THP-1 cells has been shown to block apoptosis by targeting FOXO mRNA and suppressing its translation (Liu et al. 2013). Another miR, called miR-155, has been shown to target mRNA of SH2 domain-containing inositol 5-phosphatase 1 (SHIP1), which is generally responsible for reduced phosphorylation of Akt and enhanced activation of Bad, FOXO-1, 3, caspase-3. Hence, the inhibition of SHIP1 by miR-155 has proved to inhibit apoptosis. Furthermore, infection with *Mycobacterium* in miR-155 knockdown murine macrophages has shown a higher level of apoptosis than wild-type infected macrophages, supporting the previous findings (Rothchild et al. 2016).

*Mycobacterium triggers* apoptosis to enhance their virulence and facilitate the spread of infection. Various virulence factors of *Mycobacterium tuberculosis* (*Mtb*), such as the 38-kDa antigen, p19, ESAT-6, HBHA, and PE-PGRS33, have been identified to induce apoptosis in macrophages through endoplasmic reticulum (ER) stress-dependent pathways (Lam et al. 2017). Studies have proposed that ER stress-related pathways can be activated, leading to apoptosis during *Mtb* infection in an ESAT-6-dependent manner (Behar et al. 2010). Additionally, these parasites promote apoptosis in host cells by inducing mitochondrial outer membrane permeabilization, which results in the release of cytochrome c, apoptosis-inducing factor, and Smac-DIABLO from the mitochondrial intermembrane space, which ultimately activates caspase 3/7 (Aguiló et al. 2013). Moreover, certain *Mycobacterium* species have been shown to induce apoptosis through TNF-mediated DR-induced apoptosis by activating caspase 8 (Stutz et al. 2021).

## 14.4 Therapeutic Targets and Strategies

Through extensive research over the decades of scientific progress, finding the "Achilles' heel" of such parasitic diseases has been possible. Finding a cure for these diseases from the network of strategies in the parasite-compromised host environment is challenging. Since the parasite confirms its stay inside the cell by affecting multiple defensive pathways of the host, targeting and blocking a single pathway does not give a total cure for these diseases but a rather effective one to a certain degree. Along with that, inducing apoptosis of infected host cells is not a perfect solution as it threatens the overall health and survivability of the patient. So, an extensive study is needed to find a cure for these parasitic diseases while focusing on both the host's wellness and the obstruction of parasite survival. Several studies are exploring the drug targets of parasitic diseases and strategies to get the optimum effect of the drugs. In the case of *Leishmania donovani* infection, induction of PD-1 signaling and subsequent inhibition of AKT signaling in macrophages would promote apoptosis of host cells and reduce parasite burden. This finding establishes PD-1 on macrophages as a future therapeutic tool to cure this parasitic disease (Roy et al. 2017). On the other hand, PD-1 signaling activated in host T cells causes apoptosis and reduced adaptive immunity against these parasites, hence allowing increased parasite burden (Jafarzadeh et al. 2022). Some studies have shown that treatment with anti-PD-1 or anti-PD-L1 reduces *L. amazonensis* parasite load in infected BALB/c mice (da Fonseca-Martins et al. 2019). A well-known drug against *Leishmania* is miltefosine, which works with the principle of inducing apoptosis via induction of DNA damage (Branco et al. 2016). One drug called vorinostat is being studied and is found to have an action mechanism of histone deacetylase inhibition, which causes the induction of apoptosis and the inhibition of cell proliferation, migration, and immune regulation. Thus, it can act as an antiparasitic drug against leishmaniasis and other parasitic diseases (Herrera-Martínez et al. 2020). The compound cordiaquinone E (CORe) has shown antileishmanial activity by inducing apoptosis (Rodrigues et al. 2021). *Embilica officinalis* L. (EO) fruit extract has been found to inhibit the growth and proliferation of promastigotes, along with the induction of oxidative stress, mitochondrial dysfunction, DNA laddering, and apoptosis-like cell death in parasites (Ali et al. 2021). Resveratrol has been found to have some direct and indirect effects against *Toxoplasma tachyzoites*, promoting apoptosis (Chen et al. 2019). One study showed that the inhibition of elevated levels of miR-20a expression in *T. gondii*-infected macrophages using locked nucleic acid (LNA) technology promoted apoptosis of host macrophages and can be a potential drug against toxoplasmosis (Rezaei et al. 2018). Similarly, there are drugs discovered against *Trypanosoma* infection that work with a similar principle. One study focused on the effect of naphthoquinone derivative and found that the treatment caused organelle disruption, intracellular ROS generation, and apoptosis, leading to the parasite's death (Rani et al. 2021). Some alkaloids, such as quinidine, berberine, and emetine, also showed their efficiency in treating *Trypanosoma evansi* infection by inducing apoptosis and ROS production (Rani et al. 2023). Several therapeutic strategies have been designed to reduce the spread of infection. Zinc and melatonin therapy increases cell viability and reduces apoptosis in *trypanosoma-infected* murine splenocytes (Brazão et al. 2015). Oxidovanadium (V) complex treatment shows a lower late apoptosis rate in *Trypanosoma cruzi* infection (Scalese et al. 2018). One recent study highlighted the therapeutic mechanism of antimalarial drug artesunate and found that a network of signaling pathways is targeted by this drug, including apoptosis (De Souza et al. 2008). In the case of cerebral malaria, neuronal apoptosis is induced, which leads to neuropathology. One study showed treatment with an aqueous extract *Phyllanthus niruri* reduces apoptosis by reducing the level of staurosporin-induced caspase 3/7 activity while also showing anti-malarial activity such as reduction of host inflammation (Montalvão et al. 2010). Another study has mentioned a natural anti-malarial compound that works by inducing apoptosis, such as the leaf extract of *Indigoferao blongifolia*, which reduces the number of apoptotic hepatic cells of mice infected with *Plasmodium chabaudi* (Wang et al. 2022). While other drugs have been found to induce parasite apoptosis, such as mefloquine (Jeje et al. 2023), alpha-2-macroglobulin has been found to regulate host cell apoptosis induced by *T. cruzi* infection (Dkhil et al. 2019). Another study showed apoptotic lymphocytes treated with purified chagasic IgG reduced *Trypanosoma* replication inside macrophages and increased proinflammatory cytokine TNF-α production (Gunjan et al. 2016). As there are a significant number of studies showing the survival and amplification strategy of *Mycobacterium* being induction of host cell apoptosis, there are studies focused on the therapeutic strategies regarding the inhibition of apoptosis. One study shows that sirtuin 1 (SIRT1) can inhibit *Mtb*-mediated macrophage apoptosis through the GSK-3β-mediated pathway. Resveratrol, an activator of SIRT1, thus inhibits *Mtb*-mediated apoptosis and infection spreading (Yang et al. 2020). One study highlights the effect of a drug complex with sparfloxacin, [Pt(DMSO)(spf)Cl]Cl, that shows the best potential against *Mycobacterium tuberculosis* (de Oliveira et al. 2018). One study has shown the effective treatment through apoptosis induction as well by first- and second-line antituberculosis drugs in experimental studies on cell lines (Vasilyeva et al. 2010). Drugs such as rifampicin activate apoptosis of THP-1 cells, epithelial cells, SPEV, and lymphocytes and can effectively treat tuberculosis (Erokhina et al. 2013). Another drug, pyrazinamide (PZA), has been found to induce apoptosis in rat liver cells by downregulating Bcl-2 and upregulating Bax and caspase-3 through the PI3K/AKT pathway (Xu et al. 2020).

## 14.5 Conclusion

In summary, the intricate interplay between parasitic infections and the apoptotic pathway underscores the vital role of apoptosis in host-parasite interactions, serving as both a defense mechanism for the host and a survival strategy for parasites. Understanding the molecular mechanisms of apoptosis and its modulation by parasites opens avenues for therapeutic interventions, including targeted antiparasitic drugs and tailored strategies. Disrupting key apoptotic pathway components targeted by parasites shows promise for impeding their evasion of host defense. Ongoing experimental analysis and in-depth studies are crucial for optimizing disease cures. Balancing effective parasite targeting with patient welfare is essential. Further research should refine therapeutic strategies, considering treatment duration, potential side effects, and the long-term impact on the host's immune system. This approach contributes to developing more robust antiparasitic therapies, prioritizing the overall well-being of individuals undergoing treatment.

## References

Aga E, Katschinski DM, Van Zandbergen G, Laufs H, Hansen B, Muller K, Solbach W, Laskay T (2002) Inhibition of the spontaneous apoptosis of neutrophil granulocytes by the intracellular parasite *Leishmania major*. J Immunol 169(2):898–905

Aguiló N, Marinova D, Martín C, Pardo J (2013) ESX-1-induced apoptosis during mycobacterial infection: to be or not to be, that is the question. Front Cell Infect Microbiol 3:88

Akarid K, Arnoult D, Micic-Polianski J, Sif J, Estaquier J, Ameisen JC (2004) *Leishmania major*-mediated prevention of programmed cell death induction in infected macrophages is associated with the repression of mitochondrial release of cytochrome c. J Leukoc Biol 76(1):95–103

Ali R, Islamuddin M, Tabrez S, Alsaweed M, Alaidarous MA, Alshehri BM, Banawas S, Dukhyil AAB, Rub A (2021) *Embilica officinalis* L. inhibits the growth and proliferation of *Leishmania donovani* through the induction of ultrastructural changes, mitochondrial dysfunction, oxidative stress and apoptosis-like cell death. Biomed Pharmacother 143:112156

Aoki MP, Guinazú NL, Pellegrini AV, Gotoh T, Masih DT, Gea S (2004) Cruzipain, a major *Trypanosoma cruzi* antigen, promotes arginase-2 expression and survival of neonatal mouse cardiomyocytes. Am J Phys Cell Phys 286(2):C206–C212

Baars I, Jaedtka M, Dewitz LA, Fu Y, Franz T, Mohr J, Gintschel P, Berlin H, Degen A, Freier S, Rygol S (2023) *Leishmania major* drives host phagocyte death and cell-to-cell transfer depending on intracellular pathogen proliferation rate. JCI Insight 8(14)

Balcewicz-Sablinska MK, Gan H, Remold HG (1999) Interleukin 10 produced by macrophages inoculated with *Mycobacterium avium* attenuates mycobacteria-induced apoptosis by reduction of TNF-α activity. J Infect Dis 180(4):1230–1237

Basmaciyan L, Casanova M (2019) Cell death in *Leishmania*. Parasite 26
Behar SM, Divangahi M, Remold HG (2010) Evasion of innate immunity by *Mycobacterium tuberculosis*: is death an exit strategy? Nat Rev Microbiol 8(9):668–674
Bergsbaken T, Fink SL, Cookson BT (2009) Pyroptosis: host cell death and inflammation. Nat Rev Microbiol 7(2):99–109
Bienvenu AL, Gonzalez-Rey E, Picot S (2010) Apoptosis induced by parasitic diseases. Parasit Vectors 3:1–9
Bosurgi L, Rothlin CV (2021) Management of cell death in parasitic infections. In: Seminars in immunopathology, vol 43. Springer, Berlin/Heidelberg, pp 481–492
Branco PVC, Soares REP, de Jesus LCL, Moreira VR, Alves HJ, de Castro Belfort MR, Silva VLM, Pereira SRF (2016) The antileishmanial drug miltefosine (Impavido®) causes oxidation of DNA bases, apoptosis, and necrosis in mammalian cells. Mutat Res/Genetic Toxicol Environ Mutagen 806:34–39
Brazão V, Filipin MDV, Santello FH, Azevedo AP, Toldo MPA, de Morais FR, do Prado JC (2015) Immunomodulatory properties and anti-apoptotic effects of zinc and melatonin in an experimental model of chronic Chagas disease. Immunobiology 220(5):626–633
Bruchhaus I, Roeder T, Rennenberg A, Heussler VT (2007) Protozoan parasites: programmed cell death as a mechanism of parasitism. Trends Parasitol 23(8):376–383
Carmen JC, Hardi L, Sinai AP (2006) *Toxoplasma gondii* inhibits ultraviolet light-induced apoptosis through multiple interactions with the mitochondrion-dependent programmed cell death pathway. Cell Microbiol 8(2):301–315
Chen QW, Dong K, Qin HX, Yang YK, He JL, Li J, Zheng ZW, Chen DL, Chen JP (2019) Direct and indirect inhibition effects of resveratrol against *Toxoplasma gondii* tachyzoites in vitro. Antimicrob Agents Chemother 63(3):10–1128
Chuenkova MV, Pereira MA (2003) PDNF, a human parasite-derived mimic of neurotrophic factors, prevents caspase activation, free radical formation, and death of dopaminergic cells exposed to the Parkinsonism-inducing neurotoxin MPP+. Mol Brain Res 119(1):50–61
Chuenkova MV, PereiraPerrin M (2004) Chagas' disease parasite promotes neuron survival and differentiation through TrkA nerve growth factor receptor. J Neurochem 91(2):385–394
Chuenkova MV, PereiraPerrin M (2005) A synthetic peptide modeled on PDNF, Chagas' disease parasite neurotrophic factor, promotes survival and differentiation of neuronal cells through TrkA receptor. Biochemistry 44(48):15685–15694
Cruz A, Ludovico P, Torrado E, Gama JB, Sousa J, Gaifem J, Appelberg R, Rodrigues F, Cooper AM, Pedrosa J, Saraiva M (2015) IL-17A promotes intracellular growth of *Mycobacterium* by inhibiting apoptosis of infected macrophages. Front Immunol 6:498
D'arcy MS (2019) Cell death: a review of the major forms of apoptosis, necrosis and autophagy. Cell Biol Int 43(6):582–592
da Fonseca-Martins AM, Ramos TD, Pratti JE, Firmino-Cruz L, Gomes DCO, Soong L, Saraiva EM, de Matos Guedes HL (2019) Immunotherapy using anti-PD-1 and anti-PD-L1 in *Leishmania amazonensis*-infected BALB/c mice reduce parasite load. Sci Rep 9(1):20275
de Oliveira LP, Carneiro ZA, Ribeiro CM, Lima MF, Paixão DA, Pivatto M, de Souza MV, Teixeira LR, Lopes CD, de Albuquerque S, Pavan FR (2018) Three new platinum complexes containing fluoroquinolones and DMSO: cytotoxicity and evaluation against drug-resistant tuberculosis. J Inorg Biochem 183:77–83
De Souza EM, Meuser-Batista M, Batista DG, Duarte BB, Araújo-Jorge TC, Soeiro MNC (2008) *Trypanosoma cruzi*: alpha-2-macroglobulin regulates host cell apoptosis induced by the parasite infection in vitro. Exp Parasitol 118(3):331–337
Decote-Ricardo D, Nunes MP, Morrot A, Freire-de-Lima CG (2017) Implication of apoptosis for the pathogenesis of *Trypanosoma cruzi* infection. Front Immunol 8:518
del Pilar Aoki M, Cano RC, Pellegrini AV, Tanos T, Guiñazú NL, Coso OA, Gea S (2006) Different signaling pathways are involved in cardiomyocyte survival induced by a *Trypanosoma cruzi* glycoprotein. Microbes Infect 8(7):1723–1731
Dhuriya YK, Sharma D (2018) Necroptosis: a regulated inflammatory mode of cell death. J Neuroinflammation 15(1):1–9

Dkhil MA, Al-Shaebi EM, Al-Quraishy S (2019) Effect of *Indigofera oblongifolia* on the hepatic oxidative status and expression of inflammatory and apoptotic genes during blood-stage murine malaria. Oxidative Med Cell Longev 2019

Du K, Lu F, Xie C, Ding H, Shen Y, Gao Y, Lu S, Zhuo X (2022) *Toxoplasma gondii* infection induces cell apoptosis via multiple pathways revealed by transcriptome analysis. J Zhejiang Univ-Sci B 23(4):315–327

Elmore S (2007) Apoptosis: a review of programmed cell death. Toxicol Pathol 35(4):495–516

Erokhina MV, Kurynina AV, Onishchenko GE (2013) Mitochondria are targets for the antituberculosis drug rifampicin in cultured epithelial cells. Biochem Mosc 78:1155–1163

Esch F, Lin KI, Hills A, Zaman K, Baraban JM, Chatterjee S, Rubin L, Ash DE, Ratan RR (1998) Purification of a multipotent antideath activity from bovine liver and its identification as arginase: nitric oxide-independent inhibition of neuronal apoptosis. J Neurosci 18(11):4083–4095

Feng CG, Kullberg MC, Jankovic D, Cheever AW, Caspar P, Coffman RL, Sher A (2002) Transgenic mice expressing human interleukin-10 in the antigen-presenting cell compartment show increased susceptibility to infection with *Mycobacterium avium* associated with decreased macrophage effector function and apoptosis. Infect Immun 70(12):6672–6679

Giri J, Srivastav S, Basu M, Palit S, Gupta P, Ukil A (2016) *Leishmania donovani* exploits myeloid cell leukemia 1 (MCL-1) protein to prevent mitochondria-dependent host cell apoptosis. J Biol Chem 291(7):3496–3507

Giri J, Basu M, Roy S, Mishra T, Jana K, Chande A, Ukil A (2022) Translationally controlled tumor protein–mediated stabilization of host antiapoptotic protein MCL-1 is critical for establishment of infection by intramacrophage parasite *Leishmania donovani*. J Immunol 208(11):2540–2548

Goebel S, Lüder CG, Lugert R, Bohne W, Gross U (1998) *Toxoplasma gondii* inhibits the in vitro induced apoptosis of HL-60 cells. Tokai J Exp Clin Med 23(6):351–356

Goebel S, Gross U, Lüder CG (2001) Inhibition of host cell apoptosis by *Toxoplasma gondii* is accompanied by reduced activation of the caspase cascade and alterations of poly (ADP-ribose) polymerase expression. J Cell Sci 114(19):3495–3505

Gunjan S, Singh SK, Sharma T, Dwivedi H, Chauhan BS, Imran Siddiqi M, Tripathi R (2016) Mefloquine induces ROS mediated programmed cell death in malaria parasite: *Plasmodium*. Apoptosis 21:955–964

Gupta P, Srivastav S, Saha S, Das PK, Ukil A (2016) *Leishmania donovani* inhibits macrophage apoptosis and pro-inflammatory response through AKT-mediated regulation of β-catenin and FOXO-1. Cell Death Different 23(11):1815–1826

Gutiérrez-Kobeh L, De Oyarzabal E, Argueta J, Wilkins A, Salaiza N, Fernández E, López O, Aguirre M, Becker I (2013) Inhibition of dendritic cell apoptosis by *Leishmania mexicana* amastigotes. Parasitol Res 112:1755–1762

Hashimoto M, Nakajima-Shimada J, Aoki T (2005) *Trypanosoma cruzi* posttranscriptionally up-regulates and exploits cellular FLIP for inhibition of death-inducing signal. Mol Biol Cell 16(8):3521–3528

Herrera-Martínez M, Orozco-Samperio E, Montaño S, Ariza-Ortega JA, Flores-García Y, López-Contreras L (2020) Vorinostat as potential antiparasitic drug. Eur Rev Med Pharmacol Sci 24(13)

Hojo-Souza NS, Pereira DB, Mendes TA, Passos LS, Gazzinelli-Guimarães AC, Gazzinelli-Guimarães PH, Tada MS, Zanini GM, Bartholomeu DC, Fujiwara RT, Bueno LL (2015) CD4+ T cells apoptosis in *Plasmodium vivax* infection is mediated by activation of both intrinsic and extrinsic pathways. Malar J 14:1–7

Hwang IY, Quan JH, Ahn MH, Hassan Ahmed HA, Cha GH, Shin DW, Lee YH (2010) *Toxoplasma gondii* infection inhibits the mitochondrial apoptosis through induction of Bcl-2 and HSP70. Parasitol Res 107:1313–1321

Jafarzadeh A, Kumar S, Bodhale N, Jafarzadeh S, Nemati M, Sharifi I, Sarkar A, Saha B (2022) The expression of PD-1 and its ligands increases in *Leishmania* infection and its blockade reduces the parasite burden. Cytokine 153:155839

Jeje TO, Bando H, Azad MTA, Fukuda Y, Oluwafemi IE, Kato K (2023) Antiplasmodial and interferon-gamma-modulating activities of the aqueous extract of stone breaker (*Phyllanthusniruri* Linn.) in malaria infection. Parasitol Int 97:102789

Kakani P, Suman S, Gupta L, Kumar S (2016) Ambivalent outcomes of cell apoptosis: a barrier or blessing in malaria progression. Front Microbiol 7:302

Kerr JF, Wyllie AH, Currie AR (1972) Apoptosis: a basic biological phenomenon with wide-ranging implications in tissue kinetics. Br J Cancer 26(4):239–257

Kim HJ, So HS, Lee JH, Lee JH, Park C, Park SY, Kim YH, Youn MJ, Kim SJ, Chung SY, Lee KM (2006a) Heme oxygenase-1 attenuates the cisplatin-induced apoptosis of auditory cells via down-regulation of reactive oxygen species generation. Free Radic Biol Med 40(10):1810–1819

Kim JY, Ahn MH, Jun HS, Jung JW, Ryu JS, Min DY (2006b) *Toxoplasma gondii* inhibits apoptosis in infected cells by caspase inactivation and NF-κB activation. Yonsei Med J 47(6):862–869

Lam A, Prabhu R, Gross CM, Riesenberg LA, Singh V, Aggarwal S (2017) Role of apoptosis and autophagy in tuberculosis. Am J Phys Lung Cell Mol Phys 313(2):L218–L229

Li J, Cao F, Yin HL, Huang ZJ, Lin ZT, Mao N, Sun B, Wang G (2020) Ferroptosis: past, present and future. Cell Death Dis 11(2):88

Liu Y, Jiang J, Wang X, Zhai F, Cheng X (2013) miR-582-5p is upregulated in patients with active tuberculosis and inhibits apoptosis of monocytes by targeting FOXO1. PLoS One 8(10):e78381

Lüder CG, Campos-Salinas J, Gonzalez-Rey E, van Zandbergen G (2010) Impact of protozoan cell death on parasite-host interactions and pathogenesis. Parasit Vectors 3:1–11

Molestina RE, Payne TM, Coppens I, Sinai AP (2003) Activation of NF-κB by *Toxoplasma gondii* correlates with increased expression of antiapoptotic genes and localization of phosphorylated IκB to the parasitophorous vacuole membrane. J Cell Sci 116(21):4359–4371

Montalvão F, Almeida GM, Silva EM, Borges VM, Vasconcellos R, Takiya CM, Lopes MF, Nunes MP, DosReis GA (2010) Apoptotic lymphocytes treated with IgG from *Trypanosoma cruzi* infection increase TNF-α secretion and reduce parasite replication in macrophages. Eur J Immunol 40(2):417–425

Moore KJ, Matlashewski G (1994) Intracellular infection by *Leishmania donovani* inhibits macrophage apoptosis. J Immunol (Baltimore, Md.: 1950) 152(6):2930–2937

Moore KJ, Labrecque S, Matlashewski G (1993) Alteration of *Leishmania donovani* infection levels by selective impairment of macrophage signal transduction. J Immunol (Baltimore, Md.: 1950) 150(10):4457–4465

Mucci J, Risso MG, Leguizamón MS, Frasch AC, Campetella O (2006) The trans-sialidase from *Trypanosoma cruzi* triggers apoptosis by target cell sialylation. Cell Microbiol 8(7):1086–1095

Nash PB, Purner MB, Leon RP, Clarke P, Duke RC, Curiel TJ (1998) *Toxoplasma gondii*-infected cells are resistant to multiple inducers of apoptosis. J Immunol 160(4):1824–1830

Norbury CJ, Hickson ID (2001) Cellular responses to DNA damage. Annu Rev Pharmacol Toxicol 41(1):367–401

Patel NR, Swan K, Li X, Tachado SD, Koziel H (2009) Impaired M. tuberculosis-mediated apoptosis in alveolar macrophages from HIV+ persons: potential role of IL-10 and BCL-3. J Leukoc Biol 86(1):53–60

Petersen CA, Krumholz KA, Carmen J, Sinai AP, Burleigh BA (2006) *Trypanosoma cruzi* infection and nuclear factor kappa B activation prevent apoptosis in cardiac cells. Infect Immun 74(3):1580–1587

Rani R, Narasimhan B, Varma RS, Kumar R (2021) Naphthoquinone derivatives exhibit apoptosis-like effect and anti-trypanosomal activity against *Trypanosoma evansi*. Vet Parasitol 290:109367

Rani R, Sethi K, Gupta S, Virmani N, Kumar S, Kumar R (2023) Intracellular ROS production and apoptotic effect of quinoline and isoquinoline alkaloids on the growth of *Trypanosoma evansi*. Acta Trop 245:106980

Rezaei F, Daryani A, Sharifi M, Sarvi S, Sattar Pagheh A, Hashemi N, Hejazi SH (2018) miR-20a inhibition using locked nucleic acid (LNA) technology and its effects on apoptosis of human macrophages infected by *Toxoplasma gondii* RH strain. Microb Pathog 121:269–276

Rodrigues RRL, Nunes TAL, de Araujo AR, MarinhoFilho JDB, da Silva MV, de AmorimCarvalho FA, Pessoa ODL, Freitas HPS, da Franca Rodrigues KA, Araújo AJ (2021) Antileishmanial activity of cordiaquinone E towards *Leishmania (Leishmania) amazonensis*. Int Immunopharmacol 90:107124

Rodríguez-González J, Wilkins-Rodríguez AA, Gutiérrez-Kobeh L (2022) Involvement of Akt and the antiapoptotic protein Bcl-xL in the inhibition of apoptosis of dendritic cells by *Leishmania mexicana*. Parasite Immunol 44(7):e12917

Rothchild AC, Sissons JR, Shafiani S, Plaisier C, Min D, Mai D, Gilchrist M, Peschon J, Larson RP, Bergthaler A, Baliga NS (2016) MiR-155–regulated molecular network orchestrates cell fate in the innate and adaptive immune response to *Mycobacterium tuberculosis*. Proc Natl Acad Sci 113(41):E6172–E6181

Roy S, Gupta P, Palit S, Basu M, Ukil A, Das PK (2017) The role of PD-1 in regulation of macrophage apoptosis and its subversion by *Leishmania donovani*. Clin Trans Immunol 6(5):e137

Ruhland A, Leal N, Kima PE (2007) *Leishmania* promastigotes activate PI3K/Akt signalling to confer host cell resistance to apoptosis. Cell Microbiol 9(1):84–96

Saha S, Basu M, Guin S, Gupta P, Mitterstiller AM, Weiss G, Jana K, Ukil A (2019) *Leishmania donovani* exploits macrophage heme oxygenase-1 to neutralize oxidative burst and TLR signaling–dependent host defense. J Immunol 202(3):827–840

Saraste A, Pulkki K (2000) Morphologic and biochemical hallmarks of apoptosis. Cardiovasc Res 45(3):528–537

Scalese G, Machado I, Fontana C, Risi G, Salinas G, Pérez-Díaz L, Gambino D (2018) New heterolepticoxidovanadium (V) complexes: synthesis, characterization and biological evaluation as potential agents against *Trypanosoma cruzi*. JBIC J Biol Inorg Chem 23:1265–1281

Srivastav S, Ball WB, Gupta P, Giri J, Ukil A, Das PK (2014) *Leishmania donovani* prevents oxidative burst-mediated apoptosis of host macrophages through selective induction of suppressors of cytokine signaling (SOCS) proteins. J Biol Chem 289(2):1092–1105

Sturm A, Amino R, Van de Sand C, Regen T, Retzlaff S, Rennenberg A, Krueger A, Pollok JM, Menard R, Heussler VT (2006) Manipulation of host hepatocytes by the malaria parasite for delivery into liver sinusoids. Science 313(5791):1287–1290

Stutz MD, Allison CC, Ojaimi S, Preston SP, Doerflinger M, Arandjelovic P, Whitehead L, Bader SM, Batey D, Asselin-Labat ML, Herold MJ (2021) Macrophage and neutrophil death programs differentially confer resistance to tuberculosis. Immunity 54(8):1758–1771

Valdés-Reyes L, Argueta J, Morán J, Salaiza N, Hernández J, Berzunza M, Aguirre-García M, Becker I, Gutiérrez-Kobeh L (2009) *Leishmania mexicana*: inhibition of camptothecin-induced apoptosis of monocyte-derived dendritic cells. Exp Parasitol 121(3):199–207

Van De Sand C, Horstmann S, Schmidt A, Sturm A, Bolte S, Krueger A, Lütgehetmann M, Pollok JM, Libert C, Heussler VT (2005) The liver stage of *Plasmodium berghei* inhibits host cell apoptosis. Mol Microbiol 58(3):731–742

Vasilyeva OA, Serebryakova VA, Urazova OI, Novitsky VV, Kononova TE, Naslednikova IO (2010) Apoptosis-inducing action of antituberculosis drugs of the main group in vitro. Antibiotikii Khimioterapiia= Antibiotics and Chemoterapy [sic] 55(11–12):25–29

Wang H, Li C, Ye W, Pan Z, Sun J, Deng M, Zhan W, Chu J (2021) *Toxoplasma gondii* induces apoptosis via endoplasmic reticulum stress-derived mitochondrial pathway in human small intestinal epithelial cell-line. Korean J Parasitol 59(6):573

Wang F, Song J, Yan Y, Zhou Q, Li X, Wang P, Yang Z, Zhang Q, Zhang H (2022) Integrated network pharmacology analysis and serum metabolomics to reveal the anti-malaria mechanism of artesunate. ACS Omega 7(35):31482–31494

Wei M, Lu L, Sui W, Liu Y, Shi X, Lv L (2018) Inhibition of GLUTs by WZB117 mediates apoptosis in blood-stage *Plasmodium* parasites by breaking redox balance. Biochem Biophys Res Commun 503(2):1154–1159

Xiao GH, Jeffers M, Bellacosa A, Mitsuuchi Y, VandeWoude GF, Testa JR (2001) Anti-apoptotic signaling by hepatocyte growth factor/Met via the phosphatidylinositol 3-kinase/Akt and mitogen-activated protein kinase pathways. Proc Natl Acad Sci 98(1):247–252

Xu Y, Jiang Y, Li Y (2020) Pyrazinamide enhances lipid peroxidation and antioxidant levels to induce liver injury in rat models through PI3k/Akt inhibition. Toxicol Res 9(3):149–157

Yang H, Chen J, Chen Y, Jiang Y, Ge B, Hong L (2020) Sirtuin inhibits *M. tuberculosis*-induced apoptosis in macrophage through glycogen synthase kinase-3β. Arch Biochem Biophys 694:108612

Zuniga E, Motran C, Montes CL, Diaz FL, Bocco JL, Gruppi A (2000) *Trypanosoma cruzi*-induced immunosuppression: B cells undergo spontaneous apoptosis and lipopolysaccharide (LPS) arrests their proliferation during acute infection. Clin Exp Immunol 119(3):507–515

# Role of Apoptosis in Viral Infections with Special Reference to COVID-19: Therapeutic Targets and Strategies

# 15

Shraman Jana, Satyajit Halder, Ankita Bhattacharya, Mihir K. Bhattacharya, and Kuladip Jana

**Abstract**

Viruses are intracellular obligate parasites that heavily rely on host machinery to undergo replication and survival. They activate and block almost all known types of programmed cell death. These modulate viral pathogenesis and host antiviral immunity to a great extent. Severe acute respiratory syndrome coronavirus 2 (SARS-CoV-2) had resulted in a recent global pandemic that caused millions of deaths worldwide. Similar to other viral infections, it has also been reported that SARS-CoV-2 induces cell death in infected cells. This leads to severe inflammation, contributing to viral pathogenicity and poor prognosis. Biopsy samples from postmortem patients indicate the occurrence of cell death. But, the connection between SARS-CoV-2 and mechanisms of cell death is yet to be elucidated. Among the cell death pathways, apoptosis, necroptosis, pyroptosis, and ferroptosis majorly contribute to disease pathogenesis. So, this chapter will focus on current evidence and research progress in the modulation of cell death due to SARS-CoV-2 infection. This will aid in uncovering novel therapeutic strategies in treating COVID-19 patients.

S. Jana
Campus Rd, Mohanpur, Haringhata Farm, West Bengal, India

S. Halder · K. Jana (✉)
Division of Molecular Medicine, Bose Institute, Unified Academic Campus, Kolkata, West Bengal, India
e-mail: kuladip@jcbose.ac.in

A. Bhattacharya
Department of Biochemistry, K.P.C Medical College & Hospital, Kolkata, India

M. K. Bhattacharya
ICMR-National Institute of Cholera & Enteric Diseases (NICED), Kolkata, India



K. Jana (ed.), *Apoptosis and Human Health: Understanding Mechanistic and Therapeutic Potential*, https://doi.org/10.1007/978-981-97-7905-5_15

**Keywords**

Apoptosis · Viral infections · COVID-19 · Necroptosis · Pyroptosis · Ferroptosis therapeutic targets

## 15.1 Introduction

Programmed cell death (PCD) plays a vital role in normal homeostasis and development, regulation of the immune system, and pathogenesis of infectious diseases. Numerous types of diseases like developmental disorders, neurodegenerative diseases, and cancer are linked with aberration of PCD pathways. Similarly, in viral infections, infected cells can activate PCD by nucleic acid-mediated antiviral responses, along with ER and oxidative stress (Reshi et al. 2014). Severe acute respiratory syndrome coronavirus 2 (SARS-CoV-2) is responsible for the recent pandemic COVID-19 and has become a global health priority within a few years. The SARS-CoV-2 virus is a member of the β-coronavirus family, which bears a single-stranded and positive-sense RNA genome. The genome length is around 30,000 base pairs, which encodes 16 nonstructural proteins (NSP1-16), four structural proteins-spike (S), membrane (M), envelope (E), and nucleocapsid (N), along with nine accessory proteins. Major virus targets are the cells and organs with surface angiotensin-converting enzyme 2 (ACE2) receptors (Merad and Martin 2020). Clinical manifestations include acute respiratory distress syndrome, lung injury, and organ failure. All these can be attributed to host cell death due to severe inflammatory response (Carvalho et al. 2021). Major types of cell death, such as apoptosis, necroptosis, pyroptosis, and ferroptosis, are involved in host defense against microbial infections. However, it has been demonstrated that cell death has a dual role in viral pathogenesis. Normally, cell death has a protective effect on the host, which limits virus spread and replication. However, dysregulated cell death causes cytokine storm, aberrant immune response, organ damage, and host death (Tan et al. 2007). Interestingly, the cell death pathways were believed initially to operate independently. However, recent studies have highlighted a significant crosstalk and connections among several of those pathways. Cell death caused by SARS-CoV-2 may help in viral elimination or accelerate tissue damage and disease progression (Donia and Bokhari 2021). So, understanding host-virus interaction, as well as the role of individual proteins of SARS-CoV-2 in cell death, will accelerate the discovery of novel anti-SARS-CoV-2 drugs. These drugs will only induce death of the infected cells, limiting replication and spread of the virus. In this chapter, we will focus on four cell death pathways (apoptosis, necroptosis, pyroptosis, and ferroptosis) in SARS-CoV-2 infection, the role of SARS-CoV-2 proteins in inducing cell death and relevant targeted therapies in the context of each of the pathways.

## 15.2 Identification of Viral-Associated Nonself-Nucleic Acids

Our body's innate immune system encodes host sensors to recognize nonself-viral nucleic acids in infected cells selectively. These include cell surface-expressed (Toll-like receptor 3 (TLR3)), endosome-associated (TLR3, TLR7, TLR8, and TLR9), and several cytoplasmic sensors (retinoic acid-inducible gene I (RIG-I)/ cyclic GMP-AMP synthetase (cGAS) stimulator of interferon genes (STING)/ nucleotide-binding oligomerization domain-like receptors (NLRPs) like NLRP1, NLRP6, NLRP9/absent in melanoma 2 (AIM2)). TLR3 and RIG-I mainly identify viral RNA, whereas cGAS, AIM2-like receptors (ALRs), and TLR9 mainly sense DNA. The majority of nucleic acid receptors, such as the TLR family, cGAS-STING, and the RIG-I-like helicases (RLHs), increase the expression of antiviral proteins, cytokines, and chemokines through the action of some transcription factors like IFN-regulatory factor 3 (IRF3) and nuclear factor-kB (NF-kB). These receptors induce phosphorylation and oligomerization of adaptor proteins such as mitochondrial antiviral-signaling (MAVS), STING, etc. Now, these scaffolding proteins activate several IkB kinase (IKK) related kinases, which subsequently phosphorylate transcription factors like IRF3/7 and NF-kB (DiDonato et al. 1997). These upregulate the expression of IFNs and proinflammatory cytokines, respectively. When IFNs bind to IFN receptors, they activate associated Janus kinases (JAKs), which phosphorylate both STAT1 and STAT2. These STATs form a complex that translocates to the nucleus, binds IFN-stimulated response elements (ISREs), and stimulates the expression of numerous interferon-stimulated genes (ISGs). These ISGs aid in antiviral defense and initiate adaptive immunity (Schoggins 2019).

## 15.3 Activation of Programmed Cell Death During Virus Infection

In response to virus infection, nucleic acid sensors like TLRs, RLHs, NLRPs, cGAS-STING, and Z-DNA/RNA binding protein 1 (ZBP1) not only promote the secretion of antiviral proteins but also activate different modes of PCD, like apoptosis, necroptosis, pyroptosis, and ferroptosis, with distinct signaling pathways. We will discuss the activation of various PCD pathways by viruses in the following section.

### 15.3.1 Induction of Apoptosis During Virus Infection

Virus-infected cells can undergo apoptosis via both extrinsic and intrinsic pathways. The extrinsic pathway involves transmembrane death receptors such as tumor necrosis factor receptor (TNFR), TNF-related apoptosis-inducing ligand (TRAIL) receptor, etc., which bind to their respective ligands. This leads to oligomerization of the death receptors and assembly of the death-inducing signaling complex (DISC). This consists of an adaptor molecule, i.e., FAS-associated protein containing death domain (FADD), pro-caspase-8, and pro-caspase 10. FADD activates caspase-8, which further activates the downstream executioner caspases like caspases 3/7 (Kiraz et al. 2016). The intrinsic pathway occurs in the mitochondria and is triggered by any intracellular oxidative stress or DNA damage. This results in mitochondrial outer membrane permeabilization (MOMP) and cytochrome c release in the cytosol. It is regulated by BCL-2 family proteins such as Bax, Bak, Bad, etc. (Tait and Green 2013). Afterward, cytochrome c forms a complex with apoptotic protease activating factor (APAF-1) and pro-caspase-9, known as the apoptosome. This complex activates pro-caspase-9 to form mature caspase-9. The mature caspase-9 activates pro-caspase-3 and pro-caspase-7, which perform proteolytic cleavage and subsequent cell death (Zou et al. 1997). Most viruses can induce apoptosis through one or both of these pathways. Several stages of virus replication result in apoptosis.

At first, apoptosis occurs when a virus interacts with its host cell surface receptor. For the measles virus, the viral hemagglutinin binds to the signaling lymphocytic activation molecule (SLAM) receptor of peripheral blood mononuclear cells and triggers the intrinsic apoptosis pathway (Iwasa et al. 2010). Likewise, foot and mouth disease virus binds to the integrin receptor of immature dendritic cells and induces apoptosis (Jin et al. 2007). Second, apoptosis can be induced after sensing viral replication intermediates inside a cell. For RNA viruses, cytosolic RIG-I detects viral RNA and activates IRF3. Here, IRF3 interacts with BAX and acts to induce apoptosis via a pathway commonly known as RLR-induced IRF3-mediated pathway of apoptosis (RIPA) (Chattopadhyay et al. 2010). Sendai and reovirus activate apoptosis depending on RIG-I and IRF3 (Knowlton et al. 2012). Similarly, cytoplasmic DNA sensors cGAS/STING induce apoptosis for DNA viruses. HSV and human T cell leukemia virus type 1 (HTLV-1) exhibit STING-dependent apoptosis of brain immune cells and monocytes, respectively (Reinert et al. 2021; Sze et al. 2013).

Some viruses induce unfolded protein response (UPR) and, consequently, apoptosis. There are three major pathways of UPR response, i.e., PKR-like ER kinase (PERK), inositol-requiring enzyme 1 (IRE1), and activating transcription factor (ATF6). For example, the West Nile virus triggers the PERK pathway, which phosphorylates eIF2α and activates ATF4. This promotes the expression of CCAAT-enhancer-binding homologous protein (CHOP) and initiates apoptosis (Medigeshi et al. 2007). NS4B protein of the Japanese encephalitis virus also activates the PERK pathway by inducing dimerization of PERK. Additionally, the Influenza virus alters the functionalities of several host proteins to induce apoptosis. For example, BAD is phosphorylated, activating the intrinsic apoptosis pathway (Tran et al. 2013). Finally, HIV is another pertinent example in which the TAT protein alters microtubule dynamics and translocates B cell lymphoma 2-interacting mediator of cell death (BIM) to mitochondria. This inhibits cytochrome c oxidase and induces apoptosis (Lecoeur et al. 2012).

### 15.3.2 Induction of Necroptosis During Virus Infection

Necroptosis is independent of caspases and depends mainly on TNF signaling to initiate the downstream cascade. The executioner kinases involved include receptor-interacting protein 1/3 (RIPK1/3) and mixed lineage kinase domain-like pseudokinase (MLKL), which finally trigger organelle swelling and cell lysis. Necroptosis can be activated both directly and indirectly.

Indirect initiation of necroptosis involves the formation of a TNF-receptor-associated complex, which recruits RIPK1 and TRADD with the help of death domains (Micheau and Tschopp 2003). Vaccinia virus (VACV) and murine gamma herpesvirus 68 (MHV68) trigger the cGAS-STING pathway and induce TNF-mediated necroptosis (Schock et al. 2017). Direct initiation of necroptosis is executed by ZBP1/DAI sensors or TRIF adaptor protein. During murine cytomegalovirus (MCMV) infection, necroptosis is induced by RNA synthesis and binding of Z-binding domains (ZBDs) to ZBP1/DAI (Maelfait et al. 2017). ZBP1/DAI also senses Z-RNAs during influenza A infection and gets dimerized. These now interact with and auto-phosphorylate RIPK3. After that, RIPK3 phosphorylates MLKL, which translocates to the plasma membrane and triggers membrane rupture and ion influx.

### 15.3.3 Induction of Pyroptosis During Virus Infection

Pyroptosis is a form of inflammatory cell death induced by detecting specific PAMPs by pattern recognition receptors (PRRs). Gasdermin cleavage is a critical step in pyroptosis. It inserts itself into the lipid bilayer of the cytoplasm and triggers pore formation and cell lysis. Generally, pyroptosis is activated via two pathways, canonical and noncanonical. The canonical pathway depends on caspase-1 and is activated by several viruses during infection.

Several canonical inflammasomes are involved in mediating caspase-1 dependent pyroptosis, including Nod-like receptor family pyrin domain containing 3,6 and 9 (NLRP3, NLRP6, and NLRP9) inflammasomes, AIM2 inflammasome and IFN-gamma-inducible protein-16 (IFI16) inflammasome (Wang et al. 2015). NLRP3 senses various RNA viruses like IAV, dengue virus, hepatitis C virus, etc., and DNA viruses like adenovirus and herpes virus. Two signals are required to activate the NLRP3 inflammasome. The first signal is produced when PRRs, RIG-I, and TNFR are activated, and NF-kB is produced (Bauernfeind et al. 2009). This stimulates the expression of NLRP3 and pro-IL-1β. The second signal is essential for NLRP3 to bind ASC and cause its oligomerization (Srinivasula et al. 2002). Thereby, caspase-1 is recruited, which cleaves N-terminal gastrin and leads to pore formation. NLRP6 and NLRP9 are newly discovered inflammasomes that also promote pyroptosis. AIM2 inflammasome mainly senses a wide range of cytosolic double-stranded DNA. The sugar-phosphate backbone of dsDNA activates the pyrin domain by allowing it to interact with ASC. Subsequently, pro-caspase 1 is activated to caspase-1 and leads to inflammasome assembly. Several viral dsDNA, like human papillomavirus (HPV), hepatitis B virus, and VACV, are recognized by AIM2 (Man et al. 2016). IFI16 inflammasome can bind single- and double-stranded DNA and trigger pyroptosis during viral infection. IFI16 translocates from the nucleus to cytosol to aid the antiviral immune response. IFI16 has been reported to act as a DNA sensor of Kaposi sarcoma-associated herpesvirus (KSHV), VACV, and single-stranded DNA from $CD4^+$ T cells infected with HIV (Jiang et al. 2021).

### 15.3.4 Induction of Ferroptosis During Viral Infection

Ferroptosis initiates by sensing increased lipid peroxidation and accumulation of free iron. Lipid peroxidation occurs when ROS triggers an oxidative attack on poly-unsaturated fatty acids. One of the ways to initiate ferroptosis is to inhibit cystine/glutamate antiporter, which imports cystine into the cell to reduce it to cysteine. This cysteine is required for glutathione (GSH) synthesis to reduce ROS in the cell with the help of glutathione peroxidases (GPXs) (Banjac et al. 2008). Ferroptosis is characterized by the shrinkage of mitochondria with a reduction in mitochondrial cristae. It has been reported that the Newcastle disease virus (NDV) induces ferroptosis via inhibition of the GSH-GPX4 pathway and increases ROS production (Kan et al. 2021). A classic ferroptosis inducer called erastin increases ROS production and lipid oxidation by accumulating STING in the mitochondria (Li et al. 2021). Still, more studies are required to understand how other viruses activate or inhibit ferroptosis.

## 15.4 Inhibition of Apoptosis During Virus Infection

As host cells try to counteract viral infection by activating PCD, viruses also employ various strategies to inhibit the activation and execution of PCD. Herpes and pox viruses use one-third of their genome to protect against host antiviral response. Several strategies are utilized by viruses to inhibit apoptosis. Poxviridae, Iridoviridae, and Herpesviridae possess viral BCL-2 homologs to inhibit apoptosis. These inhibit BAX and BAD oligomerization, preventing mitochondrial pore formation and cytochrome c release (Banjara et al. 2020). VACV encodes F1L, which interacts with BAX or BAK to inhibit mitochondrial pore formation (Taylor et al. 2006). Viruses can inhibit caspases and block apoptosis. CrmA was first found to inhibit caspase-1, caspase 8, and caspase-10 during cowpox virus (CPXV) infection (Zhou et al. 1997). Members of KSHV have viral FLIPs (vFLIPs) to completely inhibit the proteolytic activity of caspase-8 (Ruder et al. 2020). TNFR homologs in poxviruses deplete proapoptotic TNF along with other immunomodulatory cytokines. Some VACV strains have viral Golgi antiapoptotic protein (vGAAP) that inhibits the intrinsic apoptosis pathway via a Ca2+-dependent mechanism. US3, present in α-herpesviruses, phosphorylates BAD and caspase-3, inhibiting them (Benetti et al. 2003). So, several DNA viruses inhibit apoptosis through distinct mechanisms.

## 15.5 Cell Death in SARS-CoV-2 Infection

SARS-CoV-2 infection induces diverse inflammatory responses, including host cell death. Most studies on SARS-CoV-2 have focused on preventive and direct therapeutic strategies. But, understanding the role of SARS-CoV-2 and its encoded proteins in triggering cell death is also vital for better assessment of disease pathophysiology. Here, we will focus on the roles of cell death pathways in SARS-CoV-2 infection and potential targeted treatments.

### 15.5.1 Apoptosis in SARS-CoV-2 Infection

The structural proteins encoded by SARS-CoV-2 can induce apoptosis via multiple mechanisms. M protein has been reported to directly inhibit the ubiquitination of Bcl-2 ovarian killer BOK and promote its mitochondrial translocation—this induced apoptosis of H292 cells via an intrinsic pathway (Yang et al. 2022). Moreover, the M protein inhibits the pyruvate dehydrogenase kinase 1 (PDK1)-protein kinase B (PKB) axis, which reduces the activities of the forkhead transcription factor (FKHRL1). This upregulates the FAS level and promotes apoptosis. Another SARS-CoV-2 protein, N protein, has been reported to enhance M protein-mediated apoptosis by further inhibiting the PDK1-PKB axis (Ren et al. 2021). Caspase 6 cleaves the N protein, producing a fragment that antagonizes IFN and triggers viral replication. Jiang et al. reported that SARS-CoV-2 S protein could activate ROS production and inhibit the PI3K/AKT/mTOR pathway (Jiang et al. 2017). This inhibits the Bcl-2 protein and activates the intrinsic apoptosis pathway. S protein infection upregulates caspase-3 and caspase-6 levels and induces apoptosis of THP-1-like macrophages via overexpression of ROS and calcium production (Barhoumi et al. 2021).

ORF3a, an accessory protein of SARS-CoV-2, forms the ion channel on the cell membrane for triggering apoptosis and virus release. Recent studies suggest that ORF3a is found in the endoplasmic reticulum (ER) compartment and stimulates reticulophagy regulator 1 (RETREG1)-dependent reticulophagy. This leads to ER stress, proinflammatory responses, and cell apoptosis, which promotes viral replication (Zhang et al. 2022). ORF-7a is another transmembrane protein that recruits Bcl-XL to ER, activates the cellular stress response, and promotes cell apoptosis. Similarly, ORF-7b upregulates the expression of IL-6, TNF-α, and IFN-β via phosphorylation of IRF3 and STAT1. This drives TNF-dependent apoptosis in VeroE6 and HEK293T cells (Yang et al. 2021b). ORF-9 interacts with translocase of the outer membrane 70 (TOM70) and thereby inhibits mitochondrial antiviral signaling (MAVS) protein activation (Gordon et al. 2020). This finally triggers apoptosis during viral infection (Fig. 15.1).

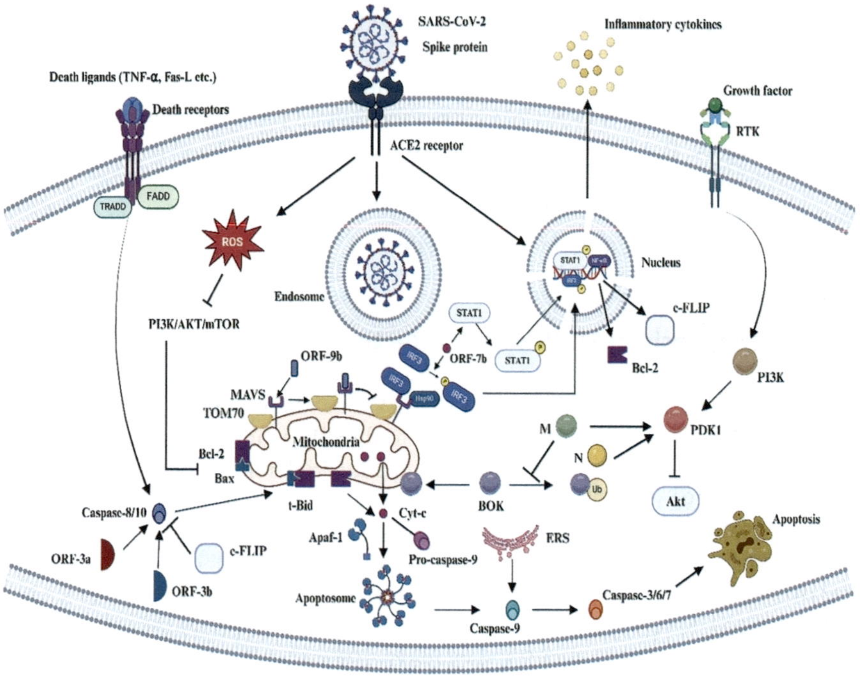


**Fig. 15.1** Apoptosis pathway in SARS-CoV-2 infection. Spike protein of SARS-CoV-2 regulates Bcl-2 and Bax to induce intrinsic apoptosis pathway. M protein protects BOK from ubiquitination and triggers intrinsic apoptosis. It further interacts with PDK1 and inhibits PKB/Akt signaling to induce caspase-dependent apoptosis. N protein further strengthens M-protein mediated inhibition of PKB/Akt signaling. ORF-3a and ORF-3b also trigger apoptosis pathways by activating caspase-8. ORF-7b phosphorylates and stimulates the nuclear translocation of STAT1 and IRF3, thereby resulting in apoptosis. ORF-9b targets TOM70 and inhibits the downstream signaling of MAVS. SARS-CoV-2 prevents apoptosis in infected cells by activating c-FLIP and NF-kB signaling

### 15.5.2 Targeted Therapies for Apoptosis in SARS-CoV-2 Infection

Inhibiting host cell apoptosis has emerged as a therapeutic target in treating COVID-19. SARS-CoV-2 infection induces cytokine storm, which is a main factor of severe mortality. Therefore, the inhibition of TNF-α and IFN-γ mediated pathways will be a better strategy to tackle this cytokine storm (Karki et al. 2021). c-FLIP restrains the apoptotic potential of DISC and inhibits caspase-8 activation. So, using small molecules which target c-FLIP will activate and induce extrinsic apoptosis via caspase-8. This is a critical method of contrasting viral replication (Hillert et al. 2020). Infliximab, an anti-TNF-α human-mouse chimeric IgG1 monoclonal antibody, has reduced inflammatory cytokine levels among severe COVID-19 patients in a phase-II trial (Hachem et al. 2021). Administration of adalimumab, a totally humanized anti-TNF-α IgG monoclonal antibody, remdesivir, and dexamethasone, has no positive results among severe COVID-19 patients.

During the outbreak of COVID-19, it was under research whether Lianhua Qingwen (LHQW) was safe for treating COVID-19 patients. It showed a shorter recovery time, inhibiting IL-6 and TNF-α and promoting infiltration of inflammatory cells. It was revealed that LHQW inhibits the upregulation of p53 and impedes p53-dependent apoptosis pathways through downregulation of caspase-3, caspase-9and Bax (Yang et al. 2021a, b). Another drug, pirfenidone, reduces oxidative stress inflammation and ACE receptor levels. This prevents COVID-19 invasion and cytokine storm in pneumocytes.

### 15.5.3 Necroptosis in SARS-CoV-2 Infection

Based on the pathogenic mechanism of COVID-19, necroptosis is thought to be increasingly associated. SARS-CoV-2 infected human lung organoids were cultivated from pathological lung tissues to demonstrate the effect of RIPK1 in SARS-CoV-2 infection. It was revealed that NSP12 interacts and activates RIPK1 and induces transcription of proinflammatory factors and host factors like ACE2 to promote virus entry into cells (Xu et al. 2021). SARS-CoV-2 immune complexes trigger necroptosis of monocytes, which depends on RIPK3 and MLKL. By blocking proteins related to necroptosis, i.e., RIPK1, RIPK3, and MLKL, the level of the SARS-CoV-2 N gene has been found to drop within monocytes (Santos et al. 2023). ORF-3a and E lead to the formation of a nonselective $Ca^{2+}$ permeable cation channel, which triggers necroptosis and apoptosis. Moreover, Ferren et al. reported that hamsters' necroptosis occurs within the lungs and brainstem during SARS-CoV-2 infection (Ferren et al. 2021).

### 15.5.4 Targeted Therapies for Necroptosis in SARS-CoV-2 Infection

Targeting components of necroptosis pathways like RIPK1, RIPK3, and MLKL has proved to be an important anti-COVID-19 strategy, which is further validated by proteomic and RNA-sequencing data. Primidone can reduce TNF-α-mediated inflammation in vivo and RIPK-1-mediated necroptosis in vitro, thereby proving useful against COVID-19 (Riebeling et al. 2021). Necrostatin-1 inhibits the necroptosis pathway and reduces systemic inflammation, viral load, and cytokine storm associated with SARS-CoV-2 infection in both in vitro and in vivo mice models.

### 15.5.5 Pyroptosis in SARS-CoV-2 Infection

SARS-CoV-2 proteins regulate inflammasome action via several mechanisms. S protein activates NLRP3 inflammasome, which is required for secretion of IL-1β. This creates hyperinflammatory response and secretion of cytokines in macrophages among COVID-19 patients (Theobald et al. 2021). NSP6 interacts with the

ATPase proton pump component (ATP6AP1) and inhibits lysosomal acidification. This activates NLRP3 inflammasome and pyroptosis of lung epithelial cells. As reported by Bertoni et al. ORF3a protein induces ASC speck formation by activating NLRP3. N protein accelerated the assembly of NLRP3 inflammasome and induced proinflammatory responses within A549 and HEK293T cells (Pan et al. 2021). Moreover, it interacts with I-kappa-b-kinase (IKK) and TGF-β activated kinase 1 (TAK1) complexes and activates the NF-kB pathway. This triggers the production of NLRP3, pro-caspase-1, and pro-IL-1β, resulting in pyroptosis (Wu et al. 2021). Also, the E protein stimulates the NF-kB pathway by triggering the generation of proinflammatory factors. Additionally, once protected from UPR stress induced by apoptosis, the E protein also helps in host translation inhibition and viral replication.

### 15.5.6 Targeted Therapies for Pyroptosis in SARS-CoV-2 Infection

MCC950 is a specific inhibitor of NLRP3 that prevents NLRP3 inflammasome assembly and might prove to be one of the possible ways to manage COVID-19-related immunopathology. Lung damage and cytokine formation associated with SARS-CoV-2 are inhibited by MCC950 treatment (Zeng et al. 2022). Disulfiram, an FDA-approved agent, inhibits GSDMD pore formation in cells and animal models, suppressing IL-1β production and pyroptosis (Hu et al. 2020). According to a phase-IV randomized study, an angiotensin receptor blocker, telmisartan, inhibits angiotensin binding to its receptors, reducing ventilation requirements and death 15/30 days during hospitalization. Anakinra is an inhibitor of IL-1 receptor, which restricts the amplification loop of inflammasome activation mediated by IL-1β. It is also approved by the FDA for treating adult COVID-19 patients who require supplementary oxygen due to pneumonia. It has been reported to suppress NLRP3-inflammasome mediated pyroptosis and decrease mice mortality affected with SARS-CoV-2. IL-6 receptor antagonist, tocilizumab, has been widely implemented for critical COVID-19 patients at ICUs (Investigators 2021).

### 15.5.7 Ferroptosis in SARS-CoV-2 Infection

Ferroptosis is considered to be responsible for cardiac arrhythmia among COVID-19 patients. Severe COVID-19 is associated with increased serum ferritin and reduced iron levels, which correlates to the link between iron overload and disease severity. SARS-CoV-2 infection activates the hepcidin pathway, which stops $Fe^{2+}$ export and triggers ferroptosis (Banchini et al. 2020). Several proinflammatory factors like IL-1β, TNF-α, and NF-kB, which are connected to ferroptosis, also increase cytokine storm due to COVID-19 infection. Huang et al. reported that the expression of ferroptosis-related genes like GPX4, SAT1, FTH1, etc., in B cells, T cells, and PBMC increases in a diseased state and reduces in the recovery phase for COVID-19 (Huang et al. 2021). Due to a deficiency in GPX4, GSH cannot undergo peroxidation to reduce lipid ROS in severe COVID-19 cases. Furthermore, patients with no history of renal diseases may suffer renal problems in COVID-19, which suggests that ferroptosis may be related to these symptoms (Ronco et al. 2020).

### 15.5.8 Targeted Therapies for Ferroptosis in SARS-CoV-2 Infection

FDA-approved iron chelator, imatinib, inactivates iron-containing enzymes and inhibits SARS-CoV-2-associated ferroptosis by binding to free iron (Han et al. 2022). Deferoxamine downregulates IL-6 levels, thereby proving to be a possible candidate medication against COVID-19-induced liver injury. However, iron chelators are also harmful as they can worsen inflammation-associated anemia and host immunity (Garrick and Ghio 2021). Since selenium (Se) can protect GPX4 from inhibition, therefore selenium can be supplemented among COVID-19 patients. Recent clinical studies in stroke mouse models reported that selenium can elevate GPX4 levels and inhibit ferroptosis. A synthetic Se-based compound, ebselen, acts as a GPX mimic, inhibits ferroptosis, and, thereby, SARS-CoV-2 virus replication by interacting with the M protein (Bermano et al. 2021).

N-acetyl cysteine (NAC) has been shown to prevent IL-6-mediated ROS production and bronchial epithelial cell ferroptosis, suggesting its utility as a candidate treatment against COVID-19. It antagonizes ferroptosis by acting as a GSH precursor in the GPX4-GSH-cysteine axis. Further clinical studies are being conducted regarding NAC as a monotherapy or combinatorial treatment. As ferroptosis depends on lipid peroxidation, therefore inhibiting lipid peroxidation is another possible anti-COVID-19 therapeutic approach. For example, liproxstatin-1 and ferrostatin-1 efficiently trap radicals and decrease the accumulation of phospholipid hydroperoxides in PUFAs, thereby being considered ferroptosis inhibitors (Miotto et al. 2020). However, more studies are necessary to validate their roles in treating COVID-19.

## 15.6 Conclusions and Future Perspectives

It is difficult to construct a universal theoretical ground to understand the interaction between the sensing of viral ligands and PCD machinery due to the complexity of the pathways and diversity of viruses. Most studies have focused on live virus infections to understand their correlations with PCDs. Therefore, it is still unknown how inactivated and replication-defective viruses regulate PCD pathways. Moreover, further studies should comment on whether viral DNA and RNA activate distinct PCD pathways. ISGs establish an antiviral state and represent a major outcome of viral nucleic acid sensing. Still, the mechanism of action of a large fraction of ISGs in the context of viral infections remains to be elucidated. On one hand, cell death is a vital host defense strategy against viruses. But, when uncontrolled, systemic inflammation and ineffective immune response occur. Thus, cell death is a double-edged sword which should be carefully modulated. Several potential drugs, such as cell death inducers and other treatment strategies, are underway to obtain better efficacy in managing COVID-19 patients. Gaining more insights into the molecular mechanism of cell death induced by various SARS-CoV-2 proteins might pave the way for better therapeutic interventions.

## References


Banchini F, Vallisa D, Maniscalco P, Capelli P (2020) Iron overload and Hepcidin overexpression could play a key role in COVID infection, and may explain vulnerability in elderly, diabetics, and obese patients. Acta Bio Medica: Atenei Parmensis 91(3):e2020013

Banjac A, Perisic T, Sato H, Seiler A, Bannai S, Weiss N, Kölle P, Tschoep K, Issels RD, Daniel PT (2008) The cystine/cysteine cycle: a redox cycle regulating susceptibility versus resistance to cell death. Oncogene 27(11):1618–1628

Banjara S, Suraweera CD, Hinds MG, Kvansakul M (2020) The Bcl-2 family: ancient origins, conserved structures, and divergent mechanisms. Biomol Ther 10(1):128

Barhoumi T, Alghanem B, Shaibah H, Mansour FA, Alamri HS, Akiel MA, Alroqi F, Boudjelal M (2021) SARS-CoV-2 coronavirus spike protein-induced apoptosis, inflammatory, and oxidative stress responses in THP-1-like-macrophages: potential role of angiotensin-converting enzyme inhibitor (perindopril). Front Immunol 12:728896

Bauernfeind FG, Horvath G, Stutz A, Alnemri ES, MacDonald K, Speert D, Fernandes-Alnemri T, Wu J, Monks BG, Fitzgerald KA (2009) Cutting edge: NF-κB activating pattern recognition and cytokine receptors license NLRP3 inflammasome activation by regulating NLRP3 expression. J Immunol 183(2):787–791

Benetti L, Munger J, Roizman B (2003) The herpes simplex virus 1 US3 protein kinase blocks caspase-dependent double cleavage and activation of the proapoptotic protein BAD. J Virol 77(11):6567–6573

Bermano G, Méplan C, Mercer DK, Hesketh JE (2021) Selenium and viral infection: are there lessons for COVID-19? Br J Nutr 125(6):618–627

Carvalho T, Krammer F, Iwasaki A (2021) The first 12 months of COVID-19: a timeline of immunological insights. Nat Rev Immunol 21(4):245–256

Chattopadhyay S, Marques JT, Yamashita M, Peters KL, Smith K, Desai A, Williams BRG, Sen GC (2010) Viral apoptosis is induced by IRF-3-mediated activation of Bax. EMBO J 29(10):1762–1773

DiDonato JA, Hayakawa M, Rothwarf DM, Zandi E, Karin M (1997) A cytokine-responsive IκB kinase that activates the transcription factor NF-κB. Nature 388(6642):548–554

Donia A, Bokhari H (2021) Apoptosis induced by SARS-CoV-2: can we target it? Apoptosis 26(1):7–8

Ferren M, Favède V, Decimo D, Iampietro M, Lieberman NAP, Weickert J-L, Pelissier R, Mazelier M, Terrier O, Moscona A (2021) Hamster organotypic modeling of SARS-CoV-2 lung and brainstem infection. Nat Commun 12(1):5809

Garrick MD, Ghio AJ (2021) Iron chelation may harm patients with COVID-19. Eur J Clin Pharmacol 77(2):265–266

Gordon DE, Hiatt J, Bouhaddou M, Rezelj VV, Ulferts S, Braberg H, Jureka AS, Obernier K, Guo JZ, Batra J (2020) Comparative host-coronavirus protein interaction networks reveal pan-viral disease mechanisms. Science 370(6521):eabe9403

Hachem H, Godara A, Schroeder C, Fein D, Mann H, Lawlor C, Marshall J, Klein A, Poutsiaka D, Breeze JL (2021) Rapid and sustained decline in CXCL-10 (IP-10) annotates clinical outcomes following TNFα-antagonist therapy in hospitalized patients with severe and critical COVID-19 respiratory failure. J Clin Trans Sci 5(1):e146

Han Y, Zhu J, Yang L, Nilsson-Payant BE, Hurtado R, Lacko LA, Sun X, Gade AR, Higgins CA, Sisso WJ (2022) SARS-CoV-2 infection induces ferroptosis of sinoatrial node pacemaker cells. Circ Res 130(7):963–977

Hillert LK, Ivanisenko NV, Busse D, Espe J, König C, Peltek SE, Kolchanov NA, Ivanisenko VA, Lavrik IN (2020) Dissecting DISC regulation via pharmacological targeting of caspase-8/c--FLIPL heterodimer. Cell Death Different 27(7):2117–2130

Hu JJ, Liu X, Xia S, Zhang Z, Zhang Y, Zhao J, Ruan J, Luo X, Lou X, Bai Y (2020) FDA-approved disulfiram inhibits pyroptosis by blocking gasdermin D pore formation. Nat Immunol 21(7):736–745

Huang L, Shi Y, Gong B, Jiang L, Zhang Z, Liu X, Yang J, He Y, Jiang Z, Zhong L (2021) Dynamic blood single-cell immune responses in patients with COVID-19. Signal Transduct Target Ther 6(1):110

Investigators R-C (2021) Interleukin-6 receptor antagonists in critically ill patients with Covid-19. N Engl J Med 384(16):1491–1502

Iwasa T, Suga S, Qi L, Komada Y (2010) Apoptosis of human peripheral blood mononuclear cells by wild-type measles virus infection is induced by interaction of hemagglutinin protein and cellular receptor, SLAM via caspase-dependent pathway. Microbiol Immunol 54(7):405–416

Jiang Y, Kou J, Han X, Li X, Zhong Z, Liu Z, Zheng Y, Tian Y, Yang L (2017) ROS-dependent activation of autophagy through the PI3K/Akt/mTOR pathway is induced by hydroxysafflor yellow A-sonodynamic therapy in THP-1 macrophages. Oxidative Med Cell Longev 2017(1):8519169

Jiang Z, Wei F, Zhang Y, Wang T, Gao W, Yu S, Sun H, Pu J, Sun Y, Wang M (2021) IFI16 directly senses viral RNA and enhances RIG-I transcription and activation to restrict influenza virus infection. Nat Microbiol 6(7):932–945

Jin H, Xiao C, Zhao G, Du X, Yu Y, Kang Y, Wang B (2007) Induction of immature dendritic cell apoptosis by foot and mouth disease virus is an integrin receptor mediated event before viral infection. J Cell Biochem 102(4):980–991

Kan X, Yin Y, Song C, Tan L, Qiu X, Liao Y, Liu W, Meng S, Sun Y, Ding C (2021) Newcastle-disease-virus-induced ferroptosis through nutrient deprivation and ferritinophagy in tumor cells. Iscience 24(8)

Karki R, Sharma BR, Tuladhar S, Williams EP, Zalduondo L, Samir P, Zheng M, Sundaram B, Banoth B, Malireddi RKS (2021) Synergism of TNF-α and IFN-γ triggers inflammatory cell death, tissue damage, and mortality in SARS-CoV-2 infection and cytokine shock syndromes. Cell 184(1):149–168

Kiraz Y, Adan A, Kartal Yandim M, Baran Y (2016) Major apoptotic mechanisms and genes involved in apoptosis. Tumour Biol 37:8471–8486

Knowlton JJ, Dermody TS, Holm GH (2012) Apoptosis induced by mammalian reovirus is beta interferon (IFN) independent and enhanced by IFN regulatory factor 3-and NF-κB-dependent expression of Noxa. J Virol 86(3):1650–1660

Lecoeur H, Borgne-Sanchez A, Chaloin O, El-Khoury R, Brabant M, Langonné A, Porceddu M, Briere JJ, Buron N, Rebouillat D (2012) HIV-1 tat protein directly induces mitochondrial membrane permeabilization and inactivates cytochrome c oxidase. Cell Death Dis 3(3):e282–e282

Li C, Liu J, Hou W, Kang R, Tang D (2021) STING1 promotes ferroptosis through MFN1/2-dependent mitochondrial fusion. Front Cell Develop Biol 9:698679

Maelfait J, Liverpool L, Bridgeman A, Ragan KB, Upton JW, Rehwinkel J (2017) Sensing of viral and endogenous RNA by ZBP 1/DAI induces necroptosis. EMBO J 36(17):2529–2543

Man SM, Karki R, Kanneganti T (2016) AIM2 inflammasome in infection, cancer, and autoimmunity: role in DNA sensing, inflammation, and innate immunity. Eur J Immunol 46(2):269–280

Medigeshi GR, Lancaster AM, Hirsch AJ, Briese T, Lipkin WI, DeFilippis V, Früh K, Mason PW, Nikolich-Zugich J, Nelson JA (2007) West Nile virus infection activates the unfolded protein response, leading to CHOP induction and apoptosis. J Virol 81(20):10849–10860

Merad M, Martin JC (2020) Pathological inflammation in patients with COVID-19: a key role for monocytes and macrophages. Nat Rev Immunol 20(6):355–362

Micheau O, Tschopp J (2003) Induction of TNF receptor I-mediated apoptosis via two sequential signaling complexes. Cell 114(2):181–190

Miotto G, Rossetto M, Di Paolo ML, Orian L, Venerando R, Roveri A, Vučković A-M, Travain VB, Zaccarin M, Zennaro L (2020) Insight into the mechanism of ferroptosis inhibition by ferrostatin-1. Redox Biol 28:101328

Pan P, Shen M, Yu Z, Ge W, Chen K, Tian M, Xiao F, Wang Z, Wang J, Jia Y (2021) SARS-CoV-2 N protein promotes NLRP3 inflammasome activation to induce hyperinflammation. Nat Commun 12(1):4664

Reinert LS, Rashidi AS, Tran DN, Katzilieris-Petras G, Hvidt AK, Gohr M, Fruhwürth S, Bodda C, Thomsen MK, Vendelbo MH (2021) Brain immune cells undergo cGAS/STING-dependent apoptosis during herpes simplex virus type 1 infection to limit type I IFN production. J Clin Invest 131(1)

Ren Y, Wang A, Fang Y, Shu T, Wu D, Wang C, Huang M, Min J, Jin L, Zhou W (2021) SARS-CoV-2 membrane glycoprotein M triggers apoptosis with the assistance of nucleocapsid protein N in cells. Front Cell Infect Microbiol 11:706252

Reshi ML, Su Y-C, Hong J-R (2014) RNA viruses: ROS-mediated cell death. Int J Cell Biol 2014(1):467452

Riebeling T, Jamal K, Wilson R, Kolbrink B, von Samson-Himmelstjerna FA, Moerke C, Ramos Garcia L, Dahlke E, Michels F, Lühder F (2021) Primidone blocks RIPK1-driven cell death and inflammation. Cell Death Different 28(5):1610–1626

Ronco C, Reis T, Husain-Syed F (2020) Management of acute kidney injury in patients with COVID-19. Lancet Respir Med 8(7):738–742

Ruder B, Günther C, Stürzl M, Neurath MF, Cesarman E, Ballon G, Becker C (2020) Viral FLIP blocks Caspase-8 driven apoptosis in the gut in vivo. PLoS One 15(1):e0228441

Santos LD, Antunes KH, Cassão G, Gonçalves JI, Abbadi BL, Bizarro CV, Basso LA, Machado P, de Souza APD, Porto BN (2023) SARS-CoV-2 immune complex triggers human monocyte necroptosis. Int Immunopharmacol 117:109954

Schock SN, Chandra NV, Sun Y, Irie T, Kitagawa Y, Gotoh B, Coscoy L, Winoto A (2017) Induction of necroptotic cell death by viral activation of the RIG-I or STING pathway. Cell Death Different 24(4):615–625

Schoggins JW (2019) Interferon-stimulated genes: what do they all do? Annu Rev Virol 6:567–584

Srinivasula SM, Poyet J-L, Razmara M, Datta P, Zhang Z, Alnemri ES (2002) The PYRIN-CARD protein ASC is an activating adaptor for caspase-1. J Biol Chem 277(24):21119–21122

Sze A, Belgnaoui SM, Olagnier D, Lin R, Hiscott J, van Grevenynghe J (2013) Host restriction factor SAMHD1 limits human T cell leukemia virus type 1 infection of monocytes via STING-mediated apoptosis. Cell Host Microbe 14(4):422–434

Tait SWG, Green DR (2013) Mitochondrial regulation of cell death. Cold Spring Harb Perspect Biol 5(9):a008706

Tan Y-X, Tan THP, Lee MJ-R, Tham P-Y, Gunalan V, Druce J, Birch C, Catton M, Fu NY, Yu VC (2007) Induction of apoptosis by the severe acute respiratory syndrome coronavirus 7a protein is dependent on its interaction with the Bcl-XL protein. J Virol 81(12):6346–6355

Taylor JM, Quilty D, Banadyga L, Barry M (2006) The vaccinia virus protein F1L interacts with Bim and inhibits activation of the pro-apoptotic protein Bax. J Biol Chem 281(51):39728–39739

Theobald SJ, Simonis A, Georgomanolis T, Kreer C, Zehner M, Eisfeld HS, Albert M, Chhen J, Motameny S, Erger F (2021) Long-lived macrophage reprogramming drives spike protein-mediated inflammasome activation in COVID-19. EMBO Mol Med 13(8):e14150

Tran AT, Cortens JP, Du Q, Wilkins JA, Coombs KM (2013) Influenza virus induces apoptosis via BAD-mediated mitochondrial dysregulation. J Virol 87(2):1049–1060

Wang P, Zhu S, Yang L, Cui S, Pan W, Jackson R, Zheng Y, Rongvaux A, Sun Q, Yang G (2015) Nlrp6 regulates intestinal antiviral innate immunity. Science 350(6262):826–830

Wu Y, Ma L, Cai S, Zhuang Z, Zhao Z, Jin S, Xie W, Zhou L, Zhang L, Zhao J (2021) RNA-induced liquid phase separation of SARS-CoV-2 nucleocapsid protein facilitates NF-κB hyper-activation and inflammation. Signal Transduct Target Ther 6(1):167

Xu G, Li Y, Zhang S, Peng H, Wang Y, Li D, Jin T, He Z, Tong Y, Qi C (2021) SARS-CoV-2 promotes RIPK1 activation to facilitate viral propagation. Cell Res 31(12):1230–1243

Yang R, Yang H, Wei J, Li W, Yue F, Song Y, He X, Hu K (2021a) Mechanisms underlying the effects of lianhuaqingwen on sepsis-induced acute lung injury: a network pharmacology approach. Front Pharmacol 12:717652

Yang R, Zhao Q, Rao J, Zeng F, Yuan S, Ji M, Sun X, Li J, Yang J, Cui J (2021b) SARS-CoV-2 accessory protein ORF7b mediates tumor necrosis factor-α-induced apoptosis in cells. Front Microbiol 12:654709

Yang Y, Wu Y, Meng X, Wang Z, Younis M, Liu Y, Wang P, Huang X (2022) SARS-CoV-2 membrane protein causes the mitochondrial apoptosis and pulmonary edema via targeting BOK. Cell Death Different 29(7):1395–1408

Zeng J, Xie X, Feng X-L, Xu L, Han J-B, Yu D, Zou Q-C, Liu Q, Li X, Ma G (2022) Specific inhibition of the NLRP3 inflammasome suppresses immune overactivation and alleviates COVID-19 like pathology in mice. EBioMedicine 75

Zhang X, Yang Z, Pan T, Long X, Sun Q, Wang P-H, Li X, Kuang E (2022) SARS-CoV-2 ORF3a induces RETREG1/FAM134B-dependent reticulophagy and triggers sequential ER stress and inflammatory responses during SARS-CoV-2 infection. Autophagy 18(11):2576–2592

Zhou Q, Snipas S, Orth K, Muzio M, Dixit VM, Salvesen GS (1997) Target protease specificity of the viral serpin CrmA: analysis of five caspases. J Biol Chem 272(12):7797–7800

Zou H, Henzel WJ, Liu X, Lutschg A, Wang X (1997) Apaf-1, a human protein homologous to *C. elegans* CED-4, participates in cytochrome c–dependent activation of caspase-3. Cell 90(3):405–413

# Role of Apoptosis in Male Infertility: Therapeutic Targets and Strategies

# 16

Dibya Pal, Satyajit Halder, Debidas Ghosh,
and Kuladip Jana

**Abstract**

Apoptosis, or programmed cell death, plays a significant role in spermatogenesis and sperm quality. Environmental factors, xenobiotics, chemicals, stress, diabetes, etc., persuade oxidative stress (OS) and reactive oxygen species (ROS) production, which can induce apoptosis in testicular cells, leading to impaired sperm production and functions and ultimately exacerbating male infertility. Apoptosis markers such as caspase activation, phosphatidyl serine externalization, mitochondrial membrane potential changes, and DNA fragmentation are evident in ejaculated spermatozoa of infertile males. ROS significantly impact male infertility through various signaling pathways. Apoptosis also regulates male germ cells from early embryonic stages through fertilization, maintaining Sertoli to germ cell ratios and ensuring testicular quality control. Understanding the molecular pathways linking apoptosis, OS, excess ROS production, and male infertility is crucial for developing targeted interventions to mitigate reproductive complications in men. This chapter highlights the intricate interplay between apoptosis, OS, and male infertility, emphasizing the need for comprehensive management strategies.

**Keywords**

Apoptosis · Reproduction · Male infertility · ROS · DNA fragmentation · Therapeutic strategies

D. Pal · D. Ghosh
Molecular Medicine, Nutrigenomics and Public Health Research Laboratory, Department of Bio-Medical Laboratory Science and Management, Vidyasagar University, Midnapore, West Bengal, India

S. Halder · K. Jana (✉)
Division of Molecular Medicine, Bose Institute, Unified Academic Campus, Kolkata, West Bengal, India
e-mail: kuladip@jcbose.ac.in



K. Jana (ed.), *Apoptosis and Human Health: Understanding Mechanistic and Therapeutic Potential*, https://doi.org/10.1007/978-981-97-7905-5_16

## 16.1 Introduction

Infertility is a significant social and medical concern affecting numerous individuals. Apoptosis, a crucial physiological process marked by chromatin disintegration during programmed cell death, plays a pivotal role in germ cell formation and maintaining the germ cell-Sertoli ratio in the testis (Bansal and Singh 2017). Reactive oxygen species (ROS), identified in the 1940s, were early implicated in male infertility, leading to oxidative stress (OS) that impairs sperm function. Studies suggest that ROS may significantly contribute to infertility in males (Jana and Sen 2012; Bisht et al. 2017), reflecting an imbalance between antioxidants and ROS production in different pathophysiological conditions (Jana and Sen 2012; Bisht et al. 2017). DNA damage, primarily caused by aberrant apoptosis and OS, commonly manifests as DNA fragmentation in spermatozoa of infertile men. ROS-induced damage may lead to single- or double-stranded DNA breaks and induction of apoptosis (Sharma et al. 2015). Semen analysis, following the World Health Organization (WHO) criteria, is pivotal for evaluating volume, sperm count, motility, and morphology, serving as a fundamental step in diagnosing male infertility (World Health Organization 1999). Environmental pollutants, i.e., car exhaust, industrial emissions, pesticides, cigarette smoke, etc., also adversely impact sperm motility, vitality, and acrosome reaction, primarily by increasing ROS production (Brahem et al. 2011).

### 16.1.1 Spermatogenesis

Spermatogenesis is an exceptionally active self-renewal process in the body, generating approximately $10^7$ sperm/day/gram of testicular tissue. This cycle, unique and unalterable for each mammalian species, regulates the duration and organization of cell cycles through interactions with germ cells, supported by Sertoli cells (Goossens and Tournaye 2017; Kuchakulla et al. 2020). The process of spermatogenesis unfolds in three distinct stages such as (i) duplication and differentiation of spermatogonia, (ii) meiosis, and (iii) spermiogenesis, which intricately transforms into round spermatids into complex spermatozoa. In humans, spermatogenesis commences at puberty and persists during a man's lifetime (Goossens and Tournaye 2017). During this intricate process, early apoptosis ensues after the initial round of spermatogenesis in the testes, eliminating excess spermatogonia (Aitken et al. 2011). Dysregulation of apoptosis leads to male infertility by disrupting the balance between germ and Sertoli cell numbers (Rodriguez et al. 1997). Later in life, apoptosis is critical in eliminating germ cells damaged by exposure to environmental toxins, chemotherapy drugs, or heat (Wang et al. 2007). Nearly 75% of spermatogonia undergo apoptosis before reaching maturity, underscoring apoptosis's crucial role in regulating spermatogenesis across mammalian species, including humans (Print and Loveland 2000).

The hypothalamic-pituitary-testicular hormonal axis intricately regulates spermatogenesis. This axis begins with the hypothalamus producing gonadotropin-releasing hormone (GnRH), which helps to secrete follicle-stimulating hormone (FSH) and luteinizing hormone (LH) by stimulating the pituitary gland. LH, in turn, triggers the Leydig cells to release testosterone through steroidogenesis. Testosterone supports the function of Sertoli cells and germ cells. FSH primarily acts on Sertoli cells, playing a vital role in their function. In addition to these hormonal inputs, spermatogenesis is influenced by locally produced factors such as inhibin and activin, steroids like estradiol-17β, and various paracrine and autocrine factors, including cytokines, laminin fragments, and collagen fragments. These components collectively create a specialized microenvironment within the seminiferous epithelium essential for maintaining spermatogenesis (Goossens and Tournaye 2017; Kuchakulla et al. 2020).

While the hypothalamic-pituitary-testicular axis primarily governs spermatogenesis, external factors such as exposure to heavy metals, cigarette smoking, radiation, chemicals/herbs, and medications or infections can impair spermatogenesis through sperm cell damage and infertility (Jana and Sen 2012; Aitken and Koppers 2011).

### 16.1.2 Apoptosis

Apoptosis is a programmed cell death process that eliminates genetically damaged cells. The factors capable of directly entering cells and triggering the apoptotic cascade without specific cell surface receptors include stressors, ROS, heat shock, ultraviolet radiation, drugs, synthetic peptides, and toxins (Shukla et al. 2012). Nowadays, the presence and activation of apoptotic signals in human sperm in response to various stimuli are well established (Grunewald et al. 2009).

Apoptosis can be initiated through two distinct mechanisms such as the extrinsic pathway, involving receptor-mediated apoptosis, and the intrinsic pathway, involving mitochondrial pathways. Specific pathways include the perforin-granzyme A and B and p53 pathways, which induce apoptosis (Paasch et al. 2004). Biochemical hallmarks of apoptosis include the externalization of phosphatidylserine on the plasma membrane, activation of caspases, and DNA fragmentation (Salvesen and Dixit 1997). The caspase activity has been linked to immature sperm, low sperm counts, reduced motility, reduced fertilization rates, and compromised plasma membrane integrity, indicated by phosphatidylserine externalization and apoptosis. Caspases are cysteine proteases crucial for apoptosis in mammals (Shaha et al. 2010). The apoptosis cycle is vital for germ cell development as it facilitates the removal of cells with genetic defects during mitosis and meiosis. In ejaculated semen samples, the proportion of apoptotic sperm from infertile men is reported to be higher compared to healthy individuals (Taylor et al. 2004).

## 16.2 Molecular Pathways of Apoptosis

### 16.2.1 Intrinsic Pathway (Mitochondrial-Mediated)

Several intracellular stress signals, such as OS, DNA damage, or growth factor withdrawal, initiate the intrinsic apoptosis pathway. These stress signals converge on the mitochondria, resulting in mitochondrial outer membrane permeabilization (MOMP) and the release of pro-apoptotic proteins into the cytoplasm. The key regulators of the intrinsic pathway include members of the Bcl-2 family proteins, which function as either proapoptotic (e.g., Bax, Bak) or antiapoptotic (e.g., Bcl-2, Bcl-xL) factors. MOMP leads to the release of cytochrome c from the mitochondria into the cytoplasm, where it interacts with Apaf-1 (apoptotic protease-activating factor 1) to form the apoptosome, a multiprotein complex that activates caspase-9. The activated caspase-9 then initiates a caspase cascade, ultimately leading to the cleavage and activation of downstream effector caspases, such as caspase-3, -6, and -7, which execute the apoptotic program by cleaving various cellular substrates and inducing apoptotic morphological changes (Chinnaiyan et al. 1995; Czabotar et al. 2014). Mitochondria are essential organelles involved in ATP synthesis, calcium signaling, generation of ROS, and regulation of apoptosis. Abnormalities in mitochondrial function can lead to various physiological disorders, including infertility (Rajender et al. 2010).

### 16.2.2 Extrinsic Pathway (Death Receptor-Mediated)

The extrinsic pathway of apoptosis is triggered by the binding of extracellular death ligands, such as tumor necrosis factor-alpha (TNF-α), Fas ligand (FasL), or TNF-related apoptosis-inducing ligand (TRAIL), to their respective cell surface death receptors (e.g., TNF receptor 1, Fas, DR4, DR5). This receptor-ligand interaction induces the formation of a death-inducing signaling complex (DISC) at the cytoplasmic domain of the death receptor, which recruits and activates caspase-8. The activated caspase-8 can directly activate downstream effector caspases, such as caspase-3, -6, and -7, through the extrinsic pathway. Alternatively, caspase-8 can cleave Bid (a BH3-only pro-apoptotic protein), leading to the activation of the intrinsic pathway of apoptosis through the mitochondria-mediated pathway (Chinnaiyan et al. 1995; Czabotar et al. 2014).

## 16.3 Apoptosis in Male Germ Cell

Primordial germ cells originate from the epiblast and migrate to the gonads. Excess cells undergo apoptosis, a process balanced by Bcl-xL and Bax proteins. In transgenic mice, overexpressing Bcl-2 or Bcl-x, inhibiting early apoptosis, leads to an accumulation of spermatogonia and spermatocytes, causing infertility (Rucker et al. 2000). Conversely, Bcl-x knockout mice show severe defects in male germ cell development (Kasai et al. 2003). Transgenic mice lacking Bax or overexpressing Bcl-2 or Bcl-x do not undergo the initial wave of apoptosis, leading to the buildup of spermatogonia and spermatocytes. Consequently, these animals become infertile (Knudson et al. 1995). Bax, Bcl-w, and Bcl-2 are key regulators of germ cell development in post-birth, while Bcl-x is crucial for the survival of primordial germ cells in the embryonic gonad (Jeong and Seol 2008). Fas, a transmembrane protein, is activated by FasL and mediates apoptosis. FasL, expressed in Sertoli cells, interacts with Fas in germ cells, triggering apoptosis (Janssen et al. 2003). Mice with mutations in Fas (Faslpr) or FasL (Faslgld) genes develop lymphoproliferative and autoimmune diseases. The Fas/FasL pathway in Sertoli cells regulates germ cell numbers, leading to hypospermatogenesis and Sertoli cell syndrome (Porcelli et al. 2006). Fas expression in germ cells may be associated with their maturation and apoptosis induction to eliminate damaged cells (Francavilla et al. 2002). Gonadotropin regulates Fas/FasL expression in the human testis, suggesting a role in gamete quality control (Shaha et al. 2010).

## 16.4 Hormones and Male Germ Cell Apoptosis

In the mammalian testes, FSH, LH, GnRH, and testosterone play crucial roles in regulating germ cell proliferation, differentiation, and viability (Bansal and Singh 2017; Jana and Sen 2012). LH stimulates steroidogenesis by activating Leydig cells, while FSH acts on Sertoli cells to support various stages of spermatogenesis (Bansal and Singh 2017; Jana and Sen 2012). Imbalances in hormone levels, whether excess or deficiency, can induce cellular apoptosis in the testis (Bansal and Singh 2017; Jana and Sen 2012). Sertoli cells possess receptors for FSH and testosterone, key spermatogenesis regulators. Withdrawal of these hormones can lead to apoptosis of germ cells (Bansal and Singh 2017; Jana and Sen 2012). In conjunction with FSH and estradiol, testosterone supports germ cell survival during the maturation in seminiferous tubules. However, estradiol alone has an inhibitory effect and can induce germ-cell apoptosis (Shaha et al. 2010). Moreover, excessive testosterone could upregulate the expression of Fas/FasL in the testis (Zhou et al. 2001). Removal of testosterone also stimulates caspase activity and results in DNA fragmentation and apoptosis in Sertoli cells (Tesarik et al. 2002).

## 16.5 Toxicant-Induced Male Germ Cell Apoptosis

Various chemicals and environmental toxins can induce apoptosis in male germ cells. Phthalates, found widely in the environment, cause testicular atrophy by triggering severe apoptosis in germ cells in laboratory animals, often linked to elevated levels of FasL in Sertoli cells (Boekelheide et al. 2009). Nitrobenzene (NB), found in pesticides and pharmaceuticals, has been identified as a testicular toxicant in vivo. Ethanol exposure disrupts the hypothalamic-pituitary-gonadal axis and induces germ/sperm cell apoptosis in the testis (Jana et al. 2010a). Bisphenol-A (BPA), a building block of plastics, is classified as an endocrine disruptor and induces apoptosis in germ cells (Li et al. 2009). Lindane, an organochlorine pesticide, impairs rat testicular function and fertility by affecting NF-kB and FasL pathways (Saradha et al. 2009). However, 2-bromopropane (2-BP), used in pharmaceuticals and dyes, is recognized as a male reproductive toxin (Yu et al. 2001). However, 2,5-hexanedione (2,5-HD), a metabolite of industrial solvents, such as n-hexane, causes apoptosis in male germ cells (Bryant et al. 2008). Benzo(a)pyrene [B(a)P] and nicotine are present in cigarette smoke, industrial emissions, or car exhaust, which induced ROS-mediated male germ cell apoptosis and infertility (Jana and Sen 2012; Banerjee et al. 2016; Jana et al. 2010b).

## 16.6 Heavy Exercise-Induced Germ Cell Apoptosis

Prolonged and strenuous exercise is suggested as a potential contributor to male infertility issues (Safarinejad et al. 2009). Intensive swimming under duress has been identified as a factor causing dysfunction in the male reproductive system. This exercise directly influences the generation of ROS in the testes, potentially disrupting spermatogenesis, steroidogenesis, and the expression of steroidogenic enzymes. As a result, this process may lead to the depletion of germ cells through apoptosis (Jana et al. 2014).

## 16.7 Role of Apoptosis in Male Infertility

Under specific pathological conditions, there is a significant increase in apoptosis of germ cells, leading to idiopathic male infertility (Pareek et al. 2007). Apoptosis predominantly affects spermatocytes, with a lesser impact on spermatogonia and minimal occurrence in spermatids. Fujisawa et al. have reported notable apoptosis, specifically in spermatocytes within the testes (Fujisawa et al. 1999). Research by Martincic et al. has identified substantial germ cell apoptosis in infertile males while noting that Sertoli cells remain unaffected by this process (Aitken and Koppers 2011). The Fas-system is crucial in regulating germ cell apoptosis under normal conditions. The Sertoli cells express FasL, which induces apoptosis in Fas-positive germ cells, highlighting a paracrine interaction between germ and Sertoli cells within the testis (Lee et al. 1997). Apoptosis rates in the testes tend to increase with age, contributing to a decline in germ cells, possibly linked to decreased androgen levels or increased OS and ROS production (Shukla et al. 2012). The monitoring of apoptosis rates can serve as an indicator for guiding the treatment plans for male infertility.

## 16.8 The Role of ROS in Male Germ Cell Apoptosis

In recent years, there has been extensive research into the role of ROS as a significant sponsor of male infertility. ROS production can disrupt semen quality by creating OS, which results from a discrepancy between ROS production and antioxidant defenses (Bansal and Singh 2017; Jana and Sen 2012; Banerjee et al. 2016; Jana et al. 2010b). ROS exposure to mitochondria can trigger apoptosis and lead to DNA fragmentation and apoptosis in a caspase-dependent and independent manner (Banerjee et al. 2016; Jana et al. 2010b; Vashisht and Gahlay 2020) (Fig. 16.1). Research validated a direct link between increased sperm damage from ROS and elevated levels of cytochrome C, caspase 9, and caspase 3 in males, indicating heightened apoptosis (Banerjee et al. 2016; Jana et al. 2010b). Age-related accumulation of ROS can induce lipid peroxidation, further exacerbating oxidative damage to DNA and apoptosis (Gunes et al. 2016). Clinical studies have confirmed that antioxidant supplementation can improve sperm DNA integrity, motility, and fertilizing capacity. For instance, a combination of vitamin C, E, and glutathione over two months significantly reduced the levels of 8OHdG, a marker of ROS-mediated sperm DNA damage (Agarwal and Said 2005). Overproduction of ROS in human semen is strongly associated with male factor infertility (Agarwal et al. 2006). ROS, oxygen-derived free radicals, are necessary at low levels for motility, hyperactivation, capacitation, and the acrosome reaction (Bui et al. 2018). However, elevated ROS levels can lead to OS, resulting in DNA damage, lipid peroxidation, telomere shortening, epigenetic changes, Y chromosome microdeletions, and apoptosis induction in sperm (Bui et al. 2018; Agarwal et al. 2006). Moreover, high ROS concentrations adversely affect sperm count, motility, and male fertility (Agarwal et al.

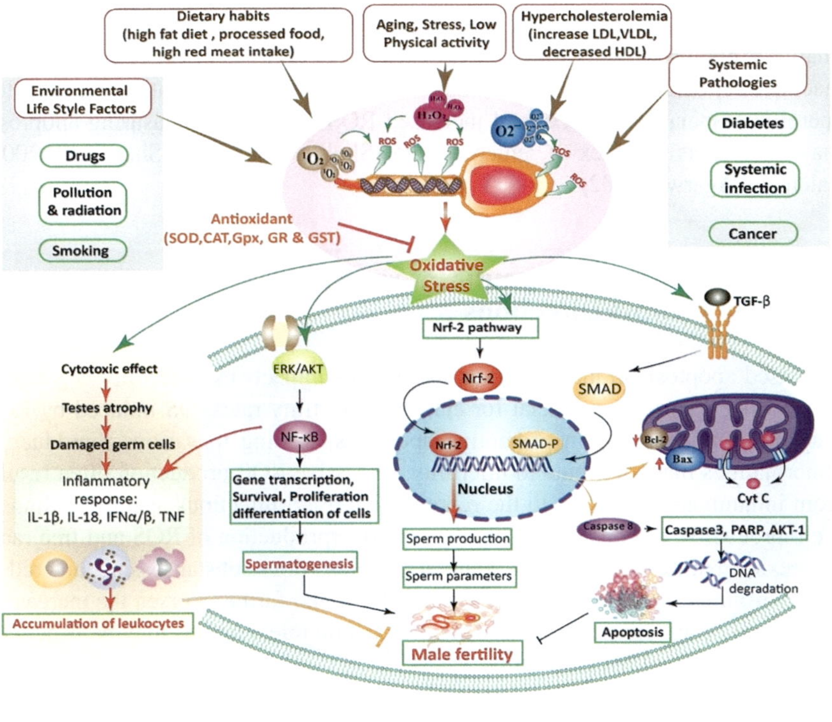


**Fig. 16.1** ROS-induced apoptosis in germ/sperm cells and male infertility

2006; Sawyer et al. 2003). Endogenous sources of ROS include immature sperm and leukocytes present in semen. At the same time, external factors such as genitourinary infections, varicocele, cigarette smoking, alcohol consumption, recreational drug use, ionizing radiation, cell phone use, mental stress, excessive exercise, spinal cord injury, parental age, and environmental pollution, etc., also contribute to ROS generation (Aitken and Clarkson 1987; Agarwal et al. 2014; Sharma et al. 2016). Sperm are predominantly vulnerable to ROS due to their high content of polyunsaturated fatty acids, which are prone to lipid peroxidation. By-products of lipid oxidation such as acrolein, malondialdehyde, and 4-hydroxy-nonenal (4-HNE) indirectly damage DNA (Aitken et al. 2012; Esterbauer et al. 1991). Specifically, 4-HNE and acrolein induce apoptosis and DNA fragmentation (Aitken et al. 2012; Esterbauer et al. 1991). Malondialdehyde is commonly considered as a biochemical marker to assess the extent of peroxidative damage to sperm (Esterbauer et al. 1991; Venkatesh et al. 2009).

## 16.9 Apoptosis as a Marker of Semen Quality

Sperm apoptosis has been identified as a potentially helpful indicator of male fertility. The loss of germ cells via apoptosis is a predominant mechanism during spermatogenesis, regulated by the expressions of p53, p21, Bcl-2, FAS, and caspases (Banerjee et al. 2016; Fujisawa et al. 1999; Lee et al. 1997; Tesarik et al. 1998). Some studies have reported increased apoptosis rates in semen samples of poor quality. Apoptosis occurs during spermatogenesis and has also been observed in ejaculated sperm. Research has demonstrated that DNA fragmentation in ejaculated sperm is associated with OS and increased ROS formation. Measuring apoptosis may be used as an index of semen quality (Shukla et al. 2012; Shen et al. 2002; Saleh and Agarwal 2002).

## 16.10 Infertility: Therapeutic Intervention and Future Directions

Increased apoptosis in male germ or sperm cells impacts male fertility negatively. Controlling apoptosis is crucial for enhancing fertility rates. OS induced by ROS plays a significant role in promoting apoptosis, leading to various reproductive abnormalities that contribute to infertility. Excessive ROS production often results from inflammatory damage to the reproductive tract. Infectious agents residing in the reproductive tract can also stimulate the overproduction of ROS and free radicals, exacerbating inflammation. Xenobiotics, hazardous substances encountered in the environment, pose a risk factor for infertility and further worsen the increasing prevalence of this issue. The following therapeutic interventions may be beneficial in the prevention of male infertility.

**Antioxidant Therapy** Antioxidant supplements, such as vitamin E/C, zinc, selenium, and coenzyme Q10, may help to reduce OS and ROS formation as well as mitigate testicular damage and apoptosis by scavenging ROS and augmenting antioxidant defense mechanisms.

**Antiinflammatory Agents** Antiinflammatory agents, including nonsteroidal anti-inflammatory drugs (NSAIDs), corticosteroids, and cytokine inhibitors, may attenuate inflammation and immune dysregulation in the testes, reducing germ cell apoptosis and preserving spermatogenesis.

**Hormonal Replacement Therapy** Hormonal replacement therapy, such as testosterone replacement therapy (TRT), restores hormonal balance and alleviates hypogonadism-associated symptoms in men, preventing germ/sperm cell apoptosis and improving sexual function and quality of life.

**Pharmacological Modulators of Apoptosis** Pharmacological agents targeting apoptotic pathways, such as caspase inhibitors, BH3 mimetics, and modulators of Bcl-2 family proteins, hold promise as potential therapeutic interventions to prevent or attenuate germ/sperm cell apoptosis in the testes.

**Lifestyle Modifications** Lifestyle modifications, including changes in dietary habits, quitting smoking and alcohol consumption, regular exercise, and weight management, can help to improve testicular functions and decrease germ/sperm cell apoptosis in men.

## 16.11 Conclusion

Apoptosis is a very familiar quality control mechanism in the testis. However, increased apoptosis can adversely affect sperm production, ultimately compromising male fertility. Therefore, regulating the apoptosis rate is crucial for maintaining fertility in men. Additionally, ROS and the resulting OS play a significant role in germ cell apoptosis. ROS have been shown to cause abnormalities in semen analysis and sperm concentration, leading to infertility in men. The ROS and OS contribute to the growing problem of infertility. Antioxidants play a protective role by neutralizing excess ROS generated by environmental toxin exposure. By regulating ROS overproduction, antioxidants may help to maintain sperm cell activity and support reproductive health.

## References

Agarwal A, Said TM (2005) Oxidative stress, DNA damage and apoptosis in male infertility: a clinical approach. BJU Int 95:503–507

Agarwal A, Sharma RK, Nallella KP, Thomas AJ Jr, Alvarez JG, Sikka SC (2006) Reactive oxygen species as an independent marker of male factor infertility. Fertil Steril 86:878–885

Agarwal A, Mulgund A, Sharma R, Sabanegh E (2014) Mechanisms of oligozoospermia: an oxidative stress perspective. Syst Biol Reprod Med 60:206–216

Aitken RJ, Clarkson JS (1987) Cellular basis of defective sperm function and its association with the genesis of reactive oxygen species by human spermatozoa. J Reprod Fertil 81:459–469

Aitken RJ, Koppers AJ (2011) Apoptosis and DNA damage in human spermatozoa. Asian J Androl 13:36–42

Aitken RJ, Findlay JK, Hutt KJ, Kerr JB (2011) Apoptosis in the germ line. Reproduction 141:139–150

Aitken RJ, Whiting S, De Iuliis GN, McClymont S, Mitchell LA, Baker MA (2012) Electrophilic aldehydes generated by sperm metabolism activate mitochondrial reactive oxygen species generation and apoptosis by targeting succinate dehydrogenase. J Biol Chem 287:33048–33060

Banerjee B, Nandi P, Chakraborty S, Raha S, Sen PC, Jana K (2016) Resveratrol ameliorates benzo(a)pyrene induced testicular dysfunction and apoptosis through p38 MAPK/ATF2/iNOS signalling. J Nutri Biochem 34:17–29

Bansal SK, Singh R (2017) Specific and generalized treatments of male infertility. In: SINGH R, Singh K (eds) Male infertility: understanding, causes and treatment. Springer, Singapore, pp 449–464

Bisht S, Faiq M, Tolahunase M, Dada R (2017) Oxidative stress and male infertility. Nat Rev Urol 14:470–485

Boekelheide K, Kleymenova E, Liu K, Swanson C, Gaido KW (2009) Dose-dependent effects on cell proliferation, seminiferous tubules, and male germ cells in the fetal rat testis following exposure to di-butyl phthalate. Microsc Res Tech 72:629–638

Brahem S, Mehdi M, Elghezal H, Saad A (2011) The effects of male aging on semen quality, sperm DNA fragmentation and chromosomal abnormalities in an infertile population. J Assist Reprod Genet 28:425–432

Bryant BH, Yamasaki H, Sandrof MA, Boekelheide K (2008) Spermatid head retention as a marker of 2, 5-hexanedione-induced testicular toxicity in the rat. Toxicol Pathol 36:552–559

Bui AD, Sharma R, Henkel R, Agarwal A (2018) Reactive oxygen species impact on sperm DNA and its role in male infertility. Andrologia 50:e13012

Chinnaiyan AM, O'Rourke K, Tewari M, Dixit VM (1995) FADD, a novel death domain-containing protein, interacts with the death domain of Fas and initiates apoptosis. Cell 81(4):505–512

Czabotar PE, Lessene G, Strasser A, Adams JM (2014) Control of apoptosis by the BCL-2 protein family: implications for physiology and therapy. Nat Rev Mol Cell Biol 15(1):49–63

Esterbauer H, Schaur RJ, Zollner H (1991) Chemistry and biochemistry of 4-hydroxynonenal, malonaldehyde and related aldehydes. Free Radic Biol Med 11:81–128

Francavilla S, D'Abrizio P, Cordeschi G, Pelliccione F, Necozione S, Ulisse S et al (2002) Fas expression correlates with human germ cell degeneration in meiotic and post-meiotic arrest of spermatogenesis. Mol Hum Reprod 8:213–220

Fujisawa M, Hiramine C, Tanaka H, Okada H, Arakawa S, Kamidono S (1999) Decrease in apoptosis of germ cells in the testes of infertile men with varicocele. World J Urol 17:296–300

Goossens E, Tournaye H (2017) Spermatogenesis: clinical and experimental considerations. In: De Jonge CJ, Barratt CLR (eds) The sperm cell: production, maturation, fertilization, regeneration. Cambridge University Press, Cambridge, pp 1–20

Grunewald S, Sharma R, Paasch U, Glander HJ, Agarwal A (2009) Impact of caspase activation in human spermatozoa. Microsc Res Tech 72:878–888

Gunes S, Hekim GNT, Arslan MA, Asci R (2016) Effects of aging on the male reproductive system. J Assist Reprod Genet 33:441–454

Jana K, Sen PC (2012) Environmental toxicants induced male reproductive disorders: identification and mechanism of action. In: Acree W (ed) Toxicity & drug testing. In Tech Publication, Croatia, pp 473–518

Jana K, Jana N, De DK, Guha SK (2010a) Ethanol induces mouse spermatogenic cell apoptosis in vivo through over expression of Fas/Fas-L, p53 and caspase-3 along with cytochrome c translocation and glutathione depletion. Mol Reprod Dev 77:820–833

Jana K, Samanta PK, De DK (2010b) Nicotine diminishes testicular gametogenesis, steroidogenesis and steroidogenic acute regulatory protein expression in adult albino rats: possible influence on pituitary gonadotropins and alters testicular antioxidant status. Toxicol Sci 116:647–659

Jana K, Dutta A, Chakraborty P, Manna I, Firdaus SB, Bandyopadhyay D, Chattopadhyay R, Chakravarty B (2014) α- lipoic acid and N-acetylcysteine protects intensive swimming exercise mediated germ cell depletion, pro-oxidant generation and alteration of steroidogenesis in rat testis. Mol Reprod Dev 8:833–850

Janssen O, Qian J, Linkermann A, Kabelitz D (2003) CD95 ligand—death factor and costimulatory molecule? Cell Death Differ 10:1215–1225

Jeong SY, Seol DW (2008) The role of mitochondria in apoptosis. BMB Rep 41:11–22

Kasai S, Chuma S, Motoyama N, Nakatsuji N (2003) Haploinsufficiency of Bcl-x leads to male-specific defects in fetal germ cells: differential regulation of germ cell apoptosis between the sexes. Dev Biol 264:202–216

Knudson CM, Tung KS, Tourtellotte WG, Brown GA, Korsmeyer SJ (1995) Bax-deficient mice with lymphoid hyperplasia and male germ cell death. Science 270:96–99

Kuchakulla M, Narasimman M, Khodamoradi K, Khosravizadeh Z, Ramasamy R (2020) How defective spermatogenesis affects sperm DNA integrity. Andrologia 53:13615

Lee J, Richburg JH, Younkin SC, Boekelheide K (1997) The Fas system is a key regulator of germ cell apoptosis in the testis. Endocrinology 138:2081–2088

Li YJ, Song TB, Cai YY, Zhou JS, Song X, Zhao X, Wu XL (2009) Bisphenol A exposure induces apoptosis and upregulation of Fas/FasL and caspase-3 expression in the testes of mice. Toxicol Sci 108:427–436

Paasch U, Grunewald S, Agarwal A, Glandera HJ (2004) Activation pattern of caspases in human spermatozoa. Fertil Steril 81:802–809

Pareek TK, Joshi AR, Sanyal A, Dighe RR (2007) Insights into male germ cell apoptosis due to depletion of gonadotropins caused by GnRH antagonists. Apoptosis 12:1085–1100

Porcelli F, Meggiolaro D, Carnevali A, Ferrandi B (2006) Fas ligand in bull ejaculated spermatozoa: a quantitative immunocytochemical study. Acta Histochem 108:287–292

Print CG, Loveland KL (2000) Germ cell suicide: new insights into apoptosis during spermatogenesis. BioEssays 22:423–430

Rajender S, Rahul P, Mahdi AA (2010) Mitochondria, spermatogenesis and male infertility. Mitochondrion 10:419–428

Rodriguez I, Ody C, Araki K, Garcia I, Vassalli P (1997) An early and massive wave of germinal cell apoptosis is required for the development of functional spermatogenesis. EMBO J 16:2262–2270

Rucker EB, Dierisseau P, Wagner KU, Garrett L, Wynshaw-Boris A, Flaws JA, Hennighausen L (2000) Bcl-x and Bax regulate mouse primordial germ cell survival and apoptosis during embryogenesis. Mol Endocrinol 14:1038–1052

Safarinejad MR, Azma K, Kolahi AA (2009) The effects of intensive, long-term treadmill running on reproductive hormones, hypothalamus-pituitary-testis axis, and semen quality: a randomized controlled study. J Endocrinol 200:259–271

Saleh RA, Agarwal A (2002) Oxidative stress and male infertility: from research bench to clinical practice. J Androl 23:737–752

Salvesen GS, Dixit VM (1997) Caspases: intracellular signaling by proteolysis. Cell 91:443–446

Saradha B, Vaithinathan S, Mathur PP (2009) Lindane induces testicular apoptosis in adult Wistar rats through the involvement of Fas–FasL and mitochondria-dependent pathways. Toxicology 255:131–139

Sawyer DE, Mercer BG, Wiklendt AM, Aitken RJ (2003) Quantitative analysis of gene-specific DNA damage in human spermatozoa. Mutat Res 529:21–34

Shaha C, Tripathi R, Mishra DP (2010) Male germ cell apoptosis: regulation and biology. Philos Trans R Soc B Biol Sci 365:1501–1515

Sharma R, Agarwal A, Rohra VK, Assidi M, Abu-Elmagd M, Turki RF (2015) Effects of increased paternal age on sperm quality, reproductive outcome and associated epigenetic risks to offspring. Reprod Biol Endocrinol 13:35

Sharma R, Harlev A, Agarwal A, Esteves SC (2016) Cigarette smoking and semen quality: a new meta-analysis examining the effect of the 2010 World Health Organization laboratory methods for the examination of human semen. Eur Urol 70:635–645

Shen HM, Dai J, Chia SE, Lim A, Ong CN (2002) Detection of apoptotic alterations in sperm in subfertile patients and their correlations with sperm quality. Hum Reprod 17:1266–1273

Shukla KK, Mahdi AA, Rajender S (2012) Apoptosis, spermatogenesis and male infertility. Front Biosci 4:746–754

Taylor SL, Weng SL, Fox P, Duran EH, Morshedi MS, Oehninger S, Beebe SJ (2004) Somatic cell apoptosis markers and pathways in human ejaculated sperm: potential utility as indicators of sperm quality. Mol Hum Reprod 10(11):825–834

Tesarik J, Greco E, Cohen-Bacrie P, Mendoza C (1998) Germ cell apoptosis in men with complete and incomplete spermiogenesis failure. Mol Hum Reprod 4:757–762

Tesarik J, Martinez F, Rienzi L, Iacobelli M, Ubaldi F, Mendoza C, Greco E (2002) In-vitro effects of FSH and testosterone withdrawal on caspase activation and DNA fragmentation in different cell types of human seminiferous epithelium. Hum Reprod 17:1811–1819

Vashisht A, Gahlay GK (2020) Using miRNAs as diagnostic biomarkers for male infertility: opportunities and challenges. Mol Hum Reprod 26:199–214

Venkatesh S, Riyaz AM, Shamsi MB, Kumar R, Gupta NP, Mittal S et al (2009) Clinical significance of reactive oxygen species in semen of infertile Indian men. Andrologia 41:251–256

Wang C, Cui YG, Wang XH, Jia Y, Sinha Hikim A, Lue YH et al (2007) Transient scrotal hyperthermia and levonorgestrel enhance testosterone-induced spermatogenesis suppression in men through increased germ cell apoptosis. J Clin Endocrinol Metab 92:3292–3304

World Health Organization (1999) WHO laboratory manual for the examination of human semen and sperm-cervical mucus interaction. Cambridge University Press, Cambridge

Yu X, Kubota H, Wang R, Saegusa J, Ogawa Y, Ichihara G, Takeuchi Y, Hisanaga N (2001) Involvement of Bcl-2 family genes and Fas signaling system in primary and secondary male germ cell apoptosis induced by 2-bromopropane in rat. Toxicol Appl Pharmacol 174:35–48

Zhou XC, Wei P, Hu ZY, Gao F, Zou RJ, Liu YX (2001) Role of Fas/FasL genes in azoospermia or oligozoospermia induced by testosterone undecanoate in rhesus monkey. Acta Pharmacol Sin 22:1028–1033

# Role of Apoptosis in Aging: Therapeutic Targets and Strategies

17

Anirban Manna, Chirantan Majumder, Satyajit Halder, and Kuladip Jana

**Abstract**

Aging, a natural phenomenon, is characterized by a gradual deterioration of the physiological functions of an organism. The increase in life expectancy of humans has given rise to numerous aging-associated diseases. A detailed understanding of the molecular mechanisms of aging might pave the way for combating such diseases. In this chapter, we discuss the role of programmed cell death (PCD) pathways in aging, with special emphasis on apoptosis. These pathways are required for maintaining normal turnover of cells and tissue homeostasis. Improper regulation of PCD is reported to be responsible for aging and its associated diseases. Aging is characterized by an enhancement of PCD in certain cell types and a diminution in others. Aging-associated enhancement of PCD is responsible for neurodegenerative diseases, sarcopenia, loss of cardiomyocytes, and a weakening of the immune system. On the other hand, senescent cells exhibit resistance to PCD and increase in numbers during aging.

**Keywords**

Aging · Apoptosis · Senescence · Neurodegenerative diseases · Sarcopenia · Immunity

## 17.1 Introduction

An introduction to different forms of cell death:

A. Manna · C. Majumder · S. Halder · K. Jana (✉)
Division of Molecular Medicine, Bose Institute, Unified Academic Campus, Kolkata, West Bengal, India
e-mail: kuladip@jcbose.ac.in





K. Jana (ed.), *Apoptosis and Human Health: Understanding Mechanistic and Therapeutic Potential*, https://doi.org/10.1007/978-981-97-7905-5_17

Programmed cell death (PCD) and necrosis are the two broad classes of cell death. The occurrence of the former depends on energy, while the latter is known to elicit a powerful immune response (Elmore 2007; Proskuryakov and Gabai 2010; Taylor et al. 2008). However, the concept of a complete lack of regulation in necrosis, as suggested previously, has been refuted by researchers, and the mechanisms of the two forms of cell death have been shown to overlap under certain conditions (Galluzzi et al. 2012).

Apoptosis, called type-I PCD, is the canonical form of PCD (Elmore 2007; Kerr et al. 1972). During apoptosis, cells are known to exhibit characteristic morphological and biochemical changes. Movement of phosphatidyl serine from the inner to the outer leaflet of the cell's plasma membrane is considered to be a hallmark of apoptosis, which can be easily studied via Annexin-V staining of the cells (Hankins et al. 2015; Leventis and Grinstein 2010). Macrophages in the vicinity of the apoptotic cells can recognize the externalized phosphatidyl serine as a signal to engulf the cells. The earliest morphological changes associated with apoptotic cells are cell shrinkage and nuclear and chromatin condensation, followed by subsequent nuclear and chromatin fragmentation. The formation of cell membrane blebs and, subsequently, multiple membrane-bound apoptotic bodies (that bud off from the cell) also characterize apoptosis. These apoptotic bodies readily get engulfed by the macrophages present nearby. As a result, the contents of the apoptotic cells are not released in the interstitial space. The macrophages also do not release proinflammatory cytokines. Hence, no immune response is induced by cells undergoing apoptosis.

A class of cysteine proteases, known as caspases, is found in the cytoplasm of cells as inactive procaspases (Shalini et al. 2015). These undergo self-activation under the right conditions via autocatalytic cleavage. The self-activated caspases can cleave and activate other caspases, initiating a caspase activation chain. Caspase-9, known as the "initiator caspase," can cause the cleavage and activation of several "effector caspases," such as caspase-3. These effector caspases target and cleave multiple cellular proteins, including cytoskeletal proteins. Mitochondria play a crucial role in regulating the apoptotic process in mammalian cells. The members of the Bcl-2 family of cellular proteins are both pro- and antiapoptotic, and there exists a balance of these two kinds of proteins in cells. Any disturbance in this balance can lead the cells down the apoptotic pathway. The proapoptotic proteins, like Bid, Bim, and Puma, have been shown to facilitate the permeabilization of the mitochondrial membrane via the activation of two other members of the Bcl-2 family: Bad and Bax. These two proteins are involved in pore formation in the mitochondrial membrane (Westphal et al. 2014). The antiapoptotic Bcl-2 family members, Bcl-2, Bcl-x, and Bcl-xL, interfere with mitochondrial membrane pore formation by binding with the proapoptotic proteins (Fig. 17.1).

Regulation of morphological alterations of the mitochondria is partly attributed to DRP1, a protein involved in mitochondrial fission (Otera et al. 2013). In apoptotic cells, DRP1-driven fission is responsible for mitochondrial membrane permeabilization (Frank et al. 2001). A family of caspase inhibitors (Inhibitors of Apoptosis Proteins (IAPs)) is also found in the cytosol. It has been reported that rather than

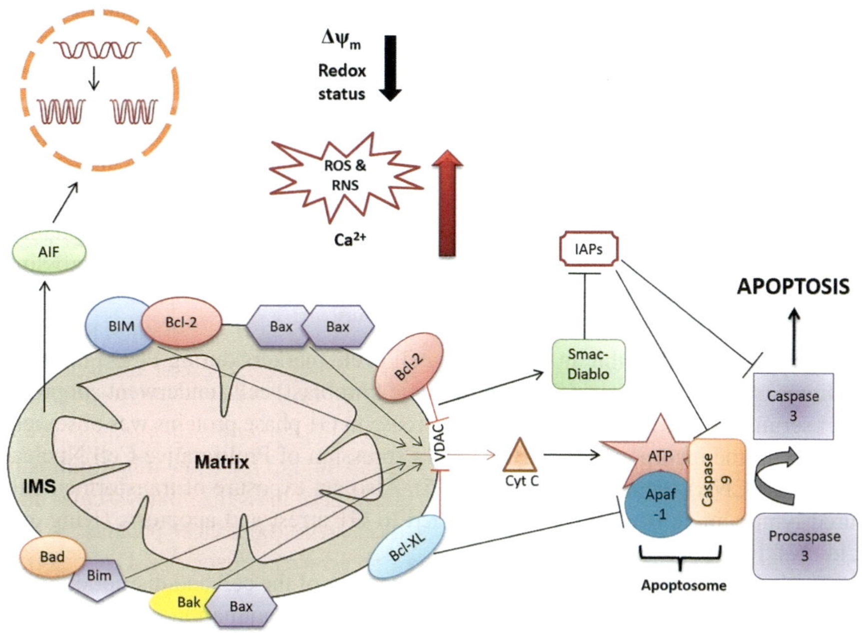


**Fig. 17.1** A depiction of the mitochondria-mediated apoptotic pathway. Several factors are known to cause improper functioning of mitochondria. These are a collapse of the mitochondrial membrane potential ($\Delta\psi_m$), increased calcium ions, increased ROS and RNS production, and lowered redox status (glutathione, NADH, and ATP). The mitochondrial outer membrane voltage-dependent anion channel (VDAC) is influenced by multiple mitochondrial proteins, such as Bcl-2, Bcl-$X_L$, Bid, and Bax, and their specific ratios. This causes the release of cytochrome c from the mitochondria to the cytosol, which leads to the formation of the apoptosome, and subsequently, apoptosis occurs. Apoptosis-inducing factor (AIF), present in the mitochondrial intermembrane space (IMS), undergoes nuclear translocation upon its release from the mitochondria and causes DNA fragmentation. Smac-Diablo is released into the cytosol along with cytochrome c; it inhibits the Inhibitors of Apoptosis (IAPs), promoting apoptosis. Notably, different cell types exhibit differences in the signals and proteins that cause apoptosis (Pollack and Leeuwenburgh 2001)

preventing cell death, in vivo caspase inhibition in mice alters the death mechanism to a more necrosis-like death (Rabinovich 2015).

Both intrinsic and extrinsic death signals are known to initiate apoptosis. The intrinsic apoptotic pathway is prompted by stressful conditions like oxidative stress, irradiation, DNA damage, hypoxia, and activation of oncogenes. These cues signal changes in the expression and/or stability of proapoptotic factors, ultimately causing permeabilization of the mitochondrial membrane and subsequent movement of mitochondrial proapoptotic factors (like cytochrome c, Smac/DIABLO, HtrA2/Omi) to the cytosol. Cytochrome c associates with two cytosolic proteins, Apaf-1 and procaspase-9, to form the apoptosome, which in turn causes proteolysis-mediated activation of procaspase-9 to caspase-9. Apoptosis is further facilitated by the Smac/DIABLO and HtrA2/Omi mediated inactivation of IAPs. DNA fragmentation and nuclear condensation are caused by the mitochondrial proteins Apoptosis

Inducing Factor (AIF), Endonuclease G (EndoG), and Caspase-Activated DNAse (CAD), which exert their functions in the nucleus. However, EndoG and AIF are also known to be involved in caspase-independent PCD. Hormone-mediated repression of apoptosis has been reported in multiple cell types, and withdrawal of these hormones is known to prompt the intrinsic apoptotic pathway (Raff et al. 1994).

The Tumor Necrosis Factor (TNF) receptor family is a class of transmembrane receptors that share an intracellular "death domain," for which they are also known as "death receptors." The extrinsic apoptotic pathway is mediated via signaling through these receptors. TNF-α activates the TNF receptor-1 in the cells expressing this receptor. This, in turn, leads to the activation of initiator caspase-8 and, ultimately, apoptotic cell death.

Some cell types exhibit an increase in cell cycle markers during PCD initiation. Cultures of quiescent mouse 3 T3 (embryonic fibroblast) cells underwent apoptosis after serum deprivation. During this, an increase in G1 phase proteins was observed, along with the incorporation of BrdU and expression of Proliferative Cell Nuclear Antigen (PCNA) (Pandey and Wang 1995). Also, an exposure of transformed cultured lymphoid cells to glucocorticoids led to G1 arrest and apoptosis (King and Cidlowski 1995, 1998).

The hallmarks of necrotic cell death are swelling of the cell and disruption of its plasma membrane (Proskuryakov and Gabai 2010). Inflammation is caused when cytoplasmic contents of the cell undergoing necrosis are released into the interstitial space. While exposure of certain cell types to moderate heat and radiation results in apoptosis, extreme levels of these stimuli lead to necrotic death. Researchers have, however, shown that unless a cell is exposed to stressful conditions of extreme nature, necrosis is caused by certain physiological and pathological stimuli and is a regulated process. Hence, it is called type-III PCD or necroptosis (Degterev et al. 2014). Necroptosis is a caspase-independent process. Stress cues such as hypoxia, ischemia, and TNFR family signaling have been shown to cause necroptosis in some cell types, where the RIP1 kinase (Receptor-Interacting serine/threonine protein kinase), JNK (c-Jun N-terminal Kinase) signaling pathway and the mitochondria play vital roles (Galluzzi et al. 2009; Pasparakis and Vandenabeele 2015). Some forms of necroptosis are characterized by cathepsin and/or calpain (cysteine proteases) mediated cleavage of mitochondrial AIF and its subsequent release from the mitochondria. This AIF then moves to the nucleus, the movement aided by cyclophilin A and inhibited by Hsp70 (Delavallée et al. 2011; Gurbuxani et al. 2003; Zhu et al. 2007). The interaction of AIF with EndoG (translocated from mitochondria) in the nucleus causes nuclear DNA fragmentation. Thus, several types of apoptosis and necroptosis may have overlapping mechanisms in the context of different cell types and stress cues.

Type-II PCD or autophagic cell death is another form of PCD (Galluzzi et al. 2012; Kroemer et al. 2009). In macroautophagy, cellular organelles, and the bulk cytoplasm are enclosed inside double membrane-bound vesicles (autophagosomes). Fusion of the autophagosomes with lysosomes results in clearance of the enclosed cellular contents. Autophagy is not associated with condensation of the chromatin. The formation of autophagic vacuoles in the cytoplasm is a hallmark of autophagy.

Macroautophagy plays a protective role in cells and prevents apoptosis. Therefore, autophagy-mediated cell death could result from the hyperactivation of this useful process.

A form of PCD known as aponeurosis is characterized by overlapping features of both apoptotic and necrotic cell death mechanisms (Papucci et al. 2004; Yakovlev and Faden 2004). Another form of PCD, called paraptosis, does not depend on the activation of caspases. Vacuole formation in the cytoplasm is a hallmark of paraptosis; it also requires gene transcription and translation. In cultured mammalian cells, it can be prompted by Insulin-like Growth Factor-1(IGF-1) signaling; it is also known to occur in vivo (Sperandio et al. 2004). Upon losing contact with the extracellular matrix, cells undergo a kind of apoptotic death known as anoikis, which can follow intrinsic or extrinsic pathways depending on the cell type. Metastatic cancer cells exhibit resistance to anoikis (Paoli et al. 2013; Taddei et al. 2012; Zhong and Rescorla 2012). Keratinocytes in the epidermis exhibit a special caspase-14-dependent PCD known as cornification. Here, the dead cells help to form the skin's outermost layer (Eckhart et al. 2013; Galluzzi et al. 2012) (Fig. 17.2).

## 17.2 PCD and Aging

As is well known, PCD plays a pivotal role in human development. Elimination of cells via PCD is responsible for the creation of interdigital spaces in the hands and feet (Chimal-Monroy et al. 2011; Hernández-Martínez and Covarrubias 2011). PCD is also crucial for brain development (Burek and Oppenheim 1996; Oppenheim 1991). In various adult tissues, the harmony of cell division and PCD is of utmost importance for the normal turnover of cells. These tissues include the intestinal epithelium, blood, epidermis, kidney, and lungs (Elmore 2007; Günther et al. 2013; Kile 2014). To replenish the cells destroyed by PCD, roughly 10 billion cells are formed daily in an adult individual (Renehan et al. 2001). During aging, PCD is known to be misregulated: suboptimal occurrence of PCD in some cell types and higher-than-optimal occurrence in others characterize aging (Gupta 2005; Higami and Shimokawa 2000; Lu et al. 2012; Salminen et al. 2011; Shen and Tower 2009; Warner 1997). In human adipose mesenchymal stem cells (MSCs), a lower expression of apoptosis genes and enhanced expression of senescence-related genes are hallmarks of aging (Alt et al. 2012). Mouse bone marrow MSCs exhibited lower expression of cell cycle- and PCD-related genes during aging (Wilson et al. 2010). When aged rats were exposed to methyl methane sulfonate (a DNA-damaging agent), the liver exhibited diminished PCD. This was characterized by enhanced ERK phosphorylation in the aged rat tissues (Suh et al. 2002; Suh 2002). Peripheral blood lymphocytes from aged mice displayed significantly diminished apoptosis when exposed to radiation; this also holds for cells from aged human patients (Polyak et al. 1997; Camplejohn et al. 2003). As expected, normal aging is characterized by a diminution of apoptotic markers in human serum (Kavathia et al. 2009).

The formation of neointima as a consequence of vascular injury is increased in aged patients. This is partly attributed to decreased apoptosis and increased

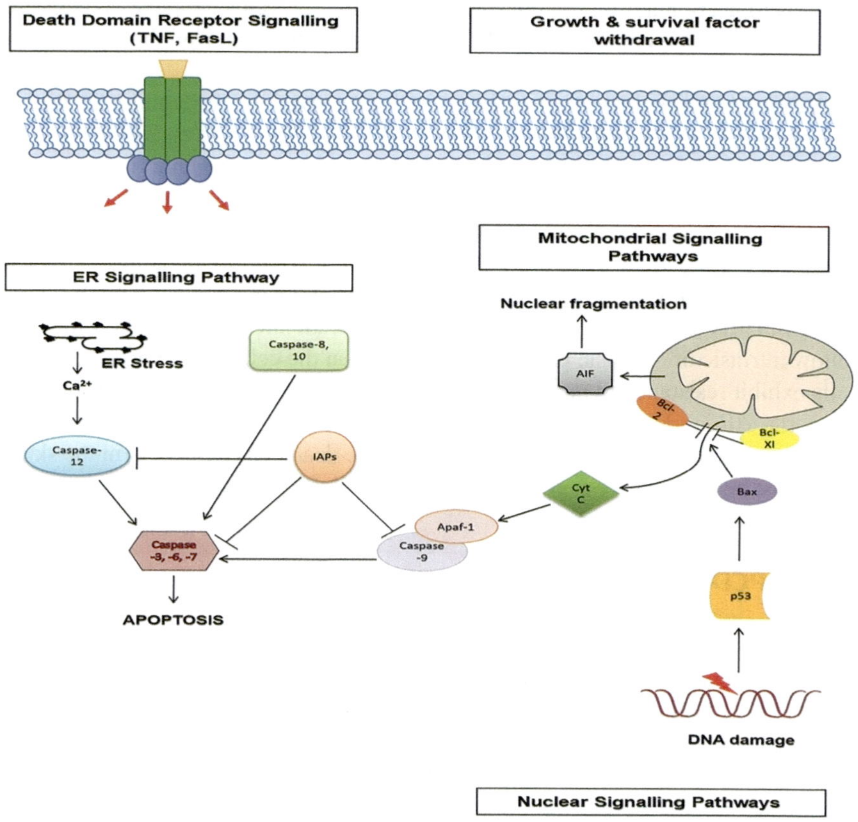


**Fig. 17.2** A simplified depiction of the pathways involved in apoptotic cell signaling. The mitochondria-mediated pathway has been described in Fig. 17.1. Endoplasmic Reticulum (ER) stress activates the ER-mediated pathway, causing a rise in intracellular calcium concentration, leading to caspase-12 activation. Ligand binding and withdrawal of growth and survival factors can activate the receptor-mediated pathways, activating caspase-8 & -10. Nuclear signaling pathways promote apoptosis by DNA damage-dependent p53 signaling (Pollack and Leeuwenburgh 2001)

proliferation of vascular smooth muscle cells. This phenomenon was also observed in aging mice (Vazquez-Padron et al. 2004). Of note, the observed decrease in apoptosis of the vascular smooth muscle cells could be associated with reduced signaling from endothelial cells in the vicinity (Qian et al. 2011).

Dysregulation of PCD and the phagocytosis-mediated removal of the cellular debris results in an inflammatory response in aged individuals. Here, boosting the NF-κB signaling pathway is known to occur (Gupta et al. 2006; Salvioli et al. 2013; Shaw et al. 2010).

Apoptosis has been shown to play a pivotal role in developing the immune system and maintaining immune homeostasis in adults. It aids in the clearance of self-reactive T and B cells. It is also essential for the regulation of immunological memory and for decreasing the number of effector T-cells postimmune responses.

Target cells, such as virus-infected ones, are cleared through apoptosis by cytotoxic T-lymphocytes and natural killer cells. The number of apoptotic lymphocytes in the thymus and bone marrow is enhanced in aged individuals, resulting in the loss of cells in these sites (Sainz et al. 2003). CD4+ and CD8+ T cells exhibit increased apoptotic death in aged individuals (Gupta 2005). However, in vitro studies revealed some T cell subtypes (CD8+ CD28−) to show senescence characteristics, like diminution of stress response and resistance to apoptosis (Chou and Effros 2013). The hallmarks of T-cell replicative senescence are lack of regulation in immune functions, downregulation of the costimulatory molecule CD28, reduction in telomere length, and enhanced proinflammatory cytokine production. Senescent T-cells in aged individuals commonly show specificity against cytomegalovirus (CMV); CMV infection could be responsible for immune senescence. Notably, senescent T-cells are also found in patients suffering from certain cancers, autoimmune disorders, and chronic infections (Chou and Effros 2013).

The death of heart myocytes (involving both apoptotic and necrotic mechanisms) and hypertrophy of the remaining myocytes are the hallmarks of aging of the mammalian heart (Olivetti et al. 1991; Kajstura et al. 1996; Kung et al. 2011).

The wound healing process, where apoptosis plays a vital role, is compromised in aged individuals, sometimes involving local inflammation (Hiebert and Granville 2012; Rai et al. 2005). Age-related increase in the occurrence of fibrotic lung disease has been attributed to the lack of proper regulation of apoptosis. While alveolar epithelial cells exhibit greater sensitivity to apoptosis, causing increased injury, fibroblasts and myofibroblasts show lowered sensitivity to apoptosis, contributing to increased fibrosis (Kapetanaki et al. 2013).

Several independent studies have reported a reduction in the normal cellular turnover rate in various organs like skin, heart, pituitary, and breast epithelium with increasing age (Rinnerthaler et al. 2015; Bergmann et al. 2009; Misell et al. 2005; Nolan et al. 1999; Richardson et al. 2014). This could be partly responsible for the general age-associated reduction in PCD markers.

## 17.3 PCD and Cellular Senescence

A senescent cell is a cell that has reached a state of irreversible cell cycle arrest. This could be due to several reasons, such as repeated cycles of cell division and a decrease in the length of the telomeres, structural disorganization of the chromatin, oxidative stress, and activation of oncogenes. A cell in which p53 has been activated can either die by apoptosis or reach the senescent state; its fate depends on the cell type and the nature of the stress to which it is exposed (Zuckerman et al. 2009). An assemblage of senescent cells in multiple tissues of aged individuals can be attributed to the cells' resistance to apoptosis (Wang 1995; Kirkland and Tchkonia 2015). Senescence-associated secretory phenotype (SASP) is a hallmark of senescent cells. It is characterized by releasing proinflammatory cytokines and ECM-degrading enzymes from such cells (Tchkonia et al. 2013). Thus, an assemblage of senescent cells in aged individuals might contribute to the degradation of the ECM and aging-associated inflammation. Studies on the senescent human fibroblasts in culture have revealed their inability to decrease Bcl-2 upon exposure to apoptotic stimuli, such as serum withdrawal and oxidative stress (Wang 1995, 1997; Ryu et al. 2007; Yeo et al. 2000). These cells exhibited a decrease in the expression of caspase-3 and an enhancement in the expression of the antiapoptotic protein Bcl-xL (Marcotte et al. 2004; Rochette and Brash 2008). A decrease in the cell cycle factors (contributing to cell division and death) might also be associated with the senescent cells' resistance to apoptosis (King and Cidlowski 1995). Interestingly, the selective destruction of senescent cells in mice by genetic and chemical means was found to improve the health of the animals with a higher level of such cells (Zhu et al. 2015).

## 17.4 Role of PCD in Aging-Associated Neurodegenerative Diseases

No overall decrease in neurons has been recorded during the normal aging process in human beings and other mammals. Possibly, PCD does not usually occur during normal aging of the brain (Andersen et al. 2003; Giannaris and Rosene 2012; Long et al. 1999; Merrill et al. 2001; Rapp and Gallagher 1996). But, PCD does play a significant role in aging-associated neurodegenerative disorders (Ghavami et al. 2014). Synaptic plasticity and some forms of memory in the normal adult brain require the activation of caspase-3 (Snigdha et al. 2012). Stress-exposed neurons may be susceptible to improper PCD induction caused by active caspase-3. Furthermore, the iron content of nervous tissue has been reported to be enhanced during aging and in patients suffering from neurodegenerative disorders (Oshiro et al. 2011). Excessive iron leads to an abundance of ROS, causing oxidative stress. Partial repression of PCD in neurons occurs via the formation of reducing equivalents (employing glucose metabolism). These reducing equivalents block the oxidation and activation of cytochrome c (Vaughn and Deshmukh 2008). Thus, interruptions in the neuronal metabolism could render these cells vulnerable to PCD activation.

The formation and piling up of the aggregates of certain proteins is the hallmark of aging-associated neurodegenerative disorders, such as Parkinson's, Alzheimer's, and Huntington's diseases. Cells usually get rid of such aggregates via macroautophagy. Macroautophagy is also implicated in the usual turnover of cellular mitochondria. Disturbances in mitochondrial functions and consequent oxidative stress play significant roles in these disorders. Thus, interruptions in the autophagic process lead to the accumulation of damaged and improperly functioning mitochondria, which cause PCD (Ghavami et al. 2014; Hroudová et al. 2014; Nixon and Yang 2012).

The exact mechanisms of cell death are debatable in the above-mentioned neurodegenerative disorders. Enhancement or changes in the activity of p53 have been implicated in Alzheimer's disease (Lanni et al. 2012). A programmed necrosis pathway involving PARP is responsible for Parkinson's disease in animal model-based studies (Lee et al. 2014). A significant upregulation of cell cycle proteins has been found in the brains of patients suffering from neurodegenerative disorders. Oxidative stress and DNA damage occur in neuronal cells in vitro, leading to an abortive cell cycle (ACC) entry and, ultimately, apoptosis; this phenomenon could also happen in vivo. Thus, the cell cycle reentry could be an important mechanism in the etiology of these diseases (Folch et al. 2012).

## 17.5 Cell Death in Various Tissues of Aged Animals

PCD or similar processes are implicated in aging-associated cell loss in various tissues. Sarcopenia, an age-related wasting away of skeletal muscles in mammals, is characterized by the loss of muscle fibers and a diminution in the size of the muscle fibers (Marzetti et al. 2008). While this phenomenon's exact mechanism is unclear, studies in rodent models reveal improper mitochondrial functioning and aberrant PCD signaling as the key contributing factors (Marzetti et al. 2013). In certain rodent muscle tissues, aging and sarcopenia are associated with increased expression of PCD markers. With age, the rat gastrocnemius muscle (the main muscle of the calf of the leg) exhibits an enhancement in Bax levels and a reduction in Bcl-2 levels. An activation of caspase-3 and DNA fragmentation were also recorded in the said muscle (Song et al. 2006). However, with age, the rat soleus muscle (which lies right beneath the gastrocnemius muscle in the calf of the leg) exhibits a caspase-independent process, where the movement of EndoG from mitochondria to the nucleus has been recorded (Leeuwenburgh et al. 2005). In aged rats, deletions in the mitochondrial DNA were associated with improper functioning of these organelles in skeletal muscle fibers, suggesting a role of mitochondrial mutations in sarcopenia (Wanagat et al. 2001). Genetically engineered mice with increased mitochondrial mutations showed increased PCD-like events in their skeletal muscle with age (Kujoth et al. 2005).

Age-dependent changes in systemic signaling can also contribute to the aging of muscles. Upon fusion of the circulatory systems of young and old mice, proliferation and regeneration of the old mouse's skeletal muscle were boosted (Conboy et al. 2005). Blood levels of the growth hormone GDF-11 (Growth Differentiation Factor-11) were diminished in older mice, and replenishment of this hormone was found to bring back the proper functioning of muscles (Sinha et al. 2014).

In vivo studies have implicated PCD in the loss of myocytes in the aging heart. Increased Bax: Bcl-2 ratio, release of cytochrome c from mitochondria to the cytosol, and activation of caspase-3 were recorded (Kwak 2013). Malfunctioning mitochondria and oxidative stress were found to play significant roles in the heart of aging mice, as revealed by a decrease in PCD upon overexpression of manganese-superoxide dismutase (MnSOD), an antioxidant enzyme (Kwak et al. 2015). Also, knockout of the lysosomal cysteine protease cathepsin-K improved cardiac function in aged mice hearts. Compared to old wild-type mice, markers of senescence (cardiac lipofuscin, p21, and p16) were diminished in old cathepsin-K knockout mice. A significant decrease in cardiomyocyte apoptosis and nuclear translocation of mitochondrial AIF was recorded in aged cathepsin-K knockout mice. Thus, both caspase-dependent and independent pathways were affected by the knockout of cathepsin-K in aged mice (Hua et al. 2015).

Kidneys have also been found to be increasingly susceptible to stress and induction of PCD with age, characterized by an increase in Bax, caspase-3, and cytochrome c and a decrease in Bcl-2 in rodent models (Wang et al. 2014). Mutations in the mitochondrial DNA lead to PCD of sensory hair cells, causing age-related hearing loss (Fujimoto and Yamasoba 2014). An enhancement in ROS levels and diminution of Bcl-2 levels cause PCD of melanocyte stem cells, ultimately leading to age-related graying of hair (Seiberg 2013). In the rodent model, caspase-3 activation is the cause of an age-associated increase in the susceptibility of gastric mucosal cells to stress and PCD (Tarnawski et al. 2014).

## 17.6 Apoptosis Dysregulation and Cellular Metabolism

The stimuli- and cellular environment-specific downregulation of mTOR signaling causes apoptosis induction in some mitotic (like endothelial progenitor cells) as well as postmitotic cells (like cardiomyocytes) (Mazelin et al. 2016; Miriuka et al. 2006). mTOR signaling downregulation positively impacts longevity and homeostasis by eliminating damaged cells and promoting apoptosis in age-accumulated senescent cells. When improper downregulation of mTOR signaling and induction of apoptosis occurs in mature nonsenescent postmitotic cells, there is excess loss of cells, improper functioning of tissues, and worsening of pathological conditions linked to postmitotic cells. During cardiac ischemia/reperfusion damage, mTOR signaling is enhanced, and this blocks unrestricted apoptosis or autophagy induction, offering protection to cardiomyocytes and ensuring proper functioning of the heart (Fan et al. 2016). Of note, a decrease in mTOR causes a diminution of apoptosis in tubular epithelial and mesangial cells (smooth muscle-like cells that control capillary lumen size and glomerular blood flow).

The insulin/IGF-1 pathway is known to promote aging. It exerts antiapoptotic effects by causing the activation of RAF/MEK/ERK and PI3K/Akt pathways. IGF-1 is known to prevent apoptosis in neuronal cells by increasing the antiapoptotic Bcl-2 levels and hindering the activity of caspase-3 (Jover-Mengual et al. 2007). Also, netrin-1 boosts the ERK pathway and prevents apoptosis of cortical neurons (Chen et al. 2017). An activation of ERK, however, can bring about oxidative stress-mediated cell death in neurons (He and Aizenman 2010). Furthermore, insulin/IGF-1 has been reported to cause the activation of PI3K/Akt. This, in turn, can spur the activity of mTOR and lead to NF-κB activation. NF-κB is a well-known regulator of inflammation and can exert cellular context-dependent pro- and antiapoptotic effects (Gudkov et al. 2011).

Depleting nutrients can result in AMPK-induced expression of genes involved in cellular energy generation or apoptosis induction. Of note, activating AMPK for an extended time can lead to cellular apoptosis. This happens due to spurring the activities of JNK, p53, FOXO-3, and caspase-3/7. There is an increase in the Bax/Bcl-2 ratio and a decline in the activities of mTORC1 and Bcl-2 (Chen et al. 2013; Queiroz et al. 2014). Moderate AMPK activation can, however, diminish the expression of cleaved caspase-3 and the activity caspase-3/7, thereby interfering with apoptosis (Dai et al. 2015). It has been proposed by Argüelles et al. (2019) that apoptosis induced by long-term AMPK activation plays a protective role by clearing the age-accumulated mitotic senescent cells and damaged postmitotic cells, thereby restoring homeostasis and ensuring proper functioning of tissues. Also, the repression of apoptosis caused by moderate AMPK activation may help to prevent cellular damage-induced pathological conditions by hindering improper activation of apoptosis (Argüelles et al. 2019).

Various stress cues can cause SIRT1 to hinder apoptosis, boost cell survival, and extend lifespan. This is achieved through the deacetylation of proteins like p53, the FOXO family, Ku70, and RelA/p65 subunit of NF-κB, which in turn causes caspase-3/9 inhibition, increase in Bcl-2 and a decrease in Bax (Hori et al. 2013). FOXO3 (mainly upon stress exposure) activates death receptor ligands like Fas ligand and TRAIL and causes upregulation of Bim and PUMA, promoting apoptosis and encouraging longevity (Zhang et al. 2011). The SIRT1-mediated inhibition of apoptosis may offer protection from tissue injury (Argüelles et al. 2019). Intriguingly, SIRT1 can promote apoptosis in some forms of cancer (Chao et al. 2017).

## 17.7 Oxidative Stress, Apoptosis, and Autophagy in Aging

An increase in oxidative stress is detrimental to cellular biomolecules and causes the death of cells via the mitochondrial apoptotic pathway (Redza-Dutordoir and Averill-Bates 2016). A heightened endoplasmic reticulum (ER) stress and consequent elicitation of unfolded protein response (UPR) was observed under the following conditions: physiological oxidative stress levels and a temporary decrease in mitochondrial functions. The UPR helps in boosting cellular longevity. Small molecules like rapamycin and resveratrol, known to increase lifespan, have been implicated in boosting UPR (Mouchiroud et al. 2013). Cells are, however, known to undergo apoptotic death due to their inability to cope with high ER stress levels under very high oxidative stress conditions (Karthikeyan et al. 2017). Inhibition of apoptosis is achieved by AMPK via SIRT1 upon reduction of energy metabolism. While functional impairment of mitochondria and consequent ATP depletion and cellular apoptosis are induced by high ROS and RNS (Reactive Nitrogen Species) levels, a decrease in ATP generation under certain conditions has been shown to boost a cell's longevity. When ATP levels are depleted (as in the case of nutrient starvation), antiaging pathways like AMPK and SIRT1 and autophagy receive a boost; proaging pathways (e.g., mTOR) are, however, blocked. A temporary loss of mitochondrial function has been reported to be associated with longevity. Aging-associated apoptosis of some long-lived postmitotic cells (like osteocytes) is caused by reduced connexin-43 levels. The PTEN/pAkt pathway is impeded by lowered connexin-43, which removes the antiapoptotic effect of IGF-1 (Davis et al. 2017). Furthermore, age-dependent enhancement in ROS generation is known to boost the activity of MAP kinases, which in turn cause mitochondrial functional impairment, ER stress, and an excess of calcium ions, ultimately facilitating the intrinsic apoptotic pathway in osteoblasts (Zhu et al. 2018). The loss of skeletal muscle mass with age (sarcopenia) is caused by impaired mitochondrial functioning and the consequent oxidative stress-driven apoptosis of skeletal muscle cells via NF-κB and FOXO (Buonocore et al. 2011).

The processes of apoptosis and autophagy are known to play opposite roles during the aging process. The apoptosis-promoting signaling pathways cause an inflammatory response that speeds up the aging process and brings about age-associated disorders. Autophagy induction, on the other hand, boosts longevity. Age-related accumulation of protein aggregates is associated with impairment in autophagy, which hampers proper mitochondrial functioning and ultimately leads to apoptosis. Caspase-mediated cleavage of Beclin-1 renders it unfit for initiation of autophagy. The C-terminal fragment of Beclin-1 can promote apoptosis by releasing mitochondrial cytochrome c into the cytosol. It is mentioned here that spermidine, a known inducer of autophagy, can repress the activation of caspase-3 and consequently block Beclin-1 cleavage. Thus, cytochrome c is not released into the cytosol, and neurons are protected from damage (Yang et al. 2017). Intriguingly, Bax has been shown to both decrease and induce autophagy under different contexts. In cells lacking antiapoptotic capacity, Bax induced the caspase-mediated cleavage of Beclin-1 and thereby caused a diminution of autophagy. However, in cells possessing significant antiapoptotic capacity, Bax can induce autophagy by interfering with the interaction of Bcl-xL and Beclin (Luo and Rubinsztein 2010). Autophagy can promote both death and survival of MSCs under different conditions. Under starvation conditions, MSCs' survival is enhanced by autophagy. However, autophagy has been shown to promote apoptosis of the said cells in an inflammatory microenvironment via the activation caspase-3 and the inhibition of Bcl-2 (Dang et al. 2015). Autophagy generally protects cells from apoptosis, which is known to be diminished with age. Indeed, the activation of autophagy achieved through pharmacological and genetic means can hinder aging and extend the lifespan of various organisms (Leon and Gustafsson 2016).

## 17.8 Context-dependent Inhibition and Induction of Apoptosis in Stem Cells

Repairing and regenerating damaged tissues employing stem cell therapy is currently an area of intense research. Unfortunately, the transplanted stem cells undergo apoptotic death due to their susceptibility to the aggressive microenvironment triggered by tissue damage and proinflammatory modulators. Thus, the survival of stem cells can be boosted by the inhibition of apoptosis. Neural stem cells (NSCs) are reported to differentiate into neuronal and glial cells and are thus indispensable for normal neurogenesis. Apoptosis of NSCs is triggered by exposure to ketamine, which diminishes neurogenesis and causes cognitive impairment. NSCs can be safeguarded from ketamine-induced damage by treatment with dexmedetomidine. This molecule suppresses apoptosis by boosting the activity of the PI3K/Akt/GSK-3β signaling pathway (Lu et al. 2018a, b). Apoptosis of mesenchymal stem cells (MSCs) caused by nutrient depletion could be reversed by cyclic helix B peptide treatment. This was achieved by promoting the activation of the Nrf2/SIRT3/FOXO3a pathway and ensuring the proper functioning of mitochondria (Wang et al. 2017). Serum depletion and hypoxia are known to cause apoptosis in MSCs. Adrenomedullin can reverse the effect via upregulation of pAkt and pGSK3β levels, boosting Bcl-2 signaling pathways and inhibiting caspase-3 activation (Si et al. 2018).

Cancer stem cells (CSCs) are the principal cause of cancer recurrence. Thus, the promotion of apoptosis in these cells can be beneficial to cancer patients. Certain molecules reported in the scientific literature can achieve this end, and these are discussed in the following section. Quercetin-3-methyl ether can block the expression of NOTCH1 and trigger Akt and PI3K phosphorylation, thereby promoting apoptosis (Cao et al. 2018). Induction of apoptosis in liver cancer stem cells can be achieved by curcumin treatment. This involves the upregulation of caspase3/9 and Bax, and the downregulation of Bcl-2. The PI3K/ Akt/ mTOR pathway is also blocked (Wang et al. 2018).

## 17.9 Telomerase Loss, Senescence, and Apoptosis

Mitotic cells' resistance to apoptosis is boosted upon overexpression of human telomerase reverse transcriptase (hTERT) in them (Crea et al. 2006). Furthermore, hTERT can protect postmitotic cells, such as neurons, from injury-induced apoptosis (Li et al. 2013). As is well known, telomerase activity boosts cellular resistance to apoptosis. Thus, inhibiting hTERT makes a cell susceptible to apoptosis. The activation of the mitochondrial apoptotic pathway was achieved by inhibiting hTERT, involving activation of caspase-3/9, upregulating Bax, and downregulating Bcl-2 (Li et al. 2018). Notably, telomere damage can result in apoptosis or senescence. However, the mechanisms that dictate the cell's fate remain elusive. Panneer Selvam and colleagues have reported that p16 (an inducer of cell cycle arrest and senescence) associates with caspase-3 and induces telomere damage-dependent senescence, as this association was found to block the activation of caspase-3 and hence apoptosis. In p16 deficient cells, however, telomere damage (promoted by the instability of telomerase) leads to the activation of caspase-3 and apoptosis. In this case, senescence does not occur. Congruently, a p16 mutant lacking the capability of association with caspase-3 failed to block telomere-damage-dependent apoptosis (Panneer Selvam et al. 2018).

## 17.10 Apoptosis in the Context of Exceptional Aging

Several studies suggest that, while apoptosis decreases with age, it receives a significant boost in exceptionally long-lived individuals (i.e., centenarians). This is exemplified by the fact that Fas and FasL are known to prevent chronic inflammation and age-associated inflammatory disorders by promoting the removal of stimulated T-cells (Argüelles et al. 2019). An increase in circulating levels of soluble Fas or sFas (an Inhibitor of Apoptosis Protein or IAP); as well as a decrease in circulating levels of sFasL (which promotes apoptosis) and total cytochrome c released from apoptotic cells, was observed with increasing age. This decrease in global apoptosis with age, as observed in normal individuals, could be associated with an increase in diseases involving improper regulation of apoptosis (Kavathia et al. 2009). Pinti and colleagues compared the mRNA levels of total Fas (tFas) and membrane Fas (mFas) as well as FasL in the peripheral blood lymphocytes collected from centenarians, young and middle-aged individuals. An age-dependent increase in the mRNA level of tFas (but not mFas) was recorded in resting lymphocytes. FasL mRNA level was, however, found to diminish with age. No significant alteration in the in vitro production of Fas/FasL mRNA was observed upon exposure of cells from all three groups to various stimuli. Centenarian peripheral blood lymphocytes exhibited normal Fas-induced apoptosis, although these individuals had a higher percentage of Fas-positive cells than other groups. The in vitro production of sFas was found to be normal in cells from centenarians. Compared to other groups, plasma sFas level in centenarians was appreciably higher. Thus, centenarians exhibited an efficient immune system functioning by balancing their lymphocytes' pro- and antiapoptotic molecules (Pinti et al. 2004). When subjected to functional transcriptomic analysis, peripheral blood mononuclear cells revealed better apoptosis maintenance in exceptionally long-lived individuals compared to normal-aged ones. Notably, an upregulation of Bcl-xL (an inhibitor of the intrinsic apoptotic pathway) as well as a decrease in cytochrome c (a marker of intrinsic apoptosis) was recorded in centenarians, indicating a protective role of Bcl-xL from age-related damage. An increase in Bcl-xL possibly contributed to maintaining the strength of the mitochondrial membrane. It ensured the proper functioning of lymphocytes and the stress response in centenarians (Borras et al. 2016).

Longevity in long-lived species has been attributed to the following molecular features: (a) unusual resistance to cancer, (b) an efficient DNA damage/ repair system, (c) a decrease in circulating IGF-1 levels and brain IGF-1 receptors, (d) increase in telomerase binding proteins (like Ku80) to ensure maintenance of telomeres, and (e) proper maintenance of protein homeostasis (Tian et al. 2017). Of note, all the above-mentioned features are known to regulate apoptosis. Thus, it seems possible that promoting apoptosis under normal conditions causes the death of nonsenescent mature cells and tissue damage, ultimately giving rise to various pathological conditions. But, postinjury induction of apoptosis is critical to survival and longevity. Apoptosis removes the cells bearing deleterious mutations that can cause genomic instability, thereby delaying aging (Argüelles et al. 2019). Indeed, apoptosis could play a pivotal role in the observed 30-fold difference in longevity between mice and humans. Compared to short-lived species like mice, an enhanced expression of Ku80 and the catalytic subunit of DNA-dependent protein kinase has been reported in species with greater longevity, such as humans (Lorenzini et al. 2009).

## 17.11 Conclusion

Modern science has yet to predict the longevity of an individual. There was a glaring disparity between the increase in human life expectancy and the quality of human life in the last century. An increased life expectancy coupled with the emergence of various age-related diseases is indeed a burden for society. Therefore, it is extremely important to delve deeper into the biology of aging for effective mitigation and/or prevention of such age-related pathological conditions. To properly maintain the body's functions as a prerequisite for healthy aging, a living organism must give rise to new cells and destroy old and damaged ones. Here, the process of apoptosis plays a pivotal role. In the present chapter, we have discussed the various forms of PCD with special emphasis on apoptosis, referred to as type-I PCD. The imbalance of apoptosis in various tissues and organs during aging leads to age-associated pathological conditions, and the mechanisms involved therein have been elaborated. Finally, we have discussed apoptosis in the context of extraordinary aging.

## References


Alt EU, Senst C, Murthy SN, Slakey DP, Dupin CL, Chaffin AE, Kadowitz PJ, Izadpanah R (2012) Aging alters tissue resident mesenchymal stem cell properties. Stem Cell Res 8(2):215–225

Andersen BB, Gundersen HJG, Pakkenberg B (2003) Aging of the human cerebellum: a stereological study. J Comp Neurol 466(3):356–365

Argüelles S, Guerrero-Castilla A, Cano M, Muñoz MF, Ayala A (2019) Advantages and disadvantages of apoptosis in the aging process. Ann N Y Acad Sci 1443(1):20–33

Bergmann O, Bhardwaj RD, Bernard S, Zdunek S, Barnabé-Heider F, Walsh S, Zupicich J, Alkass K, Buchholz BA, Druid H, Jovinge S (2009) Evidence for cardiomyocyte renewal in humans. Science 324(5923):98–102

Borras C, Abdelaziz KM, Gambini J, Serna E, Inglés M, de la Fuente M, Garcia I, Matheu A, Sanchís P, Belenguer A, Errigo A (2016) Human exceptional longevity: transcriptome from centenarians is distinct from septuagenarians and reveals a role of Bcl-xL in successful aging. Aging (Albany NY) 8(12):3185

Buonocore D, Rucci S, Vandoni M, Negro M, Marzatico F (2011) Oxidative system in aged skeletal muscle. Muscles Ligaments Tendons J 1(3):85

Burek MJ, Oppenheim RW (1996) Programmed cell death in the developing nervous system. Brain Pathol 6(4):427–446

Camplejohn RS, Gilchrist R, Easton D, McKenzie-Edwards E, Barnes DM, Eccles DM, Ardern-Jones A, Hodgson SV, Duddy PM, Eeles RA (2003) Apoptosis, ageing and cancer susceptibility. Br J Cancer 88(4):487–490

Cao L, Yang Y, Ye Z, Lin B, Zeng J, Li C, Liang T, Zhou K, Li J (2018) Quercetin-3-methyl ether suppresses human breast cancer stem cell formation by inhibiting the Notch1 and PI3K/Akt signaling pathways. Int J Mol Med 42(3):1625–1636

Chao SC, Chen YJ, Huang KH, Kuo KL, Yang TH, Huang KY, Wang CC, Tang CH, Yang RS, Liu SH (2017) Induction of sirtuin-1 signaling by resveratrol induces human chondrosarcoma cell apoptosis and exhibits antitumor activity. Sci Rep 7(1):3180

Chen MB, Zhang Y, Wei MX, Shen W, Wu XY, Yao C, Lu PH (2013) Activation of AMP-activated protein kinase (AMPK) mediates plumbagin-induced apoptosis and growth inhibition in cultured human colon cancer cells. Cell Signal 25(10):1993–2002

Chen J, Du H, Zhang Y, Chen H, Zheng M, Lin P, Lan Q, Yuan Q, Lai Y, Pan X, Chen R (2017) Netrin-1 prevents rat primary cortical neurons from apoptosis via the DCC/ERK pathway. Front Cell Neurosci 11:387

Chimal-Monroy J, Abarca-Buis RF, Cuervo R, Díaz-Hernández M, Bustamante M, Rios-Flores JA, Romero-Suárez S, Farrera-Hernández A (2011) Molecular control of cell differentiation and programmed cell death during digit development. IUBMB Life 63(10):922–929

Chou JP, Effros RB (2013) T cell replicative senescence in human aging. Curr Pharm Des 19(9):1680–1698

Conboy IM, Conboy MJ, Wagers AJ, Girma ER, Weissman IL, Rando TA (2005) Rejuvenation of aged progenitor cells by exposure to a young systemic environment. Nature 433(7027):760–764

Crea F, Sarti D, Falciani F, Al-Rubeai M (2006) Over-expression of hTERT in CHO K1 results in decreased apoptosis and reduced serum dependency. J Biotechnol 121(2):109–123

Dai YL, Huang SL, Leng Y (2015) AICAR and metformin exert AMPK-dependent effects on INS-1E pancreatic β-cell apoptosis via differential downstream mechanisms. Int J Biol Sci 11(11):1272

Dang S, Yu ZM, Zhang CY, Zheng J, Li KL, Wu Y, Qian LL, Yang ZY, Li XR, Zhang Y, Wang RX (2015) Autophagy promotes apoptosis of mesenchymal stem cells under inflammatory microenvironment. Stem Cell Res Ther 6:1–9

Davis HM, Pacheco-Costa R, Atkinson EG, Brun LR, Gortazar AR, Harris J, Hiasa M, Bolarinwa SA, Yoneda T, Ivan M, Bruzzaniti A (2017) Disruption of the Cx43/miR21 pathway leads to osteocyte apoptosis and increased osteoclastogenesis with aging. Aging Cell 16(3):551–563

Degterev A, Zhou W, Maki JL, Yuan J (2014) Assays for necroptosis and activity of RIP kinases. Methods Enzymol 545:1–33

Delavallée L, Cabon L, Galán-Malo P, Lorenzo HK, Susin SA (2011) AIF-mediated caspase-independent necroptosis: a new chance for targeted therapeutics. IUBMB Life 63(4):221–232

Eckhart L, Lippens S, Tschachler E, Declercq W (2013) Cell death by cornification. Biochimica et Biophysica Acta (BBA)-Mol Cell Res 1833(12):3471–3480

Elmore S (2007) Apoptosis: a review of programmed cell death. Toxicol Pathol 35(4):495–516

Fan G, Yu J, Asare PF, Wang L, Zhang H, Zhang B, Zhu Y, Gao X (2016) Danshensu alleviates cardiac ischaemia/reperfusion injury by inhibiting autophagy and apoptosis via activation of mTOR signaling. J Cell Mol Med 20(10):1908–1919

Folch J, Junyent F, Verdaguer E, Auladell C, Pizarro JG, Beas-Zarate C, Pallas M, Camins A (2012) Role of cell cycle re-entry in neurons: a common apoptotic mechanism of neuronal cell death. Neurotox Res 22:195–207

Frank S, Gaume B, Bergmann-Leitner ES, Leitner WW, Robert EG, Catez F, Smith CL, Youle RJ (2001) The role of dynamin-related protein 1, a mediator of mitochondrial fission, in apoptosis. Dev Cell 1(4):515–525

Fujimoto C, Yamasoba T (2014) Oxidative stresses and mitochondrial dysfunction in age-related hearing loss. Oxid Med Cell Longev 2014

Galluzzi L, Kepp O, Kroemer G (2009) RIP kinases initiate programmed necrosis. J Mol Cell Biol 1(1):8–10

Galluzzi L, Vitale I, Abrams JM, Alnemri ES, Baehrecke EH, Blagosklonny MV, Dawson TM, Dawson VL, El-Deiry WS, Fulda SJCD, Gottlieb E (2012) Molecular definitions of cell death subroutines: recommendations of the Nomenclature Committee on Cell Death 2012. Cell Death Different 19(1):107–120

Ghavami S, Shojaei S, Yeganeh B, Ande SR, Jangamreddy JR, Mehrpour M, Christoffersson J, Chaabane W, Moghadam AR, Kashani HH, Hashemi M (2014) Autophagy and apoptosis dysfunction in neurodegenerative disorders. Prog Neurobiol 112:24–49

Giannaris EL, Rosene DL (2012) A stereological study of the numbers of neurons and glia in the primary visual cortex across the lifespan of male and female rhesus monkeys. J Comp Neurol 520(15):3492–3508

Gudkov AV, Gurova KV, Komarova EA (2011) Inflammation and p53: a tale of two stresses. Genes Cancer 2(4):503–516

Günther C, Neumann H, Neurath MF, Becker C (2013) Apoptosis, necrosis and necroptosis: cell death regulation in the intestinal epithelium. Gut 62(7):1062–1071

Gupta S (2005) Molecular mechanisms of apoptosis in the cells of the immune system in human aging. Immunol Rev 205(1):114–129

Gupta S, Agrawal A, Agrawal S, Su H, Gollapudi S (2006) A paradox of immunodeficiency and inflammation in human aging: lessons learned from apoptosis. Immun Ageing 3:1–8
Gurbuxani S, Schmitt E, Cande C, Parcellier A, Hammann A, Daugas E, Kouranti I, Spahr C, Pance A, Kroemer G, Garrido C (2003) Heat shock protein 70 binding inhibits the nuclear import of apoptosis-inducing factor. Oncogene 22(43):6669–6678
Hankins HM, Baldridge RD, Xu P, Graham TR (2015) Role of flippases, scramblases and transfer proteins in phosphatidylserine subcellular distribution. Traffic 16(1):35–47
He K, Aizenman E (2010) ERK signaling leads to mitochondrial dysfunction in extracellular zinc-induced neurotoxicity. J Neurochem 114(2):452–461
Hernández-Martínez R, Covarrubias L (2011) Interdigital cell death function and regulation: new insights on an old programmed cell death model. Develop Growth Differ 53(2):245–258
Hiebert PR, Granville DJ (2012) Granzyme B in injury, inflammation, and repair. Trends Mol Med 18(12):732–741
Higami Y, Shimokawa I (2000) Apoptosis in the aging process. Cell Tissue Res 301:125–132
Hori YS, Kuno A, Hosoda R, Horio Y (2013) Regulation of FOXOs and p53 by SIRT1 modulators under oxidative stress. PLoS One 8(9):e73875
Hroudová J, Singh N, Fišar Z (2014) Mitochondrial dysfunctions in neurodegenerative diseases: relevance to Alzheimer's disease. BioMed Res Int 2014
Hua Y, Robinson TJ, Cao Y, Shi GP, Ren J, Nair S (2015) Cathepsin K knockout alleviates aging-induced cardiac dysfunction. Aging Cell 14(3):345–351
Jover-Mengual T, Zukin RS, Etgen AM (2007) MAPK signaling is critical to estradiol protection of CA1 neurons in global ischemia. Endocrinology 148(3):1131–1143
Kajstura J, Cheng W, Sarangarajan R, Li P, Li B, Nitahara JA, Chapnick S, Reiss K, Olivetti G, Anversa P (1996) Necrotic and apoptotic myocyte cell death in the aging heart of Fischer 344 rats. Am J Phys Heart Circ Phys 271(3):H1215–H1228
Kapetanaki MG, Mora AL, Rojas M (2013) Influence of age on wound healing and fibrosis. J Pathol 229(2):310–322
Karthikeyan B, Harini L, Krishnakumar V, Kannan VR, Sundar K, Kathiresan T (2017) Insights on the involvement of (−)-epigallocatechin gallate in ER stress-mediated apoptosis in age-related macular degeneration. Apoptosis 22:72–85
Kavathia N, Jain A, Walston J, Beamer BA, Fedarko NS (2009) Serum markers of apoptosis decrease with age and cancer stage. Aging 1(7):652
Kerr JF, Wyllie AH, Currie AR (1972) Apoptosis: a basic biological phenomenon with wideranging implications in tissue kinetics. Br J Cancer 26(4):239–257
Kile BT (2014) The role of apoptosis in megakaryocytes and platelets. Br J Haematol 165(2):217–226
King KL, Cidlowski JA (1995) Cell cycle and apoptosis: common pathways to life and death. J Cell Biochem 58(2):175–180
King KL, Cidlowski JA (1998) Cell cycle regulation and apoptosis. Annu Rev Physiol 60(1):601–617
Kirkland JL, Tchkonia T (2015) Clinical strategies and animal models for developing senolytic agents. Exp Gerontol 68:19–25
Kroemer G, Galluzzi L, Vandenabeele P, Abrams J, Alnemri ES, Baehrecke EH, Blagosklonny MV, El-Deiry WS, Golstein P, Green DR, Hengartner M (2009) Classification of cell death: recommendations of the Nomenclature Committee on Cell Death 2009. Cell Death Different 16(1):3–11
Kujoth GC, Hiona A, Pugh TD, Someya S, Panzer K, Wohlgemuth SE, Hofer T, Seo AY, Sullivan R, Jobling WA, Morrow JD (2005) Mitochondrial DNA mutations, oxidative stress, and apoptosis in mammalian aging. Science 309(5733):481–484
Kung G, Konstantinidis K, Kitsis RN (2011) Programmed necrosis, not apoptosis, in the heart. Circ Res 108(8):1017–1036
Kwak HB (2013) Effects of aging and exercise training on apoptosis in the heart. J Exer Rehabil 9(2):212

Kwak HB, Lee Y, Kim JH, Van Remmen H, Richardson AG, Lawler JM (2015) MnSOD overexpression reduces fibrosis and pro-apoptotic signaling in the aging mouse heart. J Gerontol A Biol Sci Med Sci 70(5):533–544

Lanni C, Racchi M, Memo M, Govoni S, Uberti D (2012) p53 at the crossroads between cancer and neurodegeneration. Free Radic Biol Med 52(9):1727–1733

Lee Y, Kang HC, Lee BD, Lee YI, Kim YP, Shin JH (2014) Poly (ADP-ribose) in the pathogenesis of Parkinson's disease. BMB Rep 47(8):424

Leeuwenburgh C, Gurley CM, Strotman BA, Dupont-Versteegden EE (2005) Age-related differences in apoptosis with disuse atrophy in soleus muscle. Am J Phys Regul Integr Comp Phys 288(5):R1288–R1296

Leon LJ, Gustafsson ÅB (2016) Staying young at heart: autophagy and adaptation to cardiac aging. J Mol Cell Cardiol 95:78–85

Leventis PA, Grinstein S (2010) The distribution and function of phosphatidylserine in cellular membranes. Annu Rev Biophys 39:407–427

Li J, Qu Y, Chen D, Zhang L, Zhao F, Luo L, Pan L, Hua J, Mu D (2013) The neuroprotective role and mechanisms of TERT in neurons with oxygen–glucose deprivation. Neuroscience 252:346–358

Li Y, Pan G, Chen Y, Yang Q, Hao T, Zhao L, Zhao L, Cong Y, Diao A, Yu P (2018) Inhibitor of the human telomerase reverse trancriptase (hTERT) gene promoter induces cell apoptosis via a mitochondrial-dependent pathway. Eur J Med Chem 145:370–378

Long JM, Mouton PR, Jucker M, Ingram DK (1999) What counts in brain aging? Design-based stereological analysis of cell number. J Gerontol Ser A: Biomed Sci Med Sci 54(10):B407–B417

Lorenzini A, Johnson FB, Oliver A, Tresini M, Smith JS, Hdeib M, Sell C, Cristofalo VJ, Stamato TD (2009) Significant correlation of species longevity with DNA double strand break recognition but not with telomere length. Mech Ageing Dev 130(11–12):784–792

Lu B, Chen HD, Hong-Guang HG (2012) The relationship between apoptosis and aging. Adv Biosci Biotechnol 3(06):705

Lu P, Lei S, Li W, Lu Y, Zheng J, Wang N, Xia Y, Lu H, Chen X, Liu Y, Zhang PB (2018a) Dexmedetomidine protects neural stem cells from ketamine-induced injury. Cell Physiol Biochem 47(4):1377–1388

Lu Q, Zhou Y, Hao M, Li C, Wang J, Shu F, Du L, Zhu X, Zhang Q, Yin X (2018b) The mTOR promotes oxidative stress-induced apoptosis of mesangial cells in diabetic nephropathy. Mol Cell Endocrinol 473:31–43

Luo S, Rubinsztein DC (2010) Apoptosis blocks Beclin 1-dependent autophagosome synthesis: an effect rescued by Bcl-xL. Cell Death Different 17(2):268–277

Marcotte R, Lacelle C, Wang E (2004) Senescent fibroblasts resist apoptosis by downregulating caspase-3. Mech Ageing Dev 125(10–11):777–783

Marzetti E, Lawler JM, Hiona A, Manini T, Seo AY, Leeuwenburgh C (2008) Modulation of age-induced apoptotic signaling and cellular remodeling by exercise and calorie restriction in skeletal muscle. Free Radic Biol Med 44(2):160–168

Marzetti E, Calvani R, Cesari M, Buford TW, Lorenzi M, Behnke BJ, Leeuwenburgh C (2013) Mitochondrial dysfunction and sarcopenia of aging: from signaling pathways to clinical trials. Int J Biochem Cell Biol 45(10):2288–2301

Mazelin L, Panthu B, Nicot AS, Belotti E, Tintignac L, Teixeira G, Zhang Q, Risson V, Baas D, Delaune E, Derumeaux G (2016) mTOR inactivation in myocardium from infant mice rapidly leads to dilated cardiomyopathy due to translation defects and p53/JNK-mediated apoptosis. J Mol Cell Cardiol 97:213–225

Merrill DA, Chiba AA, Tuszynski MH (2001) Conservation of neuronal number and size in the entorhinal cortex of behaviorally characterized aged rats. J Comp Neurol 438(4):445–456

Miriuka SG, Rao V, Peterson M, Tumiati L, Delgado DH, Mohan R, Ramzy D, Stewart D, Ross HJ, Waddell TK (2006) mTOR inhibition induces endothelial progenitor cell death. Am J Transplant 6(9):2069–2079

Misell LM, Hwang ES, Au A, Esserman L, Hellerstein MK (2005) Development of a novel method for measuring in vivo breast epithelial cell proliferation in humans. Breast Cancer Res Treat 89:257–264

Mouchiroud L, Houtkooper RH, Moullan N, Katsyuba E, Ryu D, Cantó C, Mottis A, Jo YS, Viswanathan M, Schoonjans K, Guarente L (2013) The NAD+/sirtuin pathway modulates longevity through activation of mitochondrial UPR and FOXO signaling. Cell 154(2):430–441
Nixon RA, Yang DS (2012) Autophagy and neuronal cell death in neurological disorders. Cold Spring Harb Perspect Biol 4(10):a008839
Nolan LA, Lunness HR, Lightman S (1999) The effects of age and spontaneous adenoma formation on trophic activity in the rat pituitary gland: a comparison with trophic activity in the human pituitary and in human pituitary adenomas. J Neuroendocrinol 11(5):393–401
Olivetti G, Melissari M, Capasso JM, Anversa P (1991) Cardiomyopathy of the aging human heart. Myocyte loss and reactive cellular hypertrophy. Circ Res 68(6):1560–1568
Oppenheim RW (1991) Cell death during development of the nervous system. Annu Rev Neurosci 14(1):453–501
Oshiro S, Morioka MS, Kikuchi M (2011) Dysregulation of iron metabolism in Alzheimer's disease, Parkinson's disease, and amyotrophic lateral sclerosis. Adv Pharmacol Pharma Sci 2011
Otera H, Ishihara N, Mihara K (2013) New insights into the function and regulation of mitochondrial fission. Biochimica et Biophysica Acta (BBA)-Mol Cell Res 1833(5):1256–1268
Pandey S, Wang E (1995) Cells en route to apoptosis are characterized by the upregulation of c-fos, c-myc, c-jun, cdc2, and RB phosphorylation, resembling events of early cell-cycle traverse. J Cell Biochem 58(2):135–150
Paoli P, Giannoni E, Chiarugi P (2013) Anoikis molecular pathways and its role in cancer progression. Biochimica et Biophysica Acta (BBA)-Mol Cell Res 1833(12):3481–3498
Papucci L, Formigli L, Schiavone N, Tani A, Donnini M, Lapucci A, Perna F, Tempestini A, Witort E, Morganti M, Nosi D (2004) Apoptosis shifts to necrosis via intermediate types of cell death by a mechanism depending on c-myc and bcl-2 expression. Cell Tissue Res 316:197–209
Pasparakis M, Vandenabeele P (2015) Necroptosis and its role in inflammation. Nature 517(7534):311–320
Pinti M, Troiano L, Nasi M, Bellodi C, Ferraresi R, Mussi C, Salvioli G, Cossarizza A (2004) Balanced regulation of mRNA production for Fas and Fas ligand in lymphocytes from centenarians: how the immune system starts its second century. Circulation 110(19):3108–3114
Pollack M, Leeuwenburgh C (2001) Apoptosis and aging: role of the mitochondria. J Gerontol Ser A Biol Med Sci 56(11):B475–B482
Polyak K, Wu TT, Hamilton SR, Kinzler KW, Vogelstein B (1997) Less death in the dying. Cell Death Different 4(3):242–246
Proskuryakov SY, Gabai VL (2010) Mechanisms of tumor cell necrosis. Curr Pharm Des 16(1):56–68
Qian D, Wu X, Jiang H, Gao P, Kuang C, Wang K, Huang L (2011) Aging reduces susceptibility of vascular smooth muscle cells to $H_2O_2$-induced apoptosis through the down-regulation of Jagged1 expression in endothelial cells. Int J Mol Med 28(2):207–213
Queiroz EA, Puukila S, Eichler R, Sampaio SC, Forsyth HL, Lees SJ, Barbosa AM, Dekker RF, Fortes ZB, Khaper N (2014) Metformin induces apoptosis and cell cycle arrest mediated by oxidative stress, AMPK and FOXO3a in MCF-7 breast cancer cells. PLoS One 9(5):e98207
Rabinovich GA (2015) Essential versus accessory aspects of cell death: recommendations of the NCCD 2015
Raff MC, Barres BA, Burne JF, Coles HSR, Ishizaki Y, Jacobson MD (1994) Programmed cell death and the control of cell survival. Philos Trans R Soc Lond B Biol Sci 345(1313):265–268
Rai NK, Tripathi K, Sharma D, Shukla VK (2005) Apoptosis: a basic physiologic process in wound healing. Int J Low Extrem Wounds 4(3):138–144
Rapp PR, Gallagher M (1996) Preserved neuron number in the hippocampus of aged rats with spatial learning deficits. Proc Natl Acad Sci 93(18):9926–9930
Redza-Dutordoir M, Averill-Bates DA (2016) Activation of apoptosis signaling pathways by reactive oxygen species. Biochimica et Biophysica Acta (BBA)-Mol Cell Res 1863(12):2977–2992
Renehan AG, Booth C, Potten CS (2001) What is apoptosis, and why is it important? Education and debate. BMJ 322(7301):1536–1538

Richardson RB, Allan DS, Le Y (2014) Greater organ involution in highly proliferative tissues associated with the early onset and acceleration of ageing in humans. Exp Gerontol 55:80–91
Rinnerthaler M, Streubel MK, Bischof J, Richter K (2015) Skin aging, gene expression and calcium. Exp Gerontol 68:59–65
Rochette PJ, Brash DE (2008) Progressive apoptosis resistance prior to senescence and control by the anti-apoptotic protein BCL-xL. Mech Ageing Dev 129(4):207–214
Ryu SJ, Oh YS, Park SC (2007) Failure of stress-induced downregulation of Bcl-2 contributes to apoptosis resistance in senescent human diploid fibroblasts. Cell Death Different 14(5):1020–1028
Sainz RM, Mayo JC, Reiter RJ, Tan DX, Rodriguez C (2003) Apoptosis in primary lymphoid organs with aging. Microsc Res Tech 62(6):524–539
Salminen A, Ojala J, Kaarniranta K (2011) Apoptosis and aging: increased resistance to apoptosis enhances the aging process. Cell Mol Life Sci 68:1021–1031
Salvioli S, Monti D, Lanzarini C, Conte M, Pirazzini C, Giulia Bacalini M, Garagnani P, Giuliani C, Fontanesi E, Ostan R, Bucci L (2013) Immune system, cell senescence, aging and longevity-inflamm-aging reappraised. Curr Pharm Des 19(9):1675–1679
Seiberg M (2013) Age-induced hair greying–the multiple effects of oxidative stress. Int J Cosmet Sci 35(6):532–538
Selvam SP, Roth BM, Nganga R, Kim J, Cooley MA, Helke K, Smith CD, Ogretmen B (2018) Balance between senescence and apoptosis is regulated by telomere damage–induced association between p16 and caspase-3. J Biol Chem 293(25):9784–9800
Shalini S, Dorstyn L, Dawar S, Kumar S (2015) Old, new and emerging functions of caspases. Cell Death Different 22(4):526–539
Shaw AC, Joshi S, Greenwood H, Panda A, Lord JM (2010) Aging of the innate immune system. Curr Opin Immunol 22(4):507–513
Shen J, Tower J (2009) Programmed cell death and apoptosis in aging and life span regulation. Discov Med 8(43):223–226
Si H, Zhang Y, Song Y, Li L (2018) Overexpression of adrenomedullin protects mesenchymal stem cells against hypoxia and serum deprivation-induced apoptosis via the Akt/GSK3β and Bcl-2 signaling pathways. Int J Mol Med 41(6):3342–3352
Sinha M, Jang YC, Oh J, Khong D, Wu EY, Manohar R, Miller C, Regalado SG, Loffredo FS, Pancoast JR, Hirshman MF (2014) Restoring systemic GDF11 levels reverses age-related dysfunction in mouse skeletal muscle. Science 344(6184):649–652
Snigdha S, Smith ED, Prieto GA, Cotman CW (2012) Caspase-3 activation as a bifurcation point between plasticity and cell death. Neurosci Bull 28:14–24
Song W, Kwak HB, Lawler JM (2006) Exercise training attenuates age-induced changes in apoptotic signaling in rat skeletal muscle. Antioxid Redox Signal 8(3-4):517–528
Sperandio S, Poksay K, De Belle I, Lafuente MJ, Liu B, Nasir J, Bredesen DE (2004) Paraptosis: mediation by MAP kinases and inhibition by AIP-1/Alix. Cell Death Different 11(10):1066–1075
Suh Y (2002) Cell signaling in aging and apoptosis. Mech Ageing Dev 123(8):881–890
Suh Y, Lee KA, Kim WH, Han BG, Vijg J, Park SC (2002) Aging alters the apoptotic response to genotoxic stress. Nat Med 8(1):3–4
Taddei ML, Giannoni E, Fiaschi T, Chiarugi P (2012) Anoikis: an emerging hallmark in health and diseases. J Pathol 226(2):380–393
Tarnawski AS, Ahluwalia A, Jones MK (2014) Increased susceptibility of aging gastric mucosa to injury: the mechanisms and clinical implications. World J Gastroenterol: WJG 20(16):4467
Taylor RC, Cullen SP, Martin SJ (2008) Apoptosis: controlled demolition at the cellular level. Nat Rev Mol Cell Biol 9(3):231–241
Tchkonia T, Zhu Y, Van Deursen J, Campisi J, Kirkland JL (2013) Cellular senescence and the senescent secretory phenotype: therapeutic opportunities. J Clin Invest 123(3):966–972
Tian X, Seluanov A, Gorbunova V (2017) Molecular mechanisms determining lifespan in short- and long-lived species. Trends Endocrinol Metab 28(10):722–734

Vaughn AE, Deshmukh M (2008) Glucose metabolism inhibits apoptosis in neurons and cancer cells by redox inactivation of cytochrome c. Nat Cell Biol 10(12):1477–1483

Vazquez-Padron RI, Lasko D, Li S, Louis L, Pestana IA, Pang M, Liotta C, Fornoni A, Aitouche A, Pham SM (2004) Aging exacerbates neointimal formation, and increases proliferation and reduces susceptibility to apoptosis of vascular smooth muscle cells in mice. J Vasc Surg 40(6):1199–1207

Wanagat J, Cao Z, Pathare P, Aiken JM (2001) Mitochondrial DNA deletion mutations colocalize with segmental electron transport system abnormalities, muscle fiber atrophy, fiber splitting, and oxidative damage in sarcopenia. FASEB J 15(2):322–332

Wang E (1995) Senescent human fibroblasts resist programmed cell death, and failure to suppress bcl 2 is involved. Cancer Res 55(11):2284–2292

Wang E (1997) Regulation of apoptosis resistance and ontogeny of age-dependent diseases. Exp Gerontol 32(4-5):471–484

Wang X, Bonventre JV, Parrish AR (2014) The aging kidney: increased susceptibility to nephrotoxicity. Int J Mol Sci 15(9):15358–15376

Wang S, Zhang C, Niyazi S, Zheng L, Li J, Zhang W, Xu M, Rong R, Yang C, Zhu T (2017) A novel cytoprotective peptide protects mesenchymal stem cells against mitochondrial dysfunction and apoptosis induced by starvation via Nrf2/Sirt3/FoxO3a pathway. J Transl Med 15:1–13

Wang J, Wang C, Bu G (2018) Curcumin inhibits the growth of liver cancer stem cells through the phosphatidylinositol 3-kinase/protein kinase B/mammalian target of rapamycin signaling pathway. Exp Ther Med 15(4):3650–3658

Warner HR (1997) Aging and regulation of apoptosis. Curr Top Cell Regul 35:107–121

Westphal D, Kluck RM, Dewson G (2014) Building blocks of the apoptotic pore: how Bax and Bak are activated and oligomerize during apoptosis. Cell Death Different 21(2):196–205

Wilson A, Shehadeh LA, Yu H, Webster KA (2010) Age-related molecular genetic changes of murine bone marrow mesenchymal stem cells. BMC Genomics 11:1–14

Yakovlev AG, Faden AI (2004) Mechanisms of neural cell death: implications for development of neuroprotective treatment strategies. NeuroRx 1(1):5–16

Yang Y, Chen S, Zhang Y, Lin X, Song Y, Xue Z, Qian H, Wang S, Wan G, Zheng X, Zhang L (2017) Induction of autophagy by spermidine is neuroprotective via inhibition of caspase 3-mediated Beclin 1 cleavage. Cell Death Dis 8(4):e2738–e2738

Yeo EJ, Hwang YC, Kang CM, Choy HE, Park SC (2000) Reduction of UV-induced cell death in the human senescent fibroblasts. Mol Cells 10(4):415–422

Zhang X, Tang N, Hadden TJ, Rishi AK (2011) Akt, FoxO and regulation of apoptosis. Biochimica et Biophysica Acta (BBA)-Mol Cell Res 1813(11):1978–1986

Zhong X, Rescorla FJ (2012) Cell surface adhesion molecules and adhesion-initiated signaling: understanding of anoikis resistance mechanisms and therapeutic opportunities. Cell Signal 24(2):393–401

Zhu C, Wang X, Deinum J, Huang Z, Gao J, Modjtahedi N, Neagu MR, Nilsson M, Eriksson PS, Hagberg H, Luban J (2007) Cyclophilin A participates in the nuclear translocation of apoptosis-inducing factor in neurons after cerebral hypoxia-ischemia. J Exp Med 204(8):1741–1748

Zhu YI, Tchkonia T, Pirtskhalava T, Gower AC, Ding H, Giorgadze N, Palmer AK, Ikeno Y, Hubbard GB, Lenburg M, O'Hara SP (2015) The Achilles' heel of senescent cells: from transcriptome to senolytic drugs. Aging Cell 14(4):644–658

Zhu SY, Zhuang JS, Wu Q, Liu ZY, Liao CR, Luo SG, Chen JT, Zhong ZM (2018) Advanced oxidation protein products induce pre-osteoblast apoptosis through a nicotinamide adenine dinucleotide phosphate oxidase-dependent, mitogen-activated protein kinases-mediated intrinsic apoptosis pathway. Aging Cell 17(4):e12764

Zuckerman V, Wolyniec K, Sionov RV, Haupt S, Haupt Y (2009) Tumour suppression by p53: the importance of apoptosis and cellular senescence. J Pathol 219(1):3–15

# Apoptosis: Natural Product-Derived Small Molecules as Therapeutics

18

Srijon Gayen, Suparna Ghosh, Barun Das Gupta, and Pallab Kanti Haldar

**Abstract**

Apoptosis, a highly organized and fundamental process in cell biology, plays a pivotal role in maintaining tissue homeostasis, development, and the elimination of damaged, abnormal, or potentially harmful cells. This complex cellular phenomenon is characterized by a series of composed events that lead to a cell's controlled and programmed death. Apart from the cellular stimuli, some other exogenous factors also induce apoptosis. Various natural products (NPs) have recently gathered considerable attention due to their potential effects in modulating apoptotic pathways, thus offering a promising avenue for developing novel therapeutic strategies. The NPs mainly include polyphenols, flavonoids, terpenoids, and alkaloids derived from various plants. The present chapter emphasizes the potentiality of the natural compounds via induction or inhibition of apoptosis. It summarizes the modulation of related cellular signaling pathways, including mitochondrial-dependent apoptotic pathways (involving p21 protein and proapoptotic members of Bcl-2 protein) and death receptor-mediated apoptotic pathway (tumor necrosis factor, TNF), which ultimately activate a cysteine protease family of proteolytic protein, caspases. The significance of NPs in apoptosis lays a sure foundation of valuable resources to combat cancer, inflammation, neurodegenerative, and cardiovascular disorders. Additionally, exploring natural compounds holds the potential for discovering novel agents with improved efficacy and reduced adverse effects.

**Keywords**

Apoptosis · Natural products · Caspases · Cancer · Inflammation · Neurodegeneration

S. Gayen · S. Ghosh · B. D. Gupta · P. K. Haldar (✉)
School of Natural Product Studies, Jadavpur University, Kolkata, West Bengal, India





K. Jana (ed.), *Apoptosis and Human Health: Understanding Mechanistic and Therapeutic Potential*, https://doi.org/10.1007/978-981-97-7905-5_18

## 18.1 Introduction

Apoptosis, a morphological process observed under light and electron microscopes, presents distinct characteristics within the apoptotic cell. The process involves pyknosis by condensation of chromatin and fragmentation of nucleus, blebbing of plasma membrane, and shrinkage of cells. Ultimately, the cell undergoes disintegration and generates apoptotic bodies, which are enclosed in small membrane fragments. These apoptotic bodies undergoes phagocytosis though does not produce any inflammatory response. The term "apoptosis" obtained from the Greek word meaning "To fall away from," resembling the shedding of leaves from deciduous trees during the season of autumn (Kerr et al. 1972). Apoptosis entails a marked array of biochemical as well physical alterations affecting the nucleus, cytoplasm and plasma membrane. At the early process of apoptosis, the cells undergo rounding and detachment from neighboring cells, accompanied by shrinkage. Swelling of cisternae and formation of vacuoles occurs due to dilation of endoplasmic reticulum within the cytoplasm. Meanwhile, condensation and fragmentation occurs by endonucleases within the nucleus. The fragmentation in nucleus can also be identified through the characteristic formation of "DNA ladder." The process can be analyzed by extraction of DNA from the nucleus, using an agarose gel for separation (Vaux 2002; Xu and Shi 2007).

The nucleus undergoes convolution and divides into multiple fragments, which become enveloped within the emerging apoptotic bodies. Simultaneously, the cell undergoes a pronounced breakdown process, culminating in the shedding of numerous membrane-bound spheres which contains apoptotic bodies having diverse sizes (Kerr et al. 1994). In the plasma membrane, some specific alterations enable the phagocytic cells to identify and engulf the apoptotic bodies. As the plasma membranes remain intact under the physiological condition, the cellular contents are contained, and inflammation does not occur during the apoptosis process. This kind of physiological cell demise presents a distinct morphology compared to oncosis, wherein cells swell and disintegrate in a disorganized manner, ultimately resulting in the demolition of cellular organelles with eventual rupture in the structure of plasma membrane. This process is known as necrosis where the cellular contents are released. Necrosis may also ensue in scenarios where apoptosis overwhelms phagocytic cells, particularly within the cells, where the phagocytic cells remain away (Trump et al. 1997). Given that apoptosis describes specific cellular death morphology, the word "programmed cell death" should not be interchanged as, programmed cell death specifically denotes cell demise during the development of organism depending on time and position. Since apoptosis typically avoids inducing inflammation, it is generally regarded as cell death in a physiological form. In most of the cases, programmed cell death exhibits as the morphological characteristics of apoptosis. However, some lifesaving drugs like anticancer can also induce apoptosis. For instance, without any intervention, the process of cell death initiates and doesn't strictly qualify for programmed cell death condition as the cells wouldn't naturally die (Melino 2001). In vivo observations of infrequent dying of cells have confirmed that apoptosis happens more than 20 times faster of mitosis. Neighboring cells decomposed and swiftly engulfed the apoptotic cells, leaving little evidence behind. To sustain cell homeostasis, there must be a meticulous equilibrium between the alterations of the cell population. These alterations may occur through a distinction from the proliferation, precursors, and further cell death. Without the cell death alongside mitosis, an 80-year-old individual would harbor an excess of lymph nodes and bone marrow, with an intestine spanning 16 kilometers (Desnoyers and Hengartner 1997; Lawen 2003).

Moreover, when the pathways involved in the apoptosis become dysregulated, it can lead to the development of various pathological transformations, ultimately culminating in cancer (Koff et al. 2015). Therefore, exploring methods to trigger this process in cancer cells could yield beneficial outcomes by impeding cancer progression and proliferation. Natural compounds can influence diverse signaling pathways, thereby impacting the molecular activity within cells. Worldwide, specific plant species have been documented to offer therapeutic benefits to treat various types of cancer, including those affecting pancreas, prostate, stomach, cervix, oral cavity, colon, breast, liver, lungs, skin, neck, and blood cells (Omara et al. 2020; Aye et al. 2019; Girma et al. 2018). Nature remains an unbounded origin to generate biologically active compounds in the fields of many anticancer therapies. According to a report on the discovery of anticancer drugs, among the 185 new drugs discovered from 1981 to 2019, approximately 65% are either natural compounds or derived from natural sources (Greco et al. 2021). Also, some studies have demonstrated that natural product (NP)-derived secondary metabolites exerted anticancer activities by inducing apoptosis and DNA damage within the cancer cells (Khan et al. 2019). This chapter aims to summarize the potential effect of various NP-derived molecules to target the major pathways of apoptosis to prevent cancer.

## 18.2 Definition and Significance of Apoptosis in Cell Biology

**Overview of the Apoptotic Pathway and Its Regulation** The mechanisms governing apoptosis exhibit remarkable complexity and sophistication, orchestrating a molecular event based on energy cascades. Current reports underscore the existence of two principal apoptotic pathways known as extrinsic (mediated by death receptors) and intrinsic (centred around mitochondria). Emerging documentation highlights an intricate interconnection between these pathways, where molecules can exert activities from one pathway to another one (Igney and Krammer 2002). Also, another pathway involving cytotoxicity mediated by T-cell may cause cell death by perforin-granzyme. This pathway is known as perforin/granzyme pathway and uses granzyme B and granzyme A to trigger the apoptosis. All of the three pathways are connected through a common execution pathway. Caspase-3 cleaves this perforin pathway and results in fragmentation of DNA, cross linking of protein, degradation of nuclear and cytoskeletal protein, formation of apoptotic body, ligand expression and activity of phagocytic cell and its associated receptors. Additionally, damage of single stranded DNA may cause cell death which is activated by granzyme A (Martinvalet et al. 2005). Reports suggested that, NPs may contribute as a promising approach as an anticancer agent in modern cancer therapies. In a study by Israels and Israels 2000 gives a clear suggestion on the in vitro molecular mechanism of secondary metabolites extracted from diverse NPs as a cancer therapy (Israels and Israels 2000).

**Extrinsic Pathway** Apoptosis initiates extrinsic pathway by interacting with transmembrane receptors mediated from the superfamily of tumor necrosis factor (TNF) (Locksley et al. 2001). Feature of this superfamily is the formation of a cytoplasmic domain which are approximately 80 amino acids in length (Ashkenazi and Dixit 1998). The cytoplasmic domain acts an important role to transmit death signals to the intracellular signaling pathways from its cell surface. Some of the most extensively identified ligands and including their death receptors are FasL/FasR, Apo3L/DR3, TNF-α/TNFR1, Apo2L/DR5, and Apo2L/DR4 (Chicheportiche et al. 1997; Ashkenazi and Dixit 1998; Peter and Krammer 1998; Suliman et al. 2001; Rubio-Moscardo et al. 2005). Among them, FasL/FasR and TNF-α/TNFR1 models are used comprehensively to identify the apoptosis of extrinsic pathways by binding the receptors with homologous trimeric ligands. After the binding, cytoplasmic adapter proteins enables the associated death domains to interact with the receptors. The example can be binding of the ligand Fas with its corresponding receptor Fas leading to the resulting FADD adapter protein inclusion while, the binding of the ligand TNF with its corresponding receptor TRADD adapter protein along with the FADD and RIP (Hsu et al. 1995; Wajant 2002). Successively, dimerization in the domain of death receptor connects the FADD with pro-caspase-8. This process activates the autocatalyzation of pro-caspase-8 and forms "Death inducing signalling complex (DISC)" leading to apoptosis. c-FLIP is a protein, which can inhibit the pro-caspase-8 activation by binding with the FADD, rendering the whole apoptosis process (Kischkel et al. 1995; Kataoka et al. 1998; Scaffidi et al. 1999). Toso is another protein that may also inhibit the pro-caspase-8 activation and prevent the apoptosis process in T-cells induced by Fas (Hitoshi et al. 1998).

Brevilin A, an abundant constituent of the medicinal herb *Centipeda minima* (L.) A. Braun & Asch, exhibits anticancer activities against various cancers, including breast cancer. Analysis of its underlying mechanism on apoptotic cell death proved that the compound brevilin A upregulated the caspase-8 and caspase-3 expression to cleave and reduce the conversion of Bax into Bcl2. The finding demonstrated that brevilin A can be a potential source to induce apoptosis by acting on extrinsic pathways (Lee et al. 2022).

**Intrinsic Pathway** The intrinsic pathway triggers apoptosis by generating stimuli mediated by non-receptors within the cells, especially those involved in mitochondria. The generated stimuli may exert both positive and negative signals. Due to the absence of some specific factors like hormones, cytokines, growth factors, etc., the mechanisms of suppression of apoptosis do not work, resulting in apoptotic inhibition. Conversely, radiation, hypoxia, viral infections, toxins, free radicals, and hyperthermia are the factors involving positive signals. As a result, alteration occurs on the inner membrane of mitochondria due to the loss of transmembrane potentiality, opening of transition pore, and liberation of the primary groups of pro-apoptotic proteins in the cytosol (Saelens et al. 2004).

Among the primary groups Smac/DIABLO, serine protease HtrA2/Omi stimulates the caspase-dependent mitochondrial pathway, and the cytochrome c stimulates the Apaf-1 and procaspase-9 to generate the apoptosome process (Chinnaiyan 1999; Hill et al. 2004). This process also stimulates the caspase-9. Smac/DIABLO, serine protease HtrA2/Omi acts on the protein (inhibitors of apoptosis proteins, IAP) that inhibits the apoptosis process (van Loo et al. 2002; Schimmer 2004). Some reports identified some other mitochondrial proteins that can inhibit the action of, although the experiments on the gene knockout process suggested that binding of the mitochondrial protein only with the IAP is not sufficient to call it a proapoptotic protein (Ekert and Vaux 2005).

CAD, Endonuclease G, and AIF are also included among the primary groups that may be released during the apoptosis process after cell death. AIF condensed the nucleus by fragmenting the DNA in the peripheral nuclear chromatin into more than 250 kb pieces. This process is known as "stage I" condensation (Susin et al. 2000). On another side, endonuclease G fragments the oligonucleosomal DNA by cleaving the nuclear chromatin. Consequently, CAD is liberated towards the nucleus and cleaved by caspase-3 and induce the fragmentation of oligonucleosomal DNA to condense the chromatin more promptly. This process is known as 'Stage II' condensation in mitochondria (Enari et al. 1998; Susin et al. 2000). These whole mitochondrial proceedings are arranged through the Bcl-2 protein family. P53, a tumor suppressor protein, plays a crucial role in regulating the Bcl-2 proteins, though the mechanisms involved are completely not understood till now (Schuler and Green 2001).

Although some potent mechanisms are determined to be involved, no clear proof has been demonstrated. Apoptosis involving the Fas pathway can cause mitochondrial damage induced by cleavage of caspase-8 Bad (Li et al. 1998; Esposti 2002). This process may illustrate the "cross-talk" among the extrinsic and intrinsic pathways (Igney and Krammer 2002). Moreover, 14–3-3 is a member of the phosphoserine binding molecules associated with Bad serine phosphorylation. The Bad phosphorylation in cytosol binds with 14–3-3 for sequestration, while dephosphorylation promotes cytochrome C to release in mitochondria (Zha et al. 1996).

Puma and Noxa are the two key members of the Bcl-2 family, involved in the proapoptotic process, where both Puma and Noxa act as mediators in p53-induced apoptosis. In vitro investigation demonstrated that the increased expression of Puma is strongly associated with changes in BAX expression including overexpression, translocation, conformational changes, and release of cytochrome c and reduced potency of mitochondrial membrane (Liu et al. 2003). Noxa can activate caspase-9 by interacting with the antiapoptotic Bcl-2 family members and localization itself to the mitochondria (Oda et al. 2000).

*Annona muricata* L., commonly known as Graviola, is an antioxidant-rich evergreen plant—investigations with the leaf extract of graviola on the mechanism of its apoptosis pathways. The study showed that the leaf extract exerted its apoptotic activity by an ER-dependent intrinsic pathway (Kim et al. 2018). Another study showed the pathway responsible for induction of apoptosis by the leaf extract of the plant *Dillenia pentagyna* Roxb. The extract induced the intrinsic pathway by decreasing the potentiality of mitochondrial membrane resulting in the elevated level of Bcl-2 and cytochrome-c, fragmentation of nucleus and activated the caspase-3. Moreover, a combination of the plant leaf extract with a NF-κB inhibitor significantly improved the mechanism of intrinsic pathway to induce apoptosis (De et al. 2023).

**Execution Pathway** Execution pathway is a definitive pathway of apoptosis where both the extrinsic and intrinsic pathway meets together. In this pathway, the execution caspases are activated by triggering the cytoplasmic endonucleases to degrade the nucleic materials and proteases to degrade the cytoskeletal and nucleic proteins. The key effectors are known as "executioner" in this pathway involving caspase-3, −6, and − 7. They cleaves different substrates responsible to execute the pathway, e.g., PARP, cytokeratins, nuclear protein NuMA, alpha fordin plasma membrane cytoskeletal, etc. The execution pathway is mainly responsible for the changes of morphology as well as biochemical changes, characteristic for the apoptotic cells (Slee et al. 2001).

*Numerous plant compounds concurrently triggered the apoptosis by targeting the major pathways are listed in the following table.*

## 18.3 Cellular Signaling Pathways in Apoptosis

**Role of Bcl-2 Family Proteins** In the apoptosis process, the cells alter the propensity during the development of lymphocyte, also known as the apoptotic phenotype. These phenotypes are influenced by the overexpression of Bcl-2, Bcl-xL, Fas and ligands of Fas (Krammer 2000). Typically, developmental stages marked by selection are distinguished by decreased expression of Bcl-2 and Bcl-xL, while the proliferation stages tend to exhibit heightened levels of expression (Behrens and Mueller 1997). In some of the B-cell lymphomas, the overexpression of Bcl-2 happens due to the suppression of enhancer of immunoglobulin heavy chain. Notably, Bcl-2 was discovered as the first protooncogene that can regulate the progression of cell cycle by retarding its appearance in S-phase. The oncogenic properties stem from inhibits the apoptosis thus hampering the involvement of apoptotic signal in the development of cancer. Also, it may act through an antioxidant property (Lawen et al. 1998). Bcl-2 comprises some domains of Bcl-2 homology (BH1, BH2, BH3 and BH4) that are essential for its own survival. This function is similar to that of CED-9. Currently, 3 protein groups are identified under Bcl-2 family. They are anti-apoptotic proteins, pro-apoptotic proteins and BH3 proteins. Example of anti-apoptotic proteins are Bcl-2 and Bcl-xL. They have 4 BH domains and a membrane anchor of c-terminal. Bax and Bak are the pro-apoptotic proteins that lack the 4 BH domains and Bad, Bik, Bid and Bim are the example of BH3 proteins identified by the presence of an amphiphatic helical structure. The balance between proapoptotic and antiapoptotic proteins within a cell determines its susceptibility to apoptosis, as proposed by the rheostat hypothesis (Korsmeyer 1995). Numerous members of this protein family have the capacity to form homo- and/or heterodimers, either promoting or inhibiting cell death. A multitude of death signals converge at the mitochondria through BH3-only proteins (Borner 2003).

An investigation showed that pyranocycloartobiloxanthone A, a plant derived xanthone compound showed apoptotic mechanism by downregulating the Bcl2 which triggered the release of cytochrome c to cytosol from mitochondria (Mohan et al. 2012). Another study showed that extract of *Croton tiglium* inhibited the proliferation of a lung cancer cell by increasing the expression of mRNA of Bax and decreasing the expression of Bcl-2, though the study does not implied any changes in the mRNA expression level of Fas and caspase-8 (Li et al. 2016).

**Role of p21 Protein** Disruption of mammalian cell cycle is a critical factor in the process of cellular transformation. p21 is a cyclin-dependent kinase inhibitor that acts a pivotal role to induce the arrest of cell cycle in response to various stimuli. Although p21 is primarily familiar with the growth inhibition process, it may also exert an inhibitory effect on apoptosis, contributing to its paradoxical oncogenic activities. The impact of p21 depends on various mechanisms. For instance, the selective transcriptional repression and activation of CDKN1A and pro-apoptotic genes respectively may inhibit apoptosis. This may also happen due to alteration in the expression of p21 (Okaichi et al. 1999; Kaneuchi et al. 1999; Samuels-Lev et al. 2001). Additionally, alteration such as, phosphorylation of p21 or cleavage of p21 by caspase-3 can also inhibit apoptosis (Zhang et al. 1999). The inhibitor p21 acts on apoptosis by producing various stimuli involves overexpression of p53, deficiency of growth factor, or variation in monocytes. The antiapoptotic effect of p21 may execute from the interactions between proteins or the mechanisms of DNA repair. For this mechanism, p21 also can regulate the gene transcription factors. For instance, p21 when present in cytoplasm, it prevents the protein activity of caspase-8, caspase-10, and pro-caspase-3, apoptosis signal-regulating kinase-1 (ASK1), and stress activated protein kinases (SAPKs) which are involved in induction of apoptosis. Moreover, p21 may directly bind and inhibit the activity of proapoptotic genes by E2F1 or MYC. However, it cannot be ruled out that CDK activity might be required to induce proapoptotic genes by E2F1 or MYC (Dotto 2000).

Interestingly, p21 also can be act through p53 dependent as well p53 independent mechanisms to induce apoptosis. This mechanism may involve the upregulated activity of BAX, increased activity of the death receptors engaged in the family of TNF, or may also effects the DNA repair mechanism (Gartel 2005).

Lapiferin, a sesquiterpene alcohol plant compound, may act as a natural therapeutic treatment for gingival squamous cell carcinoma treatment. A study showed that after P21 expression depletion, lapiferin's ability to prevent cell proliferation and enhance cell apoptosis was markedly diminished which possess the robust antitumor properties (Liu et al. 2021).

Another study proved that the extract of *Juniperus communis* (JCo) induced the cell cycle arrest by the overexpression of p53-dependent apoptotic pathway and CDKs and also acting on both the extrinsic as well intrinsic pathway to exert its potential effects on oesophageal squamous carcinoma cell (Li et al. 2020).

**Role of Tumor Necrosis Factor (TNF)** In the treatment of inflammatory disease, the inhibition of TNF with its associated receptors TNF receptor 1 (TNFR1) and TNF receptor 2 (TNFR 2) has been identified as a key mechanism for achieving success. Previously, it was reported that the mechanism acts on TNFR1 and prevents the activation of nuclear factor- κB (NF- κB) and mitogen-activated protein kinase (MAPK) pathway. Failure of the mechanism may lead to increased levels of proinflammatory genes towards the inflammation pathology. Recent reports exhibited that the mechanism may also induce apoptosis, necroptosis, and also pyroptosis by promoting the inflammation. Moreover, intracellular constituents are released from death cells which further associated with the overexpression of proinflammatory genes in neighboring cells. Additionally, microbial tissue infiltration may also promote the inflammation by affecting the barrier integrity in the epithelial cells of skin and intestine and cell death occurs. To overcome the problem, some researchers investigated that targeting the genetic factors related to cell death in TNF-induced inflammatory mouse models can reverse the inflammatory condition. Therefore, drugs targeting the cell death are currently under high demand in the treatment TNF-driven cancer disorder. Example of such drug may be receptor interacting serine or threonine protein kinase 1 (RIPK1) (van Loo and Bertrand 2023; Anderton et al. 2020; Martens et al. 2020; Mifflin et al. 2020).

The extract of *Artemisia annua* was able to reverse the expression of TNF-α oncogene and significantly improved the side effects of pioglitazone in the experimental mice model associated with the greater risk factor of bladder cancer (Botrous et al. 2023).

**Role of Caspases** Previously, it has been written that the induction of apoptosis can be inhibited by various stimuli generated by external and internal pathways. Both of the pathways ultimately meet at one specific point known as caspases. According to some reports, to date, total 14 mammalian caspases have been characterized (Strasser et al. 2000). Initially, the caspases are produced as proenzymes and activated by other caspases after proteolysis. Peptide caspase inhibitors may decrease the activation of caspase thus, apoptosis (Earnshaw et al. 1999; Garcia-Calvo et al. 1998). Based on various criteria, caspases can be divided into three groups namely initiator caspases, executioner caspases, and other caspases. The initiator caspases may be identified by the presence of long prodomains attached with more than 90 amino acids and DED or CARD domains, e.g., caspase-8, caspase-10, or caspase-2, caspase-9, respectively. The executioner caspases can be identified by the presence of short prodomains, e.g., caspase-3, caspase-6, and caspase-7. The group of other caspases are primarily involved with cytokine formation instead of inducing apoptosis. Interface between the subunits of 1 Arg, 1 His, 1 Cys (large subunit), and 1 Arg (small subunit) may lead to the formation of the active site (Grutter 2000). The initiator caspases cleave the cellular substrates of another effector caspases which leads to characteristic morphological changes in apoptosis. Example of such effector caspase is caspase-3. On the activation of caspase-3, several crucial cellular substrates are cleaved, along with ICAD (inhibitor of caspase-activated DNase), poly (ADP-ribose) polymerase (PARP, a DNA repair enzyme), ROCKI (Rho-associated coiled-coil-forming kinase I), fodrin, actin, and lamin. The process can cause membrane blebbing by increased kinase activation, phosphorylation of myosin light chain, DNA fragmentation, rupture of cell structure, and finally generates CAD (caspase-activated DNAse) into nucleus to cleave DNA internucleosomally, ultimately resulting in cell death (Sebbagh et al. 2001; Nagata 2000; Coleman et al. 2001; Leverrier and Ridley 2001). Some viral and cellular proteins may also act as a caspase inhibitor. Neuronal cells may express these proteins and known as neuronal apoptosis inhibitory protein, NAIP which can protect against early apoptosis (Crocker et al. 2001). Some viruses can encode the IAPs, antiapoptotic Bcl-2 proteins, and other apoptosis inhibitors to prevent infected cells from undergoing apoptosis. Structurally, caspases adopt the caspase fold, comprising the subunits of large a and small b. The heterodimers of these a/b subunits consists of total 6 a-f, b strands. Among them, a-e are arranged parallel, whether, orientation of strand f is antiparallel. On another hand, a twisted structure is generated from the b subunits surrounding the a-helices (H1, H2, H3, H4, and H5) (Salvesen and Duckett 2002).

A study demonstrated that (-)-Anonine, an alkaloid compound derived from plants, demonstrated its ability for induction of apoptosis in the cancer cells of cervical areas by the activation of caspases 3, 7, 8, and 9, suggesting its potential chemopreventive activity (Chen et al. 2008).

A recent report suggested that, the flower extract of Elaeagnus angustifolia induced apoptosis and cell proliferation by increasing the level of proapoptotic marker caspase-8 and Bax and decreasing the antiapoptotic marker Bcl-2. Moreover, the extract proved to significantly enhance the expression of p53 in human oral as well breast cancer cell line (Fouzat et al. 2022).

## 18.4 Natural Product-Derived Small Molecules in Apoptosis Modulation

NPs have historically been essential in drug development, particularly for infectious and cancerous disorders and other therapeutic areas, including multiple sclerosis and cardiovascular diseases (statins, for example) (fingolimod). NPs are essential in treating and mitigating disease and disorders for a significant part of the world population. NPs are crucial in maintaining specific biological functions, such as manage defense mechanisms and interacting with other organisms, often competitively. This is why NPs are highly significant in the context of infectious diseases and cancer (Atanasov et al. 2021; Harvey et al. 2015). However, due to improper scientific validation and safety evaluation studies, such advances in drug discovery from ethnopharmacology-inspired NP-derived small molecules, safety and efficacy remain a concern. Furthermore, screening specific bioactive molecules of interest might pose challenges and incur significant economic costs. Obtaining an adequate amount of biological material to isolate and analyze a bioactive natural substance might potentially present difficulties. Moreover, manufacturing phytomarkers entails several technological obstacles. Temperature, light, and solvents often lead to the deterioration and/or alteration of pure components. Additionally, isomers and conformations may also induce modifications in the markers. Nevertheless, it is crucial to gain a comprehensive grasp of the intricate complexities of medicinal plants by marker profiling and similar methodologies in order to bridge connection between ethnopharmacology-derived herbal medicine (Mukherjee et al. 2022).

## 18.5 Phytochemicals in Apoptosis

The following section highlights the mechanism of action of phytomolecules from NPs involved in inducing apoptosis (Fig. 18.1, Table 18.1).

- *Lignans*

Honokiol (**1**) a neolignan biphenolic compound isolated from *Magnolia officinalis* (Magnoliaceae) bark and seed cones induced apoptosis at a dose of 30 µg in skin tumor induced SKH-1 mice. The process of apoptosis in skin cancer was enhanced by the activation of p53, DNA fragmentation, caspase activation, and cleavage of poly (adenosine 5′-diphosphate-ribose) polymerase (PARP) (Chilampalli et al. 2010). Honokiol was also studied to trigger programmed cell death in ovarian cancer cells (SKOV3 and Caov-3) at concentrations >40 µM by activation of caspases 3, 7, and 9 and cleavage of PARP. It was also found that induction of apoptosis by honokiol relied on the AMPK/mechanistic target of rapamycin signaling pathway (Lee et al. 2019b). Balan et al., showed anticancer effect of Honokiol (20 µM) to induce cell death in kidney cancer cells by downregulating the expression of cMet and phosphorylated Ras (Balan et al. 2017). Xin et al., studied the apoptotic potential of the lignan mixture extracted from *Vitex negundo*. The apoptotic potential was quantified using the TUNEL test and DAPI labeling. Additionally, the western blot analysis was used to evaluate the impact on the expression and phosphorylation of proteins that are intimately associated with cell cycle and apoptosis. The compound EVn-50, at concentrations of 20 µg/mL and 40 µg/mL, was seen to trigger apoptosis in MDA-MB-435 cells by causing the cancer cells to be arrested in the G2/M phase of the cell cycle (Xin et al. 2013). The purified vitexins (EVn-50 at a concentration

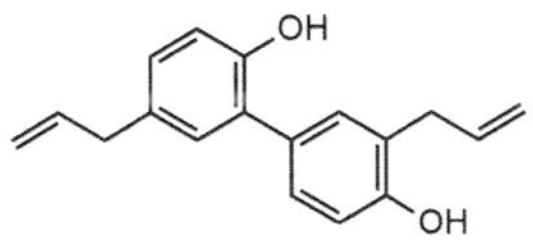

**(1)**

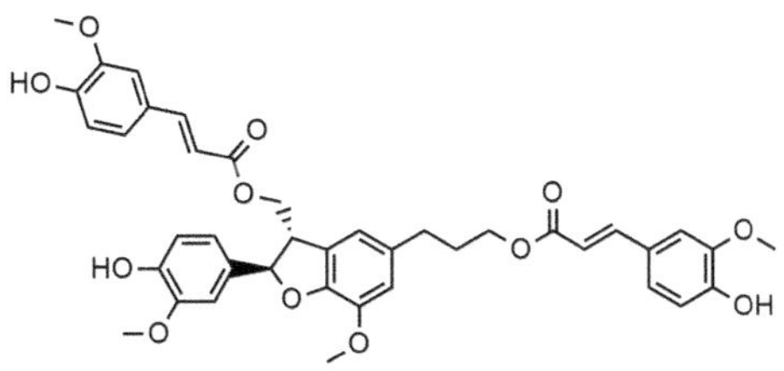

**(2)**

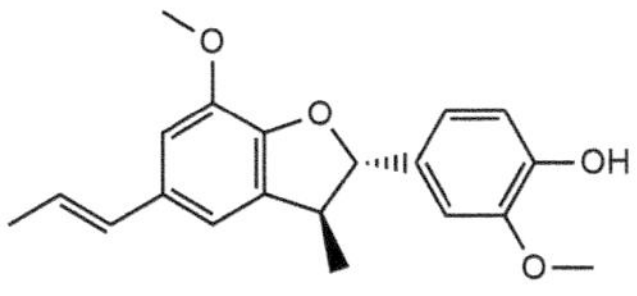

**(3)**

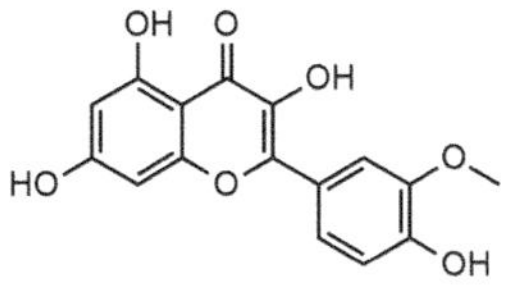

**(4)**

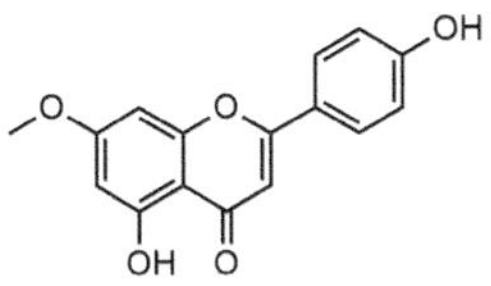

**(5)**

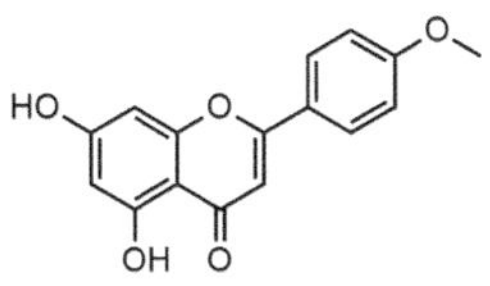

**(6)**

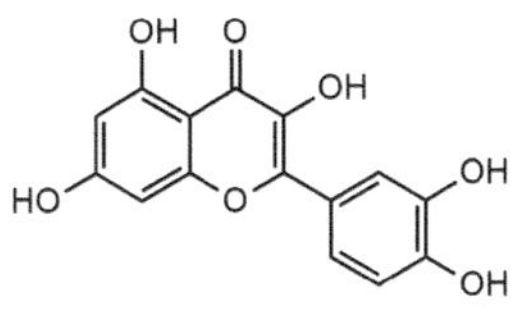

**(7)**

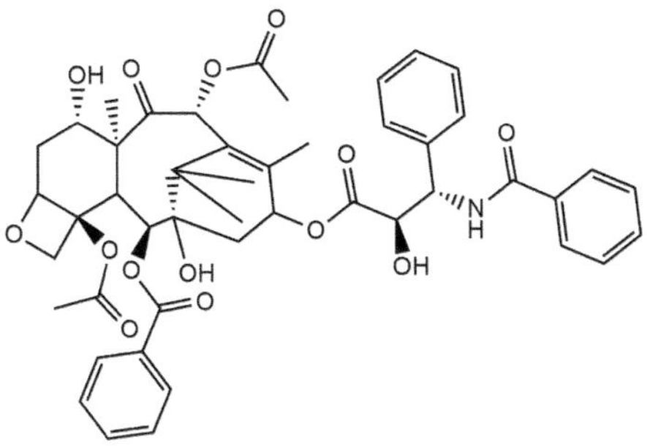

**(8)**

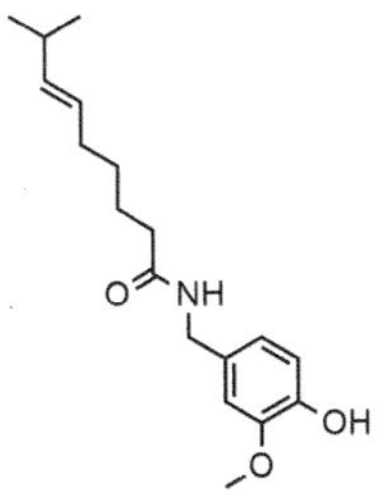

**(9)**

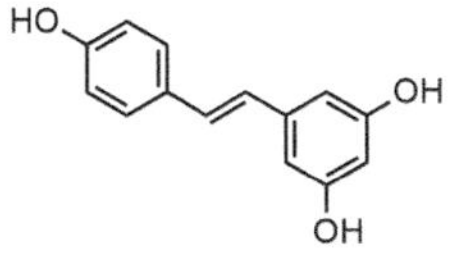

**(10)**

**Fig. 18.1** Natural product-derived phytomolecules

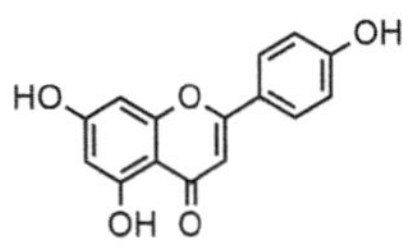

(11)

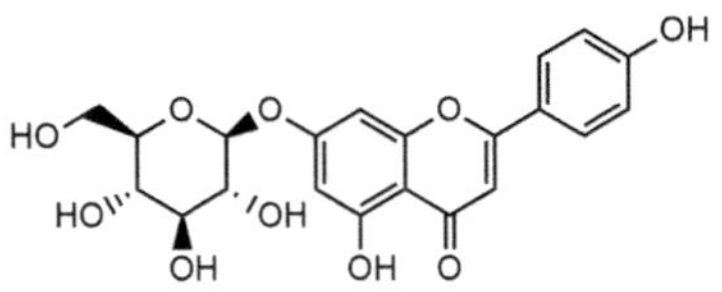

(12)

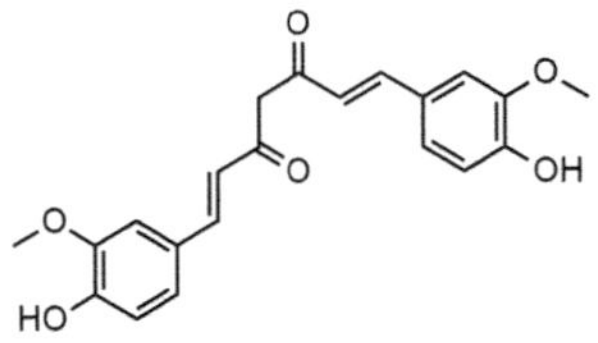

(13)

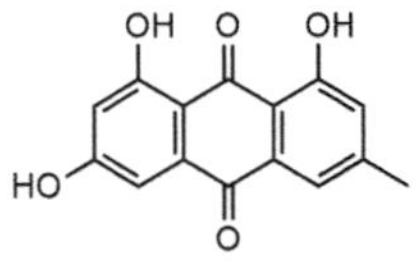

(14)

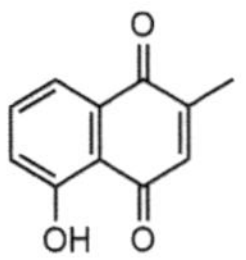

(15)

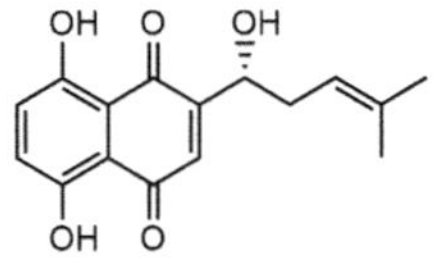

(16)

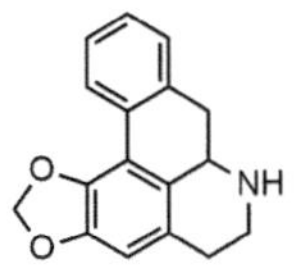

(17)

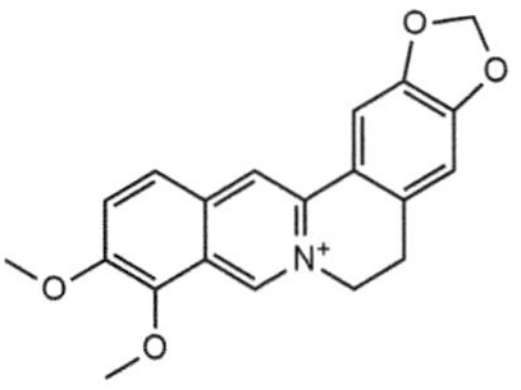

(18)

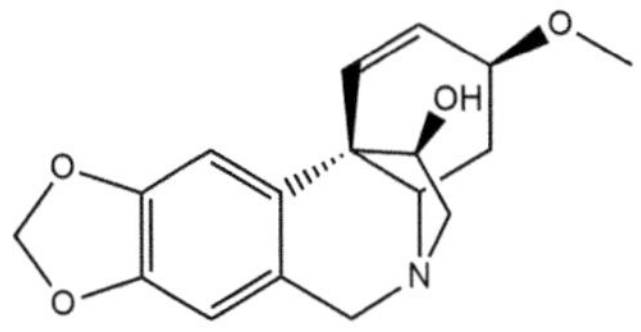

(19)

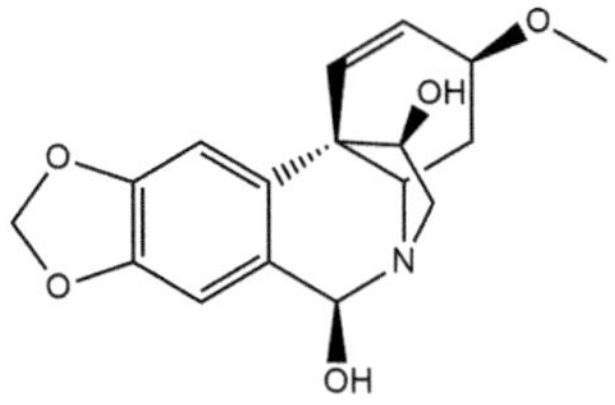

(20)

**Fig. 18.1** (continued)

of 5 μg/mL), which were obtained from air-dried seeds of *Vitex Negundo*, were discovered to have cytotoxic effects and induce apoptosis in various cancer cell lines. This was achieved through the activation of caspase-3 and caspase-9, resulting in the cleavage of poly ADP ribose polymerase protein. Additionally, the vitexins caused an increase in the expression of Bax and a decrease in the expression of Bcl-2 in breast cancer cells (MCF-7, ZR-75-1, SK-BR-7, MDA-MB-231, and MDA-MB-435 s), prostate cancer cells (PC-3 and LNCaP), and ovarian cancer cells (COC1) (Zhou et al. 2009). HJC, a new arylnaphthalene lignan isolated from

**Table 18.1** Extensive catalogue of natural compounds and their corresponding anticancer properties

| Phytomolecule | Natural products/plant scientific name (family) | Type of extract | Type of cancer | Pharmacological potential with brief mechanism of action | Reference |
|---|---|---|---|---|---|
| Formononetin | *Trifolium pratense* L. (Fabaceae) | – | Gastric cancer | • suppression of cell growth in MGC803 cells • restriction of cell mobility in SGC7901 tumor cells through the Wnt/β-catenin and Akt/mTOR signaling pathways | Yao et al. (2019) |
| Honokiol | *Magnolia officinalis* Rehder & E. H. Wilson (Magnoliaceae) | – | Colon cancer, Oral cancer, thyroid cancer | • enlisting of Jagged-1 and Hes-1 to activate genes further down the line<br>• decrease in the expression of CSC marker DCLK1<br>• depletion of the β-catenin (SP) receptor inhibitor CD44 and Jagged-1 and Hes-1 in the direction of the downstream gene target •reduction in expression of the CSC marker DCLK1 | Fried and Arbiser (2009) |
| Curcumin | *Curcuma longa* L. (Zingiberaceae) | – | Colon cancer, gastric cancer, AML, breast cancer, head and neck cancer, lung cancer | • downregulation of Gli-1, Notch-1, and cyclin D1<br>• inhibition of tumor growth of ALDH+/CD133+ cells<br>• reduction in the expression of CD44 and CD166 in chemoresistant colon cancer<br>• reduction in both the mRNA and protein levels of cyclin D1 | Kanwar et al. (2011) |

(continued)

**Table 18.1** (continued)

| Phytomolecule | Natural products/plant scientific name (family) | Type of extract | Type of cancer | Pharmacological potential with brief mechanism of action | Reference |
|---|---|---|---|---|---|
| Ursolic acid and Koetjapic acid | *Malus pumila* mill. (Malpighiaceae), *Pyrus communis* L. (Rosaceae), and *Prunus domestica* L. (Rosaceae) | – | Colon cancer | • induction of caspase-3 cleavage and inhibition of STAT3 phosphorylation in ALDH(+)/CD133(+) colon cancer-initiating cells | Wang et al. (2013b) |
| Hesperidin | *Calotropis gigantea* (L.) W. T. Aiton (Rutaceae) | Ethanolic extract | Nonsmall-cell lung carcinoma | • by modulating NF-B signal transduction, apoptosis is utilized to inhibit cell growth | Lee et al. (2019a) |
| Isobutyrophenone and Arnicolide | *Centipeda minima* (L.) A. Braun & Asch. (Compositae) | Aqueous and hydroalcoholic extract | Colon cancer | • facilitation of intracellular reactive oxygen species (ROS) synthesis • ligation of NF-κB | Taylor and Towers (1998) |
| Epigallocatechin gallate | *Camellia sinensis* (L.) Kuntze (Theaceae) | Crude green tea polyphenol extract | Skin cancer | • inhibition of PKD1 activation | Du et al. (2012) |
| Gossypol | *Gossypium herbaceum* L. (Malvaceae) | – | Prostate cancer | • induction of DNA damage, stimulation of p53, and initiation of apoptosis | Qian and Wang (1984) |
| Ethoxy Mansonone G | *Mansonia gagei* J. R. Drumm. (Malvaceae) | Hydroalcoholic extract of the heartwood | Breast cancer | • inhibition of estrogen-induced cell proliferation | Chonsut et al. (2019) |
| Genistein | Soybeans and soy products (legumes) | Gamma-tocotrienol extracted from palm oil and PSP extracted from the mushroom Turkey tail | Prostate cancer, cervical cancer, colorectal cancer, non-small-cell lung carcinoma | • hedgehog-Gli signaling inhibition and CD44 expression of CSC markers<br>• stabilization of p53 via the APE1 pathway<br>• suppression of the Akt pathway | Zhang et al. (2012a) |

| | | | | | |
|---|---|---|---|---|---|
| Sulforaphane | Cruciferous vegetables (Brassicaceae) | – | Prostate and pancreatic cancer | • inhibition of hedgehog-gli signaling and expression of CD44, which are markers of CSCs;<br>• P53 stabilization via the APE1 pathway; • suppression of the Akt pathway | Kallifatidis et al. (2011) |
| Proanthocyanidins | *Vitis vinifera* L. (Vitaceae) | Hydroalcoholic extract of grape seeds | Pancreatic adenocarcinoma | • NF-B expression is decreased and the epithelial-mesenchymal transition process is reversed | Katiyar (2008) |
| Methyl antcinate A | *Antrodia camphorate* | – | Breast cancer | • heat shock protein 27 is downregulated, whereas IkBα and p53 expression are upregulated | Hsieh et al. (2010) |
| Oridonin | *Rabdosia rubescens* (Hemsl.) H. Hara (Lamiaceae) | – | Breast cancer | • inhibition of Notch1 activity and downregulation of Jagged2 expression • induction of signaling via the PI3K/PTEN/Akt/mTORC1 and WNT/β-catenin pathways | Dong et al. (2014) |
| Resveratrol | *Vitis vinifera* L. (Vitaceae) | – | Head and neck cancer, gastric cancer, thyroid cancer, lung cancer | • inhibition of ALDH1 and CD44 in HNC-TICs in a dose-dependent manner | Hu et al. (2012) |

(continued)

**Table 18.1** (continued)

| Phytomolecule | Natural products/plant scientific name (family) | Type of extract | Type of cancer | Pharmacological potential with brief mechanism of action | Reference |
|---|---|---|---|---|---|
| Guggulsterone | *Commiphora mukul* (hook. Ex stocks) Engl. (Burseraceae) | Guggul extract | Colon cancer, prostate cancer, head, and neck cancer | • induction of cell death in PC-3 cells<br>• downregulation of ALDH1 and CD44 in HNC • activation of caspase-9, caspase-8, and caspase-3<br>• HT-29 cell apoptosis induction via activation of caspases 3 and 8<br>• dose-dependent downregulation of cIAP1, cIAP2, and Bcl-2 levels and upregulation of truncated bid, Fas, p-JNK, and p-c-Jun -TICs | Deng (2007) |
| Emodin | *Rheum palmatum* L. (Polygonaceae) | Hydroalcoholic extract | Prostate cancer | • activation of notch signaling pathway | Deng et al. (2015) |
| Artesunate | *Artemisia annua* L. (Compositae) | – | Chronic myeloid leukemia | • downregulation of VEGF expression | Du et al. (2010) |
| Allicin | *Allium sativum* L. (Liliaceae) | – | Ovarian cancer | • activation of JNK pathway | Xu et al. (2014) |
| Caffeic acid phenethyl ester (CAPE) | Propolis of honeybee hives | Caffeic acid phenethyl ester (CAPE) is a main bioactive component extracted from honeybee hive propolis | Prostate cancer | • induction of cell cycle arrest (G1/G2/M) | Lin et al. (2015) |

| | | | | | |
|---|---|---|---|---|---|
| Erinacine A (ditherpenoid) | *Hericium erinaceus* (Bull.: Fr.) Pers. | Ethanolic extract | Gastrointestinal cancer, colorectal cancer | • cell cycle arrest caused by ROS and decreased NF-κB activation in cancer stem cells | Lee et al. (2017) |
| Morusin | *Morus alba* L. (Moraceae) | – | Cervical cancer | • modulating the Bcl-2, Bax, and STAT3 proteins to induce apoptosis and limit growth | Wang et al. (2013a) |
| Deoxyshikonin | *Arnebia euchroma* (Royle) Jonst (Boraginaceae) | Hydroalcoholic extract of root | Colorectal cancer, glioma | • induction of G2/M phase cell cycle arrest | Zhu et al. (2019) |
| Apigenin | *Scutellaria baicalensis* Georgi | – | Colorectal cancer, lung cancer, osteocarcinoma, liver cancer, melanoma, prostate cancer | • support for several antiinflammatory pathways, including as PI3K/Akt and p38/MAPK<br>• Raf-1 kinase alteration causing apoptosis | Salehi et al. (2019) |
| Paclitaxel | *Taxus brevifolia* Nutt. | – | Endometrial cancer, sarcoma, cervical cancer, gastroesophageal cancer, prostate cancer, head and neck cancer | • MEK-independent signaling pathway activation | Blagosklonny et al. (1996) |

(continued)

**Table 18.1** (continued)

| Phytomolecule | Natural products/plant scientific name (family) | Type of extract | Type of cancer | Pharmacological potential with brief mechanism of action | Reference |
|---|---|---|---|---|---|
| Tannins/tannic acid | *Syzygium guineense*,Gall nuts | Extracts of *Syzygium guineense* from Cameroon specifically inhibit Wnt signaling | Triple-negative breast cancer, colon cancer, lung cancer, liver cancer, ovarian cancer, prostate cancer, pancreatic cancer, embryonic carcinoma | • reduces TGF-β-induced EMT and NF-κB activation in cells;<br>• promotes TRAIL-induced extrinsic apoptosis;<br>• downregulates JAK/STAT pathway;<br>• inhibits cell cycle progression by G1/S phase arrest;<br>• acts in concert with cisplatin and oxaliplatin | Koval et al. (2018) |
| Calotropin<br>Quercetin | *Asclepiadoideae* family nuts, tea, onions, apple and in plant sources | Methanol extract of *Calotropis gigantea* (Asclepiadaceae) | Colon cancer colorectal cancer, triple-negative breast cancer | β-Catenin inhibition, GSK-3β phosphorylation inhibition, and E-cadherin upregulation | Park et al. (2014) |
| Lycopene | *Genista tinctoria* L. | Hydroalcoholic extract | Prostate cancer | • reduce the expression of cyclin E;<br>• block β-catenin;<br>• prevent GSK-3β phosphorylation | Sferrazza et al. (2020) |

*Justicia procumbens*, causes apoptosis and caspase activation in K562 leukemia cells. HJC was isolated from dried plants of *Justicia procumbens* and treated for growth inhibition in leukemia K562 cells (8.41, 33.7 and 135 μM) in dose-dependent manner (Luo et al. 2014). Pan et al., studied the effects of Boehmenan (**2**) isolated from *Clematis armandii* dried stems to exhibit apoptosis in A549 cells. It was found to modulate apoptosis-related protein (p53, p21, cleaved caspase 3, and cleaved PARP) and cyclin D1 expression inducing $\Delta\Psi_m$ collapse in a concentration-dependent manner (Pan et al. 2016). Licarin A (**3**) has the ability to cause programmed cell death (apoptosis) and halt the progression of CCRF-CEM cells in the sub-G0/G1 phase. These effects are associated with the activation of caspase 3/7, 8, and 9, changes in the mitochondrial membrane potential (MMP), and an increase in the formation of reactive oxygen species (ROS) (Mbaveng et al. 2021).

- *Flavonoids*

Isorhamnetin (**4**), genkwanin (**5**), and acacetin (**6**) isolated from the hydroalcoholic extract of *Tephroseris kirilowii* (Turcz.) Holub. were found to cause cell cycle arrest in the G2/M phase and resulted in the induction of apoptosis and autophagy in human breast cancer cells (MDA-MB-231 and MCF-7) in a dosage-dependent manner. Isorhamnetin and genkwanin showed more potent apoptotic potential than acacetin with decreased in the expression of PARP1 and increase in levels of cleaved caspase-3 (Zhang et al. 2018). Srivastava et al., highlighted the apoptotic potential of Quercetin (**7**) at 20 μM for 8, 16, and 24 h in leukemic cells and breast cancer cells in a dose-dependent manner and its interaction with DNA directly to activate the intrinsic pathway (Srivastava et al. 2016). Isoprenoid flavonoids extracted from air-dried velamen of *Sophora davidii* were shown to cause apoptosis in HT29 cells at doses ranging from 5 μM to 10 μM. This effect was achieved via enhancing the phosphorylation of P38 MAPK and suppressing the phosphorylation of Erk MAPK (Yuan et al. 2023). A recent study highlighted the apoptotic potential of Licorice in the TUNEL staining assay by activating the PI3K/AKT pathway (Guo et al. 2024).

- *Terpenoids*

In breast cancer cells, paclitaxel (**8**) induces apoptosis via activating the Erk, p53, and p38 MAP kinase pathway in MCF7 and its derivatives, MCF7/HER-2 and MDD2 human breast cancer cell lines (Bacus et al. 2001). Paclitaxel isolated from *Taxus brevifolia* was also found to induce apoptosis at 50 mg/ml through increasing P53, Bax, and cleaved caspase 3 synthesis and the decrease in Bcl-2 synthesis in CHMm canine mammary carcinoma cell line (Ren et al. 2018). Paclitaxel at 0.1 μg/mL was also found to be potent in combination with tanshinone I (to increase fat solubility) to induce marked process of programmed cell death in human ovarian cancer cells (A2780) and mouse ID-8 cells by increasing the expression of Bax and decreasing the expression of Bcl-2 (Zhou et al. 2020).

- *Phenylpropanoid*

Capsaicin (**9**) at variable concentrations may trigger programmed cell death, known as apoptosis, by using a mechanism that does not include caspases. Additionally, it suppresses the activity of a protooncogene called FBI-1, which in turn inhibits the NF-kB pathway in breast cancer cells (specifically MCF-7 and MDA-MB-231 cells) (Chen et al. 2021). Capsaicin was shown to trigger apoptosis in gastric cancer cells (AGS cells) by enhancing the cleavage of caspase-3 and reducing Bcl-2 levels via the manipulation of the MAPK signaling pathway. Capsaicin was discovered to hinder the growth and promote self-destruction, programmed cell death, and a halt in the cell division process in the G1 phase in nasopharyngeal cancer cells via altering the PI3K/Akt/mTOR pathway. Apoptosis is a process of programmed cell death. The upregulation of autophagy markers LC3-II and Atg5 led to an increase in p62 levels. Enhancing the degradation of Fap-1 and promoting the activation of caspase-3 (Lin et al. 2017).

- *Stilbenoid*

Resveratrol (**10**) has been shown to cause cell cycle arrest and trigger mitochondrial-mediated and caspase-independent apoptosis in transgenic adenocarcinoma of mouse prostate (TRAMP) cells. Additionally, it was discovered that it enhances the process of programmed cell death (apoptosis) in A2780 ovarian cancer stem cells, regardless of the presence of ROS (Bhaskara et al. 2020). Resveratrol stimulated apoptosis in breast tumor cells by increasing the expression of tumor suppressor p53, decreasing the expression of antiapoptotic protein Bcl-2, and initiating caspase-3 activation (Zhao et al. 2020). An investigation demonstrated the combined impact of resveratrol and docetaxel on prostate cancer cell lines (C4–2B and DU-145), revealing a synergistic effect. The combination therapy induced apoptosis in prostate cancer cells by increasing the expression of proapoptotic genes and decreasing the expression of antiapoptotic genes. In addition, the combined therapy increased the expression of tumor suppressor p53 and cell cycle inhibitors such p21 and p27, leading to a halt in the cell cycle in the G2/M phase. Resveratrol triggered programmed cell death in osteosarcoma cells by activating caspase 3, cleaving PARP, increasing the expression of Bax, and reducing the expression of Bcl-2 and Bcl-xL. Furthermore, it decreased the phosphorylation of Janus kinase 2 (JAK2), signal transducer, and activator of transcription 3 (STAT3), and suppressed the production of p-PI3K, p-Akt, and NF-κB in osteosarcoma cells (Peng and Jiang 2018).

- *Flavones*

Apigenin (**11**) was discovered to trigger apoptosis in A549 cells, a type of human lung cancer cells, via activating the Erk/MAPK pathway. Furthermore, the concurrent administration of apigenin and cisplatin triggered caspase-dependent cell death in A549 cells, leading to an increase in the levels of p53 and facilitating its phosphorylation and accumulation in A549 cells. Apigenin treatment at concentrations of 50, 100, 150, or 200 mg/ml enhanced the DNA damage caused by cisplatin and promoted the programmed cell death of tumor cells in a way that relies on the presence of p53. Cell shrinkage suggesting cell death was found in samples treated with a combination of apigenin and cisplatin. Prior research has shown that the administration of apigenin at a dosage of 50 mg/kg effectively hindered the advancement and formation of xenograft tumors. This was achieved by the induction of apoptosis, inhibition of cell proliferation, and reduction of Her2/neu expression (Mafuvadze et al. 2012). Another study reported that HeLa cells treated with Apigenin 7-O-glucoside (**12**) induced apoptosis via PTEN/PI3K/AKT pathway in dose-dependent manner with an $IC_{50}$ value of 47.26 μM (Liu et al. 2020).

- *Diarylheptanoid*

In another study, curcumin (**13**) when administered at a concentration of 10 μM, effectively suppressed cell proliferation and triggered caspase-dependent apoptosis in primary effusion lymphoma cells. This effect was seen in a dose-dependent manner and was achieved via suppressing the JAK/STAT3 pathway (Bharti et al. 2003). Curcumin at concentrations of 3 μM and 6 μM was seen to control the activity of miR-7, which is important for inhibiting cell growth and proliferation, inducing programmed cell death in pancreatic cancer cells, and decreasing the expression of histone lysine methyltransferase (SET8), a protein targeted by miR-7 (Ma et al. 2014).

- *Anthraquinone*

Emodin (**14**) obtained from the rhizome of Polygonum sachalinense as an emodin azide methyl anthraquinone derivative exhibited apoptotic activity at a concentration of 9.06 μM ($IC_{50}$) in MDA-MB-453 human breast cancer cell line and at a concentration of 0.83 μM ($IC_{50}$) in Calu-3 lung adenocarcinoma cells. The apoptotic response was triggered by an elevation in the release of cytochrome C, which subsequently led to the activation of caspase-3, −9, and − 8, as well as the activation of the Bid protein. The inhibition of MMP activity also enhanced the apoptotic response (Yan et al. 2008). A different investigation documented the ability to induce programmed cell death (apoptosis) in LS1034 colon cancer cell-based tumor xenografts in male athymic BALB/c nu/nu mice when administered a dosage of 40 mg/kg. The apoptotic response was characterized by a substantial reduction in both tumor weight and volume (Ma et al. 2012).

- *Naphthoquinone*

Plumbagin (**15**) obtained from *Plumbago zeylanica* L. at an $IC_{50}$ of 9 μM was found to induce apoptosis in human promyelocytic leukemia cells (NB4 cells) by increasing Bax and Bak protein levels and activating the caspase-3, −9, and − 8 by mitochondrial depolarization (Xu and Lu 2010). Another naphthoquinone known as

Shikonin (**16**) has shown apoptogenic properties against BEL7402 ($IC_{50}$ of 5.3 μM) and Huh7 ($IC_{50}$ of 4 μM) hepatocellular carcinoma cells. The apoptotic response is characterized by proteolytic cleavage of caspase-8, caspase-9, and Bid protein along with diminished protein levels of antiapoptotic factors, such as c-FLIP and Bcl-2. Shikonin was also found to decrease tumor volumes compared to control group in Huh7 cell induced tumor bearing nude mice at a dose range of 5 or 10 mg/kg for 30 days. The apoptotic response was again marked by increased activation of caspase-8, caspase-9, and PARP enzymes (Gong and Li 2011). Downregulation of Bcl-2 protein and caspases activation followed by the inactivation of NF-κB pathway was seen in the apoptotic response in Tca-8113 oral squamous cancer cells by the treatment of Shikonin at n $IC_{50}$ of 32.5 μM (Min et al. 2008).

- *Alkaloids*

Anonaine (**17**), a bioactive alkaloid extracted from the leaves of *M. alba*, has been shown to halt the growth of human cervical cancer (HeLa) cells by triggering apoptotic pathways in them at a concentration of 100 μM. This alkaloid exerted its effect on the process of programmed cell death (apoptosis) by enhancing the levels of ROS inside the cells, altering the MMP, raising the expression of Bax protein, and activating caspase-3, −7, −8, and − 9 in the cancer cell lines (Chen et al. 2008). Berberine (**18**) at a concentration of 75 μM ($IC_{50}$), triggered apoptosis in SCC-4 human tongue cancer cells by stimulating the production of caspase-8, −9, and − 3, AIF, and endoG. Furthermore, it elevated the Bax/Bcl-2 ratio and modified the MMP in these malignant cells (Ho et al. 2009). Hemanthamine (**19**) and hemanthidine (**20**) compounds derived from the Amaryllidaceae family, caused programmed cell death (apoptosis) in Jurkat cells. This was achieved by reducing the activity of MMP, strongly activating caspase-9 and caspase-3/7, and moderately activating caspase-8. The compounds were used at concentrations ranging from 5 to 20 μM (Havelek et al. 2014).

## 18.6 Therapeutic Applications of Natural Products in Apoptosis

Some natural compounds of plant origin, like gambogic acid, camptothecin (CPT), matrine alkaloids, annonaceous acetogenins and curcumin, can modulate the levels of key proteins involved in apoptosis signaling pathways (Figs. 18.2 and 18.3). Here, we summarize the different effects of these natural compounds on regulating apoptosis and their potential therapeutic applications in cancer therapy.

## 18.7 Gambogic Acid

Gambogic acid (GA) is the main active compound derived from gamboge (Ma et al. 2015). In 2004, the Chinese Food and Drug Administration (CFDA) granted permission for the use of GA in clinical trials as a broad-spectrum antitumor agent (Chi

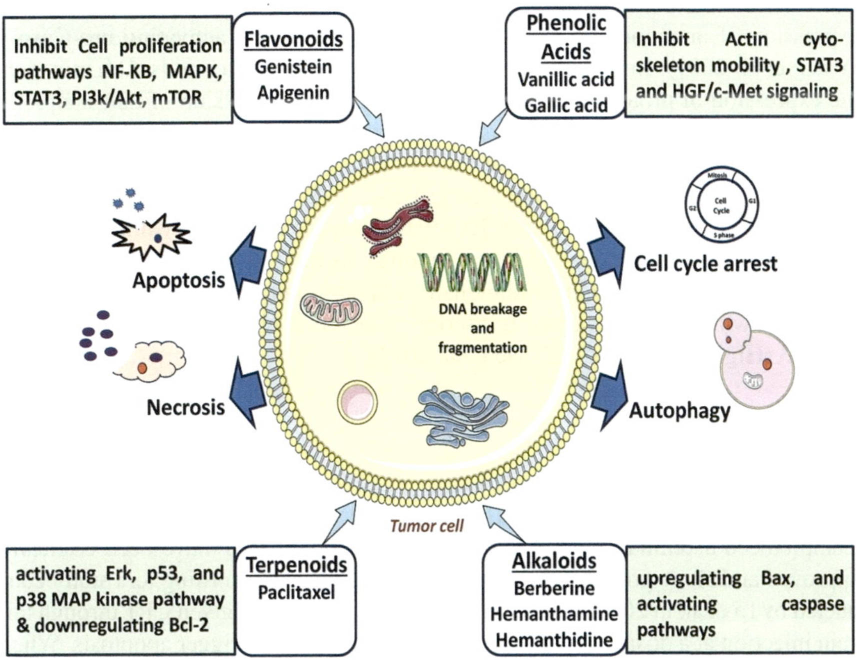


**Fig. 18.2** Induction of apoptosis through natural product-derived small molecules

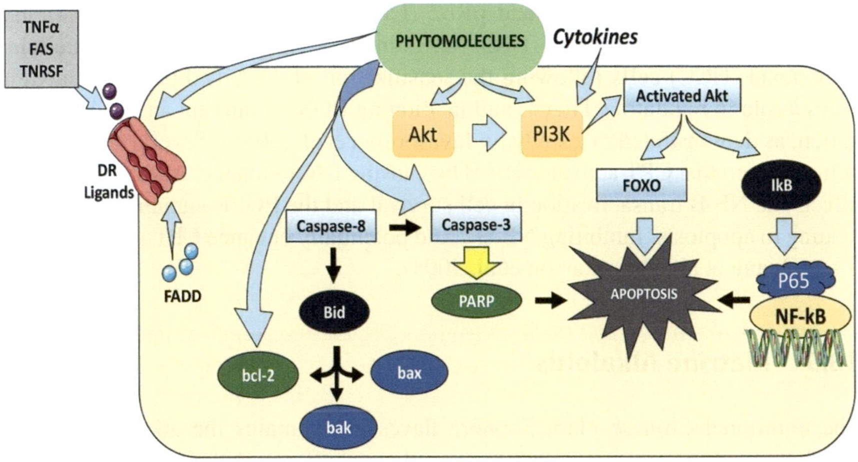


**Fig. 18.3** Modulation of apoptosis by natural products through receptors-mediated pathway

et al. 2013). Studies conducted in a laboratory setting have shown that GA has the ability to affect various proteins involved in apoptosis, such as those belonging to the Bcl-2 family, survivin, p53, pro-casp-3, NF-B, and the integrin pathway (Wang and Chen 2012; Gu et al. 2009). It has been observed that it downregulates the expression of antiapoptotic genes BclxL and Bcl-2, while activating pro-Casp-3 (Wang et al. 2009). In malignant melanoma and gastric cancer cells, GA increases the expression of proapoptotic Bax and decreases the expression of anti-apoptotic Bcl-2 (Zhong et al. 2010). According to a study by Tang et al. in 2011, GA has been found to have an impact on prostate cancer cells. It was observed that GA upregulates p53 and Bax, while downregulating Bcl-2. In B cell lymphoma, GA triggers apoptosis through the activation of PARP and caspase pathways, while simultaneously inhibiting the ERK and STAT 5 signaling pathways (Shi et al. 2013). The ROS production by GA leads to the initiation of autophagy and the activation of apoptotic pathways, which involve various transcription factors such as NF-κB, HIF-1, AP-1, and p53 (Higuchi et al. 1998; Ishaq et al. 2014).

## 18.8 Camptothecin

CPT is a quinoline alkaloid that is typically derived from the bark and stem of Camptotheca acuminate trees. Numerous derivatives of CPT have been clinically approved and show great potential in combating cancer. According to a study conducted by Li et al. in 2014, it was found that when mice were given CPT through tail vein injection at a dosage of 2.5 mg/kg, it had the ability to trigger apoptosis. When human myeloid leukemia cells are exposed to CPT, they undergo apoptosis. This occurs through the downregulation of cell cycle proteins like CRAF-1, DP1, cyclinB1, cdk4, and wee1, and the upregulation of proapoptotic proteins such as JunB, cJun, Bak, Bax, Bf11, and pRb2 (Ullmannova and Haskovec 2004). By increasing the expression of DR5, CPT induces apoptosis in Hepatocellular Carcinoma (HCC) cells following the sensitization of TRAIL. Furthermore, CPT plays a role in regulating TRAIL and facilitating ROS production and MAPK activation, as demonstrated in a study by Jayasooriya et al. (2014). Several studies have demonstrated that CPT activates NF-B by causing DNA damage. Additionally, CPT affects the NF-B transactivation in cell survival and the EGFR signaling pathways, leading to apoptosis. Inhibiting NF-B could potentially enhance CPT-induced apoptosis (Huang et al. 2000; Carson et al. 2004).

## 18.9 Matrine Alkaloids

The traditional Chinese plant Sophora flavescens contains the alkaloid matrine, which has various biological and pharmacological effects, including antiinflammatory, anticancer, and antiviral properties. Studies have shown that Matrine has demonstrated anticancer effects in various types of cancers, such as cervical, gastric, leukemia, lung, gallbladder, prostate, breast, melanoma, hepatocellular, and osteosarcoma (Wang et al. 2018). Matrine demonstrates its anticancer properties by triggering apoptosis in cancer cells. Matrine triggers apoptosis by activating the death receptor pathway, leading to the increased expression of 14 members of the TNF gene family (Zhang et al. 2009). Research has shown that matrine has the ability to trigger apoptosis in human acute myeloid leukemia cells through the mitochondrial pathway (Zhang et al. 2012b). Matrine was administered subcutaneously to mice with esophageal cancer cells at concentrations ranging from 1.25 mg/kg to 10 mg/kg in order to inhibit proliferation and induce apoptosis via the mitochondrial pathway in tumor cells (Wang et al. 2014). In addition, it has been discovered that matrine stimulates caspase-dependent apoptosis in human osteosarcoma cells through the upregulation of Fas and FasL expression, in addition to the activation of Casp-3, Casp-8, and Casp-9 (Liang et al. 2012). Matrine additionally stimulates both caspase-dependent and caspase-independent cell death in HepG2 cells via the Bid-regulated AIF nuclear translocation pathway, thereby inducing apoptosis by increasing the activation of Casp-9, while reducing the Bcl-2/Bax ratio (Zhao et al. 2012; Zhou et al. 2014). Additionally, its inhibitory effect on the PI3K/Akt pathway, which induces apoptosis in M21 cells harboring V600EBRAF, has been reported (Hui et al. 2013). Matrine inhibits the levels of phosphorylated ERK1/2 and phosphorylated Akt, increases Casp-3 activity, disrupts the MMP, decreases the Bcl-2/Bax ratio, and stimulates Cyt c release from mitochondria in human acute myeloid leukemia cells, thereby inducing apoptosis (Zhang et al. 2012a). Matrine induces caspase-dependent apoptosis and activates the p38 pathway in nonsmall cell lung carcinoma by generating ROS, according to one report (Tan et al. 2013).

## 18.10 Annonaceous Acetogenins

Annonaceae family plants elicit apoptosis in colon and lung cancer cells via the mitochondrial pathway, resulting in cell cycle arrest during the G1 phase (Moghadamtousi et al. 2014; Zorofchian et al. 2014). In addition to inhibiting colon and lung cancer cell migration and invasion, these compounds stimulate Casp3 and induce apoptosis in myelogenous leukemia K562 cells (Zirim et al. 2013). Furthermore, Annonaceae products promote apoptosis in BPH-1 cells by upregulating Bax mRNA and downregulating Bcl-2 (Asare et al. 2015). Annonaceous acetogenins inhibit multidrug-resistant MCF-7 cells and disrupt mitochondrial function by obstructing mitochondrial complex I (Oberlies et al. 1997). Annonacin, which falls under the category of annonaceous acetogenins, inhibits tumor growth in nude mice by inducing apoptosis and cell cycle arrest in MCF-7 cells via ER-related pathways (Ko et al. 2011). Squamocin, an annonaceous acetogenin composed of bis-tetrahydrofuran, activates the extrinsic and intrinsic apoptotic pathways and halts the progression of cancer cells through the G1 phase. A study has demonstrated that squamocin modulates histone H3 phosphorylation at S10 and S28 in order to induce epigenetic alterations; this establishes an entirely new mechanism of action for squamocin as an anti-cancer agent (Lee et al. 2011).

## 18.11 Quercetin

Quercetin, which is widely present in diverse dietary sources, demonstrates antioxidant and antiinflammatory characteristics that have been linked to the prevention and treatment of cancer (Ouyang et al. 2012). In vitro, it induces cell cycle arrest, inhibits cell proliferation, and promotes apoptosis, Quercetin increases Fas expression and activates Casp-3, −8, and − 9 in MDAMB-231 cells (Chien et al. 2009). Furthermore, it stimulates the expression of Cyt c, cleaved Casp-3, and PARP in cancer cells, and also, apoptosis is induced in MCF-7 cells by quercetin via a pathway dependent on p21 but independent of p53; the cell cycle is halted at the M and G2 phases (Choi et al. 2001). MCF-7 cells express a receptor for quercetin, which stimulates the firing of the initiator and effector caspases (Chou et al. 2010). Alantolactone, through its modulation of Bcl-2 family proteins, activation of caspases, and alteration of NF-κB and p53 levels, induces apoptosis in hepatoma and erythroleukemia cells, it also induces apoptosis in chronic myelogenous leukemia by impeding NF-κB activity and preventing IB-α phosphorylation induced by TNF-α (Wei et al. 2013).

## 18.12 Triptolide

Triptolide, a diterpene, was extracted from the Chinese herb Tripterygium wilfordii Hook f. (Celastraceae). It showed activity against a wide range of cancer cells, including proapoptotic and antiproliferative effects its worked By inhibiting NF-B, triptolide induces apoptosis in cells of pancreatic, lung, and colon cancer (Kye et al. 2002). Furthermore, triptolide stimulates the expression of p53 and Bax and augments the activity of caspases, thereby inducing apoptosis in gastric cancer cells (Jiang et al. 2001). Triptolide has been demonstrated to stimulate apoptosis in mouse melanoma cells through the upregulation of Casp-3 and the cleaved form of PARP (Yang et al. 2003). In addition, triptolide increased the activation of caspase-3 and caspase-9 induced by cisplatin, as well as the cleavage of PARP downstream, in gastric cancer cells (Li et al. 2012). Triptolide additionally facilitates apoptosis in HeLa cells through the activation of caspases and the induction of Cyt c release also additionally, triptolide has been observed to stimulate the release of Cyt c from mitochondria and Casp-3 activity in pancreatic cancer cells (Phillips et al. 2007).

## 18.13 Curcumin

Curcumin, an essential bioactive compound derived from the rhizome of Curcuma longa, demonstrates antitumor, antiinflammatory, and antiinfective (Nishida et al. 2008). Furthermore, epidemiological evidence suggests that elevated concentrations of curcumin may reduce the likelihood of developing various forms of cancer recurrence, including colon cancer, breast cancer, and lung cancer (Zhang et al. 2015). Curcumin, administered intravenously at a dosage of 10 mg/kg, has been utilized to treat colon tumors in mice (Zhang et al. 2015). Curcumin inhibits both COX-2 and EGFR in lung and pancreatic adenocarcinoma cells; this is associated with the enhanced induction of apoptosis and decreased survival of lung and pancreatic adenocarcinoma cell lines (Chen and Wang 2002). Curcumin elicits apoptosis in human lung cancer cells through diverse mechanisms, including the inhibition of cell cycle progression during the G2/M phase and activation of the p38, JNK, ERK, and MAPK signaling pathways (Chou et al. 2008). Also, curcumin inhibits cell proliferation, upregulates miR-192-5p, and downregulates PI3K/Akt signaling in order to induce apoptosis in human nonsmall cell lung cancer cells (Chen and Wang 2002). Curcumin can inhibit the growth of PC-9 cells by increasing the expression of the cyclin-dependent kinase inhibitor genes p21 and p27 and by increasing the expression of growth arrest and DNA damage-inducible genes 45 and 153 (GADD45 and 153) in a manner independent of p53. Additionally, it suppresses the activity of various genes, including CDK2, CDK4, CDK6, Bcl-2, and cyclin D1. This is succeeded by a robust induction of apoptosis (Saha et al. 2010). Curcumin induces apoptosis in HL-60 human leukemia cells via ceramide, a potent tumor suppressor that drives autophagic responses, apoptosis, and cell cycle arrest (Morad and Cabot 2013), according to one study. Curcumin has been shown to facilitate cross-communication between Casp-3, Casp-8, and antiapoptotic members of the Bcl family, all of which are involved in apoptosis induced by curcumin (Sun 2011). Curcumin has been documented to induce apoptosis via the mitochondrial pathway and induce endoplasmic reticulum stress and Casp-3 activation in human lung adenocarcinoma cells (Tait and Green 2010). Curcumin induces apoptosis by activating Casp-3 and -9, depleting MMP, and releasing Cyt c in human small cell lung cancer (SCLC) (Ghosh et al. 1998). It inhibits sarcoendoplasmic reticulum calcium adenosine (SERCA) activity, inducing apoptosis in ovarian cancer cells, and elevates cytosolic Ca2+ concentration, regardless of p53 status (Wang et al. 2009). In Panc-1 cells, curcumin activates Casp-9/3, reduces the Bcl-2/Bax ratio, and enhances sensitivity to detachment-induced apoptosis (Ma et al. 2015). Several studies have documented that curcumin, when combined with other agents, induces apoptosis. Inducing apoptosis in tumor cells, curcumin in combination with oxaliplatin upregulates Bax and causes cell cycle arrest, downregulates pro-Casp-3, Bcl-2, pro-PARP, survivin, c-myc, and HSP70, and cleaves PARP and activates Casp-3 (Rosenberg-Lee 2004). The induction of apoptosis in human lung cancer cells was achieved through the synergistic inhibition of cell invasion and metastasis by curcumin and carboplatin. NF-B mediates the apoptosis-inducing effects of curcumin by activating ERK1/2, caspases, and p53 and inhibiting the Akt/IKK pathway (Riedl and Shi 2004).

## 18.14 Emerging Trends in Natural Product-Based Apoptosis Therapeutics

The enchantment with natural resources and their potential for medicinal uses, along with the complex structures and synthetic challenges they present, have greatly contributed to the advancement of this field. This has significantly impacted the pharmaceutical industry and related economic activities. Utilizing the expertise of a pharmacologist, researchers are actively pursuing the discovery of small molecule drugs. These drugs are being developed using templates and designs derived from a wide range of biodiversified and chemically diversified NP pools (O'Keefe 2001; Mabry 2001). In addition, the traditional herbals have gained recognition in the field of ethnobotany and ethnopharmacology, presenting chemists with exciting opportunities and obstacles in the realm of drug discovery and the development of health-related products (Tan et al. 2015). NPs are bioactive substances derived from microbes, plants, and animals. These substances possess important pharmacological or biological properties that are harnessed in preventing and treating various diseases (Zhu et al. 2022).

Their long-standing use in traditional medicine continues, and their significant impact on modern drug development is undeniable. Current studies highlight the enduring importance of NPs in pharmaceutical research. A study by David J. Newman and Gordon M. Cragg found that a significant portion of small-molecule drugs approved from 1981 to 2019 were either natural or derived from them. Additionally, between 1981 and 2014, more than half of all new small molecule drugs worldwide were sourced from NPs (Newman and Cragg 2019). According to a study conducted by Eric Patridge and colleagues, it was found that the US Food and Drug Administration has approved a total of 547 NPs and their derivatives for medicinal use from 1827 to 2013 (Naeem et al. 2022). These products and derivatives were used to address a variety of health issues, such as hypertension, cancer, and bacterial infections. Plants are well known for their rich collection of biologically active compounds found in nature. With over 350,000 registered vascular plant species worldwide, the potential for drug discovery using compounds derived from plants is immense (Naeem et al. 2022). Phytochemicals like terpenes, alkaloids, flavonoids, and essential oils have garnered significant attention for their therapeutic potential, particularly in the field of cancer treatment (Naeem et al. 2022). When it comes to drug discovery, NPs come with their own set of advantages and challenges. Their wide range of chemical scaffolds and intricate structures often make them valuable sources of potential drugs. It is interesting to highlight that NPs have significantly advanced anticancer medications, making up around 50% of antitumor drugs (Dey et al. 2020). Notable examples include vinca alkaloids derived from Catharanthus roseus, such as vincristine and vinblastine, as well as taxanes like paclitaxel (Taxol®) and docetaxel (Taxotere®) derived from Taxus brevifolia and Taxus baccata, respectively (Dey et al. 2020). Two additional noteworthy examples are etoposide and teniposide. These antitumor agents are derived from podophyllotoxin discovered in Podophyllum species. NPs have shown promising potential in preventing and treating cancer by inhibiting cell growth, controlling the cell cycle, and disrupting various signaling pathways associated with tumor development. Furthermore, NPs have the potential to activate DNA repair mechanisms by influencing various gene products and proteins involved in the process. These include Bax, Bak, and Bid proteins, which trigger the production of protective enzymes like caspases 3, 7, 8, 9, 10, 12, as well as the modulation of antioxidant enzymes such as GST, GSH, and GPx. The chemopreventive effects happen by blocking key processes in tumor formation, such as getting rid of harmful substances, improving DNA repair, and eliminating abnormal cells through the immune system. In addition, these effects are achieved by inhibiting cell growth and targeting abnormal cells in the immune system, while also slowing down the advancement of tumors by encouraging cell specialization, preventing the formation of new blood vessels, and triggering cell death (Gaman et al. 2020).

There has been limited focus on the apoptotic effects of NPs so far. However, conducting more comprehensive studies on these effects could provide valuable insights into the exact mechanisms by which these compounds function as anti-inflammatory agents. Most studies in this field have primarily examined the apoptotic effects of phytochemicals in cancer research, aiming to discover potential anti-cancer substances. However, a significant portion of this research could also be relevant to investigating the anti-inflammatory properties of NPs. These compounds, like alkaloids and phenolics such as flavonoids, have been extensively researched in the fields of phytochemistry and pharmacology. For this review, we have chosen various NPs that have been recognized for their antiinflammatory properties. We have, however, also included several compounds that may have connections to the inflammatory process based on their mechanisms and known properties. This review can be divided into two main groups: compounds that induce apoptosis and compounds with antiapoptotic effects.

## 18.15 Conclusion and Outlook for the Field

NPs are an invaluable resource in the management of cancer due to the vast variety of compounds they contain that possess distinct structures and mechanisms of action. Utilized individually or in combination therapies, their potential for cancer prevention and treatment is evident. Screening programmes are uncovering novel prototypes of pharmacologically active compounds derived from natural sources as technological progress continues, thereby expediting the investigation of cancer therapeutics. The increasing prevalence of cancer and the inadequacies of traditional treatments emphasize the necessity for alternative approaches. Due to their superior biodegradability, biocompatibility, and efficacy, NPs have the potential to significantly transform the field of cancer treatment over the next decade. Notwithstanding obstacles such as inadequate solubility and bioavailability, formulations based on nanotechnology exhibit potential in surmounting these challenges, for NPs to reach their full potential, a greater understanding of their molecular interactions and signaling pathways is essential. The identification of biomarkers linked to carcinogenesis and the comprehension of these interactions can be facilitated with the aid of in silico methods such as molecular docking. Prior to clinical application, large-scale, well-controlled clinical trials are necessary to validate anticancer activities and ensure safety. The implementation of contemporary methodologies is crucial in order to adhere to global quality benchmarks. A partnership between practitioners of traditional medicine and the pharmaceutical industry can accelerate the development and distribution of natural compounds for patient use. Incorporating interdisciplinary endeavor's and ethnopharmacological insights into the drug discovery process is critical for maximizing the therapeutic capabilities of NPs in the context of cancer treatment. In summary, NPs continue to be critical sources of drug leads, propelling advancements in the field of cancer drug discovery and development. Their therapeutic efficacy and varied chemical profiles demonstrate their enduring significance in contemporary oncology.

## References

Anderton H, Wicks IP, Silke J (2020) Cell death in chronic inflammation: breaking the cycle to treat rheumatic disease. Nat Rev Rheumatol 16:496–513

Asare GA, Afriyie D, Ngala RA, Abutiate H, Doku D, Mahmood SA, Rahman H (2015) Antiproliferative activity of aqueous leaf extract of Annona muricata L. on the prostate, BPH-1 cells, and some target genes. Integr Cancer Ther 14(1):65

Ashkenazi A, Dixit VM (1998) Death receptors: signaling and modulation. Science 28:1305–1308

Atanasov AG, Zotchev SB, Dirsch VM et al (2021) Natural products in drug discovery: advances and opportunities. Nat Rev Drug Discov 20:200–216. https://doi.org/10.1038/s41573-020-00114-z

Aye MM, Aung HT, Sein MM, Armijos C (2019) A review on the phytochemistry, medicinal properties and pharmacological activities of 15 selected Myanmar medicinal plants. Molecules 24:293

Bacus S, Gudkov A, Lowe M, Lyass L, Yung Y, Komarov AP, Keyomarsi K, Yarden Y, Seger R (2001) Taxol-induced apoptosis depends on MAP kinase pathways (ERK and p38) and is independent of p53. Oncogene 20:147–155. https://doi.org/10.1038/sj.onc.1204062

Balan M, Chakraborty S, Flynn E, Zurakowski D, Pal S (2017) Honokiol inhibits c-Met-HO-1 tumor-promoting pathway and its cross-talk with calcineurin inhibitor-mediated renal cancer growth. Sci Rep 7:5900. https://doi.org/10.1038/s41598-017-05455-1

Behrens TW, Mueller DL (1997) Bcl-x and the regulation of survival in the immune system. Immunol Res 16:149–160

Bharti AC, Donato N, Aggarwal BB (2003) Curcumin (diferuloylmethane) inhibits constitutive and IL-6-inducible STAT3 phosphorylation in human multiple myeloma cells 1. J Immunol 171(7):3863–3871. https://doi.org/10.4049/jimmunol.171.7.3863

Bhaskara VK, Mittal B, Mysorekar VV, Amaresh N, Simal-Gandara J (2020) Resveratrol, cancer and cancer stem cells: a review on past to future. Curr Res Food Sci 3:284–295. https://doi.org/10.1016/j.crfs.2020.10.004

Blagosklonny MV, Schulte T, Nguyen P, Trepel J, Neckers LM (1996) Taxol-induced apoptosis and phosphorylation of Bcl-2 protein involves c-Raf-1 and represents a novel c-Raf-1 signal transduction pathway. Cancer Res 56(8):1851–1854

Borner C (2003) The Bcl-2 protein family: sensors and checkpoints for life-or death decisions. Mol Immunol 39:615–647

Botrous S, Elmaghraby A, El-Achy S, Mustafa Y, Badr E, Haggag A, Abdel-Rahman S (2023) Inhibition of TNF-α oncogene expression by Artemisia annua L. extract against pioglitazone side effects in male albino mice. Mol Biotechnol 66:432–441

Carson JP, Zhang N, Frampton GM, Gerry NP, Lenburg ME, Christman MF (2004) Pharmacogenomic identification of targets for adjuvant therapy with the topoisomerase poison camptothecin. Cancer Res 64(6):2096–2104

Chen M, Wang J (2002) Initiator caspases in apoptosis signaling pathways. Apoptosis 7(4):31

Chen CY, Liu TZ, Tseng WC, Lu FJ, Hung RP, Chen CH, Chen CH (2008) (−)-Anonaine induces apoptosis through Bax- and caspase-dependent pathways in human cervical cancer (HeLa) cells. Food Chem Toxicol 46:2694–2702

Chen M, Xiao C, Jiang W, Yang W, Qin Q, Tan Q, Lian B, Liang Z, Wei C (2021) Capsaicin inhibits proliferation and induces apoptosis in breast cancer by down-regulating FBI-1-mediated NF-κB pathway. Drug Des Devel Ther 15:125–140

Chi Y, Zhan XK, Yu H, Xie GR, Wang ZZ, Xiao W, Cui CX (2013) An open-labeled, randomized, multicenter phase IIa study of gambogic acid injection for advanced malignant tumors. Chin Med J 126(9):1642–1646

Chicheportiche Y, Bourdon PR, Xu H, Hsu YM, Scott H, Hession C, Garcia I, Browning JL (1997) TWEAK, a new secreted ligand in the tumor necrosis factor family that weakly induces apoptosis. J Biol Chem 272:32401–32410

Chien SY, Wu YC, Chung JG, Yang JS, Lu HF, Tsou MF, Wood WG, Kuo SJ, Chen DR (2009) Quercetin-induced apoptosis acts through mitochondrial- and caspase-3-dependent pathways in human breast cancer MDA-MB-231 cells. Hum Exp Toxicol 28(8):493–503

Chilampalli S, Zhang X, Fahmy H, Kaushik RS, Zeman D, Hildreth MB, Dwivedi C (2010) Chemopreventive effects of honokiol on UVB-induced skin cancer development. Anticancer Res 30(3):777–783

Chinnaiyan AM (1999) The apoptosome: heart and soul of the cell death machine. Neoplasia 1:5–15

Choi JA, Kim JY, Lee JY, Kang CM, Kwon HJ, Yoo YD, Kim TW, Lee YS, Lee SJ (2001) Induction of cell cycle arrest and apoptosis in human breast cancer cells by quercetin. Arch Pharm Res 19(4):837–844

Chonsut P, Mahalapbutr P, Pradubyat N, Chavasiri W, Wonganan P, Ketchart W (2019) Ethoxy mansonone G as an anticancer agent in estrogen receptor-positive and endocrine-resistant breast cancer. J Pharm Pharmacol 71(12):1839–1853

Chou CC, Hsiao HY, Hong QS, Chen CH, Peng YW, Chen HW, Yang PC (2008) Single-walled carbon nanotubes can induce pulmonary injury in mouse model. Nano Lett 8(2):437

Chou CC, Yang JS, Lu HF, IP SW, Lo C, Wu CC, Lin JP, Tang NY, Chung JG, Chou MJ, Teng YH, Chen DR (2010) Quercetin-mediated cell cycle arrest and apoptosis involving activation of a caspase cascade through the mitochondrial pathway in human breast cancer MCF-7 cells. Arch Pharm Res 33(8):1181–1191

Coleman ML, Sahai EA, Yeo M, Bosch M, Dewar A, Olson MF (2001) Membrane blebbing during apoptosis results from caspase-mediated activation of ROCK I. Nat Cell Biol 3:339–345

Crocker SJ, Wigle N, Liston P, Thompson CS, Lee CJ, Xu D, Roy S, Nicholson DW, Park DS, MacKenzie A, Korneluk RG, Robertson GS (2001) NAIP protects the nigrostriatal dopamine pathway in an intrastriatal 6-OHDA rat model of Parkinson's disease. Eur J Neurosci 14:391–400

De D, Panda SK, Ghosh U (2023) Induction of apoptosis by ethanolic extract of leaf of Dillenia pentagyna Roxb. In A549 cells via NF-κB pathway. Phytomed Plus 3(1):100368

Deng R (2007) Therapeutic effects of guggul and its constituent guggulsterone: cardiovascular benefits. Cardiovasc Drug Rev 25(4):375–390

Deng G, Ju X, Meng Q, Yu Z, Ma LB (2015) Emodin inhibits the proliferation of PC3 prostate cancer cells in vitro via the Notch signaling pathway. Mol Med Rep 12(3):4427–4433

Desnoyers S, Hengartner MO (1997) Genetics of apoptosis. Adv Pharmacol 41:35–56

Dey P, Kundu A, Kumar A, Gupta M, Lee BM, Bhakta T, Dash S, Kim HS (2020) Analysis of alkaloids (indole alkaloids, isoquinoline alkaloids, tropane alkaloids). Recent Adv Nat Prod Anal. https://doi.org/10.1016/B978-0-12-816455-6.00015-9

Dong Y, Zhang T, Li J, Deng H, Song Y, Zhai D, Peng Y, Lu X, Liu M, Zhao Y, Yi Z (2014) Oridonin inhibits tumor growth and metastasis through anti-angiogenesis by blocking the Notch signaling. PLoS One 9(12):e113830

Dotto (2000) p21(WAF1/Cip1): more than a break to the cell cycle? Biochim Biophys Acta 1471:M43–M56

Du JH, Zhang HD, Ma ZJ, JI KM (2010) Artesunate induces oncosis-like cell death in vitro and has antitumor activity against pancreatic cancer xenografts in vivo. Cancer Chemother Pharmacol 65(5):895–902

Du GJ, Zhang Z, Wen XD, Yu C, Calway T, Yuan CS, Wang CZ (2012) Epigallocatechin gallate (EGCG) is the most effective cancer chemopreventive polyphenol in green tea. Nutrients 4(11):1679–1691

Earnshaw WC, Martins LM, Kaufmann SH (1999) Mammalian caspases: structure, activation, substrates, and functions during apoptosis. Annu Rev Biochem 68:383–424

Ekert PG, Vaux DL (2005) The mitochondrial death squad: hardened killers or innocent bystanders? Curr Opin Cell Biol 17:626–630

Enari M, Sakahira H, Yokoyama H, Okawa K, Iwamatsu A, Nagata S (1998) A caspase-activated DNase that degrades DNA during apoptosis, and its inhibitor ICAD. Nature 391:43–50

Esposti MD (2002) The roles of bid. Apoptosis 7:433–440

Fouzat A, Hussein OJ, Gupta I, Al-Farsi HF, Khalil A, Moustafa AEA (2022) Elaeagnus angustifolia plant extract induces apoptosis via p53 and signal transducer and activator of transcription 3 signaling pathways in triple-negative breast cancer cells. Front Nutr 9. https://doi.org/10.3389/fnut.2022.871667

Fried LE, Arbiser JL (2009) Honokiol, a multifunctional antiangiogenic and antitumor agent. Antioxid Redox Signal 11(5):1139–1148

Gaman AM, Egbuna C, Gaman MA (2020) Natural bioactive lead compounds effective against haematological malignancies. In: Phytochemicals as lead compounds for new drug discovery. Elsevier, Amsterdam

Garcia-Calvo M, Peterson EP, Leiting B, Ruel R, Nicholson DW, Thornberry NA (1998) Inhibition of human caspases by peptide-based and macromolecular inhibitors. J Biol Chem 273:32608–32613

Gartel AL (2005) The conflicting roles of the cdk inhibitor p21CIP1/WAF1 in apoptosis. Leuk Res 29:1237–1238

Ghosh S, May MJ, Kopp EB (1998) NF-KB and Rel proteins: evolutionarily conserved mediators of immune responses. Annu Rev Immunol 16:225–260

Girma B, Mulisa E, Tessema S, Amelo W (2018) Ethnomedicine claim directed in silico prediction of anticancer activity. Ethiop J Health Sci 28:83–92

Gong K, Li W (2011) Shikonin, a Chinese plant-derived naphthoquinone, induces apoptosis in hepatocellular carcinoma cells through reactive oxygen species: a potential new treatment for hepatocellular carcinoma. Free Radic Biol Med 51(12):2259–2271. https://doi.org/10.1016/j.freeradbiomed.2011.09.018

Greco G, Catanzaro E, Fimognari C (2021) Natural products as inducers of non-canonical cell death: a weapon against cancer. Cancer 13(2):304

Grutter MG (2000) Caspases: key players in programmed cell death. Curr Opin Struct Biol 10:649–655

Gu HY, Rao SY, Zhao J, Wang J, Mu R, Rong JJ, Tao L, Qi Q, You QD, Guo QL (2009) Gambogic acid reduced bcl-2 expression via p53 in human breast MCF-7 cancer cells. J Cancer Res Clin 135(12):1777–1782

Guo Y, Wu Y, Huang T, Huang D, Zeng Q, Wang Z, Hu Y, Liang P, Chen H, Zheng Z, Liang T, Zhai D, Jiang C, Li L, Zhu H, Liu Q (2024) Licorice flavonoid ameliorates ethanol-induced gastric ulcer in rats by suppressing apoptosis via PI3K/AKT signaling pathway. J Ethnopharmacol 325. https://doi.org/10.1016/j.jep.2024.117739

Harvey AL, Edrada-Ebel R, Quinn RJ (2015) The re-emergence of natural products for drug discovery in the genomics era. Nat Rev Drug Discov 14(2):111–129

Havelek R, Seifrtova M, Kralovec K, Bruckova L, Cahlikova L, Dalecka M, Vavrova J, Rezacova M, Opletal L, Bilkova Z (2014) The effect of Amaryllidaceae alkaloids haemanthamine and haemanthidine on cell cycle progression and apoptosis in p53-negative human leukemic Jurkat cells. Phytomedicine 21(4):479–490

Higuchi M, Honda T, Proske RJ, Yeh ET (1998) Regulation of reactive oxygen species-induced apoptosis and necrosis by caspase 3-like proteases. Oncogene 17(21):2753–2760

Hill MM, Adrain C, Duriez PJ, Creagh EM, Martin SJ (2004) Analysis of the composition, assembly kinetics and activity of native Apaf-1 apoptosomes. EMBO J 23:2134–2145

Hitoshi Y, Lorens J, Kitada SI, Fisher J, LaBarge M, Ring HZ, Francke U, Reed JC, Kinoshita S, Nolan GP (1998) Toso, a cell surface, specific regulator of Fas-induced apoptosis in T cells. Immunity 8:461–471

Ho YT, Lu CC, Yang JS, Chiang JH, Li TC, Ip SW, Hsia TC, Liao CL, Lin JG, Wood WG, Chung JG (2009) Berberine induced apoptosis via promoting the expression of caspase-8, -9 and -3, apoptosis-inducing factor and endonuclease G in SCC-4 human tongue squamous carcinoma cancer cells. Anticancer Res 29:4063–4070

Hsieh YC, Rao YK, Wu CC, Huang CY, Geethangili M, Hsu SL, Tzeng YM (2010) Methyl antcinate a from Antrodia camphorata induces apoptosis in human liver cancer cells through oxidant-mediated cofilin- and Bax-triggered mitochondrial pathway. Chem Res Toxicol 23(7):1256–1267

Hsu H, Xiong J, Goeddel DV (1995) The TNF receptor 1-associated protein TRADD signals cell death and NF kappa B activation. Cell 81:495–504

Hu FW, Tsai LL, Yu CH, Chen PN, Chou MY, Yu CC (2012) Impairment of tumor-initiating stem-like property and reversal of epithelial-mesenchymal transdifferentiation in head and neck cancer by resveratrol treatment. Mol Nutr Food Res 56(8):1247–1258

Huang TT, Wuerzbergerdavis SM, Seufzer BJ, Shumway SD, Kurama T, Boothman DA, Miyamoto S (2000) NF-KB activation by camptothecin a linkage between nuclear dna damage and cytoplasmic signaling events. J Biol Chem 275(13):9501

Hui J, Yu S, Wang S, Cheng X (2013) Matrine activates PTEN to induce growth inhibition and apoptosis in V600EBRAF harboring melanoma cells. Int J Mol Sci 14(8):16040

Igney FH, Krammer PH (2002) Death and anti-death: tumour resistance to apoptosis. Nat Rev Cancer 2:277–288

Ishaq M, Khan MA, Sharma K, Sharma G, Dutta RK, Majumdar S (2014) Gambogic acid induced oxidative stress dependent caspase activation regulates both apoptosis and autophagy by targeting various key molecules (NF-KB, Beclin-1, p62 and NBR1) in human bladder cancer cells. BBA-Gen Sub 1840(12):3374–3384

Israels ED, Israels LG (2000) The cell cycle. Oncologist 5(6):510–513

Jayasooriya RG, Choi YH, Hyun JW, Kim GY (2014) Camptothein sensitizes human hepatoma Hep3B cells to TRAIL- mediated apoptosis via ROS-dependent death receptor 5 upregulation with the involvement of MAPKs. Environ Toxicol Pharmacol 38(3):959–967

Jiang XH, Wong BCY, Lin MCM, Zhu GH, Kung HF, Yang D, Lam SK (2001) Functional p53 is required for triptolide- induced apoptosis and AP-1 and nuclear factor-KB activation in gastric cancer cells. Oncogene 20(55):8009

Kallifatidis G, Labsch S, Rausch V, Mattern J, Gladkich J, Moldenhauer G, Buchler MW, Salnikov AV, Herr I (2011) Sulforaphane increases drug-mediated cytotoxicity toward cancer stem-like cells of pancreas and prostate. Mol Ther 19(1):188–195

Kaneuchi M, Yamashita T, Shindoh M, Segawa K, Takahashi S, Furuta I, Fujimoto S, Fujinaga K (1999) Induction of apoptosis by the p53–273L (Arg --> Leu) mutant in HSC3 cells without transactivation of p21Waf1/Cip1/Sdi1 and bax. Mol Carcinog 26:44–52

Kanwar SS, Yu Y, Nautiyal J, Patel BB, Padhye S, Sarkar FH, Majumdar APN (2011) Difluorinated-curcumin (CDF): a novel curcumin analog is a potent inhibitor of colon cancer stem-like cells. Pharm Res 28(4):827–838

Kataoka T, Schroter M, Hahne M, Schneider P, Irmler M, Thome M, Froelich CJ, Tschopp J (1998) FLIP prevents apoptosis induced by death receptors but not by perforin/granzyme B, chemotherapeutic drugs, and gamma irradiation. J Immunol 161:3936–3942

Katiyar SK (2008) Grape seed proanthocyanidines and skin cancer prevention: inhibition of oxidative stress and protection of immune system. Mol Nutr Food Res 52(1):S71–S76

Kerr JF, Wyllie AH, Currie AR (1972) Apoptosis: a basic biological phenomenon with wide-ranging implications in tissue kinetics. Br J Cancer 26(4):239–257

Kerr JFR, Winterford CM, Harmon BV (1994) Apoptosis. Its significance in cancer and cancer therapy. Cancer 73:2013–2026

Khan T, Ali M, Khan A, Nisar P, Jan SA, Afridi S, Shinwari ZK (2019) Anticancer plants: a review of the active phytochemicals, applications in animal models, and regulatory aspects. Biomol Ther 10:47

Kim JY, Dao TTP, Song K, Park SB, Jang H, Park MY, Gan SU, Kim YS (2018) Annona muricata leaf extract triggered intrinsic apoptotic pathway to attenuate cancerous features of triple negative breast cancer MDA-MB-231 cells. Evid Based Complement Alternat Med. https://doi.org/10.1155/2018/7972916

Kischkel FC, Hellbardt S, Behrmann I, Germer M, Pawlita M, Krammer PH, Peter ME (1995) Cytotoxicity dependent APO-1 (Fas/CD95)- associated proteins form a death-inducing signaling complex (DISC) with the receptor. EMBO J 14:5579–5588

Ko YM, Wu TY, Wu YC, Chang FR, Guh JY, Chuang LY (2011) Annonacin induces cell cycle-dependent growth arrest and apoptosis in estrogen receptor-a-related pathways in MCF-7 cells. J Ethnopharmacol 137(3):1283–1290

Koff JL, Ramachandiran S, Bernal-Mizrachi L (2015) A time to kill: targeting apoptosis in cancer. Int J Mol Sci 16:2942–2955

Korsmeyer SJ (1995) Regulators of cell death. Trends Genet 11:101–105

Koval A, Pieme CA, Queiroz EF, Ragusa S, Ahmed K, Blagodatski A, Wolfender JL, Petrova TV, Katanaev VL (2018) Tannins from Syzygium guineense suppress Wnt signaling and proliferation of Wnt-dependent tumors through a direct effect on secreted Wnts. Cancer Lett 435:110–120

Krammer PH (2000) CD95's deadly mission in the immune system. Nature 407:789–795

Kye Y, Lee JSP, Young KJ, Glenn DR (2002) Triptolide sensitizes lung cancer cells to TNF—related apoptosis—inducing ligand (TRAIL)—induced apoptosis by inhibition of NF—kB activation. Exp Mol 34(6):462–468

Lawen A (2003) Apoptosis—an introduction. BioEssays 25:888–896

Lawen A, Baker MA, Malik S (1998) Apoptosis and redox homeostasis—on a possible mechanism of action of Bcl-2. Protoplasma 205:10–20

Lee J, Jang HJ, Chun H (2019a) Calotropis gigantea extract induces apoptosis through extrinsic/intrinsic pathways and reactive oxygen species generation in A549 and NCI-H1299 non-small cell lung cancer cells. BMC Complement Altern Med 19:134. https://doi.org/10.1186/s12906-019-2561-1

Lee CC, Lin YH, Chang WH, Lin PC, Wu YC, Chang JG (2011) Squamocin modulates histone H3 phosphorylation levels and induces G1 phase arrest and apoptosis in cancer cells. BMC Cancer 11(1):58

Lee KC, Kuo HC, Shen CH, Lu CC, Huang WS, Hsieh MC, Huang CY, Kuo YH, Hsieh YY, Teng CC, Lee LY, Tung SY (2017) A proteomics approach to identifying novel protein targets involved in erinacine A-mediated inhibition of colorectal cancer cells' aggressiveness. J Cell Mol Med 21(3):588–599

Lee JS, Sul JY, Park JB, Lee MS, Cha EY, Ko YB (2019b) Honokiol induces apoptosis and suppresses migration and invasion of ovarian carcinoma cells via AMPK/mTOR signaling pathway. Int J Mol Med 43:1969–1978

Lee D, Kwak HJ, Kim BH, Kim DW, Kim HY, Kim SH, Kang KS (2022) Brevilin a isolated from Centipeda minima induces apoptosis in human gastric cancer cells via an extrinsic apoptotic signaling pathway. Plan Theory 11(13):1658

Leverrier Y, Ridley AJ (2001) Apoptosis: caspases orchestrate the ROCK 'n' bleb. Nat Cell Biol 3:E91–E93

Li H, Zhu H, Xu CJ, Yuan J (1998) Cleavage of BID by caspase-8 mediates the mitochondrial damage in the Fas pathway of apoptosis. Cell 94:491–501

Li CJ, Chu CY, Huang LH, Wang MH, Sheu LF, Yeh JI, Hsu HY (2012) Synergistic anticancer activity of triptolide combined with cisplatin enhances apoptosis in gastric cancer in vitro and in vivo. Cancer Lett 319(2):203–213

Li Y, Liu R, Yang J, Ma G, Zhang Z, Zhang X (2014) Dual sensitive and temporally controlled camptothecin prodrug liposomes codelivery of siRNA for high efficiency tumor therapy. Biomaterials 35(36):9731–9745

Li C, Wu X, Sun R, Zhao P, Liu F, Zhang C (2016) Croton Tiglium extract induces apoptosis via Bax/Bcl-2 pathways in human lung cancer A549 cells. Asian Pac J Cancer Prev 17(11):4893–4898

Li CY, Lee SC, Lai WL, Chang KF, Huang XF, Hung PY, Lee CP, Hsieh MC, Tsai NM (2020) Cell cycle arrest and apoptosis induction by Juniperus communis extract in esophageal squamous cell carcinoma through activation of p53-induced apoptosis pathway. Food Sci Nutr 9(2):1088–1098

Liang CZ, Zhang JK, Shi Z, Liu B, Shen CQ, Tao HM (2012) Matrine induces caspase-dependent apoptosis in human osteosarcoma cells in vitro and in vivo through the upregulation of Bax and Fas/FasL and downregulation of Bcl-2. Cancer Chemother Pharmacol 69(2):317–331

Lin HP, Lin CY, Huo C, Hsiao PH, Su LC, Jiang SS, Chan TM, Chang CH, Chen LT, Kung HJ, Wang HD, Chuu CP (2015) Caffeic acid phenethyl ester induced cell cycle arrest and growth inhibition in androgen-independent prostate cancer cells via regulation of Skp2, p53, p21Cip1 and p27Kip1. Oncotarget 6(9):6684–6707

Lin YT, Wang HC, Hsu YC, Cho CL, Yang MY, Chien CY (2017) Capsaicin induces autophagy and apoptosis in human nasopharyngeal carcinoma cells by downregulating the PI3K/AKT/mTOR pathway. Int J Mol Sci 18(7):1343

Liu FT, Newland AC, Jia L (2003) Bax conformational change is a crucial step for PUMA-mediated apoptosis in human leukemia. Biochem Biophys Res Commun 310:956–962

Liu MM, Ma RH, Ni ZJ, Thakur K, Cespedes-Acuna CL, Jiang L, Wei ZJ (2020) Apigenin 7-O-glucoside promotes cell apoptosis through the PTEN/PI3K/AKT pathway and inhibits cell migration in cervical cancer HeLa cells. Food Chem Toxicol 146:111843

Liu L, Peng C, Ruan Y, Zhang Q (2021) The natural product lapiferin inhibits cell proliferation and promotes cell apoptosis in gingival squamous cell carcinoma via P21 regulation. Mol Med Rep 24:482

Locksley RM, Killeen N, Lenardo MJ (2001) The TNF and TNF receptor superfamilies: integrating mammalian biology. Cell 104:487–501

Luo J, Kong W, Yang M (2014) HJC, a new arylnaphthalene lignan isolated from Justicia procumbens, causes apoptosis and caspase activation in K562 leukemia cells. J Pharmacol Sci 125(4):355–363

Ma YS, Weng SW, Lin MW, Lu CC, Chiang JH, Yang JS, Lai KC, Lin JP, Tang NY, Lin JG, Chung JG (2012) Antitumor effects of emodin on LS1034 human colon cancer cells in vitro and in vivo: roles of apoptotic cell death and LS1034 tumor xenografts model. Food Chem Toxicol 50(5):1271–1278

Ma J, Fang B, Zeng F, Pang H, Zhang J, Shi Y, Wu X, Cheng L, Ma C, Xia J, Wang Z (2014) Curcumin inhibits cell growth and invasion through up-regulation of miR-7 in pancreatic cancer cells. Toxicol Lett 231(1):82–91

Ma J, Ma Y, Liu X, Chen S, Liu C, Qin A, Fan S (2015) Gambogic acid inhibits osteoclast formation and ovariectomy- induced osteoporosis by suppressing the JNK, p38 and Akt signalling pathways. Biochem J 469(3):399–408

Mabry TJ (2001) Selected topics from forty years of natural products research: betalains to flavonoids, antiviral proteins, and neurotoxic non-protein amino acids. J Nat Prod 64:1596–1604

Mafuvadze B, Liang Y, Besch-Williford C, Zhang X, Hyder SM (2012) Apigenin induces apoptosis and blocks growth of medroxyprogesterone acetate-dependent BT-474 xenograft tumors. HORM CANC 3:160–171

Martens S, Hofmans S, Declercq W, Augustyns K, Vandenabeele P (2020) Inhibitors targeting RIPK1/RIPK3: old and new drugs. Trends Pharmacol Sci 41:209–224

Martinvalet D, Zhu P, Lieberman J (2005) Granzyme A induces caspase- independent mitochondrial damage, a required first step for apoptosis. Immunity 22:355–370

Mbaveng AT, Wamba BE, Bitchagno GT, Tankeo SB, Celik I, Atontsa BC, Lonfouo AHN, Kuete V, Efferth T (2021) Bioactivity of fractions and constituents of Piper capense fruits towards a broad panel of cancer cells. J Ethnopharmacol 271:113884

Melino G (2001) The sirens' song. Nature 412:23

Mifflin L, Ofengeim D, Yuan J (2020) Receptor-interacting protein kinase 1 (RIPK1) as a therapeutic target. Nat Rev Drug Discov 19(8):553–571

Min R, Tong J, Wenjun Y, Wenhu D, Xiaojian Z, Jiacai H, Jian Z, Wantao C, Chenping Z (2008) Growth inhibition and induction of apoptosis in human oral squamous cell carcinoma Tca-8113 cell lines by Shikonin was partly through the inactivation of NF-kappaB pathway. Phytother Res 22(3):407–415

Moghadamtousi SZ, Kadir HA, Paydar M, Rouhollahi E, Karimian H (2014) Annona muricata leaves induced apoptosis in A549 cells through mitochondrial-mediated pathway and involvement of NF-KB. BMC Complement Altern Med 14(1):1–13

Mohan S, Abdelwahab SI, Kamalidehghan B, Syam S, May KS, Harmal NSM, Shafifiyaz N, Hadi AHA, Hashim NM, Rahmani M, Taha MME, Cheah SC, Zajmi A (2012) Involvement of NF-κB and Bcl2/Bax signaling pathways in the apoptosis of MCF7 cells induced by a xanthone compound Pyranocycloartobiloxanthone a. Phytomedicine 19(11):1007–1015

Morad SA, Cabot MC (2013) Ceramide-orchestrated signalling in cancer cells. Nat Rev Cancer 13(1):51–65

Mukherjee PK, Banerjee S, Gupta BD, Kar A (2022) Chapter—1. Evidence-based validation of herbal medicine: translational approach. In: Evidence-based validation of herbal medicine, 2nd edn. Elsevier. https://doi.org/10.1016/B978-0-323-85542-6.00025-1

Naeem A, Hu P, Yang M, Zhang J, Liu Y, Zhu W, Zheng Q (2022) Natural products as anticancer agents: current status and future perspectives. Molecules 27(23):8367

Nagata S (2000) Apoptotic DNA fragmentation. Exp Cell Res 256(1):12–18

Newman DJ, Cragg GM (2019) Natural products as sources of new drugs over the nearly four decades from January 1981 to September 2019. J Nat Prod 83(3):770–803

Nishida K, Yamaguchi O, Otsu K (2008) Crosstalk between autophagy and apoptosis in heart disease. Circ Res 103(4):343

O'Keefe BR (2001) Biologically active proteins from natural product extracts. J Nat Prod 64:1373–1381

Oberlies NH, Chang CJ, Mclaughlin JL (1997) Structure-activity relationships of diverse Annonaceous acetogenins against multidrug resistant human mammary adenocarcinoma (MCF7/Adr) cells. J Med Chem 40(13):2102–2106

Oda E, Ohki R, Murasawa H, Nemoto J, Shibue T, Yamashita T, Tokino T, Taniguchi T, Tanaka N (2000) Noxa, a BH3-only member of the Bcl-2 family and candidate mediator of p53-induced apoptosis. Science 288(5468):1053–1058

Okaichi K, Wang LH, Sasaki J, Saya H, Tada M, Okumara Y (1999) A point mutation of human p53, which was not detected as a mutation by a yeast functional assay, led to apoptosis but not p21Waf1/Cip1/Sdi1 expression in response to ionizing radiation in a human osteosarcoma cell line, Saos-2. Int J Radiat Oncol Biol Phys 45(4):975–980

Omara T, Kiprop AK, Ramkat RC, Cherutoi J, Kagoya S, Moraa Nyangena D, Azeze Tebo T, Nteziyaremye P, Nyambura Karanja L, Jepchirchir A, Maiyo A, Kiptui BJ, Mbabazi I, Nakiguli CK, Nakabuye BV, Koske MC (2020) Medicinal plants used in traditional management of cancer in Uganda: a review of ethnobotanical surveys, phytochemistry, and anticancer studies. Evid Based Complement Alternat Med 2020:3529081

Ouyang L, Shi Z, Zhao S, Wang FT, Zhou TT, Liu B, Bao JK (2012) Programmed cell death pathways in cancer: a review of apoptosis, autophagy and programmed necrosis. Cell Prolif 45(6):487–498

Pan LL, Wang XL, Zhang QY, Luo XL, Xu P, Liu SY, Hu JF, Liu XH (2016) Boehmenan, a lignan from the Chinese medicinal plant Clematis armandii, induces apoptosis in lung cancer cells through modulation of EGF-dependent pathways. Phytomedicine 23(5):468–476

Park HY, Toume K, Arai MA, Sadhu SK, Ahmed F, Ishibashi M (2014) Calotropin: a cardenolide from calotropis gigantea that inhibits Wnt signaling by increasing casein kinase 1α in colon cancer cells. Chembiochem 15(6):872–878

Peng L, Jiang D (2018) Resveratrol eliminates cancer stem cells of osteosarcoma by STAT3 pathway inhibition. PLoS One 13(10):e0205918

Peter ME, Krammer PH (1998) Mechanisms of CD95 (APO-1/Fas)- mediated apoptosis. Curr Opin Immunol 10:545–551

Phillips PA, Dudeja V, McCarroll JA, Borja-Cacho D, Dawra RK, Grizzle WE, Vickers SM, Saluja AK (2007) Triptolide induces pancreatic cancer cell death via inhibition of heat shock protein 70. Cancer Res 67(19):9407–9416

Qian SZ, Wang ZG (1984) Gossypol: a potential antifertility agent for males. Annu Rev Pharmacol Toxicol 24:329–360

Ren X, Zhao B, Chang H, Xiao M, Wu Y, Liu Y (2018) Paclitaxel suppresses proliferation and induces apoptosis through regulation of ROS and the AKT/MAPK signaling pathway in canine mammary gland tumor cells. Mol Med Rep 17:8289–8299

Riedl SJ, Shi Y (2004) Molecular mechanisms of caspase regulation during apoptosis. Nat Rev Mol Cell Biol 5(11):897

Rosenberg-Lee M (2004) Caspases: an ancient cellular sword of Damocles. Cell Death Differ 11(1):29

Rubio-Moscardo F, Blesa D, Mestre C, Siebert R, Balasas T, Benito A, Rosenwald A, Climent J, Martinez JI, Schilhabel M, Karran EL, Gesk S, Esteller M, deLeeuw R, Staudt LM, Fernandez-Luna JL, Pinkel D, Dyer MJ, Martinez-Climent JA (2005) Characterization of 8p21.3 chromosomal deletions in B-cell lymphoma: TRAIL-R1 and TRAIL-R2 as candidate dosage-dependent tumor suppressor genes. Blood 106(9):3214–3222

Saelens X, Festjens N, Vande Walle L, van Gurp M, van Loo G, Vandenabeele P (2004) Toxic proteins released from mitochondria in cell death. Oncogene 23:2861–2874

Saha A, Kuzuhara T, Echigo N, Fujii A, Suganuma M, Fujiki H (2010) Apoptosis of human lung cancer cells by curcumin mediated through up-regulation of “growth arrest and DNA damage inducible genes 45 and 153”. Biol Pharm Bull 33(8):1291–1299

Salehi B, Venditti A, Sharifi-Rad M, Kręgiel D, Sharifi-Rad J, Durazzo A, Lucarini M, Santini A, Souto EB, Novellino E, Antolak H, Azzini E, Setzer WN, Martins N (2019) The therapeutic potential of apigenin. Int J Mol Sci 20(6):1305

Salvesen GS, Duckett CS (2002) IAP proteins: blocking the road to death’s door. Nat Rev Mol Cell Biol 3:401–410

Samuels-Lev Y, O’Connor DJ, Beragamaschi D, Triginate G, Hsieh JK, Zhong S, Campargue I, Naumovski L, Crook T, Lu X (2001) ASPP proteins specifically stimulate the apoptotic function of p53. Mol Cell 8:781–794

Scaffidi C, Schmitz I, Krammer PH, Peter ME (1999) The role of c-FLIP in modulation of CD95-induced apoptosis. J Biol Chem 274:1541–1548

Schimmer AD (2004) Inhibitor of apoptosis proteins: translating basic knowledge into clinical practice. Cancer Res 64:7183–7190

Schuler M, Green DR (2001) Mechanisms of p53-dependent apoptosis. Biochem Soc Trans 29:684–688

Sebbagh M, Renvoize C, Hamelin J, Riche N, Bertoglio J, Breard J (2001) Caspase-3-mediated cleavage of ROCK I induces MLC phosphorylation and apoptotic membrane blebbing. Nat Cell Biol 3:346–352

Sferrazza G, Corti M, Brusotti G, Pierimarchi P, Temporini C, Serafino A, Calleri A (2020) Nature-derived compounds modulating Wnt/β-catenin pathway: a preventive and therapeutic opportunity in neoplastic diseases. Acta Pharm Sin B 10(10):1814–1834

Shi XP, Lan XY, Chen X, Wang WW, Wen CY, Huang MJ, Liu HL (2013) Effect of gambogic acid on proliferation and apoptosis of human diffuse large B-cell lymphoma cell line OCI-LY-1. Chinese J Pathophysiol 29(5):815–820

Slee EA, Adrain C, Martin SJ (2001) Executioner caspase-3, -6, and -7 perform distinct, non-redundant roles during the demolition phase of apoptosis. J Biol Chem 276:7320–7326

Srivastava S, Somasagara R, Hegde M, Nishana M, Tadi SK, Srivastava M, Choudhary B, Raghavan SC (2016) Quercetin, a natural flavonoid interacts with DNA, arrests cell cycle and causes tumor regression by activating mitochondrial pathway of apoptosis. Sci Rep 6:24049

Strasser A, O’Connor L, Dixit VM (2000) Apoptosis signaling. Annu Rev Biochem 69:217–245

Suliman A, Lam A, Datta R, Srivastava RK (2001) Intracellular mechanisms of TRAIL: apoptosis through mitochondrial-dependent and -independent pathways. Oncogene 20:2122–2133

Sun SY (2011) Understanding the role of the death receptor 5/FADD/caspase-8 death signaling in cancer metastasis. Mol Cell Pharmacol 3(1):31

Susin SA, Daugas E, Ravagnan L, Samejima K, Zamzami N, Loeffler M, Costantini P, Ferri KF, Irinopoulou T, Prevost MC, Brothers G, Mak TW, Penninger J, Earnshaw WC, Kroemer G (2000) Two distinct pathways leading to nuclear apoptosis. J Exp Med 192:571–580

Tait SW, Green DR (2010) Mitochondria and cell death: outer mem brane permeabilization and beyond. Nat Rev Mol Cell Biol 11(9):621

Tan C, Qian X, Jia R, Wu M, Liang Z (2013) Matrine induction of reactive oxygen species activates p38 leading to caspase-dependent cell apoptosis in non-small cell lung cancer cells. Oncol Rep 30(5):2529–2535

Tan AC, Konczak I, Li G, Roufogalis BD, Sekhon B, Sze-Daniel MY (2015) An ethnopharmacological approach to the preliminary screening of native Australian herbal medicines for anticancer activity. J Comp Integ Med 12:245–249

Taylor RS, Towers GH (1998) Antibacterial constituents of the Nepalese medicinal herb. Centipeda minima Phytochem 47(4):631–634

Trump BF, Berezesky IK, Chang SH, Phelps PC (1997) The pathways of cell death: Oncosis, apoptosis, and necrosis. Toxicol Pathol 25:82–88

Ullmannova V, Haskovec C (2004) Gene expression during camptothecin-induced apoptosis in human myeloid leukemia cell line ML-2. Neoplasma 51(3):175

van Loo G, Bertrand MJM (2023) Death by TNF: a road to inflammation. Nat Rev Immunol 23:289–303

van Loo G, van Gurp M, Depuydt B, Srinivasula SM, Rodriguez I, Alnemri ES, Gevaert K, Vandekerckhove J, Declercq W, Vandenabeele P (2002) The serine protease Omi/HtrA2 is released from mitochondria during apoptosis. Omi interacts with caspase-inhibitor XIAP and induces enhanced caspase activity. Cell Death Differ 9:20–26

Vaux DL (2002) Apoptosis timeline. Cell Death Differ 9:349–354

Wajant H (2002) The Fas signaling pathway: more than a paradigm. Science 296:1635–1636

Wang X, Chen W (2012) Gambogic acid is a novel anti-cancer agent that inhibits cell proliferation, angiogenesis and metastasis. Anti-Cancer Agent Med Chem 12(8):994–1000

Wang Y, Chen Y, Chen Z, Ke WJ, Wu QL, He J (2009) Mechanism of gambogic acid-induced apoptosis in Raji cells. J Experimen Hematol 17(1):88–91

Wang L, Guo H, Yang L, Dong L, Lin C, Zhang J, Lin P, Wang X (2013a) Morusin inhibits human cervical cancer stem cell growth and migration through attenuation of NF-κB activity and apoptosis induction. Mol Cell Biochem 379(1–2):7–18

Wang W, Zhao C, Jou D, Lu J, Zhang C, Lin L, Lin J (2013b) Ursolic acid inhibits the growth of colon cancer-initiating cells by targeting STAT3. Anticancer Res 33(10):4279–4284

Wang Q, Du H, Geng G, Zhou H, Xu M, Cao H, Zhang B, Song G, Hu T (2014) Matrine inhibits proliferation and induces apoptosis via BID-mediated mitochondrial pathway in esophageal cancer cells. Mol Biol Rep 41(5):3009–3020

Wang Y, Zhong J, Bai J, Tong R, An F, Jiao P, He L, Zeng D, Long E, Yan J, Yu J, Cai L (2018) The application of natural products in cancer therapy by targeting apoptosis pathways. Curr Drug Metab 19(9):739–749

Wei W, Huang H, Zhao S, Liu W, Liu CX, Chen L, Li JM, Wu YL, Yan H (2013) Alantolactone induces apoptosis in chronic myelogenous leukemia sensitive or resistant to imatinib through NF-kappaB inhibition and Bcr/Abl protein deletion. Apoptosis 18(9):1060–1070

Xin H, Kong Y, Wang Y, Zhou Y, Zhu Y, Li D, Tan W (2013) Lignans extracted from Vitex negundo possess cytotoxic activity by G2/M phase cell cycle arrest and apoptosis induction. Phytomedicine 20(7):640–647

Xu KH, Lu DP (2010) Plumbagin induces ROS-mediated apoptosis in human promyelocytic leukemia cells in vivo. Leuk Res 34(5):658–665

Xu G, Shi Y (2007) Apoptosis signaling pathways and lymphocyte homeostasis. Cell Res 17:759–771

Xu L, Yu J, Zhai D, Zhang D, Shen W, Bai L, Cai Z, Yu C (2014) Role of JNK activation and mitochondrial bax translocation in allicin- induced apoptosis in human ovarian cancer SKOV3 cells. Evid Based Complement Alternat Med:378684

Yan Y, Su X, Liang Y, Zhang J, Shi C, Lu Y, Gu L, Fu L (2008) Emodin azide methyl anthraquinone derivative triggers mitochondrial-dependent cell apoptosis involving in caspase-8-mediated bid cleavage. Mol Cancer Ther 7(6):1688–1697

Yang SM, Chen JG, Guo Z, Xu XM, Wang LP, Pei FX, Yang J, Charles B, Zhang LR (2003) Triptolide inhibits the growth and metastasis of solid tumors. Mol Cancer Ther 2(1):65–72

Yao JN, Zhang XX, Zhang YZ, Li JH, Zhao DY, Gao B, Zhou HN, Gao SL, Zhang LF (2019) Discovery and anticancer evaluation of a formononetin derivative against gastric cancer SGC7901 cells. Investig New Drugs 37(6):1300–1308

Yuan JJ, Meng YF, Zhang MS, Guo DL, Yang JW, Dong MJ, Sun CX, Xiao SJ (2023) Isoprenoid flavonoids isolated from Sophora davidii and their activities induces apoptosis and autophagy in HT29 cells. Fitoterapia 175:105945. https://doi.org/10.1016/j.fitote.2024.105945

Zha J, Harada H, Yang E, Jockel J, Korsmeyer SJ (1996) Serine phosphorylation of death agonist BAD in response to survival factor results in binding to 14-3-3 not BCL-X(L). Cell 87:619–628

Zhang Y, Fujita N, Tsuruo T (1999) Caspase-mediated cleavage of p21Waf1/Cip1 converts cancer cells from growth arrest to undergoing apoptosis. Oncogene 18:1131–1138

Zhang S, Qi J, Sun L, Cheng B, Pan S, Zhou M, Sun X (2009) Matrine induces programmed cell death and regulates expression of relevant genes based on PCR array analysis in C6 glioma cells. Mol Biol Rep 36(4):791–799

Zhang L, Li L, Jiao M, Wu D, Wu K, Li X, Zhu G, Yang L, Wang X, Hsieh JT, He D (2012a) Genistein inhibits the stemness properties of prostate cancer cells through targeting Hedgehog-Gli1 pathway. Cancer Lett 323(1):48–57

Zhang S, Zhang Y, Zhuang Y, Wang J, Ye J, Zhang S, Han Y (2012b) Matrine induces apoptosis in human acute myeloid leukemia cells via the mitochondrial pathway and Akt inactivation. PLoS One 7(10):3517

Zhang J, Li S, An FF, Liu J, Jin S, Zhang JC, Wang PC, Zhang X, Lee CS, Liang XJ (2015) Self-carried curcumin nanoparticles for in vitro and in vivo cancer therapy with real- time monitoring of drug release. Nanoscale 7(32):13503

Zhang HW, Hu JJ, Fu RQ, Liu X, Zhang YH, Li J, Liu L, Li YN, Deng Q, Luo QS, Ouyang Q, Gao N (2018) Flavonoids inhibit cell proliferation and induce apoptosis and autophagy through downregulation of PI3Kγ mediated PI3K/AKT/mTOR/p70S6K/ULK signaling pathway in human breast cancer cells. Sci Rep 8:11255

Zhao B, Li B, Bai S, Shen L, Ren R, Jonas JB, Xu X, Lu Q, Liu Q (2012) Effects of matrine on proliferation and apoptosis of cultured retinoblastoma cells. Graefes Arch Clin Exp Opthalmol 250(6):897–905

Zhao YN, Cao YN, Sun J, Liang Z, Wu Q, Cui SH, Zhi DF, Guo ST, Zhen YH, Zhang SB (2020) Anti-breast cancer activity of resveratrol encapsulated in liposomes. J Mater Chem B 8(1):27–37

Zhong N, Jiang D, Zheng SY (2010) Anticancer effect of gambogic acid in gastric cancer line SGC-7901. In: 4th international conference on bioinformatics and biomedical engineering, Chengdu, China, Jun 18-20, 2010

Zhou Y, Liu YE, Cao J, Zeng G, Shen C, Li Y, Zhou M, Chen Y, Pu W, Potters L, Shi YE (2009) Vitexins, nature-derived lignan compounds, induce apoptosis and suppress tumor growth. Clin Cancer Res 15(16):5161–5169

Zhou H, Xu M, Gao Y, Deng Z, Cao H, Zhang W, Wang Q, Zhang B, Song G, Zhan Y (2014) Matrine induces caspase- independent program cell death in hepatocellular carcinoma through bid-mediated nuclear translocation of apoptosis inducing factor. Mol Cancer 13(1):59

Zhou J, Jiang YY, Wang HP, Chen H, Wu YC, Wang L, Pu X, Yue G, Zhang L (2020) Natural compound Tan-I enhances the efficacy of paclitaxel chemotherapy in ovarian cancer. Ann Transl Med 8(12):752

Zhu Y, Zhong Y, Long X, Zhu Z, Zhou Y, Ye H, Zeng X, Zheng X (2019) Deoxyshikonin isolated from Arnebia euchroma inhibits colorectal cancer by down-regulating the PI3K/Akt/mTOR pathway. Pharm Biol 57(1):412–423

Zhu Y, Ouyang Z, Du H, Wang M, Wang J, Sun H, Kong L, Xu Q, Ma H, Sun Y (2022) New opportunities and challenges of natural products research: when target identification meets single-cell multiomics. Acta Pharm Sin B 12(11):4011–4039

Zirim AU, Okochi VI, James AB, Adebeshi OA, Ogunnowo S, Odeghe OB (2013) Induction of apoptosis in myelogenous leukemic k562 cells by ethanolic leaves extract of Annona muricata L. Indian J Drugs 2(3):142–1151

Zorofchian MS, Karimian H, Rouhollahi E, Paydar M, Fadaeinasab M, Abdul KH (2014) Annona muricata leaves induce G1 cell cycle arrest and apoptosis through mitochondria-mediated pathway in human HCT-116 and HT-29 colon cancer cells. J Ethnopharmacol 156:277–289

# 19 Apoptosis-Based Therapeutics and Inhibitors in Clinical Trials or the Preclinical State

Omkar Kailas Gaikwad, Vishakha Yogesh Joshi, Nidhi Singh, Priyanka Das, and Pallab Datta

**Abstract**

Apoptosis is a progressive cell death process that significantly preserves the homeostatic balance between cell division and death. It is an essential process that helps regenerate cells and tissues for eukaryotic development. Moreover, it inhibits unfavorable immunological reactions and is essential for phagocytosis-mediated cell removal. The apoptosis signaling pathway is driven by various interactions between multiple protein families, including caspases, B-cell lymphoma 2 (BCL-2), and other proteases, granzyme, and perforins. This pathway consists of prosurvival and proapoptotic members working together to stabilize the choice between cellular survival and death. However, this can turn into abnormal cell proliferation, which ultimately leads to cancer. Traditional chemotherapy and radiation therapy are beneficial for the treatment of cancer. Cancer cells may withstand the death process through the antiapoptotic effect of the apoptosis protein inhibitors. Both intrinsic and extrinsic pathways are involved in the process of apoptosis caused by cellular stress and DNA damage. This review elaborates on the apoptotic signaling pathways, mechanisms, interacting members, and strategies that target these pathways.

Omkar Kailas Gaikwad and Vishakha Yogesh Joshi contributed equally with all other contributors.

O. K. Gaikwad · V. Y. Joshi · N. Singh · P. Das · P. Datta (✉)
Polymer-based Medical Devices and Complex Drug Delivery Systems Laboratory, Department of Pharmaceutics, National Institute of Pharmaceutical Education and Research (NIPER), Kolkata, India
e-mail: pd@niperkolkata.ac.in





K. Jana (ed.), *Apoptosis and Human Health: Understanding Mechanistic and Therapeutic Potential*, https://doi.org/10.1007/978-981-97-7905-5_19

**Keywords**

Apoptosis · Signaling pathways · Intrinsic pathways · Extrinsic pathways · Therapeutics · Inhibitors

## 19.1 Introduction

Apoptosis, called programmed cell death, maintains a balance between cell growth and death. Abnormalities in apoptosis result in immortal cell clones leading to various cancer forms, including follicular lymphoma, human T-cell leukemia/lymphoma virus-1-infected tumors, and B-cell chronic lymphocytic leukemia (CLL). Numerous malignancies are caused by a deficiency in some apoptotic regulatory mechanisms like phosphatidylinositol 3-kinase (PI3K)/Akt, NFκB, and p53 (Cohen 1997). Cell death occurs primarily through two mechanisms, namely apoptosis and necrosis. Necrosis occurs when cells are accidentally damaged by external trauma mediated by inflammation, whereas apoptosis occurs when internal stimuli damage cells. Signaling pathways that initiate apoptosis are being overwhelmingly studied, although a variety of complex proteins are involved in controlling this process. These proteins are triggered by different stimuli and organized in successive signaling modules (Li and Yuan 2008).

There are two primary pathways (shown in Fig. 19.1) through which the process of apoptosis takes place. One is known as extrinsic pathway, initiated by tumor necrosis factor (TNF) receptor superfamily member Fas death receptor (DR). Another one is the mitochondrial or intrinsic pathway, which causes the mitochondria to release cytochrome C and activated death signals upon stimulation. These two pathways eventually come together to form a single common pathway that activates a series of proteases known as caspases, which cleave structural and regulatory components, ultimately leading to cell death. These pathways are interconnected and diagrammatically represented in Fig. 19.1 (Ashkenazi 2008). For example, TNFα boosts NFκB production and activates antiapoptotic BCL-2 family protein, whereas the overexpression of BCL-2 in the intrinsic pathway blocks extrinsic-driven apoptosis.

### 19.1.1 The Extrinsic Pathway

Various extrinsic pathways involve various proteins, like Fas DRs, Fas complexes, membrane-bound Fas ligands, caspases 8 and 10, and the Fas-associated death domain. These components stimulate the downstream caspases, resulting in programmed cell death (Hongmei 2012). The ligation of cell surface receptors, known as DRs, starts the extrinsic pathway. Fas Apo-1, or CD95, is a member of the tumor necrosis factor receptor (TNFR) superfamily. TNF R1, DR3 (Apo 2), DR4 (TRAIL R1), DR5 (TRAIL R2), and DR6 are additional TNF receptors. Fas signaling is vital in eliminating self-reactive lymphocytes and immune surveillance of virus-infected

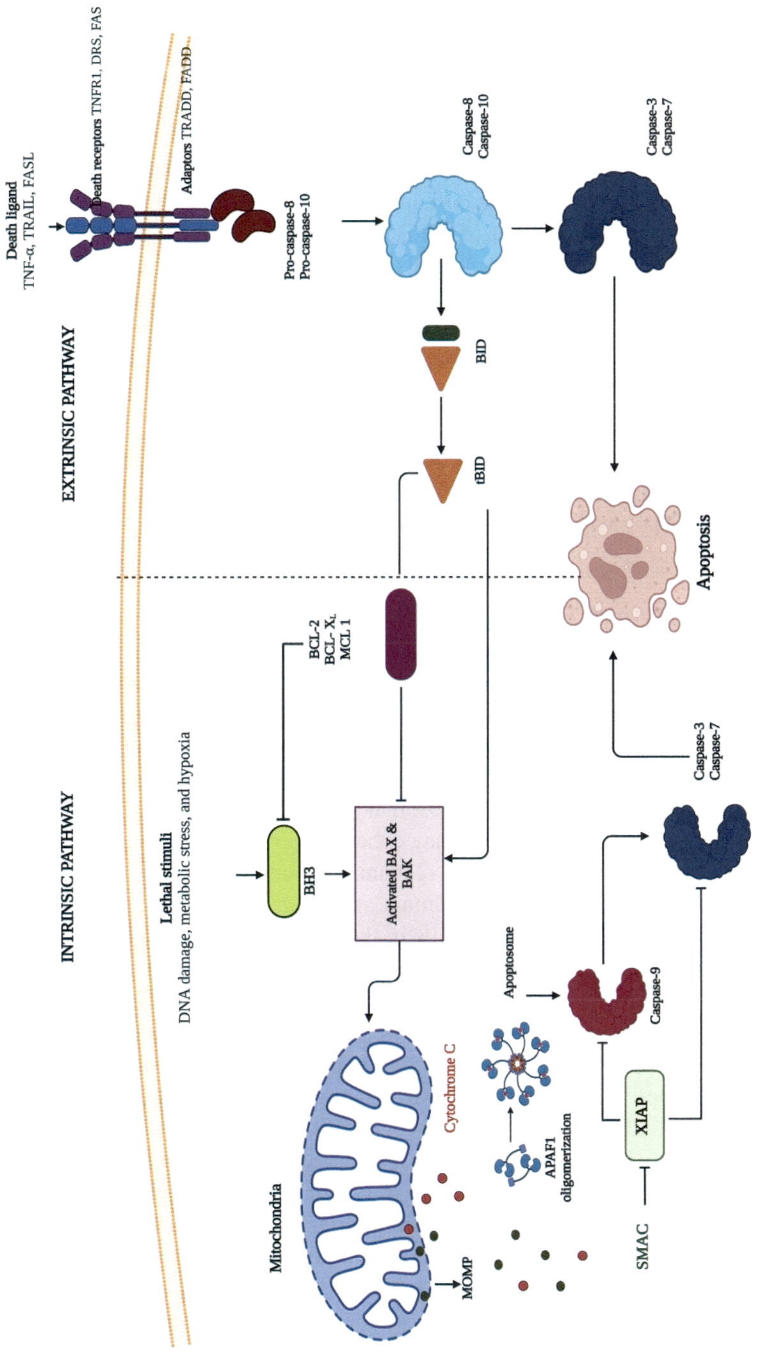


**Fig. 19.1** Diagrammatic illustration of intrinsic and extrinsic pathways involved in cell apoptosis

cells (Sayers 2011). Consequently, abnormalities in this pathway have been linked to numerous cancers and autoimmune disorders. The Fas ligand (FasL)-Fas system is primarily known as the death ligand. Still, it is also connected to several proliferative and inflammatory signaling pathways, though these are not well established (Fulda 2015). The membrane-bound FasL interacts with inactive Fas complexes by decease stimulus to generate the death-inducing signaling compound. This complex activates caspase 8, activating the downstream caspases (Elumalai et al. 2012). It also contains Fas-associated death domain protein and caspases 8 and 10. By cleaving BID, a proapoptotic member of the BCL-2 family, caspase 8 interacts with the intrinsic apoptotic pathway in some cells, releasing cytochrome C. However, caspase 8 activation may play the role of executing cell death (Kristen et al. 2013). Malignant transformation can result from dysregulation of these proteins, as mutation or deletion of the Fas gene has been found in some hematologic malignancies. Due to the transcriptional inactivity of the FasL gene, transcription factors including NFκB is the regulator of the extrinsic pathway (Soares et al. 2014). Additional pathway inhibitors are FAP-1, soluble decoy receptors like DcR3, TRAIL R-3/DcR1, TRAIL R-4/DcR2, and Fas-associated death domain protein. By competing with ligands, these decoy receptors block the activation of Fas (Tang et al. 2004).

### 19.1.2 The Intrinsic Pathway

The intrinsic pathway-mediated regulation of apoptosis is the primary function of the BCL-2 protein family. The BCL-2 gene was initially discovered near the chromosomal breakpoint of the translocation of chromosome 18 to 14 in follicular non-Hodgkin lymphoma (NHL). The BCL-2 family is overexpressed in numerous cancers, even without t (14;18) chromosomal translocations (Wu and Bratton 2013). The depletion of BCL-2 expression may enhance apoptotic responses to anticancer medications, whereas increasing BCL-2 manifestation results in resistance to radiation therapy and chemotherapy (Yanumula and Cusick 2020). The BCL-2 overexpression may lead to a build-up of cells in the G0 phase of cell cycle division and enhance chemoresistance. There are two types of members present in BCL-2 family. One is proapoptotic member including Bax, Bad, Bak, Bid, Bcl-Xs, Bik, Hrk, and Bim, and another is an antiproapoptotic member including Bcl-XL, Bfl-1, Bcl-W, and Mcl-1. Proapoptotic BCL-2 members function as promoters of apoptosis, while antiproapoptotic members suppress it by preventing the release of cytochrome C (Léveillé et al. 2010). The balance between BCL-2 and Bax is more important than BCL-2 alone. Proapoptotic proteins are subjected to posttranslational changes, such as cleavage and dephosphorylation in response to death signal. This leads to their activation and translocation to mitochondria, which ultimately causes apoptosis (Nair et al. 2014). The natural proapoptotic action of BH3-only proteins has been exhibited by multidomain BH3 proteins (Bax, Bak).

The surface mitochondrial membrane turns out to be permeable in response to apoptotic stimuli, causing the release of mitochondria-derived activator of caspase and cytochrome C. After being released into cytosol, cytochrome C interacts with

Apaf-1 to activate the proenzymes of caspase-9 (Mohana-Kumaran et al. 2014). Finally, caspase-9 activates caspase-3, which triggers the remaining caspase cascade and causes apoptosis. Through caspase-3, activated caspase causes cleavage of nuclear lamina and breakdown of the nucleus.

### 19.1.3 The Final Pathway: Caspases

The final process resulting in the implementation of a death signal is the activation of a group of proteases, also known as caspases. The caspases −3, −6, −7, −8, and − 9 have been thoroughly investigated for their role in apoptosis (Chang et al. 1993). The caspase-activated deoxyribonuclease becomes active and causes nuclear apoptosis, which cleaves the inhibitor of caspase-activated deoxyribonuclease and converges with the intrinsic and extrinsic apoptotic pathways (King and Cidlowski 1995).

In intrinsic and extrinsic pathways, caspases-8 and -9 are the upstream caspases that converge to caspase-3. The cleavage of protein kinases, DNA repair proteins, cytoskeletal proteins, inhibitory subunits of endonuclease (CIDE family) and the disruption of "housekeeping" cellular processes are initiated by downstream caspases (Wang et al. 1999). Additionally, caspases impact signaling pathways, cytoskeletal structure, and cell cycle, ultimately resulting in morphological signs of apoptosis (Mitra et al. 1998).

## 19.2 Regulation of Apoptotic Proteins

Proteins regulate the intrinsic and extrinsic pathways, including p53, NFκB, the ubiquitin-proteasome system, and PI3K. A brief description has been given in this review due to their importance in medications.

### 19.2.1 p53

p53 is a transcription factor that controls downstream genes tangled in apoptosis, DNA repair, and cell cycle arrest. In numerous malignancies, p53 loses its activity, resulting in genomic instability, compromised control of the cell cycle, and suppression of apoptosis (Wawryk-Gawda et al. 2014). p53 keeps the cell at a checkpoint following DNA damage until the impairment is restored. Apoptosis is initiated when the damage cannot be repaired (Haupt et al. 2003).

### 19.2.2 NFκB

Nuclear transcription factor (NFκB) controls the expression of several genes involved in controlling inflammation, cancer, viral replication, apoptosis, and numerous autoimmune disorders. Numerous stimuli, such as growth factors, cytokines, lymphokines, radiation, pharmaceutical drugs and stress can activate NFκB. When NFκB is not in use, it is stored in the cytoplasm, bound by IκB family inhibitor proteins (Piotrowska et al. 2008). IκB is phosphorylated by certain stimuli that activate NFκB, followed by its breakdown. As a result, the molecule is translocated to the nucleus, and the nuclear localization signals are exposed to NFκB subunits (Kucharczak et al. 2003). NFκB binds to gene sequences in the nucleus, activating the transcription of those genes. It has been demonstrated that NFκB has both proapoptotic and antiapoptotic properties, which can be calculated by the type of death stimuli (Bellas et al. 1997). Under physiological circumstances, TNF receptor-associated factor, IAP, and X-linked IAP are only a few of the complex proteins that get activated upon NFκB activation. These results in resistance to apoptotic stimuli (Stehlik et al. 1998). However, NFκB activation may result in the production of apoptosis in response to specific stimuli. It can be done by activating some proapoptotic proteins, including p53, c-myc, interferon-regulated factor-1, and caspase 1. Certain viral infections require NFκB activation to induce apoptosis (Bernal-Mizrachi et al. 2006).

### 19.2.3 The Proteasome/Ubiquitin System

It comprises a large proteinase complex. Degradation of protein involves protein recognition, attachment to several ubiquitin molecules, and digestion by 26S proteasome (Wójcik 2002). The ubiquitin/proteasome system regulates several cell cycle regulators and transcription factors, including p53, cyclins, cyclin-dependent kinase inhibitors, and NFκB. Additionally, several BCL-2 family members are ubiquitin/proteasome substrates. Proteasome inhibitors induce apoptosis by accumulating proteins like p53, proapoptotic Bad or Bax,p27or activating the stress kinase (Bazzaro et al. 2006). This causes the release of cytochrome-C and activation of the intrinsic apoptotic pathway. The relationship between apoptotic machinery and the ubiquitin/proteasome system facilitates interest in developing inhibitory drug targeting these pathways (Orlowski 1999).

### 19.2.4 PI3K

PI3K is a kinase that regulates cell survival, neovascularization, proliferation, and motility. PI3K is overexpressed in several malignancies. Cell-surface receptors produce second messengers, including phosphatidylinositol 4,5-bisphosphate 3 and phosphatidylinositol 3,4,5-trisphosphate, to transmit signals to the cytoplasm (Franke et al. 2003). Phosphatidylinositol 4,5-bisphosphate 3 signals activate 3-phosphoinositide-dependent protein kinase-1, further activating the Akt kinase. Akt activation phosphorylates proteins essential for cell survival. Phosphorylation of IkB activates NFκB, promoting survival. Phosphorylation of Bad by Akt inhibits the apoptotic signal (Duronio 2008). Phosphorylation of caspases-9 inhibits the induction of apoptosis. The PI3K pathway can cause a dysregulation of proteins linked to cancer. Epidermal growth factor stimulation activates Akt through PI3K (Patel et al. 2021).

Novel therapeutic agents aim to target the mTOR pathway, which is located downstream of Akt. Two downstream pathways, the 40S ribosomal protein S6 kinase (p70s6k) and the 4E-binding protein-1 (4E-BP1), are blocked when mTOR is inhibited. These regulate the passage of the cell cycle from G1 to S phase growth arrest. Many cancer cells can have an elevated PI3K activity due to abnormalities in cyclin D, p53, pRB, p16, and p27, which increase their susceptibility to mTOR inhibitors like CCI779 (Uno et al. 2003).

## 19.3 Strategies Involved in Apoptosis Pathways

### 19.3.1 Strategy for Intrinsic Pathway

Identifying the BCL-2 gene in follicular lymphoma patients due to the chromosomal translocation (14;18) sparked interest in focusing on this protein family to promote apoptosis and slow down the spread of malignancy. The first attempts aimed to find inhibitors of antiapoptotic family members, namely BCL-2. Among these strategies, BCL-2 expression-targeting oligonucleotides, known as BH3 mimetics, imitate the binding of BH3-only proapoptotic members to antiapoptotic BCL-2 members.

### 19.3.2 Strategy for Extrinsic Pathway

Transmembrane receptor-mediated interactions are intricated in the extrinsic signaling pathways that trigger apoptosis. These are DRs, members of the tumor necrosis factor (TNF) receptor gene superfamily.

## 19.4 Therapeutic Agents in Preclinical and Clinical Studies Targeting Apoptosis

We provided a summary of apoptotic protein regulation and pathways. Apoptosis involves numerous proteins, each of which has a different purpose. Signals from both intrinsic and extrinsic routes are present in many pathways. Therefore, specific blockers of these pathways may not be sufficient to trigger apoptosis. Several drugs have been shown to inhibit apoptosis-related proteins, either directly or indirectly. The epidermal growth factor inhibitors are not typically considered pro-apoptotic medicines. These drugs indirectly control apoptosis by the PI3K/Akt pathway. Mitogen-activated protein kinase inhibitors can trigger apoptosis by regulating the downstream protein extracellular signal-regulated kinase, which activates the intrinsic apoptosis pathway.

Here, the agents that predominantly target the apoptotic process have also been discussed. Some therapeutic medicines are now in clinical usage or studies, while others are in the preclinical stage. These agents target the extrinsic, intrinsic, common, or apoptosis-regulated pathways. Agents can be classified as either apoptosis inducers or antiapoptotic protein inhibitors.

## 19.4.1 Agents that Target the Extrinsic Pathway

These consist of all transretinoic acid (aTRA), and monoclonal antibodies agonists to DRs 4 and 5, which refer to proteins tangled in managing apoptosis. These proteins are part of the BCL-2 family, which serves as a vital role in regulating apoptosis. The chemical structure of aTRA is shown in Fig. 19.2.

### 19.4.1.1 TNF-Related Apoptosis-Inducing Ligand (TRAIL)

DRs have been investigated as possible targets for cancer treatment. TNF, Fas, and Fas L have been used as possible therapeutic targets in vivo due to their strong anticancer activity. But it was also shown that they activated nonspecific TNF receptors, which led to widespread ischemia and hemorrhagic lesions, ultimately causing fulminant hepatic failure and septic shock in animal models. APO2L, also known as TRAIL, is a novel extrinsic pathway inducer lacking TNF and Fas toxicity (Cheah et al. 2015). Five TRAIL DRs have been discovered named Decoy Receptor 1 (TRAILR-3/DcR1), Decoy Receptor 2 (TRAILR-4/DcR2), Death Receptor 4 (DR4/TRAILR-1), Death Receptor 5 (DR5/TRAILR-2), and osteoprotegerin (OPG). As a member of the tumor necrosis superfamily (TNFSF), TRAIL can link to its associated DRs in two ways: soluble or membrane-bound. The more active form of TRAIL is the membrane-bound version, often expressed on the surface of immune effector cells like CD8+ T-cells and natural killer cells. Many tumor cell types undergo apoptosis when exposed to TRAIL, inhibiting the growth of breast and colon xenografts. It is also possible to observe synergistic antitumor effects in combination with radiation or chemotherapy. It was discovered that TRAIL may cause apoptosis in human brain cells and hepatocytes in vitro but not in the animal brain cells used in preclinical research. In an in vivo model of prostate cancer, preclinical study of

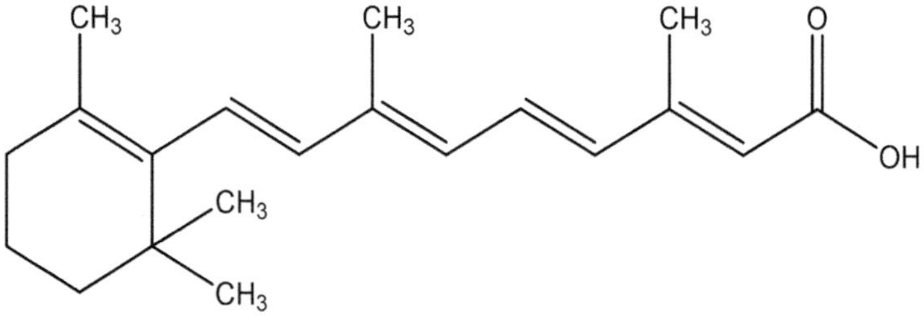


**Fig. 19.2** Chemical structure of all trans retinoic acid (aTRA)

TRAIL in conjunction with traditional chemotherapeutic drugs like doxorubicin has shown substantial suppression of tumor growth.

#### Dulanermin

Dulanermin is a recombinant soluble TRAIL variant that consists of the TNF homology province found in the outer domain of human TRAIL (amino acids 114–281). It can bind both DR4 and DR5, causing apoptosis. In preclinical studies, dulanermin showed selective cytotoxicity toward cancer cells compared to healthy cells and offered synergistic responses with several other chemotherapeutic agents. The initial human trial exhibited a satisfactory safety profile with controllable adverse events (AE). Repeated cycles showed no buildup in serum, with a short half-life (0.5–1 h) (Ouyang et al. 2018).

Unfortunately, a repeated phase II analysis revealed that dulanermin failed to improve the anticancer effect as observed with traditional chemotherapeutic agents alone. Shanghai Gebaide Biotechnology (Shanghai, China) conducted a phase III study that showed promising outcomes for dulanermin as a potential treatment option for untreated non-small cell lung cancer (NSCLC). Patients treated with dulanermin, cisplatin, and vinorelbine combinations had better progression-free survival but not overall survival (OS) rate compared to chemotherapy alone NCT03083743 (Herbst et al. 2010). Overall, dulanermin showed minimal effectiveness in clinical trials while being well-tolerated.

### 19.4.1.2 Monoclonal Antibodies

To induce apoptosis and caspase, monoclonal antibodies like HGS-ETR1, HGSETR2, and HGS-TR2J with agonist action at DR4 (TRAIL R1) and DR5 (TRAIL R2) sites have demonstrated tumor regression in colon carcinoma xenograft models. The extended half-life of the above antibodies compared to Apo2L/TRAIL may be an advantage, potentially resulting in a less frequent injection schedule. There are ongoing Phase I and Phase II clinical trials for patients with NHL, NSCLC, colon cancer, and advanced solid cancers. Studies are being conducted on the antibodies alone and with cytotoxic treatment (Trarbach et al. 2010).

#### Mapatumumab

Mapatumumab, developed by Human Genome Sciences (Rockville, MD, USA), stands out as the only human DR4 agonistic antibody that has advanced in clinical trials. In preclinical settings, mapatumumab showed a reduction in tumor growth in xenograft models and boosted the effectiveness of conventional chemotherapeutic agents (Gieffers et al. 2013). Despite its safety profile, multiple phase I/II trials revealed mapatumumab's inefficacy alone and in combination with other agents across various human cancers. For instance, a phase II trial involving NSCLC patients receiving mapatumumab alongside carboplatin and paclitaxel confirmed no added benefits compared to standard treatments (Frankfurt et al. 1996). On the other hand, several DR5 agonistic monoclonal antibodies, such as tigatuzumab, conatumumab, LBY135, drozitumab, and lexatumumab, were developed, but with limited success in clinical applications. Conatumumab, initially investigated by Amgen and later licensed to Takeda Pharmaceuticals in Tokyo, Japan, certainly targets the outer domain of DR5. Preclinical trials showcased its effectiveness in enhancing the anticancer properties of various tumor therapeutics, such as 5-fluorouracil, gemcitabine, and irinotecan. Despite the excellent safety profile demonstrated in phase I studies, mapatumumab clinical development ceased in 2011 due to the lack of clinical effectiveness (Mangiarotti et al. 1998).

#### 19.4.1.3 aTRA

aTRA was one of the initial instances of tailored treatment for human cancer. The proposed pathway of the protective effect of aTRA is illustrated diagrammatically in Fig. 19.3. During acute promyelocytic leukemia (APL), 90 mg/m$^2$ of the dose is administered intravenously every other day for nine months, using the same dose three times a week. aTRA modifies the underlying mechanism to cause differentiation in APL. It may cause the paracrine release to initiate apoptosis through the extrinsic pathway (Legler et al. 2018). Membrane-associated TRAIL causes APL, treated with aTRA to undergo apoptosis. One of the primary consequences of aTRA stands for aTRA syndrome, which is characterized by temperature, pulse, and respiratory distress. The treatment with a TRA for APL occurs either by itself or in combination with chemotherapy, which produces up to 85% to 95% full remission rate (Ratain et al. n.d.).

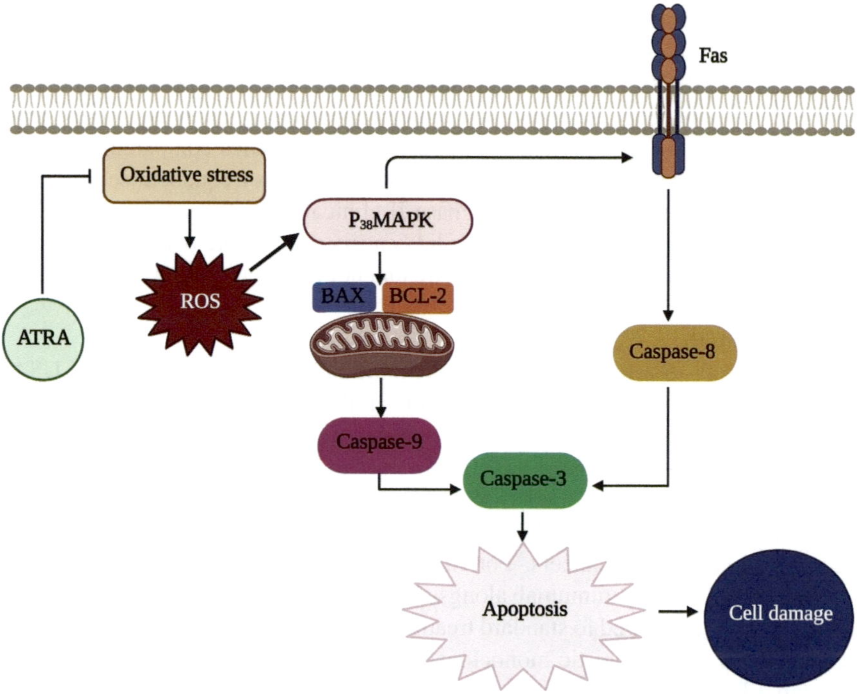


**Fig. 19.3** Proposed pathway of the protective effect of aTRA for the treatment in cancer

#### 19.4.1.4 Current TRAIL-Targeted Therapeutics Under Clinical Investigation

ABBV-621 is a fusion protein, a pioneering DR agonist developed by AbbVie in North Chicago, IL, USA.This fusion protein selectively causes tumor cells to undergo apoptosis by binding to cancer cells' DR4 and DR5 receptors. Initially developed by Apogenix (Heidelberg, Germany), a licensing contract was established for clinical expansion with AbbVie in 2014. Initial clinical trials revealed sustainable toxicity, noteworthy anticancer properties, and impacts on blood-derived pharmacodynamic indicators of programmed cell death (cleaved cytokeratin products M30/M65) (Zhang et al. 2003). In the first human trial, ABBV-621 was assessed alone and in combination with other conventional chemotherapeutic agents, such as the FOLFIRI regime and bevacizumab in colorectal cancer patients with KRAS mutations and venetoclax in hematological cancers. In individuals with relapsed or refractory AML, venetoclax was found to have preliminary anticancer efficacy (Li et al. 2006).

HexaBody-DR5/DR5, invented by Genmab (Copenhagen, Denmark), combines two humanized, noncompeting monoclonal antibodies specific to DR5 mix in 1:1 ratio. This agent exhibited activity across various cancer cell lines and in xenograft models. The safety of GEN1029 is being investigated in a two-part clinical trial, NCT03576131. It consists of a dose escalation phase (Phase 1, first-in-human) and an expansion phase (Phase 2a), starting as soon as the recommended Phase 2 dose is established.

Recently, INBRX-109, a tetravalent DR5 agonistic antibody, has gone into clinical development. Inhibrx developed it in La Jolla, CA, USA. Designed to function as a tetravalent single-domain antibody-based therapeutic agonist of DR5, it can super-cluster receptors to oppose DR5 efficiently. Its effectiveness as a single treatment in various solid human malignancies, such as colorectal and pancreatic adenocarcinomas, is assessed in a Phase 1 trial NCT03715933. The drug showed good tolerance during the first phase of dose escalation, with no notable liver toxicities identified at doses up to the maximum administered dose of 30 mg/kg. For the treatment of metastatic conventional chondrosarcoma, INBRX-109 was given fast-track designation status in 2021. A Phase 2 randomized, blinded, placebo-controlled research is slated to start later that year.

### 19.4.2 Agents Targeting Intrinsic Pathway

The substances that directly affect the intracellular membrane of the mitochondria, substances that oppose the anti-apoptotic BCL-2 protein family, and substances that increase the property of the proapoptotic BCL-2 family proteins, such as Bax.

#### 19.4.2.1 Agents Target BCL-2

Chronic lymphocytic leukemia (CLL) and B-large cell NHL, mainly the BCL-2 gene, get overexpressed in most follicular B-cell lymphomas. Cancers including prostate, breast, nonsmall cell, small cell, and melanoma that do not have the overexpressed t(14;18) translocation in the BCL-2 protein. Additionally, BCL-2 is essential for controlling how the body reacts to radiation, hormone and chemotherapy. Interestingly, modification of antiapoptotic protein through antisense treatment has been increased due to its therapeutic significance. Small compounds that recognize the surface pocket of BCL-2 or BCL-xL, antisense oligonucleotides like G3, and antisense BCL-xL are the agents that can influence the BCL-2 protein (Del Bufalo et al. 1996).

**Oblimersen Sodium**

A phosphorothioate oligonucleotide with 18 altered DNA bases is called genasense. The mRNA is the target of a single-stranded DNA molecule, which must enter the cell through endocytosis and have a complementary sequence. G3139 (Genasense) has a dose-dependent action that degrades BCL-2 mRNA before inhibiting the expression of BCL-2 protein. Studies both in vivo and in vitro have shown anticancer efficacy. Moreover, genasense increases the effectiveness of some chemotherapeutic drugs, including monoclonal antibodies, corticosteroids, alkylating agents, antimetabolites, and radiation. However, 5.3 mg/kg per day is the maximum tolerated dosage (MTD); additional Phase I investigations have been conducted on patients with CLL and solid malignancies. Numerous investigations have revealed that thrombocytopenia, tiredness, and temporary changes in liver function are the most common toxicities (Li et al. 2006). In phase I and II studies of 14 CLL patients, 7 mg/kg per day were poorly tolerated. These effects might be a by-product of G3 139's antilymphocytic action, which causes cytokine release and the onset of tumor lysis syndrome. A large-scale randomized Phase III clinical trial involved individuals with melanoma undergoing chemotherapy. Patients were randomly assigned to receive dacarbazine in addition to G3 139 or dacarbazine alone. A period of approximately one year was observed for 480 individuals. Patients in the G3139 and dacarbazine group had a median OS of 9.1 months, while those on the dacarbazine arm had a median survival of 7.9 months ($P = 0.18$). OS for patients was monitored, and it was higher in the G3139/dacarbazine arm than in dacarbazine alone. 11% of patients treated with G3139 experienced a response to therapy (CR and PR), compared to just 7% of individuals treated with dacarbazine alone. Compared to the dacarbazine group, on an average duration of 49 days, the G3139 group experienced an improvement in progression-free survival of almost 50%, lasting up to 78 days. However, the FDA advisory committee did not suggest that Genasense be approved for commercialization.

There have been further Phase III trials for multiple myeloma and CLL. A Phase III randomized trial was recently published for patients with relapsed or refractory CLL, examining the effect of fludarabine/cyclophosphamide treatment with or without G3139. Compared to 7% of the 124 patients in the control arm, major reactions happened in 16% of the G3139 arm. The G3139 arm experienced worsening of the illness progression with a median progression time of 6.1 months, in contrast to 8.9 months in the control arm. In a Phase III randomized trial including patients with advanced multiple myeloma, neither dexamethasone nor G3139 reach statistical significance for the primary objective. G3139 has often shown moderate responses in CLL and melanoma but not in multiple myeloma. The FDA has not approved this medication to treat malignant melanoma (Orlandi et al. 1998).

#### 19.4.2.2 Antisense BCL-Xl

Although BCL-xl and BCL-2 share significant sequence homology areas, BCL-xl has a distinct biological function from BCL-2. These two proteins are overexpressed in numerous cancers. Preclinical development is ongoing for antisense BCL-xl oligonucleotide and bispecific antisense BCL-2/BCL-xl oligonucleotide, which target the mRNA homology region of both BCL-2 and BCL-xl. The bispecific antisense oligonucleotide reduces the growth of several tumor types, promotes apoptosis, and simultaneously downregulates the expression of BCL-2 and BCL-xl.

#### 19.4.2.3 Antisense Clusterin

It is a glycoprotein with widespread expression in various organs whose role has been connected to how cells react in response to stress. It has also been connected to several pathogenic events, such as apoptosis prevention. Two distinct forms of the protein, a cytoplasmic and a nuclear form, may be connected to clusterin's contradictory pro- and antiapoptotic actions. Due to its significant function, it has been demonstrated in both in vitro and in vivo models that inhibiting clusterin with sequence-specific antisense oligonucleotides enhances the effects of chemotherapy in hormone-refractory prostate cancer (Kang and Reynolds 2009).

#### 19.4.2.4 Small Molecules Recognizing Surface Pocket of BCL-2 or BCL-xl

Nuclear magnetic resonance spectroscopy and X-ray crystallography have identified the outer pocket on the antiapoptotic BCL-xl protein that acts together with the BH3 domain of BCL-2 family. Both natural and synthetic organic peptides can identify the surface pockets of BCL-xl and BCL-2. Antimycin-A and other random screening procedures are used to identify the natural products. These offer crucial models for creating particular inhibitors that target BCL-2 or additional intrinsic pathway proteins. Computer-aided design of the synthetic BH3 organic peptides has shown its ability to induce apoptosis in vitro. Newer compounds with smaller molecular sizes are being developed to improve biological half-life and cell membrane permeability. For instance, preclinical development is underway for HA14–1, a protein that interacts to a surface pocket on the BCL-2 protein (Ramanarayanan et al. 2004).

#### 19.4.2.5 Agents Modulating the Proapoptotic Proteins BCL-xs and Bax

Gene therapy is involved in the introduction of Bax vectors that activate the proapoptotic Bax protein. However, it is challenging to create adenoviral vectors expressing the Bax gene due to the proapoptotic nature of the gene. Recently, vectors have been created using various molecular approaches, and this method has proved as successful in causing cancer cell lines to undergo apoptosis. BCL-xs gene therapy has the potential to cause tumor regression (Nuno Moreira et al. 2006).

### 19.4.3 Agents That Target the Mitochondrial Inner Membrane

#### 19.4.3.1 Arsenic Trioxide

The PML-RAR protein is the target of arsenic trioxide. Increased quantities cause the leukemic cells to undergo apoptosis by upsetting the inner membrane potential of the mitochondria, which suppresses the production of BCL-2 and increases caspase-3. It promotes PML-RAR protein degradation at both concentrations and differentiation at lower quantities. The FDA has approved this medication for the treatment of APL. The research on nonsmall cell lung cancer, gliomas, breast cancer, and esophageal cancer is still ongoing. In contrast, the phase II trial results for adult T-cell leukemia/lymphoma, renal cell carcinoma, and multiple myeloma have been made accessible. The main side effects are leukocytosis, hypokalemia, hypomagnesemia, torsade de pointes, QT prolongation, and congestive heart failure (Trarbach et al. 2010).

#### 19.4.3.2 Lonidamine

This substance has been derived from indazole-3-carboxylic acid, which disrupts the inherent transmembrane potential in mitochondria to cause apoptosis. By preventing oxygen consumption and interfering with the energy metabolism of tumor cells, it has a strong antiproliferative effect on them. In human breast cancer cell lines, this medication has been demonstrated to enhance the cytotoxic effect of chemotherapeutic drugs, particularly anthracyclines. It enhances the chances of radiation therapy as well (Gieffers et al. 2013). One can take 450 mg of lonidamine orally daily. Hematologic and gastrointestinal side effects are most common from this medication. As a first-line treatment for MBC, a Phase II clinical trial combining lonidamine, epirubicin, and cisplatin has demonstrated a 73% overall response rate with 13% complete response for approximately 9 months. The OS rate of patients with MBC was not improved by lonidamine in combination with high-dose epirubicin and cyclophosphamide, according to a multicenter prospective randomized trial. In nonrandomized Phase II trials, lonidamine and cytotoxic chemotherapy have shown effective responses in treating advanced non-small cell lung cancer and ovarian cancer. Lonidamine has demonstrated minimal efficacy against nonsmall cell lung cancer or glioblastoma multiforme (GBM) in the absence of cytotoxic medications (Matsubara et al. 2007).

### 19.4.4 Agents Targeting Common Pathway: Caspases Activator

Caspase activator comprises IAP targets like survivin and synthetic caspase activators such as apoptin (Mariadason et al. 2023). Caspases are the cysteine proteases that inactive zymogens in cells and act as catalytic activators when apoptosis begins. Caspases are classified into three types such as executioner caspases (caspases 3, 6, and 7), initiator caspases (caspases 2, 8, 9, and 10), and inflammatory caspases (caspases 1, 4, 5, 11, and 12). When activated, they cause a biochemical cascade of events (Vineetha et al. 2020). These medications cause intracellular protein aggregation, triggering caspase activation and subsequent apoptosis. The activation of caspases 1 and 3 may cause cell death, and these agents are being studied in preclinical research (Yadav et al. 2021).

#### 19.4.4.1 Apoptin

Apoptin is one of the substances that activate caspase. It is a small molecular weight protein (Ferreira et al. 2002), which is generated from the chicken anemia virus and cause selective apoptosis in cancer cells but not in healthy cells (Liu et al. 2022). This selectivity could be attributed to the nucleus site in cancer cells, which is required for activation (Rohn and Noteborn 2004). Healthy cells have the protein localized in the cytoplasm. This medication is still in preclinical trials.

#### 19.4.4.2 Survivin

Survivin is an IAP family member that regulates cell division and reduces apoptosis. The specific role of survivin is still unknown; however, it may prevent apoptosis by directly or indirectly inhibiting caspases, illustrated in Fig. 19.4 (van der Eb et al. 2002). Baculovirus is the active domain of survivin responsible for inhibiting apoptosis (Johnson and Howerth 2004). It is expressed in embryonic tissues, most human malignancies, and some premalignant tissues, but not in healthy tissues. Selective

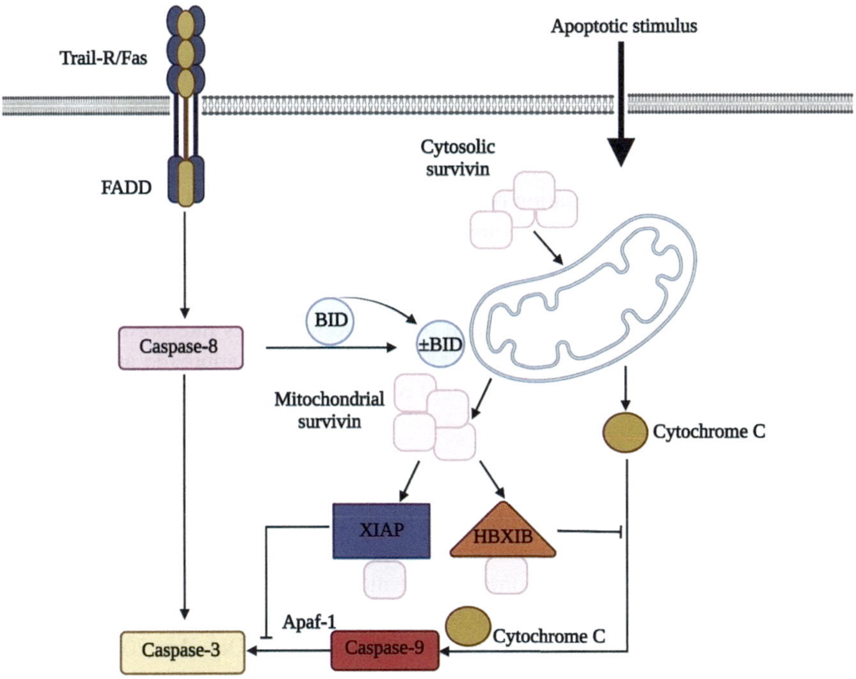


**Fig. 19.4** Diagrammatic representation of survivin as an inhibitor of apoptosis. The pathway represents how survivin is released into cytoplasm from mitochondria in response to cell death stimuli

expression and antiapoptotic action of survivin have made it a promising diagnostic marker and therapeutic target (Garg et al. 2016). It has been proven in both in vitro and in vivo experiments to cause apoptosis, inhibit tumor development, and make tumor cells more sensitive to chemotherapeutic treatments. In the past 10 years, few small-molecule survivin inhibitors have been discovered despite the exponential growth. It lacks an intrinsic catalytic activity and a drug conjugal site, making it difficult to target for direct binding (Yamamoto et al. 2008). Various preclinical and clinical studies have been summarized in Table 19.1.

### 19.4.5 Agents Targeting Modulators of the Apoptosis Pathways

These include proteasome, mTOR, and p53 inhibitors.

#### 19.4.5.1 Proteasome Inhibitors

The ubiquitin/proteasome system plays a crucial role in controlling cell proliferation and death and the turnover of regulatory proteins. Inhibitors have demonstrated in vivo anticancer activity, sensitize cells to apoptosis and potentially cause cell death (Albadari and Li 2023). Furthermore, they have been demonstrated to inhibit metastasis and angiogenesis (Rajkumar et al. 2005). As these medications promote cell apoptosis, their effects could be limited to cancer cells. For example, cancer cells appear more susceptible to apoptosis triggered by proteasome inhibitors when overexpressing the c-myc oncogene. Compared to nonproliferating cells, actively dividing cells are also more susceptible to proteasome blockage. This effect is not solely due to the rapid reproduction rate of malignant cells. Proteasome inhibitors disturb cell proceeds, causing normal cells to use checkpoint mechanisms to stop division and resume proliferation once the activity is restored. Malignant cells with defective checkpoint mechanisms are more sensitive to apoptosis triggered by proteasome inhibitors (Sanhueza et al. 2015). Proteasome inhibitors include natural products like lactacystin, peptide aldehydes like MG132, ALLN, and MG115, which are in the preclinical stages, and boric acid inhibitors like bortezomib (formerly PS341) (Murray and Norbury 2000), which is FDA-approved and currently undergoing clinical trials for the treatment of refractory multiple myeloma.

##### Bortezomib

The rate-limiting proteolytic step in the proteasome is chymotryptic threonine protease activity, specifically and potentially inhibited by the boronic acid inhibitor bortezomib. Bortezomib has been demonstrated to exhibit anticancer efficacy in various cancers, including myeloma, pancreatic cancer, CLL, prostate cancer, and colon cancer, in vitro and in vivo xenografts (Almond and Cohen 2002). Proteasome inhibition is reversible and dose-dependent (Takimoto and Diggikar 2002).

A Phase II multicenter trial in patients with relapsed and refractory multiple myeloma was conducted in response to its encouraging preclinical evidence and clinical performance in Phase I investigations. For a maximum of eight cycles, 100 Bortezomib was administered at 1.3 mg/m$^2$ on day 1, 4, 8, and 11 of 21-day cycle.

**Table 19.1** summarizes various clinical and preclinical studies related to the effectiveness and mechanism of action of IAP inhibitors. FOLFOX—chemotherapy regimen of 5-fluorouracil and oxaliplatin

| Drug | Phase | Study population | Summary | References |
|---|---|---|---|---|
| Tolinapant + pembrolizumab | Phase I | Multiple metastatic tumors | – | NCT05082259 |
| Tolinapant/Decitabine + Cedazuridine | Phase I/II | Relapsed or refractory peripheral T-cell lymphoma | – | NCT05403450 |
| Tolinapant(+capecitabine), followed by Mfolfirinox or CAPOX | Phase Ib | Metastatic rectal cancer | – | NCT05912075 |
| Xevinapant (+Avelumab) | Phase I | Multiple metastatic tumors such as NSCLC | Completed | NCT03270176 |
| Xevinapant + pembrolizumab | Phase I | Advanced pancreatic or colorectal cancer | Completed | NCT03871959 |
| Xevinapant + platinum-based chemotherapy and intensity modulated radiotherapy | Phase I | Head and neck cancer | Active, not recruiting manageable safety in combination with chemotherapy, got FDA approval in 2020 for its use in head and neck cancer in combination with cisplatin | NCT04459715 |
| Birinapant (TL32711) + intensity modulated radiation therapy | Phase I/II studies | Head and neck cancer | Terminated | NCT03803774 |
| APG-1387 (single agent) | Phase I | Advanced solid tumors and haematological cancers | Completed | NCT03386526 |
| APG-1387 + toripalimab | Phase I/II | Colorectal and NSCLC | Recruiting | NCT04284488 |
| APG-1387 + chemotherapy | Phase I/II | Advanced pancreatic cancer | Recruiting | NCT04643405 |

However, 92% of the 193 patients examined had received three or more prior treatments, and 91% had not responded to the most recent dosage of action (Herod et al. 1996). Bortezomib response rates were encouraging, with 27% of patients achieving a partial response and 4% achieving a complete response with a median response duration of 12 months. Patients with relapsed or refractory multiple myeloma were able to receive bortezomib from the FDA more quickly due to the success of clinical trials. Adverse effects were observed, including fever, tiredness, low platelet levels, peripheral neuropathy, and mild diarrhea. Patients who had relapsed following initial consolidation therapy were accepted in a second phase II trial utilizing 1.0 or 1.3 mg/m$^2$ dose. Responses were observed in 33% of patients receiving 1.0 mg/m$^2$ and 50% of those getting 1.3 mg/m$^2$ (Richardson et al. 2003). The initial studies allowed for a maximum of eight cycles of bortezomib. However, new findings indicate that providing at least five or six cycles of therapy is safe without causing significant adverse consequences. Phase III trial comparing pulsed dexamethasone treatment with bortezomib for patients with multiple myeloma was ended early due to long time to disease progression in bortezomib-treated patients. These outcomes show that bortezomib, used as a single agent, effectively treats multiple myeloma. Research combining bortezomib with other drugs is currently being under condition (Jagannath et al. 2004).

#### (κB NF) Inhibitor (IκB Kinase Inhibitor, PS-1145)

The NF-κB kinase inhibitor PS-1145 (Millenium Pharmaceuticals, Cambridge, MA) targets IKK, a key regulator of IB. PS-1145 has been shown in preclinical investigations to suppress the phosphorylation of IB in multiple myeloma cells and reduce its propagation. PS-1145 inhibits the adherence of multiple myeloma cells and the overexpression of adhesive molecules caused by NF-B (Wolf et al. 2004). Preclinical findings indicate that certain NF-B inhibitors may effectively treat multiple myeloma and other cancers with constitutively activated NF-B pathways. Furthermore, they propose that thalidomide/IMiDs can be used in combination with PS-1145 as a possible therapeutic approach to address medication resistance in multiple myeloma (Hideshima et al. 2002).

### 19.4.5.2 mTOR Inhibitors

The downstream element of the PI3K/Akt pathway is the mammalian target for mTOR. Targeted therapy may have a greater effect on cancer cells that depend on the PI3K/Akt pathway being activated or have mutations that result in constitutive activation of this system (Dai et al. 2004). For instance, overexpression of this pathway may result from the loss or inactivation of the phosphatase and tensin homolog, which has been demonstrated to raise susceptibility to the effect of CCI-779. The efficiency of mTOR inhibitors on tumor cells can also be influenced by other checkpoint regulators, including p16, p53, cyclin D, and p27 (Neshat et al. 2001).

#### CCI-779

Rapamycin is a natural compound with antibacterial, immunosuppressive, and anticancer effects and inhibits mTOR. It is an immunosuppressive medication that is licensed for use in patients undergoing kidney transplants (Dancey 2002). The National Cancer Institute tested it and discovered that it had anti-proliferative action against a range of B-lymphocytes and solid tumor cell lines. It has been demonstrated that CCI-779, a soluble rapamycin ester, is a potent cytostatic drug both in vitro and in vivo (Ghobrial et al. 2005). CCI-779 is presently being investigated in multiple clinical trials as a cancer therapy in an intravenous formulation. Since CCI-779 only inhibits tumor development in xenograft models rather than tumor regression, it could be a useful treatment to prolong the course of the disease or be used in conjunction with other anticancer therapies.

A recent Phase II research on refractory and relapsed cell lymphoma showed 38% response rate. In a different Phase II study, 26% of patients with metastatic-resistant renal cell carcinoma showed moderate responses, whereas 7% experienced a complete or partial response. Renal cell carcinoma is now under study in phase III trials (Atkins et al. 2004). Treatment for cancers of different origins, such as breast, cervical, kidney, prostate, head & neck, uterine, lungs, pancreatic, colon, ovarian, melanoma, and lymphoma, may benefit from CCI-779.

#### RAD001

Another rapamycin analog marketed as an oral formulation is RAD001 (everolimus). A Phase I trial of RAD001 showed that the drug was well tolerated orally at four doses like 5, 10, 20, and 30 mg once weekly. The adverse effects include fatigue, rash, mucositis, headache, hyperlipidemia, and gastrointestinal disturbance. Four patients experienced stable illness, while one showed improvement after three weeks of treatment (Swisher et al. 2003). When 20 mg of RAD001 is administered weekly to rodents on this schedule, the same pharmacokinetic and pharmacodynamic effects are observed, correlated with anticancer effects. Solid tumors and hematologic malignancies are the subjects of ongoing phase II clinical investigations employing RAD001.

### 19.4.5.3 P53 Inhibitors

Studies to target p53 as a therapeutic target for the treatment of cancer have been demonstrated in Fig. 19.5. This study includes antisense therapy for the protein that regulates p53 activity (MDM2) and gene therapy, such as ONYX-015 and INGN201 (Swisher et al. 2003).

#### Gene Therapy

Effective induction of apoptosis in tumor cells and surrounding cells has been demonstrated by phase I p53-based gene therapy trials that result in the return of wild-type p53 in cells. Gene therapy technology faces the challenge of infecting almost all cancer cells efficiently (Bhatt et al. 2010). Furthermore, in certain animal investigations, unchanged thymocytes perish in response to p53 activation, necessitating targeted transgene delivery. Replication-incompetent (rAd.p53, SCH5,8500) and replication-selective (ONYX-015, CG7,870) oncolytic adenoviruses are among the adenoviral agents under development. Intravascular infusion, intratumoral, or epitumoral injections are used to give agents. These drugs were reported to have an acceptable toxicity

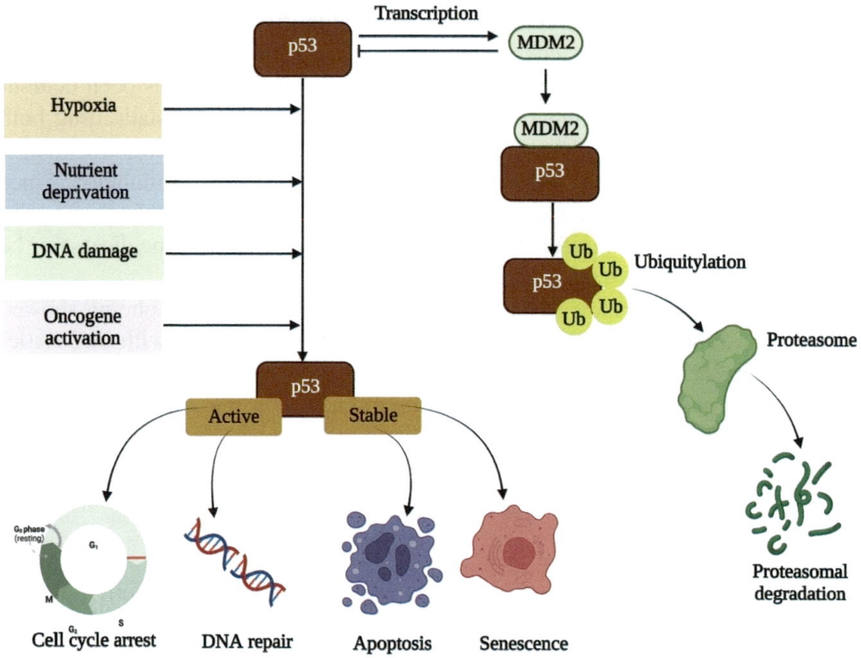


**Fig. 19.5** Diagrammatic representation of p53 as a therapeutic target for cancer treatment

profile in phase I and II studies. The toxicity profile included low-grade fever, dose-limiting cardiac output suppression (rAd.p53), and mild to moderate transaminitis in certain patients receiving intravenous or hepatic arterial infusions. Intravascular adenovirus constructs have been used to treat more than 100 cancer patients. A few patients displayed signs of viral replication or p53 gene expression (Reid et al. 2002).

### ONYX-015

Replication-competent ONYX-015 is a virus genetically engineered to multiply and lyse cancer cells lacking p53 exclusively. There is no impact of ONYX-015 on healthy cells. Both intratumoral and peritumoral injections are used to deliver it. Numerous tumor forms, including advanced head and neck cancer and pancreatic cancer, have been the subject of phase I and II trials. As of right now, this molecule is not progressing further.

### INGN 201

An inept adenovirus called INGN 201 expresses p53. Effective translation of the p53 gene into the p53 protein has been demonstrated in preclinical investigations using human cell lines and animals suffering from head and neck cancers. Numerous phase I and II clinical trials have been carried out, one of which had 19 patients with nonsmall cell, nonmetastatic lung cancer who were not candidates for surgery or chemotherapy. Patients underwent 60 Gy of radiation therapy and three intratumoral injections of INGN 201 over six weeks. The most frequent side effects of verse were chills and fevers in grades 1 and 2. Only 5% of patients had complete responses; the remaining 58% had partial responses; 16% had stable disease; 11% had progressive disease; and 11% could not provide an assessment. According to this trial, the initial injection location of INGN 201 showed tumor reduction when used in conjunction with radiation therapy, and the injection was well tolerated.

## 19.5 Nanoparticles Targeting Apoptosis

Inducing apoptosis either intrinsically through mitochondria or extrinsically by targeting cell surface receptors is one method for therapeutically destroying cancer cells (summarized in Table 19.2). In the extrinsic pathway, caspase 8 is triggered by activated cell DRs, initiating a caspase-signaling cascade. Within the mitochondrial route, procaspase 9 is activated by the multimeric Apaf-1/cytochrome c complex, which cleaves and activates caspases 3, 6, and 7. As a result, mitochondria are a desirable target for developing targeted, efficient cancer treatments that cause less collateral harm to the nearby non-cancerous cells. One current tactic under investigation is using the photothermal properties of gold nanoparticles (AuNPs) to trigger extrinsic or mitochondrial-mediated apoptosis.

Because of their surface plasmon resonance (SPR) increased light absorption and scattering, AuNPs have demonstrated enormous potential for cancer diagnosis and treatment. AuNPs effectively transform absorbed light into localized heat, a property that can be used in targeted laser photothermal cancer treatments. Moreover, the attachment of AuNPs to ligands that selectively target biomarkers on cancer cells enables photothermal therapy in addition to imaging and detection. Mkandawire et al. investigated selective targeting of AuNPs to mitochondria of Jimt-1 cancer cells to instigate apoptotic pathways. Transmission electron microscopy and fluorescence microscopy showed that AuNPs conjugates got transfected, and further guided to mitochondria, resulting in a partial breach of the outer membrane of mitochondria, which initiated cell death. The capacity to deliver AuNPs to the mitochondria of breast cancer cells and prompt apoptosis unveils a new potential use of AuNPs in cancer photothermal therapy (Mkandawire et al. 2015).

**Table 19.2** Surface functionalized nanoparticles targeting apoptosis

| Nanoparticle types | Size (nm) | Induction of cell apoptosis | Mechanism of action |
|---|---|---|---|
| AgNPs | 25–70 | Human lung fibroblast cells | Reactive oxygen species (ROS), mitochondrial dysfunction |
| Nano-$SiO_2$ | 15–30 | Human keratinocyte HaCaT cells | Oxidative stress |
| Nano-ZnO | 80 | Mice liver | Oxidative stress |
| Nano-$TiO_2$ | 21 | HaCaT cells | Inner pathway of mitochondria |
| SWCNTs | – | Rat Aorta endothelial cells (RAECs) | Mitochondrial pathways, ROS |
| Nano-Cu | 15 | Podocytes | Oxidative stress |
| $CeO_2$ NPs | 10–30 | Monocytes | Mitochondrial pathways |

## 19.6 Conclusions

Caspases are essential for inducing apoptosis, which is thought to be an energy-dependent mechanism linked to several biochemical and morphological traits. Pro- and antiapoptotic proteins, among other apoptotic proteins, are also implicated in the control of apoptosis. Knowing how apoptosis works is essential because it balances cell death and survival. It has been proposed that any abnormalities in the apoptotic pathway contribute to tumors, and a number of viable cancer treatment strategies are currently being researched to target apoptosis. Thus, investigating and learning about the cellular and molecular mechanisms of apoptosis and related proteins would be beneficial in understanding the various disease processes brought on by apoptosis and, ultimately, in developing therapeutic strategies. Apoptosis resistance appears to be one of the main causes of cancer. Therefore, targeting apoptotic pathways in cancer treatment is vital from a therapeutic standpoint. Thus, appealing anticancer treatments include blocking other related cell death mechanisms of apoptosis and focusing on the intrinsic and extrinsic routes of apoptosis. For the treatment of various cancers, several medications are presently being explored in clinical trials, such as single therapy or combined with chemotherapy, radiation therapy, or other inhibitors. However, because several pathways interfere with apoptosis and contribute to the development of cancer resistance, it is complicated to modify apoptotic pathways using targeted medicines. Therefore, applying logical combination techniques may be essential to clinical success.

## References

Albadari N, Li W (2023) Survivin small molecules inhibitors: recent advances and challenges. Molecules 28(3):1376

Almond J, Cohen G (2002) The proteasome: a novel target for cancer chemotherapy. Leukemia 16(4):433–443

Ashkenazi A (2008) Targeting the extrinsic apoptosis pathway in cancer. Cytokine Growth Factor Rev 19(3–4):325–331

Atkins MB et al (2004) Randomized phase II study of multiple dose levels of CCI-779, a novel mammalian target of rapamycin kinase inhibitor, in patients with advanced refractory renal cell carcinoma. J Clin Oncol 22(5):909–918

Bazzaro M et al (2006) Ubiquitin-proteasome system stress sensitizes ovarian cancer to proteasome inhibitor–induced apoptosis. Cancer Res 66(7):3754–3763

Bellas RE et al (1997) Inhibition of NF-kappa B activity induces apoptosis in murine hepatocytes. Am J Pathol 151(4):891

Bernal-Mizrachi L, Lovly CM, Ratner L (2006) The role of NF-κB-1 and NF-κB-2-mediated resistance to apoptosis in lymphomas. Proc Natl Acad Sci 103(24):9220–9225

Bhatt AN et al (2010) Cancer biomarkers-current perspectives. Indian J Med Res 132(2):129–149

Chang G-Q, Hao Y, Wong F (1993) Apoptosis: final common pathway of photoreceptor death in rd, rds, and mutant mice. Neuron 11(4):595–605

Cheah CY, Belada D, Fanale MA, Janikova A, Czucman MS, Flinn IW, Kapp AV, Ashkenazi A, Kelley S, Bray GL, Holden S (2015) Dulanermin with rituximab in patients with relapsed indolent B-cell lymphoma: an open-label phase 1b/2 randomised study. Lancet Haematol 2(4):e166–e174

Cohen GM (1997) Caspases: the executioners of apoptosis. Biochem J 326(1):1–16
Dai Y et al (2004) Interruption of the NF-κB pathway by bay 11-7082 promotes UCN-01-mediated mitochondrial dysfunction and apoptosis in human multiple myeloma cells. Blood 103(7):2761–2770
Dancey JE (2002) Clinical development of mammalian target of rapamycin inhibitors. Hematol/Oncol Clin 16(5):1101–1114
Del Bufalo D et al (1996) Lonidamine induces apoptosis in drug-resistant cells independently of the p53 gene. J Clin Invest 98(5):1165–1173
Duronio V (2008) The life of a cell: apoptosis regulation by the PI3K/PKB pathway. Biochem J 415(3):333–344
Elumalai P et al (2012) Induction of apoptosis in human breast cancer cells by nimbolide through extrinsic and intrinsic pathway. Toxicol Lett 215(2):131–142
Ferreira CG et al (2002) Apoptosis: target of cancer therapy. Clin Cancer Res 8(7):2024–2034
Franke TF et al (2003) PI3K/Akt and apoptosis: size matters. Oncogene 22(56):8983–8998
Frankfurt OS et al (1996) Monoclonal antibody to single-stranded DNA is a specific and sensitive cellular marker of apoptosis. Exp Cell Res 226(2):387–397
Fulda S (2015) Targeting extrinsic apoptosis in cancer: challenges and opportunities. In: Seminars in cell and developmental biology, vol 39. Elsevier, p 20
Garg H et al (2016) Survivin: a unique target for tumor therapy. Cancer Cell Int 16:1–14
Ghobrial IM, Witzig TE, Adjei AA (2005) Targeting apoptosis pathways in cancer therapy. CA Cancer J Clin 55(3):178–194
Gieffers C, Kluge M, Merz C, Sykora J, Thiemann M, Schaal R, Fischer C, Branschädel M, Abhari BA, Hohenberger P, Fulda S (2013) APG350 induces superior clustering of TRAIL receptors and shows therapeutic antitumor efficacy independent of cross-linking via Fcγ receptors. Mol Cancer Ther 12(12):2735–2747
Haupt S et al (2003) Apoptosis-the p53 network. J Cell Sci 116(20):4077–4085
Herbst RS, Kurzrock R, Hong DS, Valdivieso M, Hsu CP, Goyal L, Juan G, Hwang YC, Wong S, Hill JS, Friberg G (2010) A first-in-human study of conatumumab in adult patients with advanced solid tumors. Clin Cancer Res 16(23):5883–5891
Herod JJO et al (1996) The prognostic significance of Bcl-2 and p53 expression in ovarian carcinoma. Cancer Res 56(9):2178–2184
Hideshima T et al (2002) NF-κB as a therapeutic target in multiple myeloma. J Biol Chem 277(19):16639–16647
Hongmei Z (2012) Extrinsic and intrinsic apoptosis signal pathway review, in apoptosis and medicine. InTechOpen
Jagannath S et al (2004) A phase 2 study of two doses of bortezomib in relapsed or refractory myeloma. Br J Haematol 127(2):165–172
Johnson M, Howerth E (2004) Survivin: a bifunctional inhibitor of apoptosis protein. Vet Pathol 41(6):599–607
Kang MH, Reynolds CP (2009) Bcl-2 inhibitors: targeting mitochondrial apoptotic pathways in cancer therapy. Clin Cancer Res 15(4):1126–1132
King KL, Cidlowski JA (1995) Cell cycle and apoptosis: common pathways to life and death. J Cell Biochem 58(2):175–180
Kristen AV et al (2013) Inhibition of apoptosis by the intrinsic but not the extrinsic apoptotic pathway in myocardial ischemia-reperfusion. Cardiovasc Pathol 22(4):280–286
Kucharczak J et al (2003) To be, or not to be: NF-κB is the answer–role of Rel/NF-κB in the regulation of apoptosis. Oncogene 22(56):8961–8982
Legler K, Hauser C, Egberts JH, Willms A, Heneweer C, Boretius S, Röcken C, Glüer CC, Becker T, Kluge M, Hill O (2018) The novel TRAIL-receptor agonist APG350 exerts superior therapeutic activity in pancreatic cancer cells. Cell Death Dis 9(5):445
Léveillé F et al (2010) Suppression of the intrinsic apoptosis pathway by synaptic activity. J Neurosci 30(7):2623–2635
Li J, Yuan J (2008) Caspases in apoptosis and beyond. Oncogene 27(48):6194–6206

Li JJ et al (2006) Role of oxidative stress in the apoptosis of hepatocellular carcinoma induced by combination of arsenic trioxide and ascorbic acid 1. Acta Pharmacol Sin 27(8):1078–1084

Liu Z et al (2022) Apoptin induces pyroptosis of colorectal cancer cells via the GSDME-dependent pathway. Int J Biol Sci 18(2):717

Mangiarotti R et al (1998) All-trans retinoic acid (ATRA)-induced apoptosis is preceded by G1 arrest in human MCF-7 breast cancer cells. Br J Cancer 77(2):186–191

Mariadason J et al (2023) BCL-XL inhibitors enhance the apoptotic efficacy of BRAF inhibitors in BRAFV600E colorectal cancer

Matsubara H et al (2007) Apoptosis induction by antisense oligonucleotides against miR-17-5p and miR-20a in lung cancers overexpressing miR-17-92. Oncogene 26(41):6099–6105

Mitra D et al (1998) Role of caspases 1 and 3 and Bcl-2-related molecules in endothelial cell apoptosis associated with thrombotic microangiopathies. Am J Hematol 59(4):279–287

Mkandawire MM, Lakatos M, Springer A, Clemens A, Appelhans D, Krause-Buchholz U, Pompe W, Rödel G, Mkandawire M (2015) Induction of apoptosis in human cancer cells by targeting mitochondria with gold nanoparticles. Nanoscale 7(24):10634–10640

Mohana-Kumaran N et al (2014) Targeting the intrinsic apoptosis pathway as a strategy for melanoma therapy. Pigm Cell Melanoma Res 27(4):525–539

Murray RZ, Norbury C (2000) Proteasome inhibitors as anti-cancer agents. Anti-Cancer Drugs 11(6):407–417

Nair P et al (2014) Apoptosis initiation through the cell-extrinsic pathway. In: Methods in enzymology. Elsevier, pp 99–128

Neshat MS et al (2001) Enhanced sensitivity of PTEN-deficient tumors to inhibition of FRAP/mTOR. Proc Natl Acad Sci 98(18):10314–10319

Nuno Moreira J, Santos A, Simões S (2006) Bcl-2-targeted antisense therapy (Oblimersen sodium): towards clinical reality. Rev Recent Clin Trials 1(3):217–235

Orlandi L et al (1998) Lonidamine as a modulator of taxol activity in human ovarian cancer cells: effects on cell cycle and induction of apoptosis. Int J Cancer 78(3):377–384

Orlowski RZ (1999) The role of the ubiquitin-proteasome pathway in apoptosis. Cell Death Differ 6(4):303–313

Ouyang X, Shi M, Jie F, Bai Y, Shen P, Yu Z, Wang X, Huang C, Tao M, Wang Z, Xie C (2018) Phase III study of dulanermin (recombinant human tumor necrosis factor-related apoptosis-inducing ligand/Apo2 ligand) combined with vinorelbine and cisplatin in patients with advanced non-small-cell lung cancer. Investig New Drugs 36:315–322

Patel P et al (2021) A doxorubicin–platinum conjugate system: impacts on PI3K/AKT actuation and apoptosis in breast cancer cells. RSC Adv 11(8):4818–4828

Piotrowska A et al (2008) The structure of NF-κB family proteins and their role in apoptosis. Adv Hyg Exp Med:62

Rajkumar SV et al (2005) Proteasome inhibition as a novel therapeutic target in human cancer. J Clin Oncol 23(3):630–639

Ramanarayanan J et al (2004) Pro-apoptotic therapy with the oligonucleotide GenasenseTM (oblimersen sodium) targeting Bcl-2 protein expression enhances the biological anti-tumour activity of rituximab. Br J Haematol 127(5):519–530

Ratain MJ, Doi T, De Jonge MJ, LoRusso P, Dunbar M, Chiney M, Motwani M, Glasgow J, Petrich AM, Rasco DW, Calvo E. Phase 1, first-in-human study of TRAIL receptor agonist fusion protein ABBV-621

Reid T, Warren R, Kirn D (2002) Intravascular adenoviral agents in cancer patients: lessons from clinical trials. Cancer Gene Ther 9(12):979–986

Richardson PG et al (2003) A phase 2 study of bortezomib in relapsed, refractory myeloma. N Engl J Med 348(26):2609–2617

Rohn J, Noteborn M (2004) The viral death effector Apoptin reveals tumor-specific processes. Apoptosis 9:315–322

Sanhueza C et al (2015) The twisted survivin connection to angiogenesis. Mol Cancer 14:1–15

Sayers TJ (2011) Targeting the extrinsic apoptosis signaling pathway for cancer therapy. Cancer Immunol Immunother 60:1173–1180

Soares F et al (2014) The mitochondrial protein NLRX1 controls the balance between extrinsic and intrinsic apoptosis. J Biol Chem 289(28):19317–19330
Stehlik C et al (1998) Cytokine induced expression of porcine inhibitor of apoptosis protein (iap) family member is regulated by NF-κB. Biochem Biophys Res Commun 243(3):827–832
Swisher SG et al (2003) Induction of p53-regulated genes and tumor regression in lung cancer patients after intratumoral delivery of adenoviral p53 (INGN 201) and radiation therapy. Clin Cancer Res 9(1):93–101
Takimoto CH, Diggikar S (2002) Heat shock protein and proteasome targeting agents. Hematol/Oncol Clin 16(5):1269–1285
Tang Y et al (2004) Bax is required for resection-induced changes in apoptosis, proliferation, and members of the extrinsic cell death pathways. Gastroenterology 126(1):220–230
Trarbach T, Moehler M, Heinemann V, Köhne CH, Przyborek M, Schulz C, Sneller V, Gallant G, Kanzler S (2010) Phase II trial of mapatumumab, a fully human agonistic monoclonal antibody that targets and activates the tumour necrosis factor apoptosis-inducing ligand receptor-1 (TRAIL-R1), in patients with refractory colorectal cancer. Br J Cancer 102(3):506–512
Uno K et al (2003) TNF-related apoptosis-inducing ligand (TRAIL) frequently induces apoptosis in Philadelphia chromosome–positive leukemia cells. Blood J Am Soc Hematol 101(9):3658–3667
van der Eb MM et al (2002) Gene therapy with apoptin induces regression of xenografted human hepatomas. Cancer Gene Ther 9(1):53–61
Vineetha RC et al (2020) L-ascorbic acid and α-tocopherol synergistically triggers apoptosis inducing Antileukemic effects of arsenic trioxide via oxidative stress in human acute Promyelocytic leukemia cells. Front Oncol 10:65
Wang E, Marcotte R, Petroulakis E (1999) Signaling pathway for apoptosis: a racetrack for life or death. J Cell Biochem 75(S32):95–102
Wawryk-Gawda E et al (2014) P53 protein in proliferation, repair and apoptosis of cells. Protoplasma 251:525–533
Wójcik C (2002) Regulation of apoptosis by the ubiquitin and proteasome pathway. J Cell Mol Med 6(1):25–48
Wolf JK et al (2004) A phase I study of Adp53 (INGN 201; ADVEXIN) for patients with platinum- and paclitaxel-resistant epithelial ovarian cancer. Gynecol Oncol 94(2):442–448
Wu C-C, Bratton SB (2013) Regulation of the intrinsic apoptosis pathway by reactive oxygen species. Antioxid Redox Signal 19(6):546–558
Yadav P et al (2021) Caspase-3: a primary target for natural and synthetic compounds for cancer therapy. Chem Biol Drug Des 98(1):144–165
Yamamoto H, Ngan CY, Monden M (2008) Cancer cells survive with survivin. Cancer Sci 99(9):1709–1714
Yanumula, A. and J.K. Cusick, Biochemistry, extrinsic pathway of apoptosis. 2020
Zhang H et al (2003) All-trans retinoic acid (atRA) differentially induces apoptosis in matched primary and metastatic melanoma cells–a speculation on damage effect of atRA via mitochondrial dysfunction and cell cycle redistribution. Carcinogenesis 24(2):185–191

# Apoptosis and Cancer Immunotherapy: Current Mechanisms and Emerging Strategies

# 20

Satyajit Halder and Kuladip Jana

**Abstract**

Apoptotic pathways are essential for maintaining cellular balance and removing damaged or malignant cells. Apoptosis activation in tumor cells is a crucial process in cancer immunotherapy that leads to treatment effectiveness. Immunotherapy has significantly transformed cancer treatment by utilizing immune checkpoint blockage, bispecific antibodies, and adoptive T cell transfer. These approaches have produced remarkable clinical outcomes in both blood cancers and solid tumors. T cell-based immunotherapies seek to trigger apoptosis in cancer cells, but the ability of cancer cells to avoid apoptosis is a characteristic feature of cancer biology. Increasing the susceptibility of cancer cells to programmed cell death, known as apoptosis, is a crucial approach for enhancing the effectiveness of medical treatments. Recent progress in comprehending apoptotic signaling pathways has opened up new possibilities for innovative immunotherapeutic approaches. These improvements involve improving T cell-based therapies by making cancer cells more susceptible to death and addressing the paradox of apoptosis in T cells, which can weaken immunotherapy. This chapter examines the existing apoptosis methods in cancer immunotherapy, focusing on developing approaches such as immune checkpoint inhibitors, CAR-T cell therapy, and innovative small chemicals. Our goal is to analyze the complex relationship between immune responses and cell death signaling to present a detailed summary of the most advanced methods and potential future developments in this quickly changing area of research. Comprehending these pathways is crucial for advancing more efficient and specific cancer immunotherapies. The ultimate goal is to promote the survival of cytotoxic T cells in the

S. Halder · K. Jana (✉)
Division of Molecular Medicine, Bose Institute, Unified Academic Campus, Kolkata, West Bengal, India
e-mail: kuladip@jcbose.ac.in



K. Jana (ed.), *Apoptosis and Human Health: Understanding Mechanistic and Therapeutic Potential*, https://doi.org/10.1007/978-981-97-7905-5_20

tumor microenvironment and strengthen the body's immune response against tumors.

**Keywords**

Apoptosis · Cancer · Immunology · Immunotherapy · T cell-based therapies · Tumor microenvironment

## 20.1 Introduction

The triggering of apoptosis, or programmed cell death, is a fundamental biological process needed to maintain cellular integrity and prevent uncontrolled cell proliferation. Apoptotic pathways are highly conserved and entail a cascade of molecular events that lead to the methodical disassembly and elimination of cells. In cancer, the evasion of apoptosis is a characteristic that permits tumor cells to survive and proliferate despite genetic instability and therapeutic interventions (Hanahan and Weinberg 2011).

Cancer immunotherapies have altered immunooncology by leveraging the immune system to fight cancer and have delivered extraordinary success in patients with relapsed and resistant cancers (Chen and Mellman 2013). Immunotherapeutic techniques such as immune checkpoint inhibitors (ICIs) and chimeric antigen receptor (CAR)-T cell treatment have shown tremendous promise in treating various cancers. Central to the success of these therapies is their capacity to trigger apoptosis in cancer cells, hence increasing immune-mediated tumor elimination.

ICIs, such as those targeting the PD-1/PD-L1 and CTLA-4 pathways, relax the brakes on T cells, allowing for a more robust immune response against cancer cells (Pardoll 2012). This reactivation of T cells often induces apoptosis in tumor cells by engaging death receptors and activating proapoptotic signaling cascades (Sharma and Allison 2015). Another strategy to improve T cell detection of malignancy is using bispecific antibodies, like blinatumomab, which was approved for B-acute lymphoblastic leukemia (B-ALL) in 2014 (Topp et al. 2011). CAR-T cell therapy provides another new strategy wherein patient-derived T cells are genetically engineered to express receptors targeting tumor antigens. Upon attaching to their targets, CAR-T cells deliver lethal signals that promote apoptosis in cancer cells. Despite the extraordinary clinical responses obtained with these medicines, difficulties such as resistance to apoptosis and immune evasion mechanisms continue to hamper their efficacy. Despite their effectiveness, many individuals resist checkpoint inhibitors (Sharma and Allison 2015).

Anticancer drugs, such as immunotherapies based on T cells, try to destroy cancer cells by triggering apoptosis, a process known as programmed cell death (Kroemer et al. 2009). One crucial element of cancer progression is the development of resistance in cancer cells to apoptosis (Hanahan and Weinberg 2011). As per Tsujimoto et al. (1984), genetic alterations in antiapoptotic regulators, such as B-cell lymphoma 2 (BCL-2), are frequent in different lymphomas, allowing malignant B cells to endure harsh conditions. Apoptosis plays a vital part in cancer development since resistance to it works as a barrier against immunotherapies based on T cells (Adams and Cory 2007). According to Elmore (2007), apoptosis can be brought on by extrinsic sources such as death ligands or intrinsic factors, including DNA damage and reactive oxygen species. T lymphocytes use both methods to drive cancer cells to undergo apoptosis. The inherent apoptotic process is triggered when T cells release cytolytic granules containing granzymes and perforin in response to contact with T cell receptors. Granzymes penetrate target cells and induce proapoptotic proteins, which alter the potential of the mitochondrial membrane and activate caspases, ultimately beginning apoptosis (Trapani and Smyth 2002). In the extrinsic pathway, initiator caspases are activated, and death-inducing signaling complexes (DISCs) are generated when T cell-derived death ligands bind to receptors on target cells (Ashkenazi 2008) (Fig. 20.1).

The immune system has provided cancer cells with strategies to resist apoptosis. For instance, lower immune function genes in gastric cancer are connected to mutations in the tumor suppressor gene TP53 (Soussi 2011). Antiapoptotic and proapoptotic apoptotic regulator alteration impacts cancer immunotherapies' efficiency. For example, poor responses to CAR-T treatment are connected with genetic modifications in BCL-2 (Fisher et al. 2013). Research has shown that deleting proapoptotic

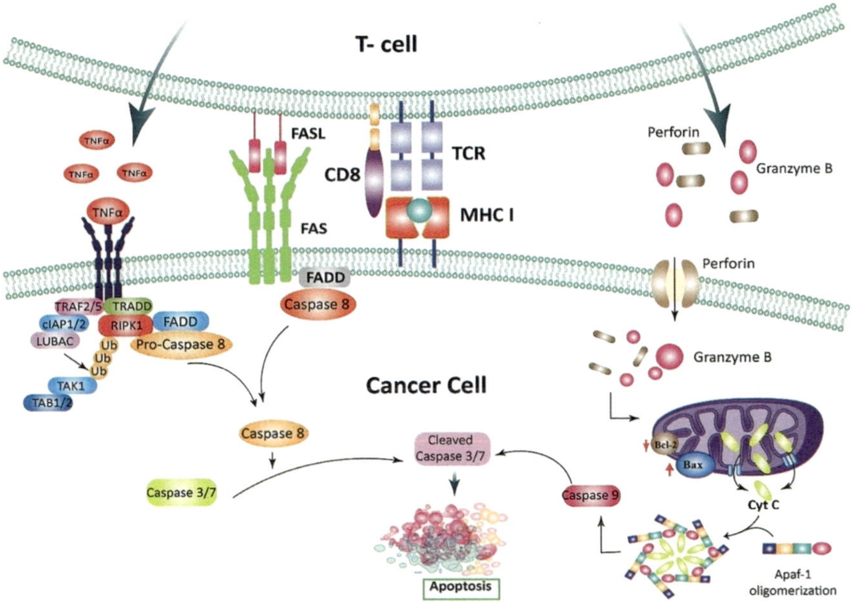


**Fig. 20.1** T cells induce apoptosis in cancer cells through intrinsic and extrinsic pathways. In the intrinsic pathway, granzymes are delivered into cancer cells via perforin. Granzymes cleave BID to produce truncated BID (tBID), which inhibits antiapoptotic proteins like BCL-2, BCL-xL, and MCL-1. This leads to BAK/BAX dimers forming on the mitochondrial membrane, releasing cytochrome C and Apaf-1 into the cytoplasm: Cytochrome C, Apaf-1, and Caspase-9 form apoptosomes, triggering apoptosis

regulators diminishes CAR-T cell efficacy, but eliminating antiapoptotic regulators boosts it (Michels et al. 2021). Emerging techniques aim to address these issues by generating innovative drugs that can directly influence apoptotic pathways. Small compounds that mimic proapoptotic proteins or inhibit antiapoptotic factors are being explored for their potential to sensitize cancer cells to immunotherapy. Additionally, combining immunotherapy with traditional chemotherapy and radiotherapy, promoting apoptosis, may have synergistic effects and enhance patient outcomes. To build logical approaches to boost the effectiveness of T cell-based immunotherapies, it is necessary to appreciate the factors in the tumor microenvironment (TME) that affect T cell death (Whiteside 2013).

In this chapter, we will analyze the existing methods by which apoptosis is employed in cancer immunotherapy and address developing options that potentially enhance the efficiency of these treatments. By understanding the relationship between apoptotic pathways and immunological effector systems, we intend to provide insights into developing more effective and targeted cancer immunotherapies.

## 20.2 Integrating Immunotherapy with Targeted Approaches

Despite the notable success of T cell-based immunotherapies, many patients do not benefit from these treatments (Charo et al. 2005). A critical factor affecting the effectiveness of these therapies is the cancer cells' resistance to apoptosis. Researchers have explored various therapeutic strategies to overcome this resistance (Ricci and Zong 2006). One approach combines conventional anticancer treatments, such as chemotherapy and radiotherapy, with immunotherapy to enhance cancer cell apoptosis. Chemotherapeutic agents (e.g., alkylating agents, anthracyclines, vinca alkaloids, and antimetabolites) and radiation (e.g., X-rays) are known to induce intrinsic apoptosis in cancer cells by causing DNA damage and disrupting the cell cycle (Ricci and Zong 2006). Moreover, certain chemotherapies (e.g., etoposide, doxorubicin, and 5-fluorouracil) and radiation have been shown to modulate death receptor-mediated extrinsic apoptosis by influencing the transcriptional activity of death receptors like Fas and TNFRSF10 families (Mattarollo et al. 2006; Todaro et al. 2013).

For instance, etoposide has been found to increase the expression of DR5 and Fas in various cancer cell lines, enhancing natural killer T cell-mediated killing (Mattarollo et al. 2006). Sublethal doses of doxorubicin upregulate TRAIL receptors on cancer cells, promoting cytotoxicity mediated by natural killer cells and tumor-infiltrating lymphocytes (TILs) (Wennerberg et al. 2013). Doxorubicin also reduces the expression of intracellular FLICE inhibitory protein (c-FLIP), sensitizing cancer cells to immune cell-mediated extrinsic apoptosis (Wennerberg et al. 2013). Similarly, 5-fluorouracil has been reported to upregulate DR5 expression in colon cancer-initiating cells, enhancing T cell-mediated cytotoxicity (Todaro et al. 2013). Radiation can induce death receptor-mediated apoptosis by causing DNA damage and intracellular stress. Sublethal radiation doses have been shown to upregulate Fas and DR5 expression in colorectal carcinoma cell lines, making them more susceptible to TRAIL-induced and Fas-induced apoptosis (Ifeadi and Garnett-Benson 2012). Additionally, sublethal radiation has been found to increase Fas expression in over 40% of various cancer cell lines, including colon, lung, and prostate cancers (Garnett et al. 2004).

These findings suggest that combining radiation with T cell-based immunotherapy could enhance the sensitivity of cancer cells to extrinsic apoptosis. Researchers have also explored the use of recombinant anti-Fas and anti-DR4/5 agonists with radiation and chemotherapeutic agents. Treatment with recombinant TRAIL, in combination with bortezomib, vorinostat (SAHA), and valproic acid, has been shown to significantly induce cancer cell apoptosis by sensitizing cancer cells to extrinsic apoptosis signals (Brooks et al. 2010; Kim et al. 2012; Koschny et al. 2014; Seol 2011).

Recent studies highlight the potential of combining T cell-based immunotherapy with conventional cancer treatments like chemotherapy and radiation to enhance cancer cell susceptibility to apoptosis. Chemotherapy agents such as etoposide, doxorubicin, and 5-fluorouracil, and radiation therapy can promote intrinsic and extrinsic apoptosis in cancer cells by inducing DNA damage and modulating death receptor activity (Mattarollo et al. 2006; Wennerberg et al. 2013; Todaro et al. 2013; Ifeadi and Garnett-Benson 2012; Garnett et al. 2004). Clinical trials have shown that combining immunotherapy with chemotherapy or radiation improves outcomes in metastatic squamous nonsmall cell lung cancer (NSCLC). This leads to FDA approval for such combinations (Paz-Ares et al. 2018; Reck et al. 2020). However, chemotherapy and radiation can also increase T cell apoptosis, with memory T cells potentially evading this effect due to high BCL-2 and low Bcl-2-like 11 (BIM) expression (Grayson et al. 2000; Sabbagh et al. 2006).

In addition to chemotherapy and radiation, targeting inhibitors of apoptosis proteins (IAPs), such as cIAP1, cIAP2, xIAP, NAIP, livin, and survivin, offers another strategy to sensitize cancer cells to apoptosis (Berthelet and Dubrez 2013; Fulda and Vucic 2012; Scott et al. 2005). SMAC mimetics like birinapant have shown promise in promoting IAP degradation, enhancing the efficacy of T cell-based therapies. Birinapant combined with ICIs or CAR-T therapy has significantly increased TNF-α-mediated apoptosis and improved survival in preclinical models of glioblastoma and other cancers (Michie et al. 2019; Beug et al. 2017; Song et al. 2022). Birinapant's ability to enhance CAR-T cell-mediated killing by sensitizing cancer cells to TNF-mediated apoptosis underscores its potential in immunotherapy combinations (Song et al. 2022). ASTX660, another IAP antagonist, has shown efficacy in head and neck squamous cell carcinoma (HNSCC) models by inducing immunogenic cell death (ICD) and enhancing antigen processing machinery (Ye et al. 2020).

Despite promising results, understanding T cell resistance mechanisms to chemotherapy and radiation remains crucial. Memory T cells' resistance to apoptosis suggests a pivotal role in developing effective combination strategies (Arina et al. 2019; Shibayama et al. 2017). Ongoing research focuses on elucidating these mechanisms to optimize immunotherapy outcomes. Additionally, the direct inhibition of apoptotic regulators, such as IAPs, continues to be an important area of investigation. The development of SMAC mimetics and other agents targeting IAPs highlights the evolving landscape of cancer treatment, where combining immunotherapy with traditional and novel approaches aims to improve patient outcomes (Morrish et al. 2020). Further preclinical and clinical studies are needed to validate these findings and ensure the safety and efficacy of such combination therapies in diverse cancer types (Fig. 20.2).

The BCL-2 family, including members such as BCL-2, BCL-XL, MCL-1, BAX, and BAK, are critical regulators of intrinsic apoptosis by modulating mitochondrial membrane permeability (Adams and Cory 2007). Overexpression or translocation of BCL-2 is a hallmark of various cancers (Correia et al. 2015). Several BCL-2 inhibitors, such as venetoclax, have been developed to counteract this effect. Venetoclax, an orally available small-molecule inhibitor specific to BCL-2, has shown significant efficacy in treating chronic lymphocytic leukemia (CLL), other lymphomas, and acute myeloid leukemia, leading to its FDA approval (Souers et al. 2013). Recent studies suggest that venetoclax can enhance the anticancer efficacy of ICIs, such as anti-PD-1 antibodies, by increasing the infiltration of effector memory T cells into tumors, indicating a potential synergistic role with proapoptotic drugs (Kohlhapp et al. 2021). Combining venetoclax with CAR-T cell therapy has demonstrated significant improvements in anticancer activity against lymphoma and

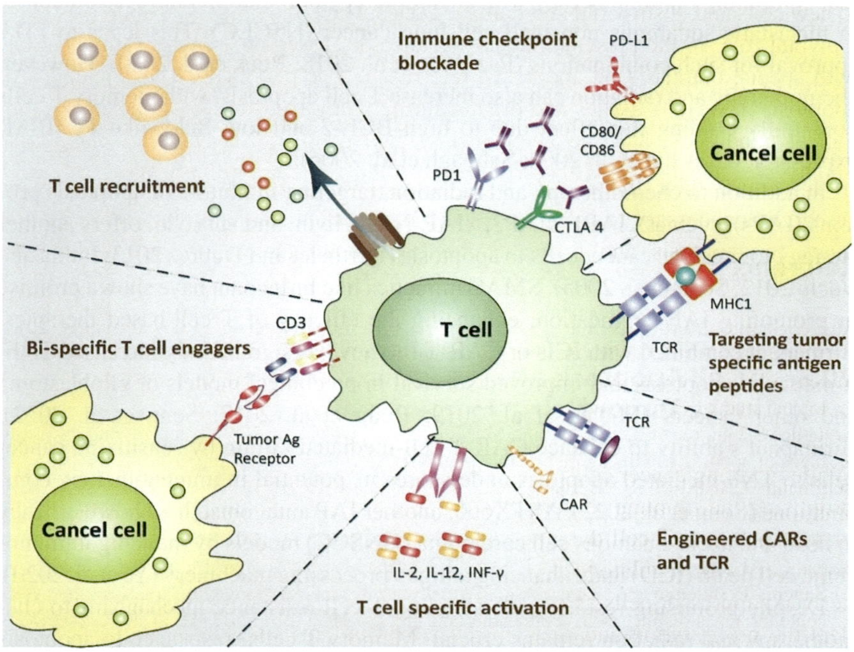


**Fig. 20.2 Advancements in T cell immunotherapy:** Schematic representation detailing the multifaceted approaches in T cell immunotherapy. The image illustrates immune checkpoint blockade, recruitment of T cells to the tumor site, targeting tumor-specific antigens with engineered CARs and TCRs, and using bispecific T cell engagers for precise immune activation against cancer cells

leukemia xenograft models, suggesting that BCL-2 inhibitors can enhance the efficacy of T cell-based immunotherapies (Yang et al. 2020). ABT737, another BCL-2 inhibitor, also showed enhanced cancer cell killing when combined with CAR-T cells (Karlsson et al. 2013). However, a significant concern with this approach is the potential toxicity of these agents on effector immune cells, such as T cells. For instance, SMAC mimetic LBW242 significantly inhibited virus-specific CD8+ T cell expansion in vivo by inducing T cell apoptosis, leading to the failure of virus replication (Gentle et al. 2014). Similarly, venetoclax reduced the viability of both genetically unmodified T cells and CAR-T cells in coculture studies, highlighting the need for careful design and sequencing of combination therapies to avoid T cell toxicity and ensure long-term therapeutic efficacy (Kohlhapp et al. 2021; Yang et al. 2020).

A promising strategy to mitigate these side effects is to pretreat cancer cells with cytotoxic drugs before administering T cell-based therapies. Clinical studies have shown that lymphoma patients who received venetoclax during bridging therapy before CAR-T cell infusion had significantly improved clinical responses compared to those who did not receive venetoclax (Yang et al. 2020)—in vitro studies corroborated these findings, demonstrating that pretreatment with venetoclax enhanced the anticancer activity of CAR-T cells while reducing their toxicity (Yang et al. 2020). These results suggest that pre-sensitizing cancer cells with anti-apoptotic inhibitors could enhance the efficacy of T cell-based immunotherapy while minimizing adverse effects on T cells. Thus, while proapoptotic molecules can significantly improve the anticancer response of T cell-based immunotherapies, the potential toxicity to effector immune cells remains a critical concern. Careful consideration of the timing and sequencing of these combination therapies is essential to maximize their therapeutic benefits and minimize adverse effects. Further research is necessary to fully understand the mechanisms underlying T cell resistance to apoptosis and optimize combination strategies for improved cancer treatment outcomes.

## 20.3 T Cell Apoptosis and Anticancer Immunity in the Tumor Microenvironment

While targeting cancer cells, cancer therapy also induces apoptosis in T cells, compromising the immune response. T cell apoptosis can result from various mechanisms, including T cell dysfunction and the immunosuppressive TME. The TME, populated with regulatory T cells, tumor-associated macrophages, and myeloid-derived suppressor cells and characterized by a scarcity of essential nutrients, hinders the proliferation and survival of cytotoxic T cells (Bates et al. 2006; Curiel et al. 2004; Rolf et al. 2013; Sinclair and Ramanathan 2013).

Cancer cells exploit T cell vulnerabilities through activation-induced cell death (AICD). Upon activation, T cells upregulate proapoptotic proteins (e.g., FasL, TRAILs, and TNF) and death receptors (e.g., Fas, TRAIL receptors, and TNF receptor), making them more susceptible to apoptosis (Croft 2009; Green et al.

2003). The FasL expression by cancer cells serves as an immunoevasion strategy, inducing apoptosis in T cells via the extrinsic pathway, as observed in various malignancies, including melanoma, colon, head/neck, liver, and lung cancers (Hahne et al. 1996; Niehans et al. 1997; O'Connell et al. 1996; Strand et al. 1996). Exosome-mediated apoptosis is another mechanism wherein cancer cells secrete exosomes expressing FasL, independently inducing T cell apoptosis and causing peripheral T cell dysfunction. This mechanism amplifies FasL-mediated T cell apoptosis and has been linked to resistance to immunotherapies in melanoma mouse models, including anti-CTLA-4 and anti-PD-1 antibodies, as well as adoptive cell transfer (ACT) (Andreola et al. 2002; Kim et al. 2005; Taylor et al. 2003; Taylor and Gerçel-Taylor 2005; Zhu et al. 2017). CAR-T cell therapies, which incorporate costimulatory domains like CD28 and 4-1BB, may increase susceptibility to apoptosis by upregulating death receptors and associated ligands. Tonic signaling from 4-1BB costimulation, driven by strong promoters such as retroviral long terminal repeats, leads to elevated FasL levels, resulting in CAR-T cell apoptosis upon activation. This underscores the need to modulate CAR activation in T cells instead of merely boosting it (June et al. 2018).

The TME contains factors that can directly trigger T cell apoptosis, including galectins. Galectin-3 (Gal-3) and Galectin-1 (Gal-1), produced by cancer and immune cells, regulate apoptosis by binding to β-galactosides on glycoproteins and glycolipids (Liu and Rabinovich 2005; Leffler et al. 2002). Gal-3 can induce T cell apoptosis in cancers such as melanoma, lung, and colorectal cancer by binding to T cell receptors like CD7, CD29, CD45, and CD71 (Fukumori et al. 2003; Li et al. 2010; Peng et al. 2008; Zubieta et al. 2006). Mechanistically, Gal-3 binds to CD45, activating pathways involving protein kinase C and ROS, leading to sustained ERK 1/2 phosphorylation, Caspase-9 activation, cytochrome c release, and Caspase-3 activation (Xue et al. 2017). Gal-1 in the TME is also linked to increased cancer progression following immune checkpoint blockade (ICB) therapy, potentially through T cell apoptosis via a CD45-dependent mechanism (Banh et al. 2011). However, this correlation varies across cancers, with elevated Gal-1 levels correlating with T cell apoptosis in pancreatic and lung cancers but not in melanoma or in vitro against activated primary T cells (Tang et al. 2012; Stowell et al. 2010).

Gangliosides, sialic acid-containing glycosphingolipids overexpressed in cancers and shed into the TME, induce T cell apoptosis by being internalized by activated T cells, leading to ROS production, cytochrome c release and activation of Caspases-8 and Caspase-9, thus promoting both intrinsic and extrinsic apoptosis (Sa and Das 2009). Gangliosides facilitate intrinsic apoptosis by inducing ROS, cytochrome c release, Caspase-9 activation, and downregulating anti-apoptotic BCL genes like BCL-XL and BCL-2. Metabolic pathways also play significant roles in T cell apoptosis. For instance, glucose deprivation reduces the proliferation of Jurkat cells and primary human T cells in vitro, potentially increasing intrinsic apoptosis (Alves et al. 2006). In the TME, competition for glucose uptake between T cells and cancer cells increases T cell susceptibility to apoptosis, impairing anticancer activity (Chang et al. 2015). Fatty acid metabolism is crucial for T cell-mediated anticancer activity, with fatty acid oxidation essential for maintaining memory phenotype. Inhibiting fatty acid synthase reduces FasL expression, preventing restimulation-induced cell death (Voss et al. 2019). Cancer cell metabolites like kynurenine, a tryptophan metabolite, can induce T cell apoptosis (Fallarino et al. 2002). Acidic and hypoxic stress in the TME, resulting from cancer cells' aerobic glycolysis (Warburg effect), also contribute to T cell apoptosis. Long-term exposure to acidic pH causes permanent damage and T cell apoptosis (Calcinotto et al. 2012). Hypoxia induces T cell apoptosis through increased NO production, Caspase-9 activation, cytochrome c release, and Caspase-3 activation (Kiang and Tsen 2008).

## 20.4 Strategies to Enhance T Cell Survival and Combat Apoptosis in the Tumor Microenvironment

Various strategies have been developed to prevent T cell apoptosis to ensure the long-term survival and functionality of T cells, which are crucial for the efficacy of cancer immunotherapies. One such approach by Yamamoto et al. (2019) involves engineering CAR-T cells to inhibit FasL-mediated T cell apoptosis by truncating the intracellular death domain of Fas or introducing a point mutation (I246N) in the Fas death domain. These modifications prevent the recruitment of FADD into the apoptotic complex, thereby inhibiting DISC formation. As a result, these engineered CAR-T cells resist cancer-induced FasL-mediated apoptosis, enhance persistence, and exhibit increased anticancer activity in murine B16 melanoma models without showing uncontrolled proliferation in vivo (Yamamoto et al. 2019).

Similarly, CRISPR-mediated knockout (KO) of Fas has been shown to reduce AICD of anti-CD19 CAR-T cells during chronic exposure to target cells, leading to increased T cell expansion (Ren et al. 2017). Generated murine T cells overexpressing BCL-2, which demonstrated superior anticancer activity compared to wild-type T cells by improving long-term survival without survival signals. Another study indicated that the constitutive overexpression of BCL-2 in CAR-T cells enhances proliferation and reduces AICD, suggesting a significant improvement in CAR-T cell persistence and overall survival in patients with lymphoma (Wang et al. 2021).

Other critical antiapoptotic regulators, such as MCL-1 and BCL-xL, also play essential roles in T cell survival and differentiation. For instance, constitutive expression of BCL-xL can rescue AICD of CD8+ T cells in viral infection models (Petschner et al. 1998). Enhanced expression of MCL-1 has been shown to promote long-term memory formation during the acute phase of vaccinia virus infections (Gui et al. 2015). Despite these benefits, the constitutive overexpression of BCL-2 in murine T cells has been associated with an increased risk of T cell lymphoma development (Linette et al. 1995).

Moreover, strategies to avoid T cell apoptosis are essential when combining immunotherapies with pro-apoptotic drugs. Our group recently developed venetoclax-resistant CAR-T cells (ven-CAR-T) by introducing a mutant form of BCL-2 with a point mutation at the 104 amino acid residue (F104L), which prevents venetoclax binding and its inhibitory function (Tahir et al. 2017; Birkinshaw et al.

2019). Consequently, overexpressing BCL-2(F104L) in ven-CAR-T resulted in significant resistance to venetoclax, enhancing the combined effects of CAR-T cells and venetoclax (Tahir et al. 2017; Birkinshaw et al. 2019).

## 20.5 Conclusion

As immunotherapy continues to advance as a frontline treatment for cancer, overcoming the limitations that hinder its efficacy in a substantial subset of patients remains paramount. This review has explored the dual role of apoptosis in T cell-based immunotherapy, addressing both cancer's resistance to apoptosis and the challenges of T cell survival. Most cancers evade apoptosis through various mechanisms, such as upregulating antiapoptotic proteins like IAPs and BCL-2 and downregulating the positive regulators of death receptor-mediated apoptosis. Enhancing cancer cell susceptibility to apoptosis is crucial for augmenting the efficacy of T cell-based therapies. Strategies combining proapoptotic agents with T cell therapies show promise. Targeting key antiapoptotic factors, such as IAPs and BCL-2, with small molecules like SMAC mimetics and ABT737 has significantly boosted CAR-T cell activity against cancer cells. Nevertheless, these agents' potential induction of T cell apoptosis necessitates careful dosage and timing optimization to avoid adverse effects.

Furthermore, the TME presents significant barriers to T cell function by secreting pro-apoptotic factors and creating a hostile milieu. Effective immunotherapy combinations are needed to counter these evasion mechanisms. CAR-T cell therapy offers a promising platform for innovative strategies, including genetic modifications such as Fas knockout, mutant Fas, or overexpression of BCL-2, to enhance T cell survival and expansion. However, these modifications raise concerns regarding lymphoproliferation and tumorigenesis, underscoring the need for safety mechanisms like the anti-inducible Caspase-9 system and antibody-mediated cellular cytotoxicity using truncated epidermal growth factor receptors.

In summary, apoptosis plays a pivotal role in T cell-based cancer immunotherapy. A comprehensive understanding of the mechanisms underlying apoptosis resistance in cancer and T cell biology is essential to promote cancer cell apoptosis while safeguarding T cell integrity. Developing novel agents, combined with advances in bioengineering, will be instrumental in the success of next-generation therapeutic strategies. Ongoing research and clinical trials will be crucial in refining these approaches, ensuring their safety, and enhancing their effectiveness in treating a broader range of cancers.

## References

Adams JM, Cory S (2007) The bcl-2 apoptotic switch in cancer development and therapy. Oncogene 26:1324–1337

Alves NL, Hooibrink B, Arosa FA, van Lier RA (2006) Glucose deprivation affects proliferation and survival of human CD8+ T cells independently of glucose uptake. J Leukoc Biol 79(4):1105–1112

Andreola G, Rivoltini L, Castelli C, Huber V, Perego P, Deho P, Squarcina P, Accornero P, Lozupone F, Lugini L, Stringaro A, Molinari A, Arancia G, Gentile M, Parmiani G, Fais S (2002) Induction of lymphocyte apoptosis by tumor cell secretion of FasL-bearing microvesicles. J Exp Med 195(10):1303–1316

Arina A, Beckett M, Fernandez C et al (2019) Tumor-reprogrammed resident T cells resist radiation to control tumors. Nat Commun 10:3959

Ashkenazi A (2008) Directing cancer cells to self-destruct with pro-apoptotic receptor agonists. Nat Rev Drug Discov 7(12):1001–1012

Banh A, Zhang J, Cao H, Bouley DM, Kwok S, Kong C, Giaccia AJ, Koong AC (2011) Tumor galectin-1 mediates tumor growth and metastasis through regulation of T-cell apoptosis. Cancer Res 71(13):4423–4431

Bates GJ, Fox SB, Han C, Leek RD, Garcia JF, Harris AL, Banham AH (2006) Quantification of regulatory T cells enables the identification of high-risk breast cancer patients and those at risk of late relapse. J Clin Oncol 24(34):5373–5380

Berthelet J, Dubrez L (2013) Regulation of apoptosis by inhibitors of apoptosis (IAPs). Cells 2:163–187

Beug ST, Beauregard CE, Healy C et al (2017) Smac mimetics synergize with immune checkpoint inhibitors to promote tumour immunity against glioblastoma. Nat Commun 8:1–15

Birkinshaw RW, Gong J-N, Luo CS et al (2019) Structures of Bcl-2 in complex with venetoclax reveal the molecular basis of resistance mutations. Nat Commun 10:2385

Brooks AD, Jacobsen KM, Li W et al (2010) Bortezomib sensitizes human renal cell carcinomas to TRAIL apoptosis through increased activation of caspase-8 in the death-inducing signaling complex. Mol Cancer Res 8:729–738

Calcinotto A, Filipazzi P, Grioni M, Iero M, De Milito A, Ricupito A, Cova A, Canese R, Jachetti E, Rossetti M, Heltai S, Marigo I, Portararo P, Parmiani G, Generoso L, Botti P, Fais S, Rivoltini L, Bellone M (2012) Modulation of microenvironment acidity reverses anergy in human and murine tumor-infiltrating T lymphocytes. Cancer Res 72(11):2746–2756

Chang CH, Qiu J, O'Sullivan D, Buck MD, Noguchi T, Curtis JD, Chen Q, Gindin M, Gubin MM, van der Windt GJ, Tonc E, Schreiber RD, Pearce EJ, Pearce EL (2015) Metabolic competition in the tumor microenvironment is a driver of cancer progression. Cell 162(6):1229–1241

Charo J, Finkelstein SE, Grewal N et al (2005) Bcl-2 overexpression enhances tumor-specific T-cell survival. Cancer Res 65:2001–2008

Chen DS, Mellman I (2013) Oncology meets immunology: the cancer-immunity cycle. Immunity 39(1):1–10

Correia C, Schneider PA, Dai H et al (2015) Bcl2 mutations are associated with increased risk of transformation and shortened survival in follicular lymphoma. Blood 125:658–667

Croft M (2009) The role of TNF superfamily members in T-cell function and diseases. Nat Rev Immunol 9(4):271–285

Curiel TJ, Coukos G, Zou L, Alvarez X, Cheng P, Mottram P, Evdemon-Hogan M, Conejo-Garcia JR, Zhang L, Burow M, Zhu Y, Wei S, Kryczek I, Daniel B, Gordon A, Myeroff L, Liu Y, Jaime-Ramirez A, David O et al (2004) Specific recruitment of regulatory T cells in ovarian carcinoma fosters immune privilege and predicts reduced survival. Nat Med 10(9):942–949

Elmore S (2007) Apoptosis: a review of programmed cell death. Toxicol Pathol 35(4):495–516

Fallarino F, Grohmann U, You S, McGrath BC, Cavener DR, Vacca C, Orabona C, Bianchi R, Belladonna ML, Volpi C, Santamaria P, Fioretti MC, Puccetti P (2002) The combined effects of tryptophan starvation and tryptophan metabolites on T and natural killer cell proliferation. J Immunol 168(10):4918–4924

Fisher R, Pusztai L, Swanton C (2013) Cancer heterogeneity: implications for targeted therapeutics. Br J Cancer 108(3):479–485

Fukumori T, Takenaka Y, Yoshii T, Kim HR, Hogan V, Inohara H, Kagawa S, Raz A (2003) CD29 and CD7 mediate galectin-3-induced type II T-cell apoptosis. Cancer Res 63(23):8302–8311

Fulda S, Vucic D (2012) Targeting IAP proteins for therapeutic intervention in cancer. Nat Rev Drug Discov 11:109–124
Garnett CT, Palena C, Chakarborty M et al (2004) Sublethal irradiation of human tumor cells modulates phenotype resulting in enhanced killing by cytotoxic T lymphocytes. Cancer Res 64:7985–7994
Gentle IE, Moelter I, Lechler N et al (2014) Inhibitors of apoptosis proteins (IAPs) are required for effective T-cell expansion/survival during antiviral immunity in mice. Blood 123:659–668
Grayson JM, Zajac AJ, Altman JD et al (2000) Cutting edge: increased expression of Bcl-2 in antigen-specific memory CD8+ T cells. J Immunol 164:3950–3954
Green DR, Droin N, Pinkoski M (2003) Activation-induced cell death in T cells. Immunol Rev 193(1):70–81
Gui J, Hu Z, Tsai C-Y et al (2015) MCL1 enhances the survival of CD8+ memory T cells after viral infection. J Virol 89:2405–2414
Hahne M, Rimoldi D, Schröter M, Romero P, Schreier M, French LE, Schneider P, Bornand T, Fontana A, Lienard D, Cerottini JC, Tschopp J (1996) Melanoma cell expression of Fas(Apo-1/CD95) ligand: implications for tumor immune escape. Science 274(5291):1363–1366
Hanahan D, Weinberg RA (2011) Hallmarks of cancer: the next generation. Cell 144(5):646–674
Ifeadi V, Garnett-Benson C (2012) Sub-lethal irradiation of human colorectal tumor cells imparts enhanced and sustained susceptibility to multiple death receptor signaling pathways. PLoS One 7:e31762
June CH, O'Connor RS, Kawalekar OU, Ghassemi S, Milone MC (2018) CAR T cell immunotherapy for human cancer. Science 359(6382):1361–1365
Karlsson H, Lindqvist AC, Fransson M et al (2013) Combining CAR T cells and the Bcl-2 family apoptosis inhibitor ABT-737 for treating B-cell malignancy. Cancer Gene Ther 20:386–393
Kiang JG, Tsen KT (2008) Biological functions and medical applications of ribonucleases. Crit Rev Oncol Hematol 68(2):97–104
Kim SH, Bianco N, Harper DR, Fewell JG (2005) Exosomes as novel therapeutic agents for drug delivery. Expert Opin Biol Ther 5(1):1159–1169
Kim H-B, Kim M-J, Lee S-H et al (2012) Amurensin G, a novel SIRT1 inhibitor, sensitizes TRAIL-resistant human leukemic K562 cells to TRAIL-induced apoptosis. Biochem Pharmacol 84:402–410
Kohlhapp FJ, Haribhai D, Mathew R et al (2021) Venetoclax increases intratumoral effector T cells and antitumor efficacy in combination with immune checkpoint blockade. Cancer Discov 11:68–79
Koschny R, Boehm C, Sprick MR et al (2014) Bortezomib sensitizes primary meningioma cells to TRAIL-induced apoptosis by enhancing formation of the death-inducing signaling complex. J Neuropathol Exp Neurol 73:1034–1046
Kroemer G, Galluzzi L, Brenner C (2009) Mitochondrial membrane permeabilization in cell death. Physiol Rev 87(1):99–163
Leffler H, Carlsson S, Hedlund M, Qian Y, Poirier F (2002) Introduction to galectins. Glycoconj J 19(7–9):433–440
Li L, Liu H, Yue GG (2010) Galectin-3 mediates oxidative stress-induced apoptosis of VSMCs through mitochondria-dependent caspase cascade. Mol Cell Biochem 344(1–2):167–173
Linette G, Hess J, Sentman C et al (1995) Peripheral T-cell lymphoma in lckpr-bcl-2 transgenic mice. Blood 86:1255–1260
Liu FT, Rabinovich GA (2005) Galectins as modulators of tumour progression. Nat Rev Cancer 5(1):29–41
Mattarollo SR, Kenna T, Nieda M et al (2006) Chemotherapy pretreatment sensitizes solid tumor-derived cell lines to V alpha 24+ NKT cell-mediated cytotoxicity. Int J Cancer 119:1630–1637
Michels J, Johnson PW, Packham G (2021) Mcl-1. Int J Biochem Cell Biol 33(1):29–49
Michie J, Beavis PA, Freeman AJ et al (2019) Antagonism of IAPs enhances CAR T-cell efficacy. Cancer Immunol Res 7:183–192
Morrish E, Brumatti G, Silke J (2020) Future therapeutic directions for Smac-mimetics. Cells 9:406

Niehans GA, Brunner T, Frizelle SP, Liston JC, Salerno CT, Knapp DJ, Green DR, Kratzke RA (1997) Human lung carcinomas express Fas ligand. Cancer Res 57(6):1007–1012
O'Connell J, O'Sullivan GC, Collins JK, Shanahan F (1996) The Fas counterattack: Fas-mediated T cell killing by colon cancer cells expressing Fas ligand. J Exp Med 184(3):1075–1082
Pardoll DM (2012) The blockade of immune checkpoints in cancer immunotherapy. Nat Rev Cancer 12(4):252–264
Paz-Ares L, Luft A, Vicente D et al (2018) Pembrolizumab plus chemotherapy for squamous non-small-cell lung cancer. N Engl J Med 379:2040–2051
Peng W, Li W, Zhao W, Gao W, Xiao X (2008) Galectin-3 expression in CD4+ T cells improves immune tolerance in liver transplantation. Transplantation 86(1):79–87
Petschner F, Zimmerman C, Strasser A et al (1998) Constitutive expression of Bcl-xL or Bcl-2 prevents peptide antigen-induced T cell deletion but does not influence T cell homeostasis after a viral infection. Eur J Immunol 28:560–569
Reck M, Ciuleanu T-E, Dols MC et al (2020) Nivolumab (nivo) + ipilimumab (ipi) + 2 cycles of platinum-doublet chemotherapy (chemo) vs 4 cycles chemo as first-line (1L) treatment (tx) for stage IV/recurrent non-small cell lung cancer (NSCLC): CheckMate 9LA. J Clin Oncol 38:9501
Ren J, Zhang X, Liu X et al (2017) A versatile system for rapid multiplex genome-edited CAR T cell generation. Oncotarget 8:17002–17011
Ricci MS, Zong W-X (2006) Chemotherapeutic approaches for targeting cell death pathways. Oncologist 11:342–357
Rolf J, Bell SE, Kovesdi D, Janas ML, Soond DR, Webb LM, Santinelli S, Saunders T, Hebeis B, Killeen N, Okkenhaug K, Turner M (2013) Phosphoinositide 3-kinase activity in T cells regulates the magnitude of the germinal center reaction. J Immunol 190(9):5136–5145
Sa G, Das T (2009) Gangliosides in cancer apoptosis. Prog Mol Biol Transl Sci 90:205–256
Sabbagh L, Srokowski CC, Pulle G et al (2006) A critical role for TNF receptor-associated factor 1 and Bim down-regulation in CD8 memory T cell survival. Proc Natl Acad Sci 103:18703–18708
Scott FL, Denault J-B, Riedl SJ et al (2005) Xiap inhibits caspase-3 and -7 using two binding sites: evolutionarily conserved mechanism of IAPs. EMBO J 24:645–655
Seol D-W (2011) P53-independent up-regulation of a TRAIL receptor DR5 by proteasome inhibitors: a mechanism for proteasome inhibitor-enhanced TRAIL-induced apoptosis. Biochem Biophys Res Commun 416:222–225
Sharma P, Allison JP (2015) The future of immune checkpoint therapy. Science 348(6230):56–61
Shibayama Y, Tsukahara T, Emori M et al (2017) Implication of chemoresistant memory T cells for immune surveillance in patients with sarcoma receiving chemotherapy. Cancer Sci 108:1739–1745
Sinclair C, Ramanathan S (2013) Immunotherapy for ovarian cancer: are we there yet? J Clin Oncol 31(16):2103–2105
Song EZ, Wang X, Philipson BI et al (2022) The IAP antagonist birinapant enhances chimeric antigen receptor T cell therapy for glioblastoma by overcoming antigen heterogeneity. Mol Ther Oncolytics 27:288–304
Souers AJ, Leverson JD, Boghaert ER et al (2013) ABT-199, a potent and selective Bcl-2 inhibitor, achieves antitumor activity while sparing platelets. Nat Med 19:202–208
Soussi T (2011) p53 alterations in human cancer: more questions than answers. Nat Rev Cancer 11(10):743–753
Stowell SR, Arthur CM, Dias-Baruffi M, Rodrigues LC, Gourdine JP, Heimburg-Molinaro J, Ju T, Molinaro RJ, Rivera-Marrero CA, Xia B, Smith DF, Cummings RD (2010) Innate immune lectins kill bacteria expressing blood group antigen. Nat Med 16(3):295–301
Strand S, Hofmann WJ, Grambihler A, Hug H, Volkmann M, Otto G, Wesch H, Mariani SM, Hack V, Neuberger M, Waldmann T, Galle PR, Krammer PH (1996) Hepatic failure and liver cell damage in acute Wilson's disease involve CD95 (APO-1/Fas) mediated apoptosis. Nat Med 2(5):512–517
Tahir SK, Smith ML, Hessler P et al (2017) Potential mechanisms of resistance to venetoclax and strategies to circumvent it. BMC Cancer 17:399

Tang D, Kang R, Coyne CB, Zeh HJ, Lotze MT (2012) PAMPs and DAMPs: signal 0s that spur autophagy and immunity. Immunol Rev 249(1):158–175
Taylor DD, Gerçel-Taylor C (2005) Tumour-derived exosomes and their role in cancer-associated T-cell signalling defects. Br J Cancer 92(2):305–311
Taylor DD, Zacharias W, Gercel-Taylor C (2003) Exosome isolation for proteomic analyses and RNA profiling. Methods Mol Biol 728:235–246
Todaro M, Orlando V, Cicero G et al (2013) Chemotherapy sensitizes colon cancer initiating cells to Vγ9Vδ2 T cell-mediated cytotoxicity. PLoS One 8:e65145
Topp MS, Gökbuget N, Stein AS, Zugmaier G, O'Brien S, Bargou RC et al (2011) Safety and activity of blinatumomab for adult patients with relapsed or refractory B-precursor acute lymphoblastic leukaemia: a multicentre, single-arm, phase 2 study. Lancet Oncol 16(1):57–66
Trapani JA, Smyth MJ (2002) Functional significance of the perforin/granzyme cell death pathway. Nat Rev Immunol 2(10):735–747
Tsujimoto Y, Finger LR, Yunis J, Nowell PC, Croce CM (1984) Cloning of the chromosome breakpoint of neoplastic B cells with the t (14; 18) chromosome translocation. Science 226(4678):1097–1099
Voss K, Li J, Ly D, Griffin R, Mansurov A, Melder R, Dranoff G, Doench JG, Root DE, Wong KK (2019) Metabolic priming by CD47 antibody leads to enhanced responses to immune checkpoint blockade. Cold Spring Harbor Laboratory
Wang H, Han P, Qi X et al (2021) Bcl-2 enhances chimeric antigen receptor T cell persistence by reducing activation-induced apoptosis. Cancers (Basel) 13:197
Wennerberg E, Sarhan D, Carlsten M et al (2013) Doxorubicin sensitizes human tumor cells to NK cell- and T-cell-mediated killing by augmented TRAIL receptor signaling. Int J Cancer 133:1643–1652
Whiteside TL (2013) The tumor microenvironment and its role in promoting tumor growth. Oncogene 27(45):5904–5912
Xue H, Liu S, Meng Q (2017) Galectin-3-induced apoptosis of activated T cells in lung cancer cell lines. Oncol Lett 13(5):3659–3665
Yamamoto TN, Lee P-H, Vodnala SK et al (2019) T cells genetically engineered to overcome death signaling enhance adoptive cancer immunotherapy. J Clin Invest 129:1551–1565
Yang M, Wang L, Ni M et al (2020) Pre-sensitization of malignant B cells through venetoclax significantly improves the cytotoxic efficacy of CD19.CAR-T cells. Front Immunol 11:608167
Ye W, Gunti S, Allen CT et al (2020) ASTX660, an antagonist of cIAP1/2 and XIAP, increases antigen processing machinery and can enhance radiation-induced immunogenic cell death in preclinical models of head and neck cancer. OncoImmunology 9:1710398
Zhu X, Lang J, Solis LM, Zimmel R, Komaki R, Cox JD, Dorsey JH, Court LE, Shen Y (2017) Lymphocyte apoptosis induced by radiation in patients with non-small-cell lung cancer. Int J Radiat Oncol Biol Phys 98(3):742–749
Zubieta MR, Furman D, Barrio M et al (2006) Galectin-3 expression correlates with apoptosis of tumor-associated lymphocytes in human melanoma biopsies. Am J Pathol 168:1666–1675

# Summary and Future Perspectives

Programmed cell death, particularly apoptosis, is a cornerstone of biology, governing vital processes from development and tissue homeostasis to the pathogenesis of various diseases. Initially characterized in *C. elegans*, the mechanisms regulating apoptosis have provided foundational insights into human health and disease. Evolutionary conserved proteins such as the caspase and BCL-2 families are central to these processes, which orchestrate the intricate pathways of cell death. In mammals, these proteins regulate developmental apoptosis and play critical roles in responding to cellular stress and disease conditions. The dysregulation of apoptosis is implicated in diverse pathological contexts, including cancer and neurodegenerative disorders, highlighting its importance in biomedical research and therapeutic development.

Apoptosis, often viewed as a double-edged sword, presents both challenges and opportunities in therapeutic targeting. In cancer, understanding and manipulating apoptotic pathways have emerged as promising strategies for treatment. Targeting cancer-favored prosurvival mechanisms through genetic interventions or pharmacological inhibition offers avenues for developing more effective therapies. However, the complexity of apoptotic pathways necessitates a nuanced approach. For instance, in cancers driven by mutations in p53 or overexpression of BCL-2, effector caspases may remain inactive despite their presence. Simply blocking caspases could lead to cellular dormancy rather than effective elimination, underscoring the need to address apoptosis not merely as a process but as a symptom of underlying cellular dysregulation.

The evolving landscape of cancer therapy highlights the challenges associated with targeting cell death pathways. Therapeutic resistance, heterogeneous responses across different cancers, and potential unintended toxicities pose significant hurdles. Overcoming these challenges requires a paradigm shift in therapeutic strategies, emphasizing integrating diverse approaches and exploring synergistic combinations based on robust preclinical and clinical data. Integrative methodologies such as liquid biopsies, single-cell analyses, advanced imaging techniques, and comprehensive molecular profiling (genomics and proteomics) are pivotal for elucidating resistance mechanisms and optimizing therapeutic outcomes.





K. Jana (ed.), *Apoptosis and Human Health: Understanding Mechanistic and Therapeutic Potential*, https://doi.org/10.1007/978-981-97-7905-5

Recently, targeting cell death pathways has expanded beyond apoptosis to include alternative forms of regulated cell death, such as necroptosis, pyroptosis, and ferroptosis. These pathways offer new avenues for therapeutic intervention, particularly in diseases where apoptosis is compromised or insufficient. Necroptosis, for example, has garnered attention for its role in inflammatory diseases and its potential as a mechanism for cancer cell eradication. Similarly, pyroptosis, driven by inflammatory caspases, contributes to immune responses and has implications for infectious diseases and cancer immunotherapy. Understanding the molecular underpinnings of these pathways opens novel opportunities for developing targeted therapies that exploit specific vulnerabilities in diseased cells while sparing normal tissues.

In the realm of neurodegenerative disorders, where cellular stress plays a central role, the focus shifts to identifying and mitigating the triggers of neuronal dysfunction. Apoptosis and related cell death pathways are implicated in the progression of diseases such as Alzheimer's and Parkinson's, where aberrant protein aggregation and mitochondrial dysfunction contribute to neuronal demise. Therapeutic strategies aim to enhance neuronal survival, promote the clearance of toxic aggregates, and restore cellular homeostasis. Innovative approaches, including developing selective pharmacological agents and biomarkers, are critical for monitoring treatment efficacy and guiding personalized therapeutic regimens.

The development and clinical translation of cell death-targeting therapies are further bolstered by advances in drug delivery systems, particularly nanoparticle-based technologies. These systems offer enhanced bioavailability, targeted delivery, and reduced systemic toxicity, addressing longstanding challenges in effectively delivering therapeutic agents. Continued refinement of these technologies, coupled with identifying biomarkers predictive of treatment response, holds promise for realizing the full therapeutic potential of cell death-targeting therapies in clinical settings.

Recent research has explored novel elements in regulating and manipulating apoptosis and other cell death pathways. For instance, studies have identified specific microRNAs that regulate apoptosis by targeting essential proteins in the path, offering the potential for new therapeutic interventions. Additionally, the role of long noncoding RNAs (lncRNAs) in apoptosis regulation is an emerging field, with evidence suggesting that lncRNAs can modulate the expression of apoptosis-related genes and influence cell death outcomes.

Another exciting area of research is the interplay between cell death pathways and the immune system. Immune checkpoint inhibitors, which have revolutionized cancer therapy, are known to influence apoptosis and other cell death pathways in tumor cells. Understanding the crosstalk between apoptosis and immune responses could lead to more effective combination therapies that enhance antitumor immunity while promoting tumor cell death.

Autophagy, a cellular process involved in the degradation and recycling of cellular components, is intricately linked to apoptosis and other cell death mechanisms. Recent studies have shown that autophagy can have both prosurvival and prodeath roles in cancer, depending on the context. Selective autophagy, such as mitophagy (the selective degradation of mitochondria), plays a critical role in maintaining cellular homeostasis and can influence the sensitivity of cancer cells to therapy. Targeting autophagy-related pathways, either alone or in combination with other medicines, holds promise for improving cancer treatment outcomes.

The tumor microenvironment, characterized by hypoxia, low pH, immunosuppression, and nutrient deprivation, plays a crucial role in shaping cell death responses. Strategies to modulate the tumor microenvironment to enhance the effectiveness of cell death-targeting therapies are under active investigation. For example, normalizing tumor vasculature or reprogramming immune cells within the tumor microenvironment can improve drug delivery and enhance the efficacy of apoptosis-inducing agents.

Emerging technologies such as CRISPR-Cas9 gene editing offer unprecedented opportunities to rapidly manipulate cell death pathways. This technology enables researchers to knock out or modify specific genes involved in apoptosis and other cell death mechanisms, providing valuable insights into their functions and potential as therapeutic targets. Additionally, CRISPR-based screens can identify novel regulators of cell death pathways, paving the way for new drug discovery.

Looking forward, the future of cell death research and therapeutic development lies in interdisciplinary collaborations and the integration of cutting-edge technologies. Advances in genomics, proteomics, and bioinformatics will enable precise characterization of cell death pathways and their dysregulation in disease states. Novel therapeutic modalities, including combination therapies that exploit synergistic interactions between different cell death pathways, offer new avenues for overcoming therapeutic resistance and improving patient outcomes. By embracing these multidimensional approaches, researchers and clinicians can pave the way toward more effective, personalized treatments for diseases characterized by dysregulated cell death processes.

In summary, studying programmed cell death and its therapeutic targeting remains dynamic and rapidly evolving. While significant progress has been made in understanding the molecular mechanisms underlying apoptosis and other cell death pathways, many challenges and opportunities lie ahead. Integrating insights from primary research with clinical applications, leveraging advanced technologies, and fostering interdisciplinary collaborations will be essential for translating these discoveries into effective therapies. As our understanding of cell death continues to deepen, the potential for developing novel treatments for various diseases, from cancer to neurodegeneration, holds great promise for improving human health.

Dr. Kuladip Jana
Editor

GPSR Compliance

*The European Union's (EU) General Product Safety Regulation (GPSR) is a set of rules that requires consumer products to be safe and our obligations to ensure this.*

*If you have any concerns about our products, you can contact us on ProductSafety@springernature.com*

In case Publisher is established outside the EU, the EU authorized representative is:

Springer Nature Customer Service Center GmbH
Europaplatz 3
69115 Heidelberg, Germany

**Batch number: 10732076**

Printed by Printforce, the Netherlands